RESONANT INFRARED DETECTORS AND EMITTERS

SERIES IN MODERN CONDENSED MATTER PHYSICS

ISSN: 2010-2119
Editors-in-charge: I. Dzyaloshinski and Yu Lu

Published

Vol. 14: *Resonant Infrared Detectors and Emitters*
by Kwong-Kit Choi

Vol. 13: *Quantum Dissipative Systems (Third Edition)*
by U. Weiss

Vol. 12: *Simple Views on Condensed Matter (Third Edition)*
by P.-G. de Gennes

Vol. 11: *Magnetism in Heavy Fermions*
by H. B. Radousky

Vol. 10: *Quantum Dissipative Systems (Second Edition)*
by U. Weiss

Vol. 9: *Theory of Critical Phenomena in Finite-Size System: Scaling and Quantum Effects*
by J. G. Brankov, D. M. Danchev and N. T. Tonchev

Vol. 8: *Simple Views on Condensed Matter (Second Edition)*
by P.-G. de Gennes

Vol. 7: *The Physics of Quantum Well Infrared Photodetectors*
by K. K. Choi

Vol. 6: *Low-Dimensional Quantum Field Theories for Condensed Matter Physicists*
eds. L. Yu, S. Lundqvist and G. Morandi

Vol. 5: *Lectures on Electron Correlation & Magnetism*
by P. Fazekas

Vol. 4: *Simple Views on Condensed Matter*
by P.-G. de Gennes

Vol. 2: *Quantum Dissipative Systems*
by U. Weiss

Vol. 1: *Exotic Properties of Superfluid Helium 3*
by G. E. Volovik

Series in Modern Condensed Matter Physics – Vol. 14

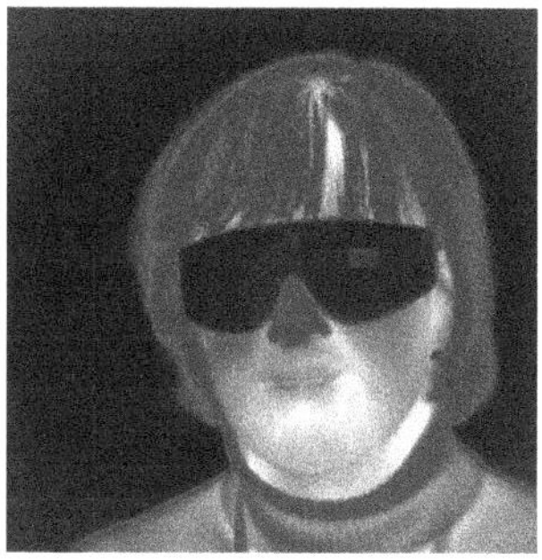

RESONANT INFRARED DETECTORS AND EMITTERS

Kwong-Kit Choi
Banpil Photonics, USA

World Scientific

NEW JERSEY • LONDON • SINGAPORE • BEIJING • SHANGHAI • HONG KONG • TAIPEI • CHENNAI • TOKYO

Published by

World Scientific Publishing Co. Pte. Ltd.
5 Toh Tuck Link, Singapore 596224
USA office: 27 Warren Street, Suite 401-402, Hackensack, NJ 07601
UK office: 57 Shelton Street, Covent Garden, London WC2H 9HE

Library of Congress Cataloging-in-Publication Data
Names: Choi, Kwong-Kit, author.
Title: Resonant infrared detectors and emitters / Kwong-Kit Choi, Banpil Photonics, USA.
Description: New Jersey : World Scientific, [2024] | Series: Series in modern condensed matter physics, 2010-2119 ; vol. 14 | Includes bibliographical references and index.
Identifiers: LCCN 2024004837 | ISBN 9789811286513 (hardcover) | ISBN 9789811286520 (ebook for institutions) | ISBN 9789811286537 (ebook for individuals)
Subjects: LCSH: Infrared detectors. | Infrared sources. | Infrared radiation--Mathematical models. | Optical resonance--Mathematical models.
Classification: LCC TA1573 .C46 2024 | DDC 681/.2--dc23/eng/20240310
LC record available at https://lccn.loc.gov/2024004837

British Library Cataloguing-in-Publication Data
A catalogue record for this book is available from the British Library.

For any available supplementary material, please visit
https://www.worldscientific.com/worldscibooks/10.1142/13684#t=suppl

Desk Editor: Shaun Tan Yi Jie

Typeset by Stallion Press
Email: enquiries@stallionpress.com

Preface

Infrared detection serves many important functions in military operations. It is used in night vision, surveillance, piloting, automatic navigation, targeting, and tracking. It also finds plenty of civilian applications in industrial processes, environmental monitoring and protection, homeland security, infrared astronomy, remote temperature sensing, medical research, etc. In Chapter 1, the fundamentals of infrared detection are introduced.

This book mainly covers a specific detector, which is known as the quantum well infrared photodetector (QWIP). QWIPs operate on particle-in-a-box quantization, which often serves as the first example in quantum mechanics. Its physical reality is not supposed to be taken seriously. It may then come as a surprise not only that such a system can be realized, but it can even be an excellent detector in many practical applications. By putting two semiconductor layers together in an alternate fashion, the difference in their bandgap energies creates just such a classic textbook example. When the layers are thin enough, the electron states are quantized into energy levels. The incident light sends the electrons from a lower-energy bound state to a higher-energy free state, converting the optical energy into electrical signal in the process. An infrared detector is thus born. In Chapter 2, we describe the design of this material to yield different desirable optoelectronic properties.

To use an array of QWIPs for infrared imaging, however, there is an unusual hurdle. The detector does not sense normal incident light. The light must travel at an angle inside the detector to induce a finite signal. Each detector pixel therefore needs an optical coupling structure to redirect light. To prepare the readers for rigorous optical design, we describe several electromagnetic modeling tools in Chapter 3. The use of these tools is mandatory because once the incident light is deflected, it will be confined in the detector volume by internal reflections and cause destructive optical interference (DOI) in general. A large internal intensity is possible only if the detector geometry is properly designed.

In Chapter 4, we introduce the first optical coupling geometry, the corrugated-QWIP. Using a pair of $\sim 45^\circ$ sidewalls to reflect light, the internal optical paths create only optical fringes and DOI can be avoided. This one-size-fits-all, wavelength-independent geometry enables the QWIP to behave like a usual detector with isotropic absorption properties.

Nevertheless, a corrugated-QWIP cannot absorb all the light with different polarizations and it needs a thick absorbing layer and a large voltage to operate. The more popular coupling scheme is grating coupling. Grating-coupled QWIPs can yield near-unity quantum efficiency. However, when the detector pixel is small, DOI occurs between the diffracted and internally reflected light, making high quantum efficiency infeasible. In Chapter 5, we introduce an alternative structure, the resonator-QWIP. This structure uses an array of rings, each acting as an individual resonator, to convert optical energy into localized surface waves, thereby avoiding DOI among traveling waves in dense arrays. Furthermore, using resonator-QWIP as an optical antenna to focus light into a small active mode volume, one can further reduce dark current to achieve higher-temperature operation.

Since an efficient receiving antenna is also a good transmitting antenna, we describe the use of the resonator-pixel as an optical transmitting antenna in Chapter 6 to achieve high power and highly directional optical and thermal emissions for active imaging and thermophotovoltaic applications. Finally, in Chapter 7, we describe an example of infrared applications, which is remote temperature sensing from space. Using direct knowledge on the detector optoelectronic properties, one can simplify camera calibration, obtain temperature reading directly from camera digital output, and account for the atmospheric effects systematically. I hope the readers find the book useful and would like to thank many of my colleagues at Army Research Laboratory, NASA Goddard Space Flight Center, L3, Teledyne, FLIR, RVS, DRS, Jet Propulsion Laboratory, SVS, Attollo, OSEMI, and Banpil Photonics for many years of collaboration.

Kwong-Kit Choi

About the Author

Dr. Kwong-Kit Choi is Chief Scientist at Banpil Photonics, Inc., USA and a senior technical consultant to government, industry, and academia. He worked with the U.S. Army Research Laboratory for 31 years and had been the Senior Research Scientist for physical sciences for the last 17 years. In this capacity, he invented the quantum well infrared photodetector (QWIP) and the resonator-pixel technology that have pivotal impacts on military, space, and scientific applications worldwide. He also participated in the design and production of QWIP sensors on board NASA Landsat-8 and Landsat-9 satellites for Earth's resource and climate observation.

Dr. Choi is an elected Fellow of the American Physical Society (APS) and the Institute of Electrical and Electronics Engineers (IEEE). He is an inductee of the NASA Space Technology Hall of Fame and a recipient of the Distinguished Presidential Rank Award. He owns 20 patents, published over 200 peer-reviewed papers, and authored the seminal text *The Physics of Quantum Well Infrared Photodetectors.* His current research interests include focal plane array demonstration, new detector design, and infrared optoelectronics. He holds a Ph.D. in physics from Yale University.

Contents

Chapter 1

Fundamentals in Infrared Detection

1.1 Infrared Radiation

Remote sensing plays an important role in scientific investigations and information gathering. Of importance is the thermal radiation emitted by physical objects pertaining to their finite temperatures. For terrestrial objects, most of the radiation falls within the infrared wavelengths between 3 and 20 μm. Objects on Earth thus appear to be brighter in these wavelengths without external light sources. Furthermore, their emission spectra often reveal the intrinsic material properties that can be utilized for characterization and identification. Infrared detection is thus an important tool for passive observation and data collection. The applications range from machine vision to natural resources exploration to astronomy.

The intensity of thermal radiation from an object is governed by Planck's law modified by the object's material properties and physical geometry. The latter two properties are characterized by the object's emissivity ε. Overall, the spectral radiance $L(\varepsilon, T, \lambda)$ of any object can be expressed as

$$L(\varepsilon, T, \lambda) = \varepsilon(\lambda) \frac{2c}{\lambda^4} \frac{1}{e^{hc/\lambda kT} - 1}, \tag{1.1}$$

where L is in the unit of number of photons per second per unit projected area per wavelength per unit projected solid angle (ph/s/cm^2/μm/sr), T is the absolute temperature, λ is the wavelength, c is the speed of light in vacuum, h is the Planck constant, and k is the Boltzmann constant.

When the radiation is Lambertian, *i.e.*, isotropic in different directions, the spectral exitance of a planar surface in a cone of half-angle θ, $M_\theta(T,\lambda)$ in ph/s/cm^2/μm is

$$\begin{aligned} M_\theta(\varepsilon,T,\lambda) &= \int_0^{2\pi}\int_0^{\theta} L(\varepsilon,T,\lambda)\sin\theta\cos\theta d\theta d\phi, \\ &= L(\varepsilon,T,\lambda)\pi\sin^2\theta, \\ &= L(\varepsilon,T,\lambda)\Omega, \end{aligned} \tag{1.2}$$

where ϕ is the azimuthal angle and θ is the polar angle in the spherical coordinates shown in Fig. 1.1. In Eq. (1.2), $\sin\theta d\theta d\phi$ is the area element of a unit sphere and $\cos\theta$ accounts for the smaller projected area in the direction of emission. $\Omega = \pi\sin^2\theta$ is the projected solid angle. It is smaller than the solid angle ω defined as the area on a unit sphere subtended by angle θ, where $\omega = \iint \sin\theta d\theta d\phi = 2\pi(1-\cos\theta)$.

The exitance $M(\varepsilon,T,\lambda)$ into the entire hemisphere is obtained by setting $\theta = 90°$, which gives

$$M(\varepsilon,T,\lambda) = \pi L(\varepsilon,T,\lambda). \tag{1.3}$$

For an ideal blackbody, $\varepsilon(\lambda) = 1$ for all λ. Figure 1.2 shows $M_{BB}(T,\lambda)$ of a blackbody at different T. The peak wavelength $\lambda_{\max}$ can be obtained by setting $dM(T,\lambda)/d\lambda = 0$, yielding

$$\lambda_{\max}T = 3670\,\mu\text{m K}. \tag{1.4}$$

Therefore, a terrestrial scene near $T = 300$ K appears to be the brightest at 12.2 μm, while a jet engine running at 2000 K will be

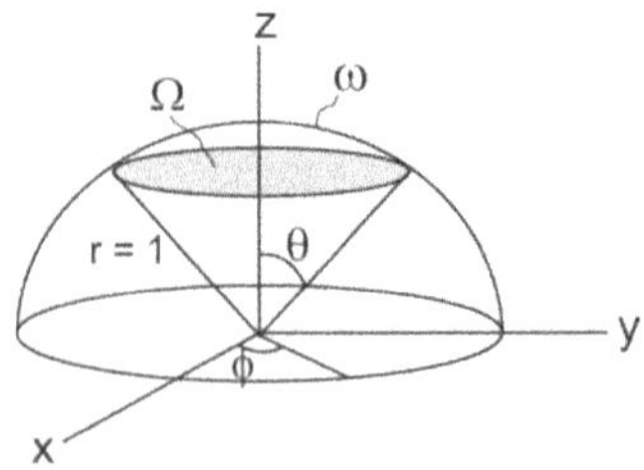

Figure 1.1. The figure shows the azimuthal angle ϕ, the polar angle θ, the solid angle ω and the projected solid angle Ω defined in a unit sphere where the radius $r = 1$.

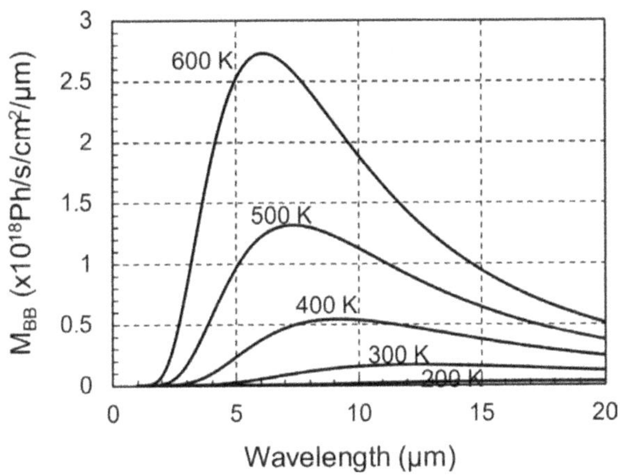

Figure 1.2. The exitance of a blackbody into the hemisphere as a function of wavelength at different temperatures.

easier to detect at $1.8\,\mu$m. The total photon flux emitted by a blackbody is

$$\begin{aligned} M &= \int_0^\infty M_{BB}(T, \lambda) d\lambda, \\ &= \frac{4\pi k^3 T^2}{c^2 h^2} \text{Zeta}(3), \\ &= 1.52023 \times 10^{11} T^3. \end{aligned} \tag{1.5}$$

At 300 K, $M = 4.105 \times 10^{18}$ ph/s/cm^2. For an ordinary object with ε less than 1, the total emission will be less. The notable materials having very low ε are metals such as 0.02 for gold and aluminum. On the other hand, natural landscapes tend to have high emissivity. Most of them exhibit an ε higher than 0.9 in the displayed infrared wavelengths in Fig. 1.3 [1]. Therefore, infrared signal from the ground is close to that of a blackbody at these wavelengths.

Besides its own emission, an object also reflects radiation. In daytime, the major infrared source is the sun. In nighttime, in addition to the occasional moonshine, recombination of oxygen atoms with other oxygen atoms and hydroxyl groups in the atmosphere radiates in the near-infrared band between 0.5 and $2\,\mu$m [2]. These

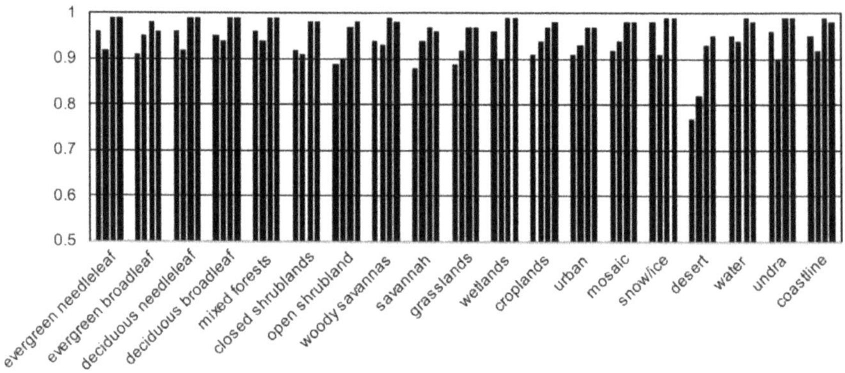

Figure 1.3. The emissivity of different landscapes measured at 3.7 μm, 8.5 μm, 11.0 μm, and 12.0 μm.

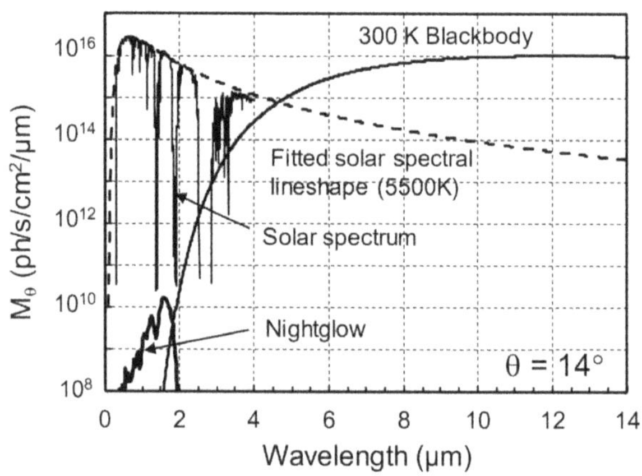

Figure 1.4. The natural infrared sources observed within a polar angle θ of 14°.

atoms and radicals are created during daytime by photodissociation of oxygen molecules and water in the upper atmosphere around 90 km altitude.

Figure 1.4 compares the spectral exitance M_θ of a typical solar spectrum in daytime and the nightglow spectrum in a moonless night with the intrinsic emission of a 300 K blackbody. The solar radiation is taken to be the standard ASTM G173-03 solar spectrum [3]. It can be approximated by that of a blackbody at $T = 5500$ K. Since the solar and nightglow spectra are not directly observed but are reflected by the object, their contributions to the detected signal are reduced by the object's reflectivity R.

By Kirchhoff's law, $R = 1-\varepsilon$, with which the displayed landscapes will have an average R of 0.08 around $\lambda = 3.7\,\mu$m. The typical reflected sunlight at this wavelength will thus be about 13 times lower than the plotted exitance level and it happens to be about the same as the 300 K blackbody radiation. Therefore, infrared detection at or around 3.7 μm will be affected by the sun in daytime operation. On the other hand, with the generally lower reflectivity at the longer wavelengths, the sun's reflection is 3 orders of magnitude below the blackbody level, and therefore the detection is much less affected. At near-infrared range below 2 μm, thermal radiation is very weak. In this case, the nightglow can provide a useful light source particularly in a moonless or overcast night.

1.2 Atmospheric Effects

For long range detection, Earth's atmosphere is a significant factor in determining the signal level. It can scatter, refract, or absorb the optical signal according to atmospheric conditions such as humidity, rain, fog, haze, dust, smoke, and chemical gases. These effects cause the atmospheric transmittance τ_{atm} to vary both in wavelength and in time. Sophisticated software tools have been developed to model some of these effects. One of the computer algorithms, ATRAN, which stands for Atmospheric TRANsmission, was developed by Lord [4]. It calculates τ_{atm} at altitudes ranging from 0 to 30 km by implementing several atmospheric models for mixed gases, water vapor, and ozone. The program covers a wide range of wavelengths from 0.8 to 10,000 μm in selectable spectral resolution and zenith angle. For example, Fig. 1.5 shows the modeled τ_{atm} from an altitude of 1.3 km to the top of the atmosphere at zenith angle $\theta = 0°$ (*i.e.*, facing vertically upward). From 2 to 14 μm, the atmosphere has high transmittance in several wavelength bands. The board maximum between 3 and 5 μm defines the mid-wavelength infrared (MWIR) atmospheric window and that between 8 and 14 μm defines the long

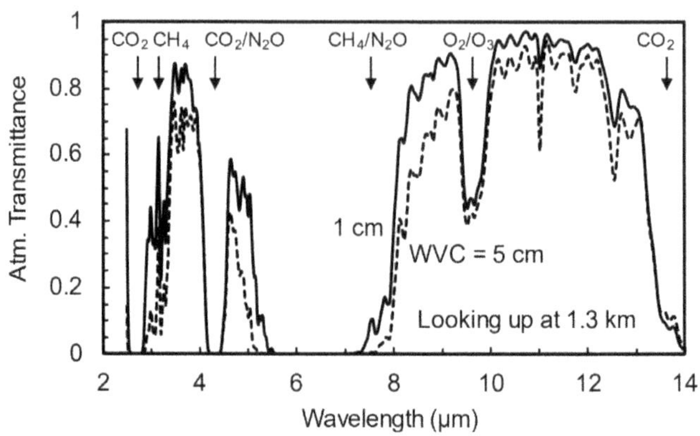

Figure 1.5. The atmospheric transmittance at 1.3 km latitude and 0° zenith angle modeled by ATRAN at two water vapor column (WVC) values. The arrows show the absorption wavelengths of different gas species.

wavelength infrared (LWIR) atmospheric window. Since these two windows locate near the emission maximum for objects on Earth, they are the primary windows used for infrared detection.

The broad minima in the transmittance are usually caused by water vapor absorption. The differential effect on τ_{atm} can be seen in Fig. 1.5 by setting the water vapor column of 1 cm and 5 cm in the model. The water vapor column is the height of the water column if all the water in the air is condensed out. Water vapor absorption is seen to affect the two atmospheric windows equally. In addition to water vapor, other chemical gases such as CO_2, O_3, CH_4, and N_2O also reduce the transmission at certain wavelengths.

Another computer code for τ_{atm} modeling is MODTRAN (MODerate resolution atmospheric TRANsmission) [5]. In addition to gaseous absorption, MODTRAN also takes into account the effects of aerosol particles in visibility and ground reflection. Figure 1.6 shows the modeled τ_{atm} from the ground vertically up to 10 km altitude at two absolute humidity values. An absolute humidity of 1.2 g/m^2 is usually encountered in the extremely dry Artic winter, while that of 18.8 g/m^2 is applicable for tropical regions. Since water moisture decreases exponentially with altitude and is mostly confined below 3 km, the water vapor absorption has a larger effect on the transmission at sea level. On the other hand, ozone absorption

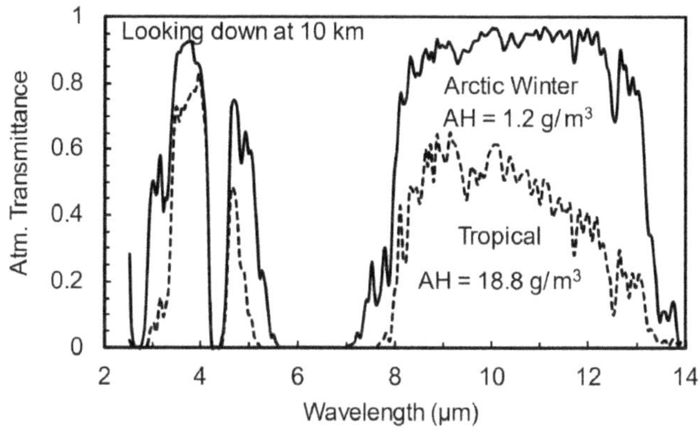

Figure 1.6. The atmospheric transmittance at 10 km latitude and 90° zenith angle modeled by MODTRAN at two absolute humidity values.

around 9.5 μm is much reduced near the ground level since the majority of ozone is situated from 15 to 35 km above ground.

1.3 Camera Optics and Modulation Transfer Function

After passing through the atmosphere, the signal reaches the point of detection. To determine the amount of radiation received by a detector, two optical configurations are considered. In Fig. 1.7, a detector element of area A_d is exposed to an extended blackbody source through a circular opening defined by the aperture stop. The opening is either unoccupied in Fig. 1.7(a) or occupied by a focusing lens in Fig. 1.7(b).

Without the lens, the photon flux in ph/s from an extended source between λ_1 and λ_2 falling on the detector is

$$\begin{aligned}\Phi_d &= \int_{A_s'} L\Omega_d \cos\theta dA_s',\\ &= \int_0^{\rho_i} L\frac{A_d\cos\theta}{r^2/\cos^2\theta}\cos\theta(2\pi\rho)d\rho,\\ &= 2\pi LA_d\int_0^{\theta_i}\sin\theta\cos\theta d\theta,\\ &= A_d(\pi L\sin^2\theta_i).\end{aligned}\tag{1.6}$$

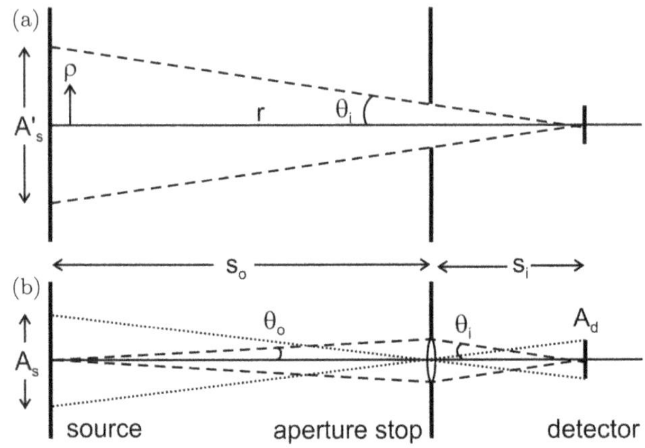

Figure 1.7. The detector faces a blackbody source through a field stop without a lens as in (a) or with a lens as in (b). The distance of the detector to the source is $r = s_o + s_i$.

In the first line of Eq. (1.6), $L = \int L(\lambda)d\lambda$, Ω_d is the projected solid angle of the detector, and $\cos\theta dA'_s$ is a projected area element of the source. In the second line, $\rho_i = r\tan\theta_i$, and θ_i is the half-cone angle of the field of view. The final result of Eq. (1.6) is similar to that of Eq. (1.2) except that both A and θ refer to the detector area and viewing angle rather than the source area and emission angle.

With the lens at the aperture stop, the source forms a focused image at s_i according to $1/s_o + 1/s_i = 1/f$, where f is the focal length of the lens. With the lens, a different source area A_s is imaged onto the detector area A_d. The flux received by the lens from A_s is

$$\Phi_{\text{lens}} = A_s L\Omega_o = A_s\pi L\sin^2\theta_o, \tag{1.7}$$

where Ω_o and θ_o are the projected solid angle and the half-cone angle of the lens subtended on the source. The flux Φ_{lens} is then focused onto the detector so that $\Phi_{\text{det}} = \Phi_{\text{lens}}$. Since $A_s = (s_o/s_i)^2 A_d$ from the lens magnification, Eq. (1.7) then becomes

$$\Phi_d = \left(\frac{s_o}{s_i}\right)^2 A_d\pi L\sin^2\theta_o,$$

$$= A_d(\pi L \sin^2\theta_i),$$

$$= A_d \frac{\pi L}{4F^2+1}, \tag{1.8}$$

where F is the f-number $\equiv f/D$ and D is the lens diameter. In the second line of Eq. (1.8), an approximation $s_o \sin\theta_o = s_i \sin\theta_i$ has been made, and in the third line, the object is assumed to be far away from the detector as in remote sensing. In this case, s_i approaches f, and the image forms on the focal plane. The value of $\sin^2\theta_i$ is then equal to $(4F^2+1)^{-1}$. The detector array placed on the focal plane is referred to as the focal plane array (FPA). A comparison between Eq. (1.6) and Eq. (1.8) shows that Φ on the detector is the same with or without the lens when viewing an extended source. The only change is the viewing area A_s being smaller than A'_s. When the image of a finite-size object is smaller than the detector size, *i.e.*, $A_s(s_i/s_o)^2 < A_d$, Eq. (1.8) is not applicable, and the original Eq. (1.7) should be used, *i.e.*, Φ falling on the detector from a small source is Φ_{lens}.

Furthermore, optical diffraction due to the finite extent of the aperture imposes a minimum image size of the object even when the object becomes a point source. If this image size is larger than the detector size, Φ_{det} will be less than Φ_{lens}. The existence of a minimum image size also limits the spatial resolution of the detection. The diffraction effect of the lens can be derived from a simplified version of the Fresnel-Kirchhoff formula [6]

$$E_p = C \int_A e^{ikr}\, dA, \tag{1.9}$$

where E_p is the optical electric field at the point p on the focal plane, C is a constant proportional to the incoming radiation, k is the optical wave vector, r is the optical pathlength along k from the aperture to point p on the focal plane, and A is the area of the aperture with diameter $D = 2R$ shown in Fig. 1.8.

Equation (1.9) describes the interference effects of different light rays after passing through the aperture and being focused by the lens onto the focal plane. This simplified formula only accounts for

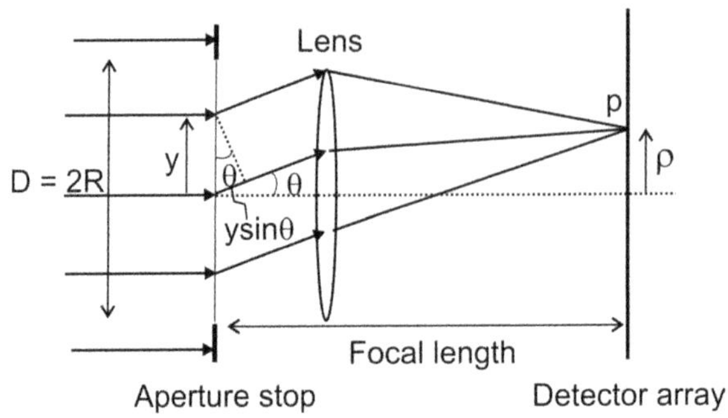

Figure 1.8. The ray diagram of a circular aperture with a focusing lens. The lens is slightly displaced from the aperture to facilitate the discussion.

the phase difference of the light rays arriving at p and ignores any intensity variation across the aperture. For a diffraction angle θ, the pathlength difference is $y \cdot \sin\theta$ and the area element dA at y is $2(R^2 - y^2)^{1/2}dy$. Equation (1.9) is then

$$\begin{aligned} E(q) &= C\int_{-R}^{R} e^{iky\sin\theta} 2\sqrt{R^2 - y^2}dy, \\ &= 2R^2C\int_{-1}^{1} e^{iuq}\sqrt{1^2 - u^2}du, \\ &= \pi R^2 C\frac{2J_1(q)}{q}, \end{aligned} \tag{1.10}$$

where $u = y/R$, $q = kR\sin\theta \cong kR\theta = 2\pi R\rho/(\lambda f) = \pi D\rho/(\lambda f) = \pi\rho/(\lambda F)$, ρ is the distance away from the focal plane center, f is the focal length, and F is the focal number. $J_1(q)$ is the Bessel function of the first kind. The irradiance distribution $I(q)$ is proportional to E^2. Therefore, the normalized $I(q)$, which is known as the point spread function (PSF), is

$$I(q) = \frac{\pi}{(\lambda F)^2}\left(\frac{J_1(q)}{q}\right)^2, \tag{1.11}$$

where $\int I(q)2\pi q dq = 1$, *i.e.*, the total power on the focal plane is normalized to unity. At $q = 3.832$, $I = 0$ which defines the bright central area, known as the Airy disk. The corresponding half-cone angle $= 1.22\lambda/D$ and the distance from the disk center $= 1.22\lambda F$ define the resolving power of the lens. They are the minimum angle and minimum distance two image points must separate for them to be distinguishable. When this criterion is satisfied, commonly known as the Rayleigh criterion, the intensity between the two PSF drops to 0.73 of the peak value. Other criteria exist such as the Houston criterion where the two PSFs are separated by the full width at half maximum (FWHM) of the central intensity, in which the intensity at the valley is 0.96 of the peak, and the Sparrow criterion where the PSFs produces a plateau in the intensity profile. To give an example, Fig. 1.9 shows the PSF with $\lambda = 9\,\mu$m and $F = 2$. The radius of the Airy disk is then $1.22\lambda F$ or $21.96\,\mu$m, which is the lens resolution under the Rayleigh criterion. Its distance at FWHM is $18.52\,\mu$m ($= 1.029\lambda F$), which is the Houston criterion. The Sparrow criterion gives a smaller distance of $17.0\,\mu$m or $0.947\lambda F$. A square detector of $20\,\mu$m will approximately match these lens resolutions.

A comparison between PSF and detector size d is shown in Fig. 1.10. It is seen that the detector will need to be larger than the Airy disk diameter of $44\,\mu$m to receive a majority (84.5%) of energy from a point source but the image will appear as a square of the detector pixel size, which is said to be pixelated. When d reduces

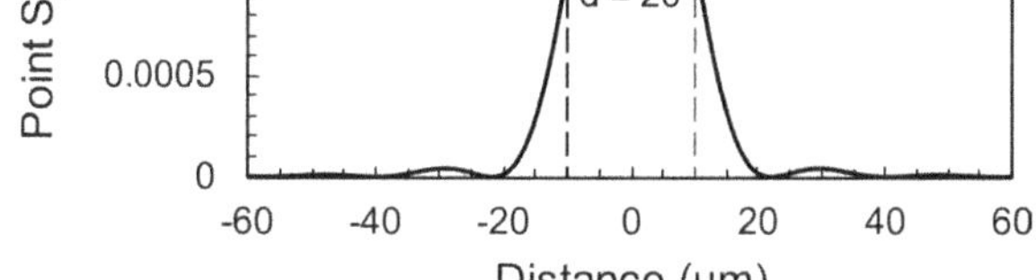

Figure 1.9. The point spread function of a lens with f-number of 2 under $9\,\mu$m incident light. A $20\,\mu$m detector detects at the same resolution as the lens.

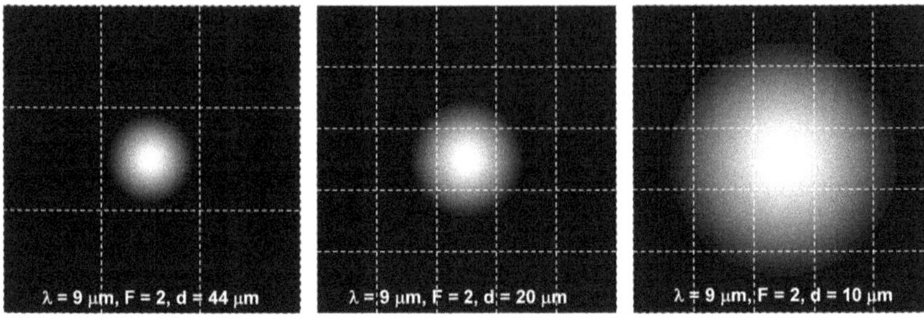

Figure 1.10. The comparison of the point spread function under $f/2$ and $9\,\mu$m incident light and the detector size.

to $20\,\mu$m, the energy received reduces to 59.8% of the total energy but the detector array begins to be able to separate individual Airy disks. When d further reduces to $10\,\mu$m, the energy received is 21.4%, and the PSF spreads to 25 detectors. The individual PSF produced by the lens can thus be well resolved. From this example, the characteristics of an imaging system consisting of a lens and a detector array clearly depends on the length ratio between the lens resolution given by λF and the signal sampling given by d. For a given ratio $r = \lambda F/d$, the fraction of energy on the detector (EoD) and the spatial resolution of the system will be fixed. The pixel sizes considered give $r = 0.41$, 0.9, and 1.8 for $d = 44$, 20 and $10\,\mu$m, respectively. Notice that by adopting a smaller optical F, one can use a smaller d in the FPA without affecting r and hence the system characteristics on EoD and spatial resolution, which drives the development of small-pixel FPAs.

A more systematic approach in modeling detector sampling is to apply Fourier optics. Here, we examine the case along a major axis, the x axis. The sampled detector signal s_d at the location x on the FPA is given by

$$s_d(x) = \int_{-\infty}^{\infty} \phi(x')h(x'-x)dx', \tag{1.12}$$

where ϕ is the PSF of the lens at $y = 0$ shown in Fig. 1.9, and $h(x'-x)$ is the detector sampling function, where $h(x'-x) = 1$ when $|x'-x| \leq d/2$ and 0 otherwise. From the convolution theorem in Fourier transform, the detector signal in each spatial frequency ν

in cycles/mm is given by

$$S_d(\nu) = \Phi(\nu)H(\nu), \tag{1.13}$$

where $\Phi(\nu)$ is the spatial frequency component at ν in $\phi(x)$, and $H(\nu)$ is the detector response at ν. Both are obtained from the Fourier transform of $\phi(x)$ and $h(x)$. $H(\nu)$ is evaluated to be

$$H(\nu) = \int_{-\frac{d}{2}}^{\frac{d}{2}} e^{i2\pi\nu x}dx = d\left(\frac{\sin \pi\nu d}{\pi\nu d}\right) = d\,\mathrm{sinc}(\pi\nu d). \tag{1.14}$$

$\Phi(\nu)$ needs to be evaluated by numerical discrete Fourier transform in terms of u, where

$$\Phi(u) = \frac{1}{N}\sum_{n=0}^{N-1}\phi(n)e^{i2\pi un}, \tag{1.15}$$

and u, n, and N are integers. The integer n is the sampling distance per $1\,\mu$m or per 10^{-3} mm increment. N is the total number of samples, which is taken to be 256 in this example. The integer u is the spatial frequency per $1000/N$ cycles/mm increment. The spatial frequency ν is given by $1000u/N$ cycles/mm. The normalized $|H(\nu)|$ and $|\Phi(\nu)|$ are shown in Fig. 1.11.

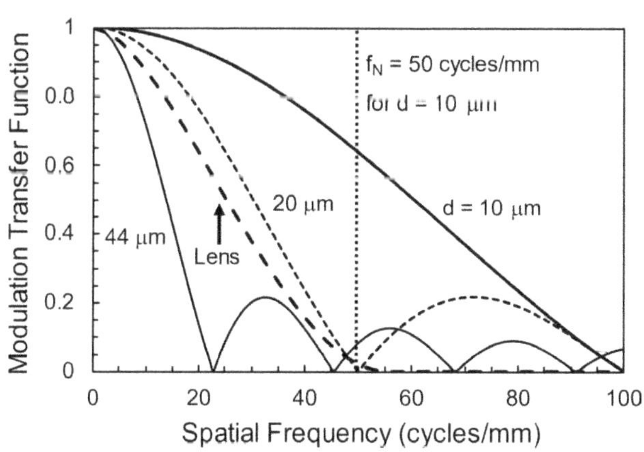

Figure 1.11. The modulation transfer functions $|\Phi(\nu)|$ of the lens with $F = 2$ and $\lambda = 9\,\mu$m and $|H(\nu)|$ of the detectors with $d =$ 10, 20, and $44\,\mu$m. The Nyquist frequency f_N of the $10\,\mu$m detector is also shown.

From $\Phi(\nu)$, the lens PSF contains spatial frequencies up to a cutoff frequency $\nu_c \sim 50$ cycles/mm or down to a period length of $20\,\mu$m, comparable to the Airy disk radius of $21.96\,\mu$m. A $20\,\mu$m detector, with its sampling frequency ν_s at $1000/d = 50$ cycles/mm, is able to sample $\Phi(\nu)$ up to ν_c but with zero response at this frequency. With the decreasing $H(\nu)$, the output signal at ν, $S_d(\nu)$, is modified from $\Phi(\nu)$. To better preserve $\Phi(\nu)$, $H(\nu)$ needs to have a wider bandwidth, which requires a smaller d. By sampling at $\nu_s = 100$ cycles/mm in a $10\,\mu$m detector array, $H(\nu)$ decreases less rapidly and $\Phi(\nu)$ is less affected. The real-space function $\phi(x)$ will be better represented. In addition, according to the Nyquist theorem, ν_s must be larger than ν_c by at least a factor of 2 to avoid aliasing, an effect causing a signal at ν to appear as a signal with $\nu_s - \nu$ when $\nu_s/2 < \nu < \nu_s$. To avoid aliasing at ν below 50 cycles/mm, ν_s must be larger than 100 cycles/mm or d less than $10\,\mu$m. The choice of a proper d depends on applications and features in the scene.

When the optical intensity from the scene is not uniform but varies as $i_{\text{obj}}(x')$ at the aperture, the sampled signal $y(x)$ on the FPA is given by

$$y(x) = \int_{-\infty}^{\infty} i_{\text{obj}}(x') s_d(x' - x) dx', \tag{1.16}$$

which again from convolution theorem yields

$$Y(\nu) = I_{\text{obj}}(\nu) S_d(\nu), \tag{1.17}$$

where $I_{\text{obj}}(\nu)$ represents the modulation of $i_{\text{obj}}(x')$ at the frequency ν. $S_d(\nu)$ can then be interpreted as the optical transfer function (OTF) of the lens/detector sampling system at ν. Since $S_d(\nu)$ is generally a complex number, where $S_d(\nu) = |S_d(\nu)| \exp(-2\pi i \varphi(\nu))$, $|S_d(\nu)|$ is referred to as the modulation transfer function and $\varphi(\nu)$ is the phase transfer function. An asymmetric transfer system can cause changes in both modulation and phase of a periodic intensity pattern. With the concept of optical transfer, one can generalize the concept to the entire optical train from the source to the eye, where the final OTF is the product of individual OTF of each component in the train. The discussed atmospheric transmittance τ_{atm}, lens PSF $\Phi(\nu)$, and the detector response function $H(\nu)$ are examples of individual OTFs.

Combining $L(\varepsilon, T, \lambda)$ in Eq. (1.1), Φ_{lens} in Eq. (1.7) and I(q) in Eq. (1.9), one can estimate the number of photons received by a detector per second from an isolated point source against a cold background such as that in space. Consider an example in which the object is a square blackbody with an area of 25 cm × 25 cm and it situates at 10 km away from the detector. The objective lens is 5 cm in diameter and the focal length ℓ is 10 cm (*i.e.*, $F = 2$). The detector size d is 20 μm. The classical image of this object will be $2.5 \times 2.5\,\mu\text{m}^2$, much smaller than the detector's collection area. Its apparent brightness would appear to be $(20/2.5)^2 = 64$ times weaker than that of an extended object filling the instantaneous field of view ($= \text{d}/\ell$). In addition, with diffraction at the lens, the flux of the image spreads out to a larger area, further reducing EoD. We consider two wavelength bands. One is the MWIR band between 3 and 5 μm, and another is the LWIR band between 8 and 10 μm. Assuming $\tau_{\text{atm}} = 1$ and the PSF is centered at the detector, the photon flux on the detector is

$$\Phi_d(T) = A_s \Omega_0 \int_{\lambda_1}^{\lambda_2} \int_{-d/2}^{d/2} \int_{-d/2}^{d/2} I(q) L(T, \lambda) dx dy d\lambda, \tag{1.18}$$

where $A_s = 625\,\text{cm}^2$, $\Omega_0 = 1.960 \times 10^{-11}$ sr, and $q = \pi(x^2 + y^2)^{1/2}/(\lambda F)$. Figure 1.12 shows the calculated Φ_d. At 300 K, Φ_d is 4.33×10^7 ph/s in MW and 6.45×10^8 ph/s in LW. (The two fluxes become equal at higher temperature of 570 K.) From Fig. 1.12, Φ_d from the point source is 71.6 times and 101.2 times lower than that of an extended light source calculated from Eq. (1.8) in the MW and LW, respectively. Therefore, the diffraction effect further reduces EoD by 1.12 times in MW and 1.58 times in LW beyond the small object's size.

The larger Φ_d of an extended light source such as that from a terrestrial scene makes it easier to detect the presence of the source. However, if all objects have exactly the same temperature (and also the same emissivity), the scene will be featureless. Therefore, in thermal imaging, one needs to detect the difference in the intensity level arising from different object temperatures. In the above point-source imaging at low background, the foreground emission such as

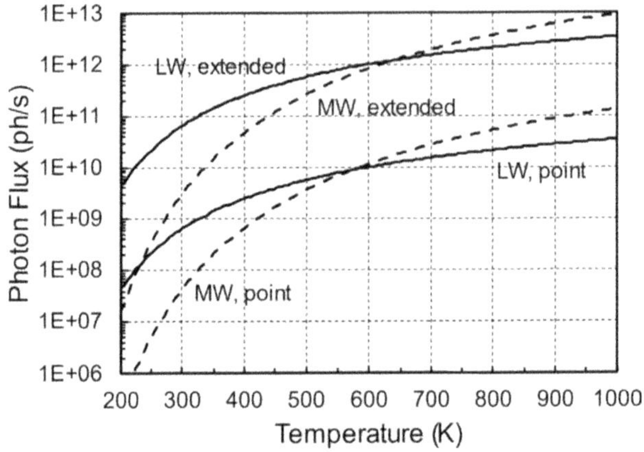

Figure 1.12. The photon flux Φ_d received by a $20\,\mu$m detector from a distant point source using Eq. (1.18) and an extended light source using Eq. (1.8), both located at 10 km away. The objective lens diameter is 5 cm and the f-number is 2.

that from the atmosphere and the camera optics ultimately limits the image contrast.

The intensity profile exhibited in an extended scene depends on the thermal contrast $C(\lambda_1, \lambda_2, T)$ of the detection band. $C(\lambda_1, \lambda_2, T)$ is the percentage change of intensity per degree change of the object temperature between λ_1 and λ_2. It is given by

$$C(\lambda_1, \lambda_2, T) = \frac{1}{\Phi_d}\frac{d\Phi_d}{dT} = \frac{1}{L}\frac{dL}{dT}, \tag{1.19}$$

where L is the shorthand for $\int L(\lambda)d\lambda$. The value of C can be obtained exactly by differentiating L numerically. On the other hand, when photon energy hc/λ is much larger than the blackbody thermal energy kT, $\exp[hc/(\lambda kT)] - 1 \approx \exp[hc/(\lambda kT)]$ and

$$C(\bar{\lambda}, T) = \frac{hc}{\bar{\lambda} kT^2}, \tag{1.20}$$

where $\bar{\lambda}$ is a fitted wavelength between λ_1 and λ_2. For the 3–5 μm band, $\bar{\lambda}$ is chosen to be 4.45 μm, and at the 8–10 μm band, it is 9 μm. In Fig. 1.13, the calculated contrast is plotted using the exact and approximate formulas, and they agree quite well. The scene has a

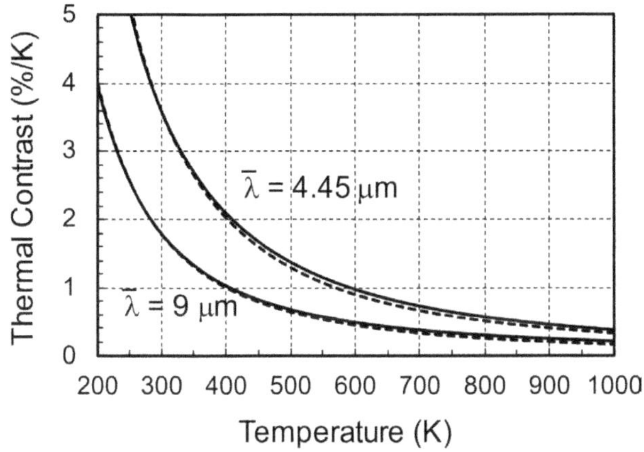

Figure 1.13. The C values from Eq. (1.19) (solid curves) and Eq. (1.20) (dashed curves).

higher contrast at lower temperature or in MW band. The MW can therefore give higher contrast images than the LW. Hotter objects in MW are then more noticeable. On the other hand, the lower contrast in the LW brings up a brighter background and thus provides better situational awareness. At 300 K, C is 3.59%/K for the MW and 1.78 %/K for the LW.

1.4 Photodetector

After the light passing through the lens, its intensity is sampled by detector elements (or pixels) in the array. There are generally two types of detectors, which are photoconductors and photodiodes. We will focus on photoconductors first. For photoconductors, their electrical conductivity increases with the optical intensity. Under a voltage bias, both photocurrent and dark current flow inside the detector. Photocurrent arises from photocarrier generation when the material absorbs optical energy. The dark current is generated from the presence of conducting carriers even without light. They can be low energy charge carriers that conduct through material defects and electronic surface states or high energy charge carriers that are generated from thermal excitations. A sensitive photoconductor should have negligible dark current in comparison with the photocurrent.

Let us assume the carriers are electrons. The photocurrent I_p can be expressed as $A_d \rho e v$, where A_d is the detector area, ρ is the photoelectron volume density, e is the electron charge, and v is the electron drift velocity under bias. The quantum efficiency η of a detector is the fraction of photon flux Φ being absorbed and converted into freely conducting electrons. For a given η, the photoelectron generation rate is $\eta\Phi$ within the detector active volume $A_d t$, where t is the thickness of the detector material. If each photoelectron survives on average τ second before recombination, the time-averaged ρ will be $\eta\Phi\tau/A_d t$. I_p is then given by

$$I_p = A_d \frac{\eta\Phi\tau}{A_d t} e v = \eta\Phi e \frac{\tau}{\tau_{tr}} = e\eta g\Phi, \tag{1.21}$$

where $\tau_{tr} = t/v$ is the electron transit time across the active layer and $g = \tau/\tau_{tr}$ is the photoconductive gain. Therefore, I_p of a photoconductor depends on two independent parameters, η and g. The former governs the generation of conducting electrons and the latter governs the distance of these electrons traveled before recombination. A large I_p requires both to be large. Very often, they are both voltage-dependent. The product of η and g is defined to be the conversion efficiency (CE). From Eq. (1.21), CE is the fraction of the photon flux being converted into electrical current, and it is equal to the number of electrons flowing out of the detector per incident photon. The fact that I_p depends on both η and g reflects infrared detection to be a two-step process: the creation of photoelectrons and their transportation. Although η and g have the same effect on the I_p magnitude, the nature of their contributions is different. Detectors with the same CE may have different η and g and thus different properties.

The quantity g can vary from less than 1 to larger than 1. If $g > 1$, the rate of electron collection ($\eta\Phi g$) will exceed the rate of photon absorption ($\eta\Phi$). This is possible because Eq. (1.21) assumes the photoelectron density ρ to be invariant under bias. It is valid only if an extracted electron from the photocurrent is replaced by another electron of the same energy from the electrical contact. In such case, each photoelectron conducting out of the active volume in the time τ_{tr} will be replaced by another electron from the opposite electrode to preserve ρ. During the photoelectron lifetime of τ, τ/τ_{tr} $(=g)$ number of replacements will have taken place, yielding g number of electrons collected externally for every photoelectron created. When $\tau > \tau_{tr}$, more electrons than the absorbed photons will be collected.

Although a large g yields a larger I_p, it does not have the same effects as η on the intrinsic sensitivity of the detector. If we treat each photon from the scene as a piece of information, the intrinsic sensitivity depends on the number of these messages being received by the detector, which depends on η. Although each message may subsequently be duplicated into g identical copies, it does not change the amount of information gathered from the scene. In this regard, since the detector produces a charge of $g \cdot e$ each time it absorbs a photon, one can treat $g \cdot e$ as the basic charge unit of the detector in the sensitivity analysis.

Equation (1.21) is valid for most photoconductors. However, there are circumstances under which ρ depends on bias. The replacing electrons may have energies lower than the photoelectrons in tunneling contacts rather than ohmic contacts or the replacing electrons do not immediately replace the outgoing electrons due to the large dielectric relaxation time. In such cases, the detector characteristics are more complicated [7,8], and they will not be considered here.

The sensitivity of a detector depends on the signal-to-noise (S/N) ratio. This ratio depends on the signal integration time τ_{int}, which determines the noise bandwidth B. To know the relationship between τ_{int} and B, we assume the fluctuation of current at each angular frequency ω is uncorrelated during signal integration. The variance P of the total current noise is the sum of the variance of each frequency component. When the noise current component at ω, $i_n(\omega)$, passes into a charge collector, usually a capacitor, the charge output $N(\omega)$ is given by

$$N(\omega) = H(\omega) i_n(\omega), \tag{1.22}$$

where $H(\omega)$ is the frequency response function of the collector, whose time response function $h(t) = 1$ if $0 \leq \tau_{\text{int}}$, and $h(t) = 0$ otherwise.

$H(\omega)$ is then given by

$$
\begin{aligned}
H(\omega) &= \int_{-\infty}^{\infty} h(t)e^{i\omega t}dt = \int_{0}^{\tau_{\text{int}}} e^{i\omega t}dt, \\
&= \tau_{\text{int}}\frac{\sin(\omega\tau_{\text{int}}/2)}{\omega\tau_{\text{int}}/2}e^{i\omega\tau_{\text{int}}/2}, \\
&= \tau_{\text{int}}\text{sinc}(\omega\tau_{\text{int}}/2)e^{i\omega\tau_{\text{int}}/2}.
\end{aligned}
\tag{1.23}
$$

$|H(\omega)|$ can be interpreted as the effective charging time at ω. For noise with $\omega = 0$, $|H(0)| = \tau_{\text{int}}$. The *dc* noise is unattenuated. When ω increases, $|H(\omega)|$ decreases as $\text{sinc}(\omega\tau_{\text{int}}/2)$. When ω are multiples of $2\pi/\tau_{\text{int}}$, $|H(\omega)| = 0$. At these frequencies, the noise fluctuation undergoes a number of complete cycles during τ_{int}, and the net noise charge is thus zero. Likewise, when a half-cycle is completed, $|H(\omega)|$ reaches a local maximum. Since the half-period time decreases with increasing ω, the local maximum is smaller at higher ω and therefore the high frequency noise is suppressed. Quantitatively, when the noise current $i_n(\omega)$ at different frequencies has a constant magnitude i_{n0} independent on ω, *i.e.*, a white noise, the variance P is given by

$$
\begin{aligned}
P &= \int_{-\infty}^{\infty} |N(\omega)|^2 d\omega = i_{n0}^2 \int_{0}^{\infty} |H(\omega)|^2 d\omega \\
&= (i_{n0}^2\tau_{\text{int}}^2)\frac{\pi}{\tau_{\text{int}}}.
\end{aligned}
\tag{1.24}
$$

On the other hand, if the noise components are not attenuated, the variance P produced by the noise within a bandwidth B in Hertz is given by

$$
P = \int_{0}^{B} |H(0)i_{n0}|^2 d(2\pi\nu) = (i_{n0}^2\tau_{\text{int}}^2)2\pi B. \tag{1.25}
$$

The equivalent noise bandwidth of the measurement is defined by equating Eqs. (1.24) and (1.25), with which $B = 1/(2\tau_{\text{int}})$. When τ_{int} is longer, the band of noise that can change the signal is thus narrower.

The noise of a detector arises from the fluctuations in the photoelectron density ρ and drift velocity v. In a simple term, the thermal fluctuations in v give rise to Johnson noise in a typical resistor. It is a form of white noise and therefore $B = 1/(2\tau_{\text{int}})$ applies. The associated noise current i_n is given by

$$i_n = \sqrt{\frac{4kTB}{R}}, \tag{1.26}$$

where T is the detector temperature and R is the detector resistance.

The fluctuations in ρ on the other hand give rise to generation-recombination (g-r) noise, which is also a form of white noise. To present a simplified derivation of the g-r noise, we denote K to be the total number of carrier generation events from both thermal and optical excitations in a time period of τ_{int}. Each event produces a charge ge in the external circuit. The average current flow I will then be Kge/τ_{int}. Typically, the carrier generation follows Poisson statistics, in which the variance of K, denoted by σ_K^2, is equal to K itself. Hence,

$$i_n^2 = \left(\sigma_K \frac{ge}{\tau_{\text{int}}}\right)^2 = \left(K \frac{ge}{\tau_{\text{int}}}\right) \frac{ge}{\tau_{\text{int}}} = I \frac{ge}{\tau_{\text{int}}} = 2egIB. \tag{1.27}$$

Equation (1.27) only considers the fluctuation in the carrier generation where g is a constant. If g fluctuates due to the presence of carrier recombination events, the variance doubles. Hence,

$$i_n = \sqrt{4egIB}. \tag{1.28}$$

The quantity i_n is referred to as g-r noise. The ratio of the noise currents from Eq. (1.26) and Eq. (1.28) is $\sqrt{(kT/eVg)} = \sqrt{(8.614 \times 10^{-5} T/gV)}$, where V is the applied voltage. At cryogenic temperatures, this current ratio is small at the normal operating V and g. Therefore, Johnson noise can usually be neglected for photoconductors. In addition to the Johnson and g-r noises, there are also other possible noise sources such as $1/f$ noise and shot noise from tunneling and leakage currents. In the following discussion, we will only consider g-r noise.

It is more convenient to express noise in the number of electrons in detector array operation. Let us denote the average number of electrons collected in each τ_{int} time frame to be N and the associated number of noise electrons to be n. The average current I will be

eN/τ_{int}. With $B = 1/(2\tau_{\text{int}})$ and i_n in Eq. (1.28),

$$n^2 = \left(\frac{i_n \tau_{\text{int}}}{e}\right)^2 = \frac{4eg}{2\tau_{\text{int}}} \frac{eN}{\tau_{\text{int}}} \frac{\tau_{\text{int}}^2}{e^2} = 2gN = 2g(N_p + N_d), \quad (1.29)$$

where N_p and N_d are the numbers of photoelectrons and dark electrons collected, respectively. Both electrons are assumed to have the same g. The fluctuation of the collected electrons in each time frame is referred to as the temporal noise. In detector arrays, random fluctuations in the material characteristics and pixel dimensions can introduce another type of noise known as the fixed pattern noise or spatial noise. Assuming the fixed pattern noise follows the normal statistical distribution, one can define the array nonuniformity σ to be $\Delta N/N$, where ΔN is the residual standard deviation of the distribution after electronic corrections. On average, the fluctuation of N from pixel to pixel is then σN. In an imaging system, other than the noise from the detectors, the supporting readout electronics (ROIC) can also incur noise such as analog-to-digital conversion (ADC) nonlinearity and digitization noise. Without specifying the noise sources of these electronics, the read noise can be parameterized by the number of noise electrons, n_r. The total number of noise electrons n_{tot} arising from the temporal, spatial, and read components is

$$n_{\text{tot}}^2 = 2g(N_p + N_d) + \sigma^2(N_p + N_d)^2 + n_r^2. \quad (1.30)$$

With the known noise in an imaging system, the S/N ratio of the system is

$$\frac{S}{N} = \frac{N_p}{n_{\text{tot}}} = \frac{N_p}{\sqrt{2g(N_p + N_d) + \sigma^2(N_p + N_d)^2 + n_r^2}}. \quad (1.31)$$

1.5 Detection Range

With Eq. (1.31), one can determine the sensitivity of an imaging system. To detect an isolated object at a distance, one is interested in the detection range of the system. On the other hand, to image an extended scene, one is interested in the smallest temperature difference one can differentiate. For the former, the quantity of interest is the noise equivalent photon flux density (NEFD) in ph/s/cm^2 at the location of the imager. NEFD is the photon flux density that gives S/N of 1. In Eq. (1.31), N_p from the object is directly proportional to the photon flux Φ_d falling on the pixel between λ_1 and λ_2, which in turn depends on the average PSF. Based on $I(q)$ in Eq. (1.11), the average fractional EoD between λ_1 and λ_2 is

$$\gamma_o = \frac{\int_{\lambda_1}^{\lambda_2} \int_{-d/2}^{d/2} \int_{-d/2}^{d/2} I(q) L(T,\lambda) dx dy d\lambda}{\int_{\lambda_1}^{\lambda_2} L(T,\lambda) d\lambda}, \tag{1.32}$$

when the diffraction spot of the image falls on the center of a detector pixel. On the other hand, when the spot falls on the corner, EoD is

$$\gamma_c = \frac{\int_{\lambda_1}^{\lambda_2} \int_0^d \int_0^d I(q) L(T,\lambda) dx dy d\lambda}{\int_{\lambda_1}^{\lambda_2} L(T,\lambda) d\lambda}. \tag{1.33}$$

If we take γ_c for a more conservative estimate,

$$N_p = \eta g \Phi_{\text{lens}} \gamma_c \tau_{\text{int}} = \eta g A_{\text{lens}} \mathcal{J}_{\text{lens}} \gamma_c \tau_{\text{int}}, \tag{1.34}$$

where $\mathcal{J}_{\text{lens}}$ is the photon flux density on the lens. If the collected photoelectrons are entirely from the object, the value of $\mathcal{J}_{\text{lens}}$ can be solved by substituting Eq. (1.34) into Eq. (1.31) and setting $S/N = 1$, which gives NEFD.

Instead of evaluating NEFD, we directly estimate the detection range r in a specific example. In this example, the object size A_s is $25 \times 25\,\text{cm}^2$, temperature is 300 K, and emissivity is 1. There is no transmission loss in the atmosphere and in the camera optics. The lens of the imager is 5 cm in diameter and f-number is 5. The detector is $20\,\mu\text{m} \times 20\,\mu\text{m}$, and it detects either at 3–5 μm MW band or 8–10 μm LW band. Therefore, the detection is at $F\lambda/d = (5 \cdot 5)/20 = 1.25$ for 5 μm in the MW and 2.5 for 10 μm in the LW. Recall that $F\lambda$ is approximately equal to FWHM of the diffraction spot. Therefore, the image is sampled appropriately in this system. The integration time τ_{int} is 5 ms. The average quantum efficiency in either band is denoted η. The detector dark current is assumed to be negligible but the foreground photocurrent I_w generated by the heated window of the imager is significant at the assumed T of 400 K. The foreground I_w carries no signal from the object and it plays the same role as the dark current. The emissivity of the window material is 0.05. I_w is usually far larger than the signal current I_p from a distant object. The collected N_w, instead of N_p, dominates the temporal noise. With $N_w \gg N_p \gg N_d$, NEFD obtained from Eq. (1.31) becomes

$$NEFD = \frac{\sqrt{2gN_w + \sigma^2 N_w^2 + n_r^2}}{\eta g A_{\text{lens}} \gamma_c \tau_{\text{int}}}. \tag{1.35}$$

The value of $N_w = I_w \tau_{\text{int}}/e = A_d \Omega \eta g \tau_{\text{int}} \int L(0.05, 400, \lambda) d\lambda = A_d \Omega \eta g \tau_{\text{int}} L_w$, where L is given by Eq. (1.1). The threshold of detection is defined to be 4 times the NEFD in this case [9]. From Eq. (1.7),

$$\mathcal{J}_{\text{lens}} = \frac{A_s}{r^2} \int_{\lambda_1}^{\lambda_2} L(T, \lambda) d\lambda = \frac{A_s}{r^2} L, \tag{1.36}$$

and

$$r^2 = \frac{A_s L A_{\text{lens}} \eta g \gamma_c \tau_{\text{int}}}{4\sqrt{2\eta g^2 \frac{A_d \pi L_w}{4F^2+1} \tau_{\text{int}} + \left(\sigma \eta g \frac{A_d \pi L_w}{4F^2+1} \tau_{\text{int}}\right)^2 + n_r^2}}. \tag{1.37}$$

In Eq. (1.37), γ_c is 0.2037 for the MW band and 0.1573 for the LW band.

Using Eq. (1.37), the impact of each detector parameter can be evaluated. For example, when the temporal noise dominates, r is proportional to $\eta^{1/4}$ but is independent of g. If the spatial noise dominates, the range is independent of both η and g, and inversely proportional to $\sqrt{\sigma}$. If the detection is read noise limited, the range is directly proportional to $\sqrt{(\eta g)}$ and inversely proportional to $\sqrt{n_r}$. The null effect of g in the first two cases is expected from the discussion in Section 1.4. Since g appears in both the numerator and denominator of Eq. (1.37), its effect cancels out when read noise is negligible. The value of n_r thus sets an acceptable value of g and vice versa.

To yield a more quantitative picture, the detection range is plotted at different η, g, σ, and n_r. In Fig. 1.14, η is varied while keeping $g = 0.2$, $n_r = 200$, and σ constant at 5 different levels from 0% to 0.1%. The detection has a longer range in LW than in MW for 300 K objects. The $\eta^{1/4}$ dependence is observed only when $\sigma = 0\%$. For a

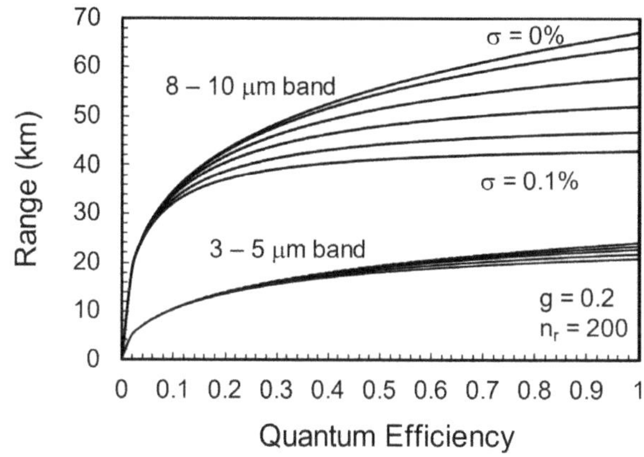

Figure 1.14. The calculated range as a function of η for different σ in 0.02% steps for $g = 0.2$ and $n_r = 200$.

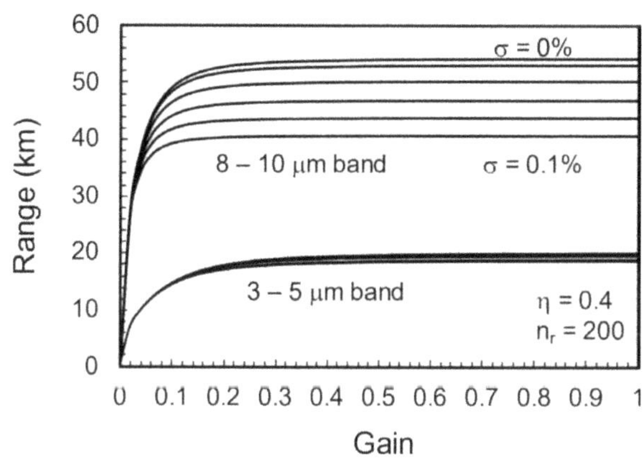

Figure 1.15. The calculated range as a function of g for different σ in 0.02% steps for $\eta = 0.4$ and $n_r = 200$.

finite σ, the dependence on η is significantly reduced. At $\sigma = 0.1\%$, r increases little when $\eta > 0.3$. The detection is thus spatial noise limited in this case.

Figure 1.15 shows the gain dependence for fixed $\eta = 0.4$ and $n_r = 200$. When g is larger than 0.2, r is practically independent of g. Only when g becomes very small, the read noise becomes dominant, and r decreases sharply. Note that the CE $= \eta g$ changes from 0 to 0.4

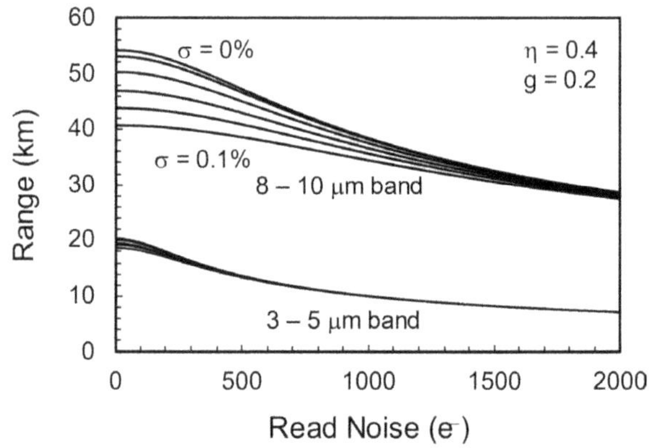

Figure 1.16. The calculated range as a function of n_r for different σ in 0.02% steps for $\eta = 0.4$ and $g = 0.2$.

in this example. Therefore, CE does not uniquely specify detector performance.

In Fig. 1.16, the effect of read noise is shown at $\eta = 0.4$ and $g = 0.2$. The read noise is seen to have significant effects on the range especially in highly uniform arrays. It needs to be less than 500 to maintain a 90% performance.

In the MW band, the detection range r is 23.99 km at $\eta = 1$ and $\sigma = 0.02\%$ in Fig. 1.14. In this case, the detector generates $N_p = 1822$ e$^-$ from the object and $N_w = 0.403$ Me$^-$ from the window. Therefore, $N_w \gg N_p$, and $n_w = 401.4$ e$^-$ from the window dominates the temporal noise as assumed. The spatial noise n_s is 80.6 e$^-$. With the additional n_r of 200 e$^-$ from the readout, the total noise $n_{\text{tot}} = \sqrt{(n_w^2 + n_s^2 + n_r^2)}$ is 455.6. S/N ratio, which is N_p/n_{tot}, is equal to 4 at 23.99 km as expected. Similarly, in the LW band, the detected range is 64.28 km at $\eta = 1$ and $\sigma = 0.02\%$. At this r, the detector generates $N_p = 4127$ e$^-$ and $N_w = 2.114$ Me$^-$, much larger than N_p. The noise n_w is 919.6 e$^-$, and n_s is 422.8 e$^-$. Combined with $n_r = 200$ e$^-$, n_{tot} is 1032 e$^-$, and S/N is again 4 at 64.28 km. To visualize the achievable detection range at different array parameters, the bars in Fig. 1.17 show some of the examples in the LW where the individual parameters are sequentially compromised.

set #	η	g	σ(%)	n_r	Range (km)	% of range	% of range
1	1	1	0	0	68.09	100.00	
2	1	0.4	0	0	68.09	100.00	
3	1	0.4	0	200	67.89	99.71	
4	1	0.4	0.02	0	64.90	95.32	
5	1	0.4	0.02	200	64.74	95.09	
6	0.5	0.4	0.02	200	55.54	81.57	
7	0.4	0.4	0.02	200	52.70	77.41	
8	0.4	0.2	0.02	200	51.70	75.94	
9	0.4	0.2	0.02	500	46.59	68.43	
10	0.4	0.4	0.04	200	50.07	73.54	
11	0.4	0.4	0.04	500	48.74	71.59	
12	0.4	0.2	0.04	500	45.10	66.24	
13	0.3	0.2	0.04	500	41.19	60.50	
14	0.2	0.2	0.04	500	35.70	52.43	
15	0.1	0.2	0.04	500	27.00	38.66	
16	0.1	0.2	0.04	1000	20.20	29.67	

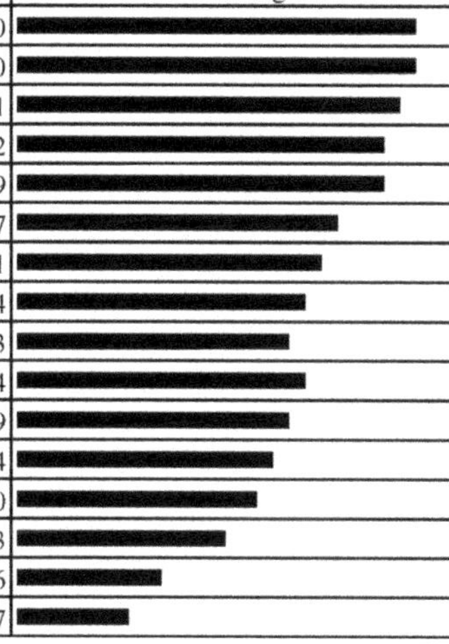

Figure 1.17. The calculated detection range in the 8–10 μm band for a selected set of imager parameters on a $25 \times 25\,\text{cm}^2$, 300 K, blackbody target. Other system parameters are: $F = 5$, $A_{\text{lens}} = \pi(2.5\,\text{cm})^2$, $A_d = (20\,\mu\text{m})^2$, $\tau_{\text{int}} = 5\,\text{ms}$, $T_s = 300\,\text{K}$, and $T_w = 400\,\text{K}$.

Another type of photodetectors is photodiode (PD). PDs are often preferred over photoconductors (PCs) in infrared detection due to their intrinsic advantages in both signal and noise. In PDs, after optical absorption, a photoelectron will either recombine in the absorbing layer resulting in no external signal or pass over the P/N junction resulting in a physical electron being collected. Therefore, for PDs, the absorption QE η is different from the external QE $\eta_{\text{ext}} = \eta f$, where f is the fraction of photoelectrons being collected. While η represents the fraction of incident photons being absorbed, η_{ext} counts only those resulted in producing external electrons. We should emphasize that η_{ext} is fundamentally different from CE $= \eta g$ of a PC. In a PD, only the photoelectrons that are collected produce signal and noise. Due to the very low resistance of the active layer, photoelectrons recombining before reaching the P/N junction produce no external effects. On the other hand, the entire active layer of a PC is resistive. Any photoelectrons generated will change the local resistence of the layer and produce external effects. Therefore,

ηf of a PD and ηg of a PC, both occasionally referred to as the external quantum efficiency, are distinct physical quantities and they should not be compared in relation to performance assessment.

In the following, we assume all photoelectrons are collected in the PD and therefore $f = 1$ and $\eta_{\rm ext} = \eta$. For the collected electrons, g is always 1. Since the recombination process does not affect the current fluctuation, there is no recombination noise. The current noise is given by

$$i_n = \sqrt{2eIB}, \tag{1.38}$$

and is referred to as the shot noise. With larger signal and lower noise, the detection range of a PD is generally larger than a PC, and it is given by

$$r^2 = \frac{A_s L A_{\rm lens} \eta \gamma_c \tau_{\rm int}}{4\sqrt{\eta \frac{A_d \pi L_w}{4F^2+1}\tau_{\rm int} + \left(\sigma \eta \frac{A_d \pi L_w}{4F^2+1}\tau_{\rm int}\right)^2 + n_r^2}}. \tag{1.39}$$

The calculated r is plotted in Fig. 1.18 and Fig. 1.19, which show a larger r and a smaller effect from n_r because of the larger collected N_w. To compare the difference in the performance more quantitatively, the range ratio of a PC over a PD is shown in

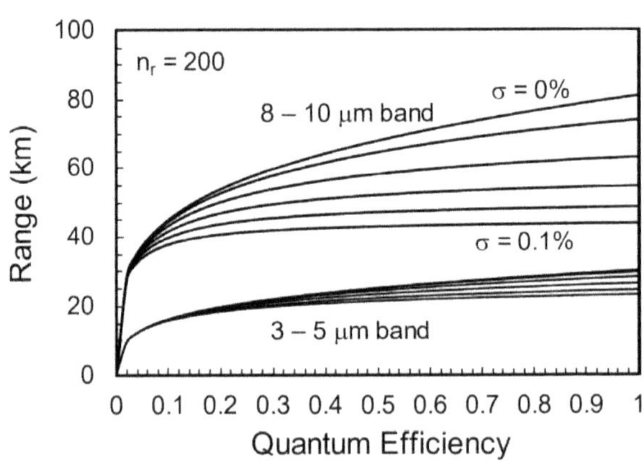

Figure 1.18. The calculated range of a photodiode as a function of η for different σ in 0.02% steps and $n_r = 200$.

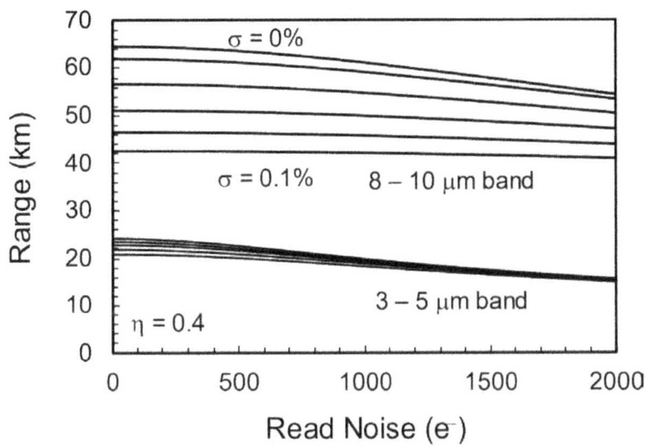

Figure 1.19. The calculated range of a photodiode as a function of n_r for different σ in 0.02% steps and $\eta = 0.4$.

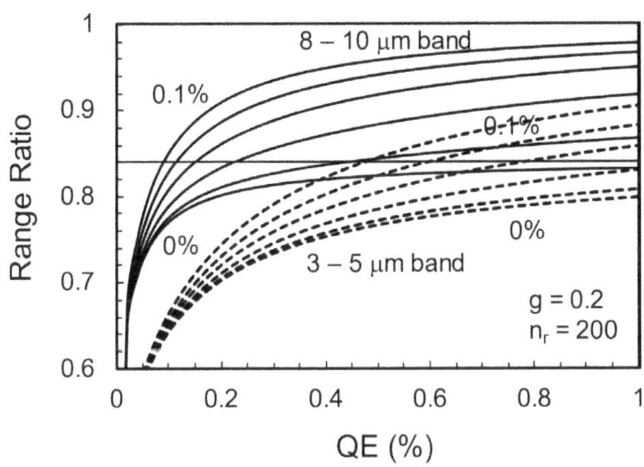

Figure 1.20. The range ratio between a photoconductor and a photodiode with different σ. The read noise is 200 and the gain of the photoconductor is 0.2. The straight line indicates the temporal noise limited ratio of 0.84.

Fig. 1.20. When the temporal noise dominates in the detection, the S/N of a PC is $\sqrt{2}$ smaller than that of a PD. The range should then be shorter by a factor of $2^{-1/4} = 0.84$. In the presence of spatial and read noises, this ratio is modified. Figure 1.20 shows that in MW, a PC in general performs worse than this factor because it is more affected by the read noise due to less charge collection.

On the other hand, a PC generally performs better than expected in LW because both the PC and PD are dominated by the spatial noise. At a very large $\sigma = 0.1\%$, the PC range can approach the PD range with a ratio larger than 0.95 when $\eta > 0.4$. To achieve the above range advantage, a PD collects 5 times more electrons because of the difference in the gain. In MW and at $r = 29.67$ km, $\eta = 1$, and $\sigma = 0.02\%$, the PD generates $N_p = 5954$ e$^-$ and $N_w = 2.014$ Me$^-$. The noise from the window n_w is 1419 e$^-$, and the spatial noise n_s is 403 e$^-$. With $n_r = 200$ e$^-$, $n_{\text{tot}} = 1489$. $S/N = N_p/n_{\text{tot}}$ is therefore 4 as designed. Similarly, in LW, at $r = 74.09$ km, $\eta = 1$, and $\sigma = 0.02\%$, $N_p = 15533$ e$^-$ and $N_w = 10.57$ Me$^-$. The value of n_w is 3251 e$^-$, n_s is 2114 e$^-$, and n_{tot} is 3883. S/N is again 4, consistent with the design.

1.6 Detection Thermal Sensitivity

In thermal imaging of an extended scene, a common measure of the detection sensitivity is the noise equivalent temperature difference (NEΔT). NEΔT is the temperature increment at the scene that produces a signal increment equal to the system noise. It is the minimum temperature variation one can detect under the constraint of the S/N ratio. It is, however, not the minimum ΔT a human can discern in the scene. The minimum resolvable ΔT also depends on the feature size and pattern of the object and the modulation transfer function of the entire optical train from the object to the eye [10]. NEΔT is only part of the determining factor and it is the one that is related to detector array performance. According to its definition and Eq. (1.31), NEΔT is

$$NE\Delta T = \frac{n_{\text{tot}}}{\frac{dN_p}{dT}} = \frac{n_{\text{tot}}}{\frac{dN_p}{d\Phi_d}\frac{d\Phi_d}{dT}} = \frac{n_{\text{tot}}}{C\Phi_d\frac{dN_p}{d\Phi_d}}$$
$$= \frac{\sqrt{2g(N_p + N_d) + \sigma^2(N_p + N_d)^2 + n_r^2}}{CN_p}, \qquad (1.40)$$

where $C = C(\lambda_1, \lambda_2, T)$ is the thermal contrast in Eq. (1.19). Here, we have used the fact that N_p is linearly proportional to Φ_d as indicated in Eq. (1.21). Since N_p appears in both the numerator and denominator of Eq. (1.40), the NEΔT dependence on the scene emissivity is much reduced.

In Section 1.5, we have discussed detecting an isolated object against a cold background where the emission from the heated camera window dominates the temporal noise. In the present case of terrestrial imaging, we are detecting objects that are part of the background and they have similar temperatures. The emission from the window and lens is usually negligible compared with the background due to their much lower ε. The photoelectrons generated by the background are both signal and a source of noise.

To gain insights into the effects of different array parameters, we again discuss a specific example, in which we assume the average temperature of the scene is 300 K, there are no atmospheric and camera optics transmission attenuation, and the f-number is 2. The detector pixel size is $20\,\mu\text{m} \times 20\,\mu\text{m}$, and the integration time τ_{int} is 2 ms. The detection is either between 3–5 μm MW band or between 8–10 μm LW band. $F\lambda/d$ is $(2 \cdot 5)/20 = 0.5$ in the MW and $(2 \cdot 10)/20 = 1$ in the LW. The detection thus tends to be undersampled. With these detection bands and temperature, the thermal contrast C are 3.59%/K and 1.78%/K for the MW and LW, respectively. Since NEΔT is inversely proportional to the contrast, thermal imaging in the MW has an intrinsic advantage of having a larger C by a factor of 2.

Because of the extended nature of the light source and the small adopted f-number, Φ_d on the detector pixel is larger than the previous point-source case by more than a factor of 600. Because of the much larger photon flux, the number of electrons collected in the ROIC is much larger and needs to be monitored. At present, the commercial ROICs with unit cell size of $20\,\mu\text{m} \times 20\,\mu\text{m}$ tend to have charge-handling capacity less than 10 Me^-. The ROIC charge capacity can pose a limit on the collected electrons and the achievable NEΔT. In anticipating technological advances that will increase this capacity, the discussion will not take the charge limitation into account. To monitor the collected charge more closely, it is preferable to specify $N = N_p + N_d$ and the photocurrent to dark current ratio $r = I_p/I_d = N_p/N_d$ in Eq. (1.40). In terms of N and r,

$$N_p = \left(\frac{r}{1+r}\right) N, \tag{1.41}$$

and

$$N_d = \left(\frac{1}{1+r}\right) N, \tag{1.42}$$

which give

$$NE\Delta T = \frac{1}{C}\left(1 + \frac{1}{r}\right)\sqrt{\frac{2g}{N} + \sigma^2 + \left(\frac{n_r}{N}\right)^2}. \tag{1.43}$$

From Eq. (1.43), for a fixed N, NEΔT doubles as I_d increases from zero to I_p, *i.e.*, $r = 1$. Therefore, I_d has to keep lower than I_p for high sensitivity. This can usually be achieved by lowering the operating temperature. When the temporal noise of the photocurrent from the scene dominates over other noise sources, background limited infrared performance is realized, and it is the highest sensitivity one can achieve in infrared detection.

When the detection is temporal noise limited, *i.e.*, the first noise term under the square root in Eq. (1.43) is largest, NEΔT is proportional to g/N, which usually means NEΔT $\propto 1/\eta$ since $N \propto \eta g \tau_{\text{int}}$ for a fixed τ_{int}. However, if τ_{int} is flexible, one can decrease g and lengthen τ_{int} to keep N constant. With a smaller g but the same N, NEΔT can be reduced. When g is less than 0.5, NEΔT of a PC can be lower than that of a PD, albeit at a longer τ_{int} by a factor of $1/g$. When the detection is spatial noise limited, *i.e.*, the second noise term is the largest, the noise is directly proportional to signal. There is no S/N improvement by having a larger η, g, or τ_{int}, and NEΔT is a constant as long as I_p/I_d is a constant. Finally, when it is read noise limited, a larger N will reduce the impact of the read noise and improve NEΔT.

To simplify further discussion, we assume $I_d = 0$ (*i.e.*, $r = \infty$), with which $N = N_p$. In this case,

$$NE\Delta T = \frac{1}{C}\sqrt{\frac{2(4F^2+1)}{\eta A_d \pi L \tau_{\text{int}}} + \sigma^2 + \left[\frac{n_r(4F^2+1)}{\eta g A_d \pi L \tau_{\text{int}}}\right]^2}. \tag{1.44}$$

Figure 1.21 shows NEΔT as a function of η at several σ steps from 0% to 0.1%. In MW, NEΔT is more dependent of η and less on σ because of the small collected N. The reverse is true for LW because of the larger N. For example, at $\eta = 1$ and $g = 0.2$, $N_{MW} = 1.24$ Me$^-$ while $N_{LW} = 26.1$ Me$^-$. The collected N in LW is thus 21 times more than in MW.

The larger N allows the achievable NEΔT$_{LW}$ being lower than NEΔT$_{MW}$ despite the lower thermal contrast C in LW. However, the high sensitivity comes with a stringent requirement on the spatial uniformity. For NEΔT lower than 20 mK, σ has to be smaller than 0.04% irrespective of η. For $\sigma > 0.02\%$, NEΔT improves very little once $\eta > 0.2$ although more photoelectrons are being collected. This may argue for the less importance of η. However, a large η is needed to keep I_p/I_d large in the assumption. Therefore, a large η is important.

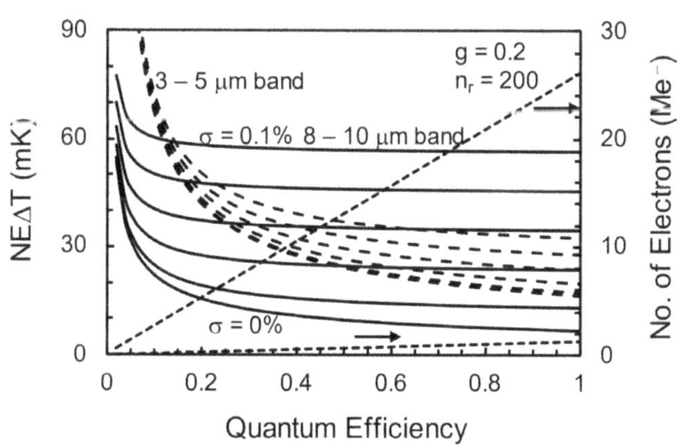

Figure 1.21. The calculated NEΔT in MW and LW for different σ in 0.02% steps for $g = 0.2$ and $n_r = 200$. The two straight lines are the collected electrons in the two bands. Other system parameters are: $F = 2$, $A_d = (20\,\mu\text{m})^2$, $\tau_{\text{int}} = 2$ ms, and $T_{\text{scene}} = 300$ K.

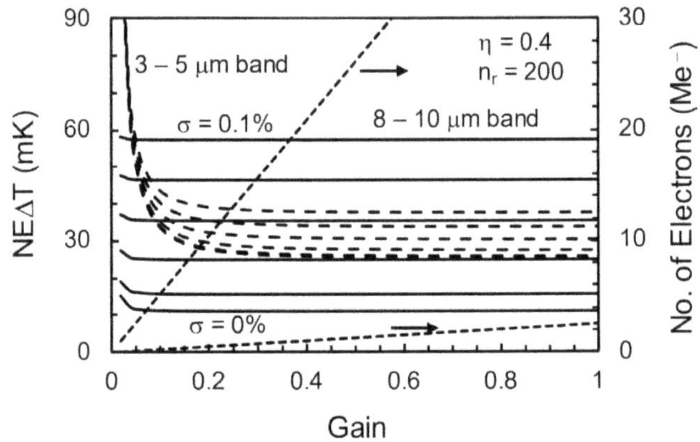

Figure 1.22. The calculated NEΔT in MW and LW for different σ in 0.02% steps for $\eta = 0.4$ and $n_r = 200$. The two straight lines are the collected electrons in the two bands.

NEΔT in both bands can be reduced by increasing τ_{int}. The feasible τ_{int} is generally limited by the ROIC charge-handling capacity and the speed of the event one wants to observe.

The effects of g are shown in Fig. 1.22. In MW, NEΔT becomes independent of g when $g > 0.2$, while it is true for LW when $g > 0.05$. Therefore, g plays a relatively minor role in affecting NEΔT. A small g lessens the ROIC charge capacity requirement but at the same time increases the read noise requirement. A desirable g will depend on the achievable η. When η is large, a small g will give a reasonable signal at a low noise. Figure 1.23 shows the effects of read noise. Because of the large collected $N = 10.4$ Me$^-$ in the LW under the present assumption of $\eta = 0.4$ and $g = 0.2$, the impact from the read noise is small up to $n_r = 2000$ e$^-$. On the other hand, because of the smaller $N = 0.5$ Me$^-$ in the MW, NEΔT tends to be linearly dependent of the read noise.

To visualize the achievable NEΔT at different array parameters, Fig. 1.24 shows some of the examples in the LW when individual parameters are sequentially compromised. Ideally, when $\eta = g = 1$, $\sigma = n_r = 0$, NEΔT $= 6.94\,\text{mK}$. When g reduces to 0.4 in set #2 and #3, NEΔT is unaffected. But when σ increases to 0.02% in set #5, NEΔT degrades by 1.9 times. When η decreases to 0.4

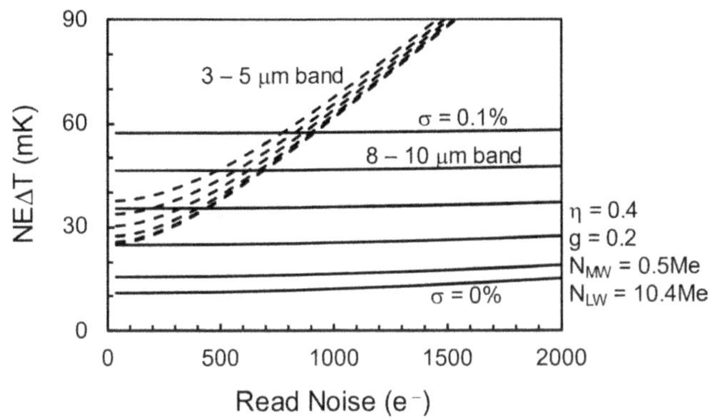

Figure 1.23. NEΔT in MW and LW for σ in 0.02% steps for $\eta = 0.4$ and $g = 0.2$.

in set #9 while having $\sigma = 0.02\%$, NEΔT degrades only by a similar factor of 2.29. However, when σ increases to 0.04% in set #11, NEΔT jumps to 3.60 times. With the same $\sigma = 0.04\%$, NEΔT gradually increases to 4.78 times in set #15 when $\eta = 0.1$, at which NEΔT = 33.19 mK.

From Fig. 1.24, pixel uniformity is seen to play a major role in determining the thermal sensitivity. The same is also true for the detection range in Fig. 1.17. The spatial nonuniformity σ is $\delta N/N$, where δN is the standard deviation of the N population after electronic correction. Even when σ increases from 0 to 0.1%, the detection range can shrink by half while NEΔT increases by a factor of 6. Therefore, the intrinsic pixel uniformity and the subsequent nonuniformity correction are important components in a detection system. The most commonly employed correction procedure is called two-point correction. In this procedure, two mean values, $\langle N_1 \rangle$ and $\langle N_2 \rangle$, of the N population are measured at two blackbody temperatures, T_1 and T_2. From these two calibration points, one can assign a common setpoint $(T_1, \langle N_1 \rangle)$ and a common slope $\langle m \rangle = (\langle N_2 \rangle - \langle N_1 \rangle)/(T_2 - T_1)$ for the entire array. The signal N of an individual pixel in a raw image is replaced by N' where $N' = \langle N_1 \rangle + (N - N_1)\langle m \rangle / m$. After this correction procedure, all the pixels in the image will have common values at the two calibration temperatures.

set #	η	g	σ(%)	n_r	NEΔT (mK)	Ratio	NE ΔT Ratio
1	1	1	0	0	6.94	1.00	
2	1	0.4	0	0	6.94	1.00	
3	1	0.4	0	200	6.95	1.00	
4	1	0.4	0.02	0	13.20	1.90	
5	1	0.4	0.02	200	13.20	1.90	
6	0.5	0.4	0.02	200	14.92	2.15	
7	0.4	0.4	0.02	200	15.71	2.26	
8	0.4	0.2	0.02	200	15.74	2.27	
9	0.4	0.2	0.02	500	15.93	2.29	
10	0.4	0.4	0.04	200	24.99	3.60	
11	0.4	0.4	0.04	500	25.02	3.60	
12	0.4	0.2	0.04	500	25.13	3.62	
13	0.3	0.2	0.04	500	26.03	3.75	
14	0.2	0.2	0.04	500	27.82	4.01	
15	0.1	0.2	0.04	500	33.19	4.78	
16	0.1	0.2	0.04	1000	38.05	5.48	

Figure 1.24. The calculated NEΔT in the 8–10 μm band for a selected set of imager parameters. Other system parameters are: $F = 2$, $A_d = (20\,\mu\text{m})^2$, $\tau_{\text{int}} = 2\,\text{ms}$, and $T_s = 300\,\text{K}$.

This two-point correction can effectively eliminate most sources of nonuniformity such as variations in detector dark current and ROIC ADC offset voltage (additive nonuniformity) or variations of detector area and bias and ROIC ADC gain (multiplication nonuniformity). However, when the spectral responses are different among detectors in an array, finite residual nonuniformity will remain. This variation can be due to the difference in the detector material composition or the mechanical stress on detectors that changes the material bandgap.

To give an example, if the detectors in an array have different cutoff wavelengths in the 8–10 μm detection band, the detectors will absorb different number of photons. In Fig. 1.25, the photon flux Φ absorbed by detectors having different λ_c between 9.5 μm and 10 μm is plotted. A detector with shorter λ_c will absorb less photons and thus produce smaller N. Assigning 10 μm to be the mean λ_c in this example, one can scale up Φ at the shorter λ_c using the two-point

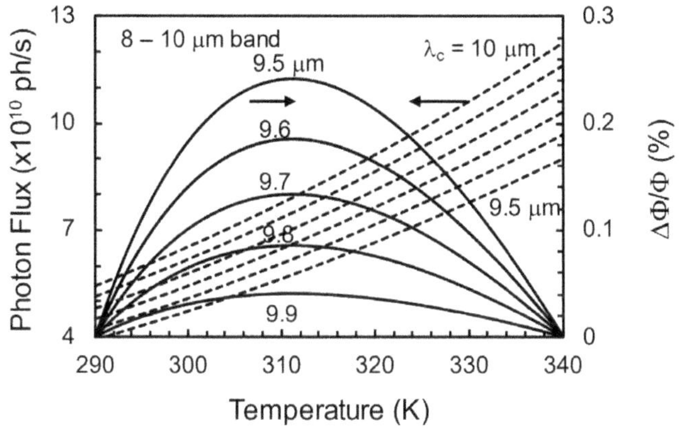

Figure 1.25. The dashed curves show the photon flux sensed by detectors with different cutoff wavelengths. The solid curves are the percentage difference when the flux levels are scaled to be the same at both ends.

correction so that all the absorbed fluxes and thus N will be the same at the two calibration points $T_1 = 290\,\text{K}$ and $T_2 = 340\,\text{K}$.

However, because of the different T dependence, the scaled Φ still have finite differences between these two points. The largest $\Delta\Phi/\Phi$ occurs at $T \sim 310\,\text{K}$ in Fig. 1.25, where $\Delta\Phi = |\Phi'(\lambda_c) - \Phi(10\,\mu\text{m})|$, $\Phi = [\Phi(\lambda_c) + \Phi(10\,\mu\text{m})]/2$, Φ is the original flux, and Φ' is the scaled flux. Specifically, at $\lambda_c = 9.5$, 9.6, 9.7, 9.8, and 9.9 μm, the maximum residual $\Delta\Phi/\Phi$ is 0.24, 0.19, 0.13, 0.086, and 0.041%, respectively. Since $\sigma = \Delta N/N = \Delta\Phi/\Phi$, the residual nonuniformity of the array depends on the λ_c variation. To maintain a σ of 0.04% so that NEΔT can be below 20 mK in Fig. 1.21, λ_c has to be controlled to within 0.1 μm in the pixel population.

The NEΔT of PDs can be similarly obtained. With a larger photocurrent by a factor of g and a smaller temporal noise by a factor of $\sqrt{2}$, NEΔT of a PD is generally smaller than the PC with the same array parameters. It is given by

$$NE\Delta T = \frac{1}{C}\left(1 + \frac{1}{r}\right)\sqrt{\frac{1}{N} + \sigma^2 + \left(\frac{n_r}{N}\right)^2}$$
$$= \frac{1}{C}\sqrt{\frac{(4F^2+1)}{\eta A_d \pi L \tau_{\text{int}}} + \sigma^2 + \left[\frac{n_r(4F^2+1)}{\eta A_d \pi L \tau_{\text{int}}}\right]^2}. \quad (1.45)$$

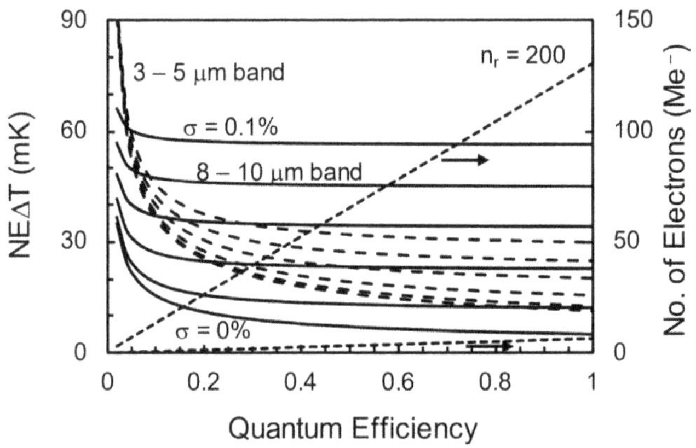

Figure 1.26. The calculated NEΔT in MW and LW of a PD for different σ in 0.02% steps for $n_r = 200$. The two straight lines are the collected electrons in the two bands. Other system parameters are: $F = 2$, $A_d = (20\,\mu\text{m})^2$, $\tau_{\text{int}} = 2\,\text{ms}$, and $T_{\text{scene}} = 300\,\text{K}$.

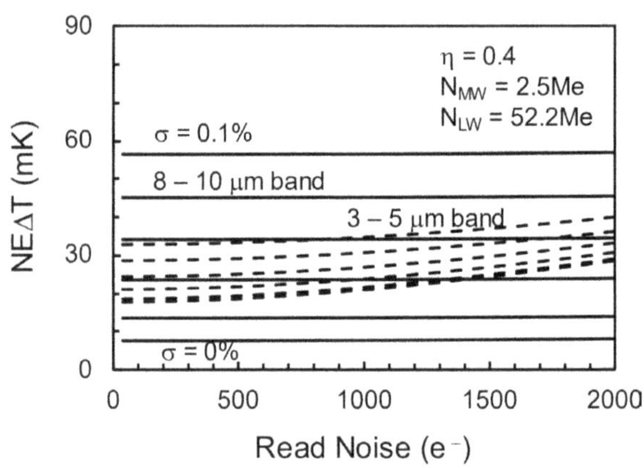

Figure 1.27. NEΔT in MW and LW for σ in 0.02% steps for $\eta = 0.4$.

Figure 1.26 and Fig. 1.27 show NEΔT of a PD with different η, σ, and n_r. To compare NEΔT of a PC and a PD quantitatively, the NEΔT ratio of PC over PD is plotted in Fig. 1.28. NEΔT of a PC is always larger than that of a PD when the two have the same array parameters, *i.e.*, the same η, σ, and n_r. When the detection is temporal noise limited, the NEΔT is expected to be larger by a

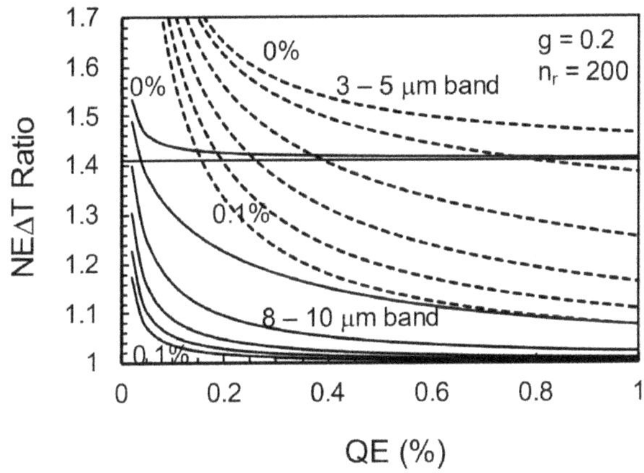

Figure 1.28. The NEΔT ratio of a photoconductor over a photodiode with the same array parameters.

factor of $\sqrt{2}$. While the factor of $\sqrt{2}$ is generally applicable for MW, the PD advantage in LW is quickly diminished at larger η and σ.

At $\sigma = 0.04\%$, the NEΔT ratio is 1.053 for $\eta = 0.4$ and 1.023 at $\eta = 1$. Therefore, the thermal sensitivity of a LW PC array can approach a PD array at a finite σ. Note that the higher sensitivity of the PD is achieved by having 5 times larger N in the present example. At $\eta = 1$, N_p of a PD is 130.6 Me$^-$, and it can far exceed the ROIC charge limit, N_c. As mentioned before, if the signal integration is limited to N_c, the maximum signal is fixed at N_c, while the temporal noise of a PD is $\sqrt{N_c}$ and that of a PC is $\sqrt{(2gN_c)}$. The S/N ratio of a PC over a PD is $\sqrt{[1/(2g)]}$. Therefore, NEΔT of a PC can be lower than that of a PD if $g < 0.5$ when the detection is readout charge limited.

1.7 Conclusion

In this chapter, we have reviewed some of the basic concepts in infrared detection. The purpose of infrared detection is to observe passive radiation emitted from the target according to its temperature and emissivity. By detecting the object's intrinsic emission, one can more reliably acquire the information of the object and avoid uncertainties related to external lighting conditions. The spectral emission characteristics of the object may also be used for target recognition and scientific studies. To detect terrestrial objects around 300 K, their thermal emission is peaked around 10 μm. Since this LWIR band falls into one of the atmospheric transmission windows, it can provide the strongest signal and is often preferred especially in high-speed applications. On the other hand, because of the smaller diffraction blur, larger thermal contrast, lower water vapor absorption, and higher detector operating temperature, 3–5 μm MWIR band also offers substantial advantages especially in detecting hotter objects. These two bands possess sufficient distinct advantages that warrant the development of two-color detection. Therefore, different detector architectures that detect two colors either simultaneously or sequentially are often sought.

The figure-of-merit in infrared detection is the detection range r in detecting isolated point objects or thermal sensitivity NEΔT in imaging extended scenes. These figures-of-merit can be concisely specified by a few array parameters: η, g, σ, and n_r. The ideal maximum range is obtained when $\eta = 1$, $g = 1$, $\sigma = 0$, and $n_r = 0$, where $r = 25.6$ km in MW and 68.1 km in LW for PCs. With $r \propto \eta^{1/4}$ in Eq. (1.37), r increases rapidly with η at small η but tends to saturate when $\eta > 0.5$. The r saturation is more pronounced when σ is substantial. Therefore, it is essential to yield $\eta \sim 0.5$. The parameter g appears in both signal and noise. Its effects are mostly canceled out in the S/N ratio. However, to minimize the effects of n_r, g should maintain to be 0.2 or larger. From Eq. (1.37), $r \propto \sigma^{-1/2}$ and $n_r^{-1/2}$, and if σ is less than 0.04% and n_r is less than 200, r will not be significantly affected. If we set $\eta = 0.5$, $g = 0.2$, $\sigma = 0.04\%$, and $n_r = 200$, the range in MW is 19.2 km, which is 75.0% of the ideal range. The range in LW is 51.6 km, which is 75.8% of the ideal range. The corresponding r of a PD is 24.4 km in MW and 58.4 km in LW, which are about 30% larger than that of a PC.

Similar observations are true for NEΔT for thermal imaging. The ideal NEΔT is 15.8 mK in MW and 6.94 mK in LW for PC. From Eq. (1.43), NEΔT $\propto \eta^{-1/2}$ and it becomes insensitive to η when $\eta > 0.4$. The photoconductive gain g cancels out in the S/N except in the read noise term and thus is not significant. Since NEΔT is linearly proportional to σ and n_r in Eq. (1.43), these two parameters have stronger effects on NEΔT than the detection range. If one can achieve $\eta = 0.4$, $g = 0.2$, $\sigma = 0.04\%$, and $n_r = 200$, NEΔT is 29.6 mK in the MW, which is 4.26 times larger than the ideal value and 25.0 mK in the LW, which is 3.60 times larger. For PDs, NEΔT for the same array parameters is 21.0 mK in the MW, which is 0.71 times lower than that of the PC, and 23.8 mK in the LW, which is 0.95 times lower.

The above discussion set the expected performance of an infrared FPA in LW and MW and the required array parameters to achieve this performance. These criteria set the stage for more detailed discussions on different detector material technologies. The discussion so far only concentrates on certain generic properties of a MW or a LW detector whether it is a PC or a PD. Each detector material will bring its own distinct attributes and the corresponding advantages and disadvantages. These characteristics are outside the scope represented by the four array parameters, η, g, σ, and n_r, which address only certain aspects on detection sensitivity. Some of these attributes may include linearity of photoresponse, optical and electrical crosstalk, $1/f$ noise, wavelength coverage, operating temperature, calibration stability, radiation hardness, producibility, and cost, etc. There is probably no single detector material that can provide the best option for all applications. The focus is often placed on choosing a technology that can fulfil the application requirement in the most cost-effective manner. This explains the presence of several technologies for MW and LW detection and their continual research and development. In this book, we concentrate on one of the technologies, the quantum well infrared photodetectors, which are commonly employed for LW infrared imaging.

References

1. Chen, Y., Sun-Mack, S., Minnis, P., Young, D. F. and Smith, W. L. (2004). Seasonal surface spectral emissivity derived from TERRA MODIS data, in *13th AMS Conference on Satellite Oceanography and Meteorology*.

2. Kalogerakis, K. S. (2019). A previously unrecognized source of the O_2 atmospheric band emission in Earth's nightglow, *Sci. Adv.*, 5, p. 9255.
3. The solar spectrum is derived from SMARTS2 version 2.9.2, which is the model used to generate the American Society for Testing and Materials (ASTM) terrestrial reference spectra for ASTM Standard G-173-03. The receiving surface is at 37° tilt toward the equator (*i.e.*, zenith angle = 53°) and facing the sun when the sun is at zenith angle of 48.19° south.
4. Lord, S. D. (1992). A new software tool for computing Earth's atmospheric transmission of near- and far-infrared radiation, Report No. NASA-TM-103957.
5. Berk, A., Conforti, P., Kennett, R., Perkins, T., Hawes, F. and van den Bosch, J. (2014). MODTRAN6: a major upgrade of the MODTRAN radiative transfer code, in *2014 6th Workshop on Hyperspectral Image and Signal Processing: Evolution in Remote Sensing (WHISPERS)*.
6. See for example: Fowles, G. R. (1975). *Introduction to Modern Optics*, 2nd Ed., Dover Publications, p. 115.
7. Arrington, D. C., Hubbs, J. E., Gramer, M. E. and Dole, G. A. (2000). Nonlinear response of QWIP detectors: summary of data from four manufacturers, *Proc. SPIE*, 4028, pp. 288–299.
8. Choi, K. K., Kennerly, S. W., Yao, J. and Tsui, D. C. (2001). Characteristics of QWIPs at low background, *Inf. Phys. & Tech.*, 42, pp. 221–235.
9. LaCroix, L. and Kurzius, S. (2005). Peeling the onion: an heuristic overview of hit-to-kill missile defense in the 21st century, *Proc. SPIE*, 5732, pp. 225–249.
10. Ugarte, A. R. (2009). Modeling for improved minimum resolvable temperature difference measurements, http://hdl.handle.net/10945/28367 (Thesis, Naval Postgraduate School, USA).

Chapter 2

Quantum Well Infrared Photodetectors

2.1 Introduction

Infrared detection usually utilizes optical transitions between valence band and conduction band as in intrinsic photodetectors, or between impurity band and conduction band as in extrinsic photodetectors. After Esaki and Tsu [1] proposed synthesized semiconductor superlattices, a new possibility of infrared detection was born. Electrons residing in a multiple quantum well (MQW) structure can be freed by photoexcitation and conduct through electron states above the quantum barriers, similar to an extrinsic photoconductor, except that the "impurity" states can now be engineered.

Chiu and Smith *et al.* [2, 3] was the first group to propose using GaAs/AlGaAs material to build MQW structures for infrared detection. Even in this early study, some of the potential advantages of the structures have been anticipated, which include the flexibility of tailoring detection wavelengths and the possibility of large-scale detector integration. However, these MQWs were not designed to have quantized energy levels. The infrared absorption relies on classical free electron absorption in bulk materials, which is a weak absorption process. The detectors cannot be made sensitive.

When each QW is thin enough to produce quantized states, the large dipole moment between envelope electron wavefunctions can lead to a much stronger optical absorption. Coon *et al.* [4, 5] proposed a detector that consists of a single QW and operates on intersubband transitions. The detector is not a photoconductor.

The detection involves charging and discharging the QW repeatedly. No experiments were reported.

West and Eglash [6] are the first researchers to observe intersubband absorption in GaAs/AlGaAs MQW structures. Subsequently, Levine and Choi *et al.* [7–12] made a series of investigations into the absorption and transport properties of these structures, based on which the first quantum well infrared photodetector (QWIP) was created. The QWIPs in these studies operated on optical transitions between the ground state and the first excited state that is slightly below the barrier. This type of structure is known as the bound-to-bound (B-B) state detector. Detectors with bound-to-continuum (B-C) transitions were later employed by Levine *et al.* [13–16] to obtain larger photocurrent. However, the detector dark current was also increased. Choi [17] made a systematic analysis on the energy level structures of QWIPs. From this analysis, the energy at the absorption maximum can be determined. To maximize the photocurrent to dark current ratio, the final state is placed approximately $\Gamma/2$ above the barrier height, where Γ is the full absorption width. This design utilizes the full absorption spectrum while keeping the thermal activation barrier as high as possible. This detector structure is referred to as the bound-to-quasi-bound (B-QB(+)) detectors. When higher operating temperature is important for an application, Gunapala *et al.* [18] recommended the strictly B-QB detectors, where the final state aligns with the barrier height. Other than photoconductive QWIPs, MQW photovoltaic detectors [19–21] and phototransistors [22] have also been proposed.

In fact, intersubband absorption and its use in infrared detection started before artificial semiconductor superlattices were proposed. Prior to the observation of intersubband transitions in MQWs, a large amount of literature on intersubband absorption has already been accumulated in the Si/SiO_2 system [23,24]. An n-channel metal-oxide-semiconductor field-effect-transistor based on intersubband absorption was studied as an infrared detector [25]. However, the photocurrent in this structure is produced to travel in the upper subband of a conducting channel between source and drain. A large dark current coexists in the lower subband of the same channel. The photoconductivity is therefore small. Furthermore, the separations of the subband energies in this system are too small for the usual mid-wave and longwave applications. Therefore, the field-effect-transistor structure is not being used.

2.2 Basic Material Structures of Quantum Well Infrared Photodetectors

A QWIP can be made either n-type or p-type. Since n-type materials are more sensitive due to the larger absorption coefficient, we will discuss these materials exclusively. A typical QWIP structure consists of a number of alternate material layers stacked together; each has a different bandgap. The material layers are grown sequentially by molecular beam epitaxy in mono-atomic layer fashion. Due to the bandgap difference, a potential well is created in the conduction band of the smaller band gap material. When each potential well is a few nanometers thick, the electron states in the well form discrete quantized levels. The smaller band gap layer is thus a quantum well. The quantization energies in the QW can be adjusted freely by changing the well width and the barrier height, and hence the energy spacing can be designed to match the detection wavelength of interest. Figure 2.1 shows a generic QWIP material structure, and Fig. 2.2 shows the corresponding conduction band diagram.

The wells are made of $In_yGa_{1-y}As$ while the barriers are made of $Al_xGa_{1-x}As$, where x and y are the Al molar ratio and In molar ratio, respectively. The inclusion of InGaAs increases the bandgap discontinuity, with which the structure can be designed to absorb at both mid-wavelength infrared (MWIR) and long wavelength infrared (LWIR). The InGaAs well includes ~5 Å of GaAs on each side to yield a better interface. The detection wavelength is determined by the material composition and the well width W. The barrier thickness B is chosen to be thick enough to suppress tunneling current along the z axis. The AlGaAs barriers at both ends are thinner to facilitate electron transport between contacts and adjacent QWs. The wells, as well as the contacts, are doped with silicon

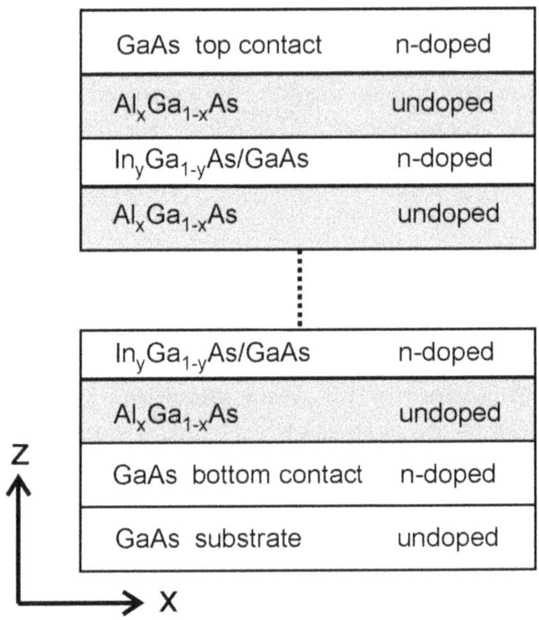

Figure 2.1. A generic QWIP material layer structure. The x axis is parallel to the layers and the z axis is perpendicular to the layers.

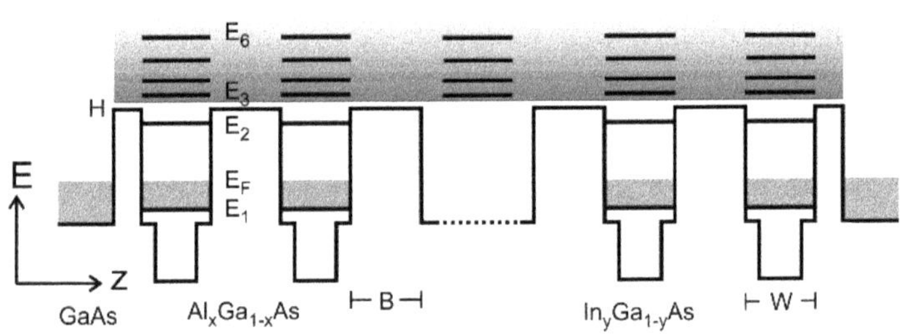

Figure 2.2. A typical energy diagram of an n-type QWIP along the z axis. The wells are made of GaAs/InGaAs/GaAs. The inclusion of InGaAs in the well is to cover both the MWIR and the LWIR applications. E_1 to E_6 are the resonant QW states. The shaded region is the global conduction band. Reproduced with permission from *IEEE J. Quant. Electr.*, 45(10), pp. 1255–1264 (2009).

to provide free electrons for optical absorption. The QWs therefore contain degenerate electron gas with a finite Fermi energy E_F.

A more accurate description of the band structure of a QWIP along the growth direction is that of minibands and the associated miniband Bloch wavefunctions. A Bloch wavefunction is a product of a periodic wavefunction $U_n(q, z)$ and a plane wave $\exp(iqz)$, where q is the miniband wavevector, and n is miniband index. The absorption spectrum is determined by miniband-to-miniband transitions in which q is conserved. However, when B is large, the miniband width is small and the number of minibands participated in the optical absorption is large. The overall absorption spectrum is determined by the variation of oscillator strengths among minibands rather than the variation within each miniband. Moreover, when the miniband width is smaller than the individual level broadening, which can be due to electron scattering or material layer nonuniformity, Bloch wavefunctions are not coherent in different layers. In this case, q is not conserved, and one can regard each QW as an isolated structure and use the single quantum well (SQW) eigenfunctions ψ_n and eigenvalues E_n to calculate its optical properties. From explicit numerical calculations, it is shown that the absorption spectrum of a SQW is practically indistinguishable from that of a MQW when $B = 300$ Å in the presence of a level broadening of 10 meV [26].

The barrier thickness B has little effects on the SQW energy structure when E is below the barrier height H. On the other hand, the resonant state energies E_n above H depends on B explicitly, *i.e.*, SQWs with different B will have different resonant energy structures. However, we will show that as long as the resonant energy level spacing is small compared with the level broadening, the absorption spectrum does not depend on B. The optical properties of a MQW structure are the same as that of a SQW regardless of their separation.

Although the energy levels in a SQW are discrete along the growth direction, the energy along the layer is still continuous. Each energy level is therefore also a subband. The total energy of an electron in a subband is continuous. Since the in-plane electron momentum is conserved during intersubband transition, the electron in-plane energy is unchanged in the process. Electrons with different in-plane momenta therefore have the same transition energy equal to the quantized level spacing. The existence of the extra dimensions is then immaterial to the optical properties of the detector. The material can thus be conveniently idealized as a one-dimensional (1D) quantum well, typically described as the first example in quantum mechanics.

With the flexibility of molecular beam epitaxy technology, a QW unit can be regarded as a 1D artificial atom. These atoms need not be identical in the structure, and a MQW can be viewed as a 1D artificial molecule. Different combinations of artificial atoms give rise to different molecular absorption spectra. This spectral versatility is a unique strength of QWIPs. Detectors of special functionality such as broadband absorption, two-color absorption, and voltage tunable absorption can be designed using certain molecular or superlattice structures. Designing a suitable QW structure for an intended application is part of the production process in QWIP technology. In detector operation, a voltage is applied between top and bottom contacts. In the presence of radiation, an electron in the ground state absorbs a photon and jumps to an excited state where the conductivity of the electron across the QWs is higher. A photocurrent is thus created.

2.3 Intersubband Transition

Normally, due to the small photon momentum, dipole transition between two electron momentum states in the same energy band is forbidden because of the conservation of momentum. In a QW structure, however, momentum conservation is relaxed in the z direction because of the lack of translational symmetry in this direction. Electrons can jump from the QW ground state E_1 to a higher eigenstate E_n if the incident photon has the matching energy equal to $E_n - E_1$. The intraband transitions between two in-plane momentum states remain forbidden. Consequently, QWs can only be responsive to the z-polarization component of the incident light. The z-component of the optical electric field drives the ground state electrons to oscillate until they jump to a higher energy state, in which optical energy is absorbed. With the uniaxial absorption properties, the QW structure behaves as an absorbing material for vertically polarized light but as a dielectric material for horizontally polarized light. Its permittivity is thus a tensor having diagonal elements $(\varepsilon, \varepsilon, \varepsilon - i\varepsilon_i)$, where ε is the relative permittivity of the QW material and ε_i is from intersubband absorption. When z-polarized light travels in the x direction, its intensity decays exponentially according to $I(x) = I_0 \exp(-\alpha x)$. The decay constant α is the absorption coefficient given by

$$\alpha = \frac{4\pi}{\lambda}\kappa = \frac{2\pi}{n\lambda}\varepsilon_i, \tag{2.1}$$

where $(n - i\kappa)$ is the z-component of the complex refractive index of the QW, and λ is the free-space incident wavelength.

The absorption coefficient $\alpha(\lambda)$ depends on the dipole oscillator strength f_n from the ground state E_1 to an excited state E_n. Each excited state is broadened by material nonuniformity and electron scattering. Therefore, the density of states of E_n is a combination of Gaussian distribution brought by the material nonuniformity and Lorentzian distribution brought by the electron scattering. In experiment, we found that the density of states can be represented adequately by a Gaussian distribution and written as

$$\rho_n(E) = \frac{1}{\sqrt{2\pi}\sigma_n} \exp\left(-\frac{(E - E_n)^2}{2\sigma_n^2}\right), \tag{2.2}$$

where σ_n is the line broadening parameter for state n. The quantity α_n can be obtained from Fermi's golden rule:

$$\begin{aligned} r_n(\hbar\omega) &= \frac{2\pi}{\hbar} \int \left| \left\langle \psi_n, N - 1 \left| \frac{e}{m^*} \boldsymbol{A} \cdot \boldsymbol{P} \right| \psi_1, N \right\rangle \right|^2 \\ &\quad \times \rho_n(E)\delta(E - \hbar\omega - E_1)dE, \\ &= \frac{\pi e^2 N \hbar^2}{m^{*2}\varepsilon_r \varepsilon_0 V \omega} \left| \left\langle \psi_n \left| \frac{\partial}{\partial z} \right| \psi_1 \right\rangle \right|^2 \rho_n(\hbar\omega + E_1), \end{aligned} \tag{2.3}$$

where r_n is the optical transition rate, N is the number of photons with energy $\hbar\omega$ within the active volume V, e is the electron charge, m^* is the electron effective mass, and $\varepsilon_r = n^2$ is the real part of the QW dielectric constant. In the first line of Eq. (2.3), the vector potential $\boldsymbol{A}$ is

$$\boldsymbol{A} = \left(\frac{\hbar}{2\varepsilon_r\varepsilon_0 V\omega}\right)^{1/2} (\boldsymbol{a}e^{-i\omega t+ikx} + \boldsymbol{a}^\dagger e^{i\omega t-ikx})\boldsymbol{u}, \tag{2.4}$$

where $\boldsymbol{u}$ is a unit-vector pointing in z, $\boldsymbol{a}$ and $\boldsymbol{a}^\dagger$ are photon creation and annihilation operators, k is the photon wavevector along x, and the momentum operator $\boldsymbol{P} = -i\hbar\partial/\partial z$ pointing in z.

In the second line of Eq. (2.3), $N/V = nI_s/(c\hbar\omega)$, where I_s is the incident intensity in $\mathrm{W/m^2}$, and c is the speed of light in vacuum. We further define the oscillator strength f_n to be

$$f_n = \frac{2\hbar}{m^*\omega_n}\left|\left\langle \psi_n \left| \frac{\partial}{\partial z} \right| \psi_1 \right\rangle\right|^2, \tag{2.5}$$

where we have approximated ω by $\omega_n = (E_n - E_1)/\hbar$ for each transition. Equation (2.3) then reads

$$r_n(\hbar\omega) = \frac{\pi e^2 I_s}{2m^* n\varepsilon_0 c\omega} f_n \rho_n(\hbar\omega + E_1). \tag{2.6}$$

The total transition rate to all n states for each electron is

$$r(\hbar\omega) = \sum_{n=2}^{\infty} r_n(\hbar\omega) = \frac{\pi e^2 I_s}{2m^* n\varepsilon_0 c\omega} \sum_{n=2}^{\infty} f_n \rho_n(\hbar\omega + E_1). \tag{2.7}$$

The number of transitions per unit volume $G(\hbar\omega)$ is then $(N_s/L)r(\hbar\omega)$, where N_s is the 2D electron density in each QW period, $L = W + B$ is the total thickness of each period, W is the well width, and B is barrier thickness. The optical transitions lead to an exponential decay of I_s so that $G(\hbar\omega)$ is also equal to

$$G(\hbar\omega) = -\frac{1}{\hbar\omega}\frac{d}{dx} I_0 e^{-\alpha x} = \frac{\alpha}{\hbar\omega} I_s. \tag{2.8}$$

Equating $G(\hbar\omega)$ in Eqs. (2.7) and (2.8) leads to

$$\alpha(\hbar\omega) = \frac{N_s}{L}\frac{\pi e^2 \hbar}{2m^* n\varepsilon_0 c} \sum_{n=2}^{\infty} f_n \rho_n(\hbar\omega + E_1). \tag{2.9}$$

The value of α can be calculated from material parameters. The quantities m^* and $n = \sqrt{\varepsilon_r}$ are taken to be average values between the InGaAs well and AlGaAs barrier material. For $In_yGa_{1-y}As$ with a small y value, we assume $\varepsilon_r(\text{InGaAs}) \cong \varepsilon_r(\text{GaAs})$, and its $\varepsilon_r(\lambda) = 14.662 - 0.466\lambda$ in the LWIR band, where λ is the incident wavelength in μm. For $Al_xGa_{1-x}As$, $\varepsilon_r(\lambda) = (1 - 0.251x)\varepsilon_{\text{GaAs}}$. For $Al_xGa_{1-x}As$, $m^* = 0.0665m_0(1 + 1.256x)$, and for $In_yGa_{1-y}As$, $m^* = 0.0665m_0(1 - 0.579y)$. The energy levels E_n and the wave functions ψ_n can be calculated using the transfer matrix method described later. Calculating f_n for all the excited states gives the overall absorption lineshape. To estimate the order of magnitude of α, let us assume the doping density N_D to be $1\times 10^{18}\,\text{cm}^{-3}$ in 50 Å QW. The 2D electron density $N_s = N_DW = 5 \times 10^{11}\,\text{cm}^{-2}$. The period length L is 550 Å, f is 1 at the absorption peak, and $\sigma = 8\,\text{meV}$ in level broadening. The value of α at the peak is calculated to be $0.24\,\mu\text{m}^{-1}$. This value is quite close to the commonly observed values of 0.15–0.22 μm^{-1}.

The oscillator strength f_n in Eq. (2.5) depends on the overlap integral between an eigenstate ψ_n and $\partial\psi_1/\partial z$. One can then adopt a physical picture in that the optical transition transforms the electron wavefunction ψ_1 in state E_1 into an excited state wavefunction $\chi = \xi\partial\psi_1/\partial z$, where ξ is a normalization quantity to keep $\int \chi^*\chi dz = 1$. The transition rate to a particular eigenstate E_n is proportional to the spectral weight of χ in ψ_n. Since χ is generally not an eigenstate of the QW, it can be expressed as a linear combination of the QW eigenstates so that

$$\chi = \sum_{n=2}^{\infty} a_n\psi_n, \tag{2.10}$$

where a_n is the coefficient of expansion. The oscillator strength to each E_n can then be expressed as

$$\begin{aligned} f_n &= \frac{2\hbar}{m^*\omega_n}\left|\left\langle \psi_n \left| \frac{\partial}{\partial z} \right| \psi_1 \right\rangle\right|^2 \\ &= \frac{2\hbar}{m^*\omega_n\xi^2}|\langle\psi_n|\chi\rangle|^2 = \frac{2\hbar}{m^*\omega_n\xi^2}|a_n|^2, \end{aligned} \tag{2.11}$$

where $|a_n|^2$ is the spectral weight. Since ψ_1 is tightly bound and remains the same as B changes, χ will also be highly localized and unchanged with B. χ is thus a property of the well. This physical picture is similar to the optical transition to one of the eigenlevels although this level is broadened by electron elastic and inelastic scattering. In this picture, there is only one excited level χ. The QW state ψ_n's merely sample different $|a_n|^2$ of χ at different E. The absorption spectrum is therefore unchanged with B. When the first excited eigenstate E_2 is a deeply bound state, ψ_2 is highly localized and resembles χ. In this case, $|a_2|^2 \approx 1$ and f_2 acquires a large value. By orthogonality of ψ_n's, $\langle\psi_n|\chi\rangle \approx 0$, if $n \neq 2$. Therefore, for a highly B-B detector, the transition is exclusively to E_2. For B-C detectors, the detection spectrum will be distributed according to the coupling of the localized χ and the barrier states that form the eigenstates of the entire quantum well structure.

2.4 Transfer Matrix Method

The eigenstates ψ_n and the eigenenergies E_n of an arbitrary QW structure can be quantitively obtained by the transfer matrix method (TMM). In this method, the electron wavefunction ψ_j in the jth material layer can be written as

$$\psi_j = A_j e^{ik_j z} + B_j e^{-ik_j z}, \tag{2.12}$$

where A_j and B_j are two constants to be determined, and k_j is the wavevector in z direction given by

$$k_j(E) = \frac{\sqrt{2m_j^*(E)}}{\hbar}\sqrt{E - \Delta E_j}, \tag{2.13}$$

where $m_j^*(E)$ is the energy-dependent effective mass, and E and ΔE_j are the electron energy and the conduction band offset measured from the GaAs conduction band edge, respectively. The constants A_j and B_j relate to A_{j+1} and B_{j+1} of the adjacent $(j+1)^{\text{th}}$ layer through boundary conditions. There is some ambiguity in the structures with different m^* in different layers. The most popular version [27] is

$$\psi_j = \psi_{j+1}, \tag{2.14}$$

and

$$\frac{1}{m_j^*}\frac{\partial \psi_j}{\partial z} = \frac{1}{m_{j+1}^*}\frac{\partial \psi_{j+1}}{\partial z}. \tag{2.15}$$

With this set of boundary conditions, the connection between A and B of two adjacent layers is given by

$$\begin{bmatrix} A_j \\ B_j \end{bmatrix} = \frac{1}{2} M_{j,j+1} \begin{bmatrix} A_{j+1} \\ B_{j+1} \end{bmatrix} \tag{2.16}$$

where

$$\begin{aligned} &M_{j,j+1} \\ &= \begin{bmatrix} (1+\gamma_{j,j+1})e^{i(k_{j+1}-k_j)d_{j,j+1}} & (1-\gamma_{j,j+1})e^{-i(k_{j+1}+k_j)d_{j,j+1}} \\ (1-\gamma_{j,j+1})e^{i(k_{j+1}+k_j)d_{j,j+1}} & (1+\gamma_{j,j+1})e^{-i(k_{j+1}-k_j)d_{j,j+1}} \end{bmatrix} \end{aligned} \tag{2.17}$$

and $\gamma_{j,j+1} = (m_j^* k_{j+1})/(m_{j+1}^* k_j)$ and $d_{j,j+1}$ is the spatial coordinate of the interface. The energy level structure of a multi-layer material system can be obtained by studying the global transmission coefficient T_G of a plane wave through the material. When the incident energy of the wave coincides with one of E_n's, strong resonant transmission occurs and T_G attains a local maximum. Assuming the plane wave incident from the left, the coefficients A_1 and B_1 for the first material layer on the left and the coefficients A_p and B_p of the last p^{th} layer on the right are connected by

$$\begin{bmatrix} A_1 \\ B_1 \end{bmatrix} = \frac{1}{2^{p-1}} M_{1,p} \begin{bmatrix} A_p \\ B_p \end{bmatrix}, \tag{2.18}$$

where $M_{1,p} = M_{1,2}M_{2,3}\ldots M_{p-1,p}$. By assuming that there is no wave traveling from the right in the last period, one can set $A_p = 1$ and $B_p = 0$, with which all the coefficients in the layers are determined. The value of T_G at a given E is then equal to

$$T_G(E) = \frac{1}{|A_1(E)|^2}\frac{v_p(E)}{v_1(E)} = \frac{2^{2p-2}}{|a_{11}(E)|^2}\frac{k_p(E)}{m_p(E)}\frac{m_1(E)}{k_1(E)} \tag{2.19}$$

where v is the electron group velocity, and a_{11} is the first diagonal element of $M_{1,p}$. The local maxima in $T_G(E)$ determine the locations of E_n. In general, these local maxima are less than unity unless the material structure is symmetrical. Calculating $A_n(E)$ and $B_n(E)$ at $E = E_n$ will yield the wavefunction ψ_n. If the first and last layers are made of the same material, $v_1 = v_p$ and T_G simply equals to $(2^{2p-2})/|a_{11}|^2$. In the presence of an electric field under detector operation, the above formulism can be modified by subdividing each material layer j into a number of sublayers n and digitizing the potential drop $V_{j,n}$ so that ΔE_j in Eq. (2.13) is replaced by $\Delta E_j - V_{j,n}$ in formatting the $p \times n$ transfer matrix. For symmetric QWs, the shift of the level spacing is small under bias because the first-order effect of ΔV is zero under parity symmetry. Therefore, the energy levels of a symmetric QW under operation are approximately equal to that at $V = 0$.

After E_n's are determined from the maxima locations of T_G, the wave amplitude A_j and B_j of individual layers can be obtained from Eq. (2.16). From the obtained wavefunctions ψ_n, the oscillator strength in Eq. (2.11) and the absorption coefficient in Eq. (2.9) can be theoretically calculated.

2.5 Material Parameters

For QWIP structures detecting at the 8–12 μm, the material is usually made of GaAs as the well layer and $Al_xGa_{1-x}As$ as the barrier layer. For this material system, it has been determined that the energy level can be accurately obtained by assuming the barrier height $\Delta E = H = 748x$ meV and the effective mass m^* is energy independent. In this structure and energy range, the band non-parabolicity is approximately cancelled out by many-body effects [28]. On the other hand, when the QWIP structure is applied to MWIR detection where the electron energy is larger, band non-parabolicity dominates and the energy dependency of m^* has to be taken into account. The many-body effects are then addressed by taking a larger barrier height, which serves as a fitting parameter. The material parameters of the $In_yGa_{1-y}As/Al_xGa_{1-x}As$ material system that can be used for both MWIR and LWIR detection were collected and summarized by Shi and Goldys [29]. The unstrained bandgap of the jth layer in the unit of meV is

$$E^u_{g,j} = 1424 + 1186x_j + 370x_j^2 - 1499y_j + 429y_j^2, \tag{2.20}$$

where x_j and y_j are the Al and In molar ratios in the jth layer, respectively. Since the material layers are under strain, the values of bandgaps are modified. Assuming the strain is accommodated exclusively in the InGaAs well layers, which are usually much thinner than the GaAs/AlGaAs layers, the modified bandgap for $y_j > 0$ is

$$E^s_{g,j} = E^u_{g,j} + 2a\left(1 - \frac{c_{12}}{c_{11}}\right)\varepsilon_{11}, \tag{2.21}$$

where $a = -8700 + 2700y_j$ meV is the deformation potential, $c_{12}/c_{11} = (5.32 - 0.79y_j)/(11.81 - 3.48y_j)$ is the ratio of the elastic constants, and $\varepsilon_{11} = 0.0078x_i - 0.405y_j/(5.6533 + 0.405y_j)$ is the hydrostatic component of the strain induced by the adjacent GaAs/AlGaAs material layers. Therefore, in the present assumption, the bandgap of InGaAs is widened while those of GaAs and AlGaAs are unchanged. The band offset ΔE_j is taken as

$$\begin{aligned} \Delta E_j &= \beta_{\text{AlGaAs}}(E^u_{g,j} - 1424) \quad \text{if } x \geq 0, \\ &= \beta_{\text{InGaAs}}(E^s_{g,j} - 1424) \quad \text{if } y > 0, \end{aligned} \tag{2.22}$$

where β_{AlGaAs} is the band offset parameter for the GaAs/AlGaAs interface and β_{InGaAs} is for the InGaAs/GaAs interface. The β parameters depend on material calibrations. Their values can vary among different material suppliers, in different ranges of x and y, in different growth reactors, and even in different wafer locations within a large growth reactor. A typical value for β_{AlGaAs} is about 0.67 for the LW detectors and 0.60 for the MW detectors. β_{InGaAs} can usually be taken to be 0.63. The effective mass at the conduction band edge is given by

$$m_j^* = 0.0665(1 + 1.256x_j - 0.570y_j). \tag{2.23}$$

Taking the band nonparabolicity into account, the effective mass $m_j^*(E)$ can be expressed in the form of

$$m_j^*(E) = m_j^* \left(1 + \frac{E - \Delta E_j}{E_{g,j}}\right) \tag{2.24}$$

where $E_{g,j}$ is the strain-modified bandgap. With these material parameters, all of the input parameters in the transfer matrix are known. Since m^* varies from layer to layer, strictly speaking, the oscillator strength of Eq. (2.5) needs to be evaluated individually in each layer. Here, for simplicity, we will use the weighted average of m_j^* in the well region according to its layer thicknesses, *i.e.*,

$$m^* = \frac{\sum_j m_j^* d_j}{\sum_j d_j}. \tag{2.25}$$

2.6 Perturbation Theory

Although the TMM method calculates the absorption spectrum accurately, it is rather lengthy. For this reason, one can employ perturbation theory to obtain the spectrum, albeit in a qualitative manner. Based on the physical picture of a single excited state χ with its spectral weight distributed across a range of energy, which is the result of the coupling of this state to the continuum states in the barrier, it acquires a finite energy width, analogous to scattering broadening. To calculate the density of states ρ_χ of χ, a model Hamiltonian is introduced [30]:

$$\mathcal{H} = E_\chi b^\dagger b + \sum_k E_k c_k^\dagger c_k + \sum_k D_k (c_k^\dagger b + b^\dagger c_k), \tag{2.26}$$

which describes a localized impurity state E_χ embedded in a continuous set of states E_k in the barrier medium. In Eq. (2.26), b and c are state operators, $D_k = |\langle E_k|H|E_\chi\rangle|$ is the coupling parameter, $|E_k\rangle$ are plane wave states with wavevector k, and H is the barrier height. It can be shown that D_k is equal to $2\hbar J_k$, where J_k is the 1D current density [31]. D_k determines the hopping rate of χ into the barrier states and can be calculated explicitly. However, we will treat it as a fitting parameter to explain the experimental data, similar to fitting the line broadening with σ in Eq. (2.2).

With the Hamiltonian $\mathcal{H}$, the first-order retarded self-energy Σ_{ret} is given by

$$\Sigma_{\text{ret}} = \sum_k \frac{D_k^2}{E - E_k + i\delta}, \tag{2.27}$$

where $\delta = 0^+$ is a positive infinitesimal. Using a free particle dispersion relation for the barrier states and assuming D_k is independent of k, the real part of Σ_{ret} is zero. The absorption peak is thus unaffected by the coupling. The imaginary part is given by

$$-Im\Sigma_{\text{ret}} = \frac{D^2 L}{\hbar}\left(\frac{m_b^*}{2}\right)^{1/2} \frac{1}{\sqrt{E}} = \frac{1}{C\sqrt{E}}, \tag{2.28}$$

where L is the period length, and C is a constant defined in Eq. (2.28). With a finite $Im\Sigma_{\text{ret}}$, the spectral function $A(E)$ is

$$A(E) = \frac{-2Im\Sigma_{\text{ret}}}{(E - E_\chi)^2 + (Im\Sigma_{\text{ret}})^2} = 2C\frac{\sqrt{E}}{1 + C^2E(E - E_\chi)^2}. \tag{2.29}$$

Here, E and E_χ are measured from the top of the barrier. By shifting the origin to the bottom of the well and equating the density of states ρ_χ to $A(E)/2\pi$, it reads

$$\rho_\chi(E) = \frac{C}{\pi}\frac{\sqrt{E - H}}{1 + C^2(E - H)(E - E_\chi)^2}. \tag{2.30}$$

Since the transition from E_1 to E_χ is the only transition in this impurity, its oscillator strength f_χ should be the sum of all f_n's in the TMM formulism. With the density of state of χ, the absorption coefficient spectrum is then given by

$$\alpha(E) = \frac{N_s}{L}\frac{\pi e^2\hbar}{2m^* n\varepsilon_0 c}\left(\sum_{n=2}^{\infty} f_n\right)\rho_\chi(E). \tag{2.31}$$

We expect $\Sigma f_n = 1$ as required by the Thomas-Reiche-Kuhn sum rule.

2.7 Bound-to-Quasi-Bound Detector Design

In this section, we consider a SQW structure made of GaAs well with width W and $Al_xGa_{1-x}As$ barrier with thickness B and barrier

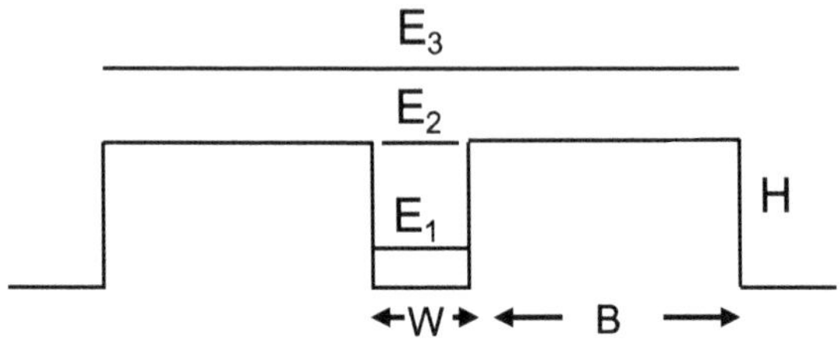

Figure 2.3. An energy diagram of a SQW structure. E_1 to E_3 are eigenstates of the structure.

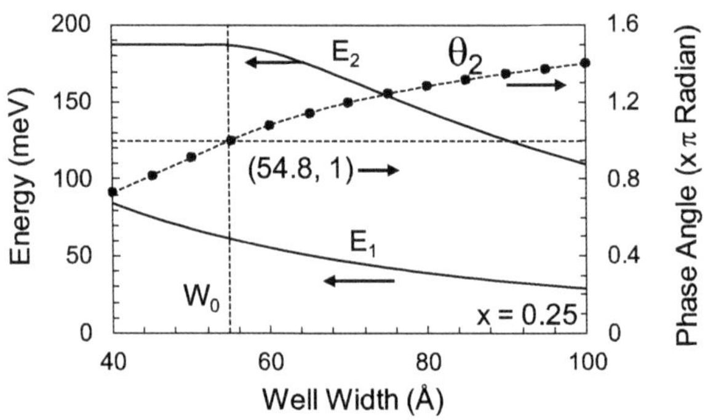

Figure 2.4. The energy level structure as a function of the well width for $x = 0.25$. The phase angle θ of E_2 is also shown. $\theta = \pi$ at $W = 54.8$ Å defines the B-QB detector structure.

height $H = 748x$ meV as shown in Fig. 2.3. It is terminated by two GaAs layers on both sides.

To understand the level structure of a SQW, we set x to be 0.25 and B to be 1000 Å and vary W. As seen in Fig. 2.4, when W decreases from 100 Å, E_1 and E_2 rise toward the top of the barrier H. When W decreases to a critical width $W_0 = 54.8$ Å, E_2 reaches $H = 187.0$ meV and remains very close to H thereafter. For example, at a small W of 40 Å, $E_2 = 187.4$ meV $\approx H$. At W_0, the energy spacing $E_2 - E_1$ is the largest, with which the thermal activation energy for the dark current will be the largest. At the same time, since E_2 is aligned with H, the photoelectrons promoted from E_1 to E_2 are free to conduct in the barrier conduction band without tunneling. Therefore, this QW structure $(W, x) = (54.8\,\text{Å}, 0.25)$

is the optimum structure in most applications. The corresponding detection wavelength $\lambda = hc/(E_2 - E_1) = 9.84\,\mu\text{m}$ in this example. This QW structure is known as the B-QB detector.

Since E_2 remains nearly the same as H once E_2 is raised above H, the unique B-QB structure cannot be defined by the position of E_2. Rather, the B-QB detector can be defined by the phase angle θ_2 of E_2. The phase angle is the total phase change from one end of the well to the other end. When W is large, E_2 is deep in the well, E_2 resembles a complete cycle of sine wave. θ_2 is then close to 2π. When E_2 increases towards the barrier top, θ_2 decreases. When ψ_2 is almost at the top of the barrier, the exponential decay length of ψ_2 in the barrier approaches ∞, with which $\partial\psi_2/\partial z = 0$ at $z = \pm W/2$. To match the boundary condition at these locations, ψ_2, which is a sine wave in the well region, must have phases equal to $\pm\pi/2$. The total phase change is then π for the B-QB detector.

For a given H (*i.e.*, a given x), there is one unique W that yields $\theta_2 = \pi$. Therefore, one can define the B-QB well structure by

$$\theta_2 = k_z W = \frac{\sqrt{2m^* E_2}}{\hbar} W = \frac{\sqrt{2m^* H}}{\hbar} W = \pi. \tag{2.32}$$

When $\theta_2 = \pi$ (or simply $\theta = \pi$ by dropping the subscript), the wave amplitude in the well is also the largest compared to other unbounded states since its amplitude reaches unity at the well edges. When W is narrower than W_0, E_2 and thus k_2 are practically unchanged as seen in Fig. 2.4. (E_2 is exactly equal to H if $B \to \infty$.) When $\theta = k_2 W$ becomes smaller, the amplitudes $\pm\sin(k_2 W/2)$ at $\pm W/2$ also become smaller, and so does the oscillator strength f_2. The absorption peak in a narrow well must then appear in the continuum above the barrier as required by the oscillator strength sum rule. A B-C detector thus forms. The absorption peak, however, does not necessarily locate at the energy E_m (with $m > 2$) where $\theta_m = \pi$, although the state E_m still has the maximum amplitude in the well. In a narrow QW, the E_1 state is also pushed up in energy and becomes less localized. The α value depends on the wavefunction in the well as well as in the barrier. If $m \gg 2$, the densely oscillatory ψ_m in the barrier can give an overall lower f_m. Based on an analytical approach, one can calculate

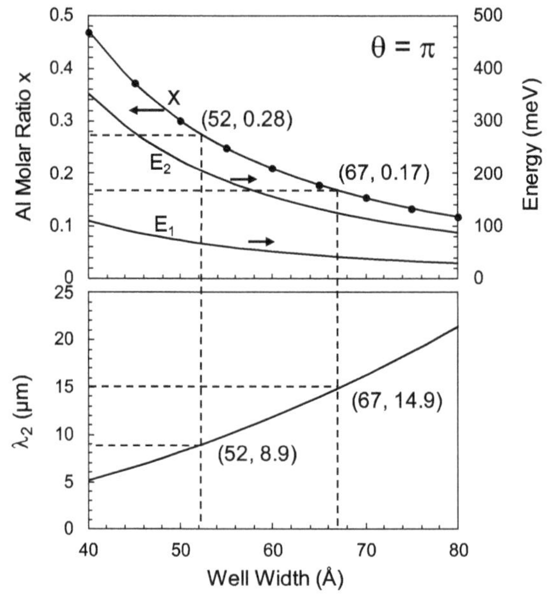

Figure 2.5. The B-QB detector design as a function of well width.

the peak absorption energy of the B-C detectors [17] but it will not be repeated here.

For the B-QB design, W and x are related by Eq. (2.32). Figure 2.5 shows x, E_1, E_2, and λ_2 as a function of W, where λ_2 is the wavelength of the transition from E_1 to E_2. For example, when W is 52 Å, x should be 0.28, and λ_2 is 8.9 μm. When W is 67 Å, x should be 0.17 and λ_2 is 14.9 μm, etc. Conversely, for a desired λ_2, the values of x and W can be read off from Fig. 2.6. From this plot, if 9 μm is required, x should be 0.272 and W should be 52.5 Å. Finally, λ_2 can also be plotted against x as shown in Fig. 2.7. In addition to the theoretical λ_2, Fig. 2.7 also shows the peak wavelengths λ_p observed in responsivity measurements. The observed λ_p generally follows the trend of the B-QB design but it is shorter than λ_2.

Firstly, these detectors were not designed strictly according to the B-QB design rule. The absorption maximum can be above the barrier height and in the continuum. Secondly, even when this design is followed, the photoelectrons created under the barrier need tunneling

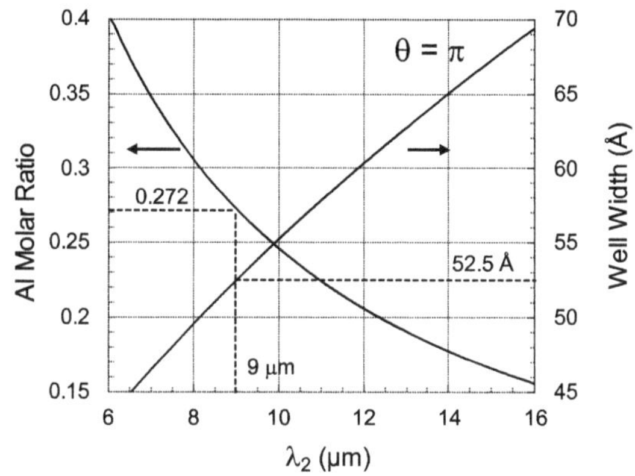

Figure 2.6. The values of x and W as functions of λ_2 for a B-QB detector.

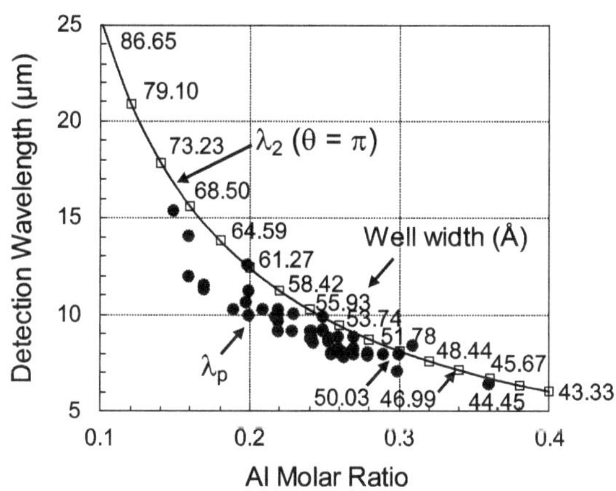

Figure 2.7. The experimental peak detection wavelength λ_p and the designed λ_2 as functions of x.

to be free as conducting electrons. The detection spectrum can be skewed by the rising tunneling probability of the barrier. Thirdly, the photoconductive gain can be larger when the photoelectrons have certain excess energy above the barrier. Finally, even with B-QB detectors, λ_2 and the maximum absorption wavelength are not the same as shown next. Nevertheless, the tuning curve in Fig. 2.7 is a useful guide to the expected λ_p when B-QB detectors are adopted.

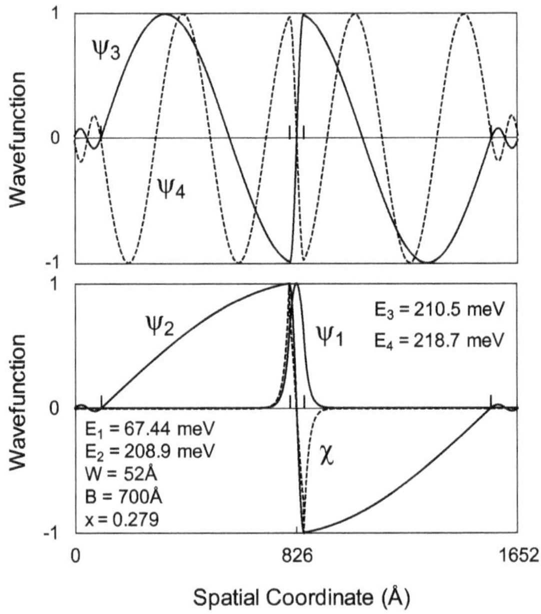

Figure 2.8. The calculated ψ_n for a B-QB detector.

To study the absorption properties of the B-QB detectors in more detail, let us consider a specific example, in which $W = 52$ Å, which gives $x = 0.279$ and $H = 208.7$ meV for $\theta = 1.002\pi$ design. The barrier width B is chosen to be 700 Å. With these QW specifications, TMM yields $E_1 = 67.44$ meV and $E_2 = 208.9$ meV. E_2 is 0.2 meV above H. The corresponding ψ's are shown in Fig. 2.8, in which ψ_1 and its normalized derivative $\psi_1' \equiv \chi$ are seen to be highly localized in the well. On the other hand, ψ_2, although its energy is only slightly higher than H, is already delocalized into the barriers. At E_2, $\theta_2 = 1.007\pi$, and its amplitude reaches ± 1 at $\pm W/2$. Meanwhile, the higher energy states ψ_3 and ψ_4 also have similar amplitudes around χ with energies $E_3 = 210.5$ meV and $E_4 = 218.7$ meV. At a result, although the oscillator strength f_2 to E_2 is the largest at 0.177 in Fig. 2.9, those at higher energy transitions such as $f_3 = 0.156$ and $f_4 = 0.142$ are not significantly less than f_2. The transition to E_2 thus does not dominate the optical absorption in B-QB detectors. In fact, $f_{15} = 0.0067$ to the 15th state at 356.3 meV is still appreciable.

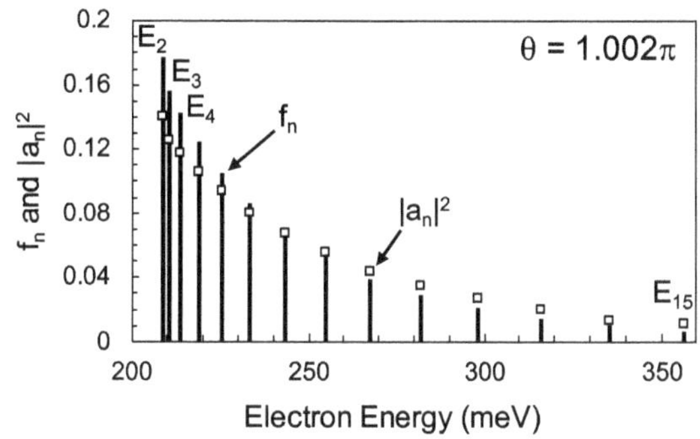

Figure 2.9. The distributions of f_n and $|a_n|^2$ as a function of electron energy.

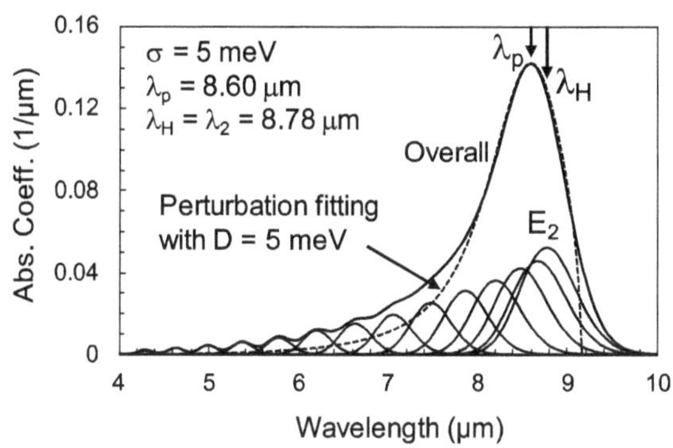

Figure 2.10. The absorption spectrum of a B-QB detector. The detection peak is 0.18 μm, shorter than the wavelength λ_2 to E_2, which is also the wavelength λ_H that can overcome the barrier H. The theoretical FWHM Γ is 1.00 μm.

The sum of f_2 to f_{15} is 1.032, which is close to the expected value from the sum rule.

If the line broadening σ_n is assumed to be the same at 5.0 meV for all energy states in Eq. (2.2), the overall spectral lineshape of the detector is shown in Fig. 2.10. It has a full width at half maximum (FWHM) Γ of 1.0 μm. The peak wavelength λ_p is 8.60 μm, which is 0.18 μm shorter than λ_2 of 8.78 μm. In Fig. 2.10, we also define λ_H, which is the wavelength that excites the ground state electrons to the top of the barrier. When E_2 is exactly aligned with H, λ_H is equal to λ_2. When the doping density N_D is 1×10^{18} cm^{-3} in the

52 Å QW, N_s is $5.2 \times 10^{11}\,\text{cm}^{-2}$. For a QW period L of 752 Å, the peak α is calculated to be $0.142\,\mu\text{m}^{-1}$.

The absorption spectrum calculated from TMM can be fitted using Eq. (2.30) obtained from the perturbation theory. In this theory, the coupling parameter D is a measure of the broadening of the virtual state χ due to the decay of χ into the barrier states, and it is the only fitting parameter. The other parameters in ρ_χ are known from the design. In Eq. (2.30), E_χ is the expectation energy of χ, which is given by

$$\langle E_\chi \rangle = \sum_{n=2}^{\infty} a_n^2 E_n \approx \sum_{n=2}^{15} a_n^2 E_n = 219.04\,\text{meV}. \tag{2.33}$$

However, to fit the TMM α curve with ρ_χ, we need to take two other factors into account. Firstly, the TMM α curve takes electron scattering and nonuniformity of material composition into account. These factors broaden the individual energy levels and make optical transitions below the barrier possible even when all the excited eigenstates are above H. Therefore, to allow for transitions below H in the perturbation theory, an effective barrier height $H' < H$ is introduced. This approach is similar to the reduction of bandgaps in the presence of band-edge impurity states in bulk semiconductors. Secondly, although $f_n = 2\hbar|a_n|^2/(m^*\omega_n\xi^2)$ and $|a_n|^2 \propto \rho_\chi$ share close similarity in Fig. 2.9, their overall energy dependencies are not exactly the same because of the additional $1/\omega_n$ dependence in f_n. Therefore, when ρ_χ is fitted to α, the apparent E'_χ can be different from $\langle E_\chi \rangle$. With these two considering factors, ρ_χ is fitted by adjusting all three parameters, D, H' and E'_χ in Eq. (2.30). The result is shown in Fig. 2.10 as the dashed curve using $D = 5.0\,\text{meV}$, $H' = 203.0\,\text{meV}$, and $E'_\chi = 206.0\,\text{meV}$. The fact that $D = \sigma$ indicates that the state χ decays into the barrier states in the similar time frame as the scattering time. H' being 5.7 meV below H is consistent with the presence of transitions 5.0 meV below the barrier. E'_χ being 13 meV below $\langle E_\chi \rangle$ is also consistent with the steeper energy dependence of f_n relative to $|a_n|^2$. With these fitting parameters, the perturbation theory matches the TMM α spectrum near the peak region satisfactorily. In addition to providing a more convenient way to fit experimental data, the perturbation theory also preserves the simple notion that an electron absorbs a photon and jumps to an upper state in a QW structure to produce a photocurrent, independent of the detector types.

2.8 Bound-to-Bound Detector

When θ_2 is substantially larger than π, the transition is bound-to-bound. An example is a detector consisting of 50 periods of QWs with $W = 70$ Å, $B = 150$ Å, and $x = 0.27$. To fit the observed λ_p, x is adjusted to 0.29, with which $H = 748x = 216.9$ meV. The calculated ψ_1 to ψ_4 are shown in Fig. 2.11 with $E_1 = 47.9$ meV and $E_2 = 177.5$ meV. θ_2 is therefore 1.244π. Notice that the derivative of ψ_1 has spikes at the well edges. It is due to m^* being different in the well and barrier so that ψ_1' is discontinuous in Eq. (2.15). This discontinuity also exists in Fig. 2.8 for the B-QB detector but it is less noticeable. Since ψ_2 and ψ_1' are very similar in shape, f_2 is large at 1.004. On the

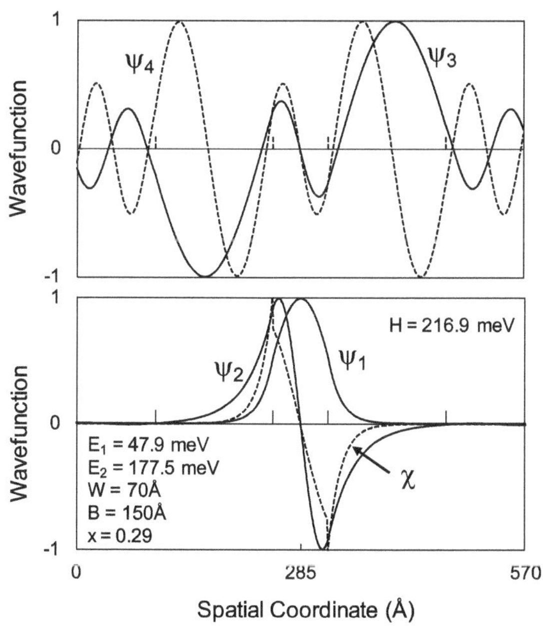

Figure 2.11. The calculated wavefunctions of a B-B detector.

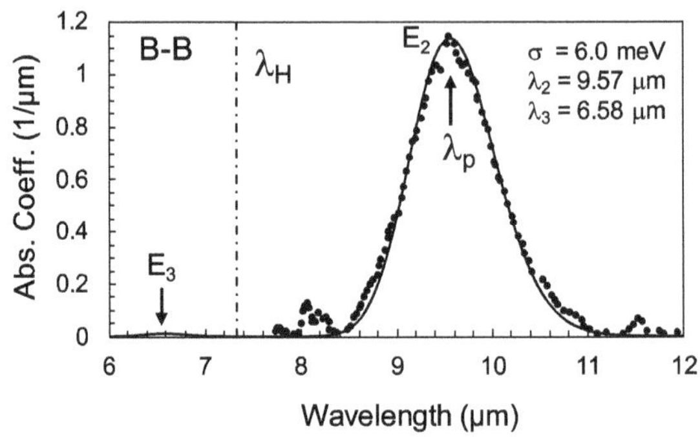

Figure 2.12. The measured and fitted α lineshape of a B-B detector. The measured FWHM Γ is $1.04\,\mu$m.

other hand, ψ_3 and ψ_4 locate away from the well, which gives small $f_3 = 0.010$ and $f_4 = 0.0028$. By fitting the individual lineshape with a Gaussian distribution, the broadening parameter σ is found to be 6.0 meV, and the fitted lineshape is shown in Fig. 2.12. The spectrum has a FWHM Γ at $1.04\,\mu$m. This absorption spectrum is measured using a Fourier transform infrared spectrometer at Brewster angle (73°). The electric field is polarized in the plane of incidence onto a detector wafer at room temperature. With the maximum $f_\rho = 0.0668$, $L = 220$ Å, and $N_D = 1 \times 10^{18}\,\text{cm}^{-3}$, the theoretical α_p is $1.137\,\mu\text{m}^{-1}$ at $9.57\,\mu$m.

2.9 Bound-to-Quasi-Bound (−) Detector

When the second level E_2 is approaching the top of the barrier from below, it forms a B-QB(−) detector structure. An example is a detector with 50 QWs. The nominal $B = 300$ Å, $x = 0.23$, and $W = 52$ Å. Based on the observed spectrum, W is adjusted to 58.2 Å and x is 0.243. The wavefunctions ψ_1, ψ_2, and the normalized derivative χ are shown in Fig. 2.13. E_2 is 180.9 meV, which is 0.9 meV below H of 181.8 meV. $\theta_2 = 1.044\pi$, yielding a weakly bound E_2 state. The oscillator strength is shown in Fig. 2.14. Although $f_2 = 0.57$ is dominant, f begins to spread to other high-energy states. The calculated and measured absorption spectrum are shown in Fig. 2.15. It exhibits a slightly asymmetric lineshape of a B-C

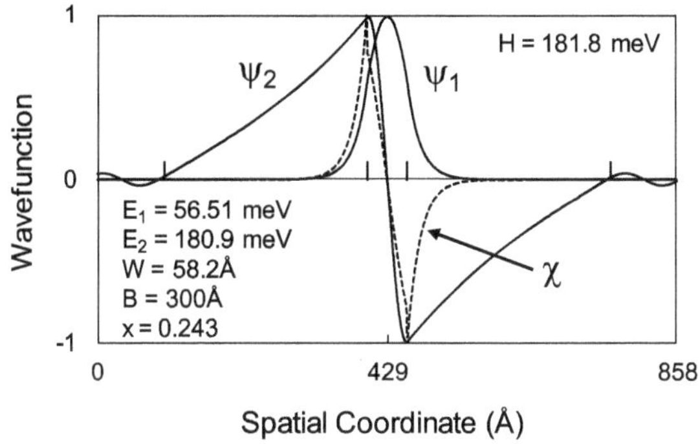

Figure 2.13. The calculated wavefunctions of a B-QB(−) detector.

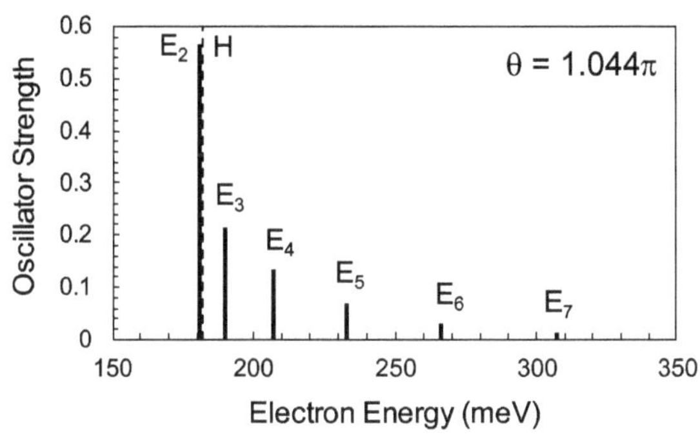

Figure 2.14. The calculated oscillator strengths of a B-QB(−) detector.

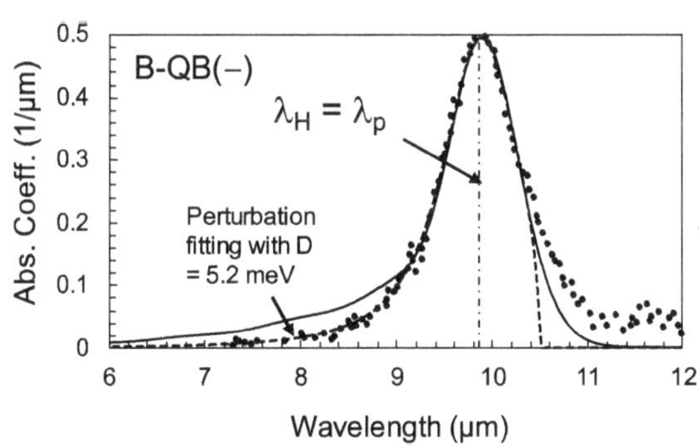

Figure 2.15. The measured and fitted α lineshapes of a B-QB(−) detector. The TMM α is shown as the solid curve and the perturbative α is shown as the dashed curve. The measured FWHM Γ is 0.91 μm.

detector. The linewidth Γ remains to be small at $0.91\,\mu$m, similar to a B-B detector. The peak wavelength λ_p aligns almost exactly with λ_H, the absorption wavelength, to reach the top of the barrier. This example shows that it needs a slightly bound state to align the absorption peak with H. Since there is an abrupt spectral change when E_2 emerges from the well, it is worth to distinguish B-QB($-$) detectors and B-QB($+$) detectors according to θ_2.

To fit the measured spectrum, σ_n's in the TMM calculation are chosen to change linearly from $\sigma_2 = 4.4$ meV to $\sigma_7 = 19.4$ meV. The larger σ at the higher energy is needed to produce a non-oscillatory α at the shorter wavelengths, and it can be attributed to the shorter lifetime at the higher energy. In the perturbation theory, the fitted H' is 174 meV and it is 7.8 meV smaller than H. $E'_\chi = 179$ meV and $D = 5.2$ meV. With these fitting parameters, the fitting agrees with the data below $10.5\,\mu$m. With the overall maximum $f_\rho = 0.0571$, $L = 350$ Å, and $N_D = 1 \times 10^{18}\,\text{cm}^{-3}$, the theoretical α_p is $0.50\,\mu\text{m}^{-1}$ at $\lambda_p = 9.9\,\mu$m for this B-QB($-$) detector.

2.10 Bound-to-Quasi-Bound (+) Detector

When E_2 is slightly above H, a B-QB($+$) detector is formed. An example is a detector with 30 QWs, $W = 50$ Å, $B = 500$ Å, and $x = 0.25$. To fit the experiment, x is adjusted to 0.265. The wavefunctions are calculated in Fig. 2.16. E_2 is 199.3 meV, and it

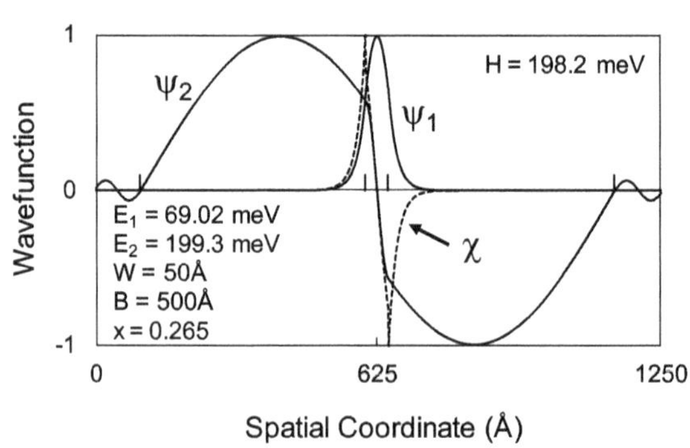

Figure 2.16. The calculated wavefunctions of a B-QB(+) detector.

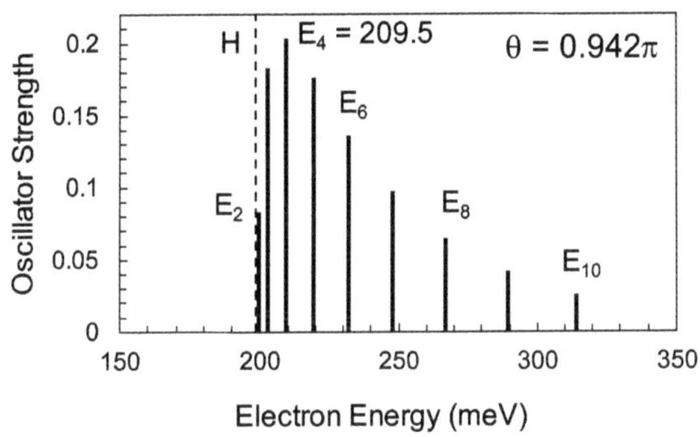

Figure 2.17. The calculated oscillator strengths of a B-QB(+) detector.

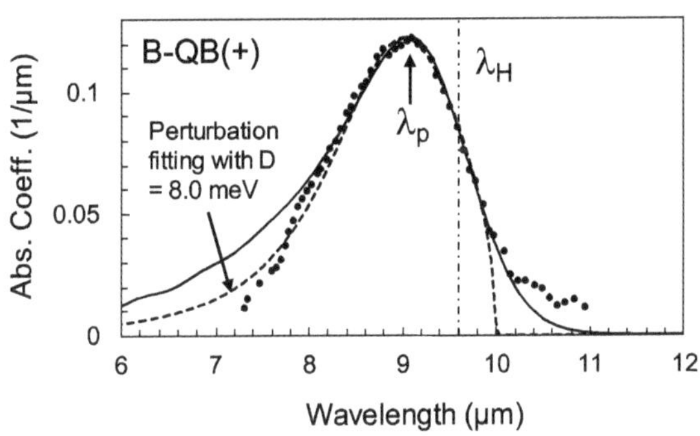

Figure 2.18. The measured and fitted α lineshape of a B-QB(+) detector. The measured FWHM Γ is $1.78\,\mu$m.

is 1.1 meV above $H = 198.2$ meV with $\theta_2 = 0.942\pi$. The oscillator strength f is shown in Fig. 2.17. Because of $\theta_2 < \pi$, ψ_2 does not reach ± 1 at the well boundaries and its amplitude reduces in the well. The resulting f_2 $(=0.083)$ is smaller than f_3 $(=0.182)$ and f_4 $(=0.203)$, which is the largest. Since E_4 is at 209.5 meV, the absorption peak is 11.3 meV above H. The absorption spectrum shows a characteristic asymmetric lineshape in Fig. 2.18 with $\lambda_p = 9.0\,\mu$m, $\lambda_H = 9.6\,\mu$m, and $\Gamma = 1.78\,\mu$m, which is substantially larger than that of a B-QB(−) detector.

This example shows a small shift of E_2 across the barrier can significantly shorten λ_p while broadening Γ. The corresponding coupling parameter D is larger at 8.0 meV in Fig. 2.18. The larger

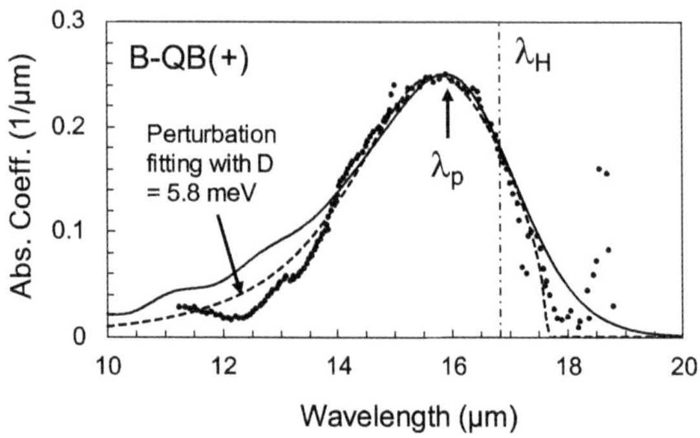

Figure 2.19. The measured and fitted α lineshape of a B-QB(+) detector. The measured Γ is 3.31 μm.

D indicates the level coupling to be stronger when χ is higher in the continuum. The fitted H' and E'_χ in the perturbation theory are 193 meV and 198 meV, respectively. With the maximum $f_\rho = 0.0252$, $L = 550$ Å, and $N_D = 1 \times 10^{18}$ cm^{-3}, the theoretical α_p is 0.123 μm^{-1} at $\lambda_p = 9.0\,\mu$m.

Another B-QB(+) detector is presented. Its absorption spectrum is shown in Fig. 2.19, which has a longer λ_p at 15.8 μm, while its λ_H is at 16.8 μm. The detector has 30 QWs with $W = 66$ Å, $B = 500$ Å, and $x = 0.15$. To fit the experimental spectrum, x changes to 0.1515. The linewidth Γ is 3.31 μm, which is substantially larger than that of 1.78 μm of the previous detector. The larger Γ of the second detector is because of its longer λ_p. Since $\partial\lambda = -\lambda^2\partial E/hc$, for the same spectral broadening in $\Delta E = \Gamma$, $\Delta\lambda \propto \lambda^2\Gamma$ will be larger when λ_p is at a longer wavelength. For the same ΔE, the bandwidth ratio of the two detectors would have been differed by $\Gamma_2/\Gamma_1 = (\lambda_{p2}/\lambda_{p1})^2 = (15.8/9.0)^2 = 3.08$. Meanwhile, the measured Γ ratio is 1.86, indicating that the second detector has a smaller ΔE by a factor of 0.60. The smaller ΔE is consistent with the smaller fitted D of 5.8 meV than the first D of 8.0 meV by a factor of 0.72. The smaller D of the second detector indicates it is closer to being a B-QB detector. With the maximum calculated $f_\rho = 0.0401$, $L = 566$ Å, and $N_D = 1 \times 10^{18}$ cm^{-3}, α_p is 0.25 μm^{-1}.

Although the α lineshape of a QWIP can be deduced from transmission measurements, it is not quantitative because of the presence of free carrier absorption (FCA) in the top and bottom contact regions. Free carrier absorption is possible with elastic impurity scattering and inelastic phonon scattering to relax the momentum conservation, and it rises as $\lambda^{3.2}$ when $\lambda > 5\,\mu$m. The absorption coefficient α of free carrier absorption is about $0.02\,\mu\text{m}^{-1}$ at $\lambda = 10\,\mu$m, $N_D = 1 \times 10^{18}\,\text{cm}^{-3}$, and $T = 300\,$K [32]. The rising absorption background at the longer λ needs to be subtracted out to yield the intersubband absorption lineshape. For detectors with $\lambda_p > 10\,\mu$m, background subtraction can introduce an error. Therefore, the absorption lineshape is more accurately measured from the photocurrent spectrum.

From the photocurrent measurement, the detector spectral responsivity $R(\lambda)$, in the unit of A/W, can be determined. From the photocurrent I_p in Eq. (1.21),

$$R(\lambda) = \frac{I_p(\lambda)}{P(\lambda)} = \frac{eg\eta\Phi}{\left(\frac{hc}{\lambda}\right)\Phi} \propto \frac{eg\lambda}{hc}(1 - e^{-\alpha(\lambda)t}) \approx \frac{egt}{hc}\lambda\alpha(\lambda), \quad (2.34)$$

where $P(\lambda)$ is the optical power in Watt, and t is the optical pathlength in the material. In Eq. (2.34), $\alpha t \ll 1$ is assumed, and in this case, $\alpha(\lambda) \propto R(\lambda)/\lambda$. Therefore, by measuring the $R(\lambda)/\lambda$ spectrum, the α lineshape can be obtained. Figure 2.20 shows the measured $R(\lambda)/\lambda$ in arbitrary units for four detectors along with the

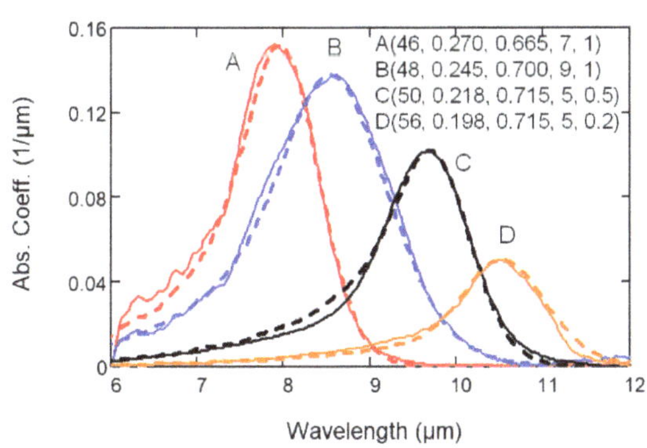

Figure 2.20. The measured $R(\lambda)/\lambda$ lineshapes (solid curves) and the calculated α for four detectors having $(W, x, \beta_x, \sigma, N_D)$ shown in the figure, where N_D is in $10^{18}\,\text{cm}^{-3}$. The barrier thickness is 500 Å for all detectors except 600 Å for detector D.

calculated $\alpha(\lambda)$. In the TMM calculation, band nonparabolicity is taken into account but without many-body effects. Therefore, the apparent β_{AlGaAs} in Eq. (2.22) needs to adjust to fit the observation. From this set of samples, β_{AlGaAs} varies from 0.665 to 0.715 when the nominal x decreases from 0.27 to 0.198. With the broadening parameter σ between 5 and 9 meV, the intrinsic absorption lineshape can be satisfactorily explained.

2.11 Coupled Ground State Detector

From Eq. (2.9), α of a QWIP depends on the electron volume density in each period $N_s/L = N_D W/L$. To increase α, one can increase the doping density N_D in the well, increase W, or decrease L. However, increasing N_D will increase the Fermi level and therefore the thermally activated dark current. Increasing W is infeasible as W controls the detection wavelength. Decreasing the period length L by decreasing the barrier thickness below 300 Å will increase the tunneling dark current. Therefore, there are limitations in increasing α of a QWIP. In the past, there were attempts in increasing α by placing more QWs in each period. Intuitively, it should increase α proportionally without increasing the dark currents. Unfortunately, explicit modeling of α below shows that it is generally not the case. In Fig. 2.21 we adopt the SQW structure mentioned in Section 2.10, which has $W = 50$ Å, $B = 500$ Å, $x = 0.265$, and $N_D = 1\times10^{18}\ \text{cm}^{-3}$.

The largest oscillator strength of this SQW structure is at E_4 with $f_4 = 0.20$, and the sum of the calculated oscillator strength $\Sigma f_n = 1.01$. Assuming a broadening parameter σ of 7 meV, the peak α is calculated to be $\alpha_p = 0.114\ \mu\text{m}^{-1}$, and the 50% cutoff wavelength λ_c is at 9.8 μm. Another measure of the absorption properties of the structure is the integrated absorption strength (IAS) defined as

$$\int \alpha(\hbar\omega)d(\hbar\omega) = \frac{N_s}{L}\frac{\pi e^2\hbar}{2m^* n\varepsilon_0 c}\sum_{n=2}^{\infty} f_n \int \rho_n(\hbar\omega + E_1)d(\hbar\omega)$$
$$= \frac{N_s}{L}\frac{\pi e^2\hbar}{2m^* n\varepsilon_0 c}\sum_{n=2}^{\infty} f_n. \tag{2.35}$$

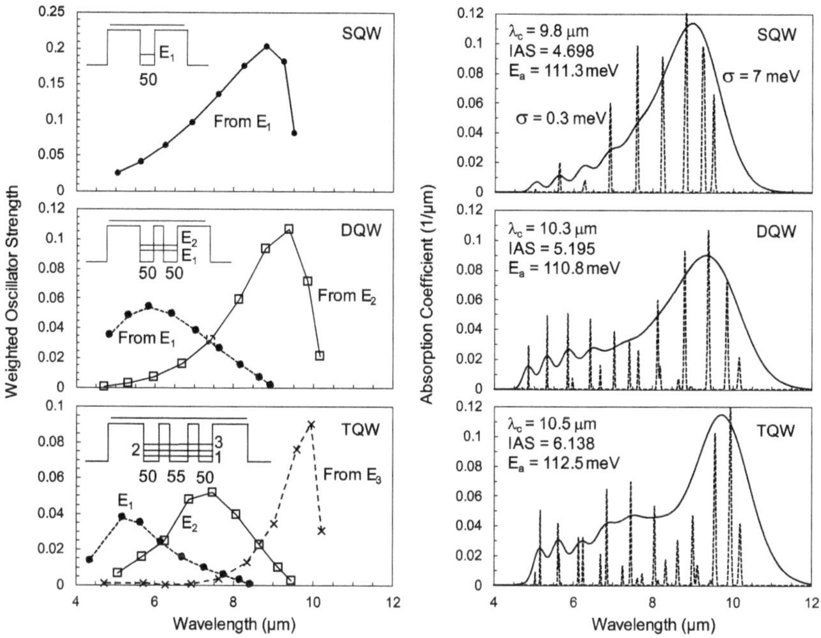

Figure 2.21. The modeled oscillator strength and absorption coefficient of single, double, and triple quantum wells. α with $\sigma = 7\,\text{meV}$ is to show the expected absorption spectrum while that with $\sigma = 0.3\,\text{meV}$ is in arbitrary units to show the underlying optical transitions.

Inputting $N_s = N_D W$ and $L = W + B$, IAS of the present SQW structure from Eq. (2.35) is $4.698\,\text{meV}/\mu\text{m}$. Directly integrating the α spectrum in Fig. 2.21 as a function of $\hbar\omega$ yields IAS $=$ $4.675\,\text{meV}/\mu\text{m}$, which shows the calculation consistency. The smaller $\sigma = 0.3\,\text{meV}$ in Fig. 2.21 is to show the transition wavelengths. To determine the Fermi level in this structure, we note that the spin-degenerated, 2D density of states per unit area in GaAs is given by

$$g_{2D}(E) = \frac{dN}{dk}\frac{dk}{dE} = 2\frac{2\pi k}{(2\pi)^2}\frac{m^*}{\hbar^2 k} = \frac{m^*}{\pi\hbar^2}, \tag{2.36}$$

which is a constant and is equal to $1.744 \times 10^{36}/\text{J/m}^2$ or $2.794 \times 10^{10}/\text{meV/cm}^2$. When the doping density N_D is measured in the unit of $10^{18}/\text{cm}^3$, and the well width W is measured in $10^{-8}\,\text{cm}$, the 2D electron density N_s is equal to $N_D W$ in $10^{10}\,\text{cm}^{-2}$. Using these units, the Fermi energy E_f is $N_s/g_{2D} = (N_D W/2.794)$ meV and the Fermi level E_F measured from the GaAs band edge is $E_f + E_1$. For the present SQW with $N_D = 1$, and $W = 50$, the Fermi energy and Fermi level are 17.90 meV and 86.92 meV, respectively. The activation energy E_a for the thermally activated dark current is $H - E_F = 111.3$ meV.

When another 50 Å QW is inserted into the period and it is separated from the first QW by a 30 Å barrier, a double QW structure is formed as shown in Fig. 2.21. Instead of simply doubling the oscillator strength at each transition, the two degenerate ground levels hybridize and form two new levels $E_1 = 61.40$ meV and $E_2 = 77.60$ meV. Therefore, the QW structure behaves as a new artificial molecule and has its own distinctive absorption spectrum. Due to parity symmetry, the even parity E_1 state only excites to the odd parity upper states and vice versa for the odd parity E_2 state. The sum of the oscillator strength $\Sigma f_n = 1.885$, indicating that there are two absorbing oscillators. All the excited states remain above the barrier. Since the two states have a common E_F, the electron population of each state will be different. The self-consistent E_F of the DQW is 87.40 meV, very similar to the 86.92 meV of the SQW, while the Fermi energy of E_1 is $E_{f1} = 26.0$ meV and that of E_2 is $E_{f2} = 9.8$ meV. Since E_{f2} is about half of the SQW $E_f = 17.90$ meV despite the total $N_s = N_{s1} + N_{s2}$ being double in the DQW, we define the weighted oscillator strengths, $f'_n \equiv N_{sn} f/N_s$, and they are shown in Fig. 2.21. The maximum f'_2 from E_2 is 0.1066, in comparison with SQW $f = 0.2033$. The lower f'_2 should be compensated exactly by the doubled N_s. However, the period length of the DQW is 80 Å larger than that of the SQW because of the added QW, which dilutes the volumetric density. In addition, the DQW upper level wavefunctions are distributed over a larger period and thus have less spectral overlap with χ. These two factors lead to a lower DQW $\alpha_p = 0.090\,\mu\text{m}^{-1}$ than the $0.114\,\mu\text{m}^{-1}$ for SQW. Therefore, adding another doped QW to a SQW period can actually reduce the peak absorption, contrary to expectation.

On the other hand, the DQW in Fig. 2.21 shows more transition peaks and a broader absorption band. The IAS calculated from the

α spectrum is 5.195 meV/μm, which is slightly larger than the SQW IAS of 4.675 meV/μm. The DQW thus has a small improvement in the overall absorption ability while keeping the activation energy E_a almost the same at 110.8 meV. To compensate for the lower α_p of a DQW, a third QW can be added to the period, and it has a slightly wider W of 55. The structure is shown in Fig. 2.21. With more transitions, $\Sigma f_n = 2.747$, and $\alpha_p = 0.115\,\mu\text{m}^{-1}$, almost the same as the $0.114\,\mu\text{m}^{-1}$ for the SQW. The IAS is larger at 6.139 meV/μm. The entire spectral detection range is from 5 μm to 11 μm, representing a 6 μm bandwidth. The self-consistent E_F of the TQW is 85.74 meV, yielding $E_a = 112.5$ meV. The activation energy of the TQW is larger than that of the SQW because the TQW E_1 (= 55.30 meV) state is lower than the SQW E_1 (= 69.02 meV) state after level hybridization. Therefore, the coupled ground state QW structure is a viable approach to widen the spectral absorption without increasing dark current. However, it cannot be used to increase the peak quantum efficiency (QE).

2.12 Coupled Excited State Detector

Instead of designing detectors with nearly degenerate ground states, one can also design detectors with coupled excited states. In Fig. 2.22,

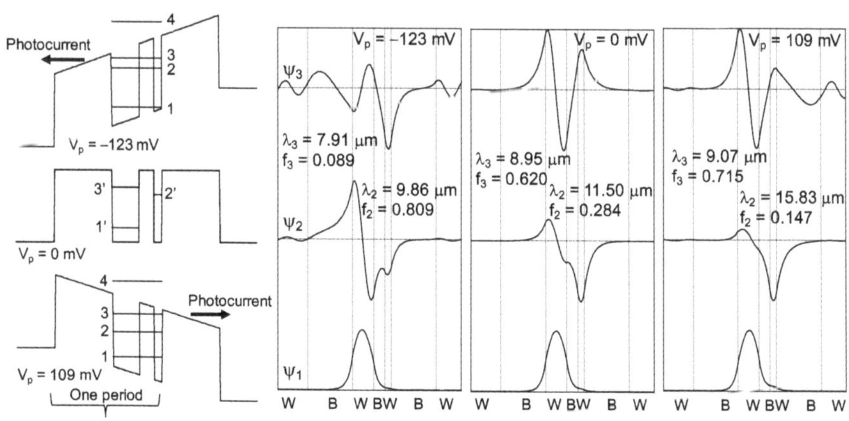

Figure 2.22. The band structure and the wavefunctions of a coupled excited state detector under positive forward bias, zero bias, and negative reverse bias.

a detector contains a thick well W_1 and a much thinner well W_2 in each unit cell. The second state of an isolated W_1, which is labeled as e'_3, is very close to the first state e'_2 of an isolated W_2 so that e'_2 and e'_3 are coupled when the two are placed together and form E_2 and E_3. In the present example, $W_1 = 72\,\text{Å}$ and $W_2 = 20\,\text{Å}$, and they are separated by a 39 Å barrier. Each coupled QW unit cell is then followed by a thicker 154 Å barrier to form a MQW structure. The barrier x is 0.31 with $H = 231.9\,\text{meV}$. Both $e'_2 = 157.9\,\text{meV}$ and $e'_3 = 177.5\,\text{meV}$ are below H. After coupling, $E_2 = 154.8\,\text{meV}$ and 185.5 meV, and the detector is a B-B detector.

The wavefunctions of the detector under different voltages are shown in Fig. 2.22. When there is no bias, *i.e.*, the voltage per period $V_p = 0$, ψ_2 is more localized in the thinner well W_2, and this localization grows stronger at forward bias such as at $V_p = 109\,\text{mV}$. In contrast, ψ_2 is more localized in the thicker well W_1 under reverse bias. Because of the changes of the wavefunctions with V_p, the corresponding oscillator strength and the absorption spectrum depend on the applied bias. Figure 2.23 shows the f spectrum at several V_p values. Under forward bias, the state E_2, which is mainly originated from e'_2, moves further away from e'_3, and thus loses the

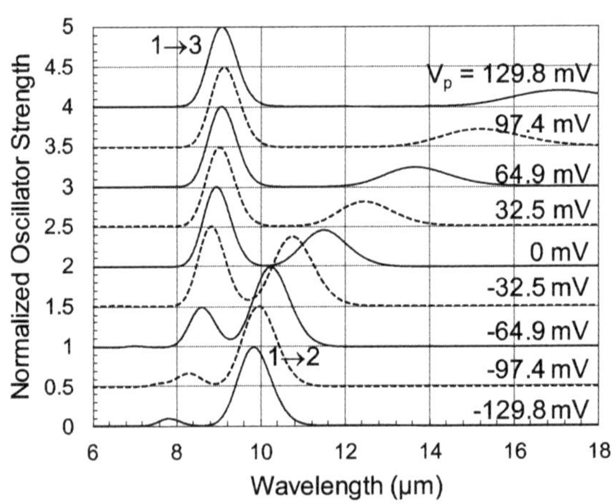

Figure 2.23. The normalized oscillator strength of the coupled QW under forward (+) bias and reverse (−) bias.

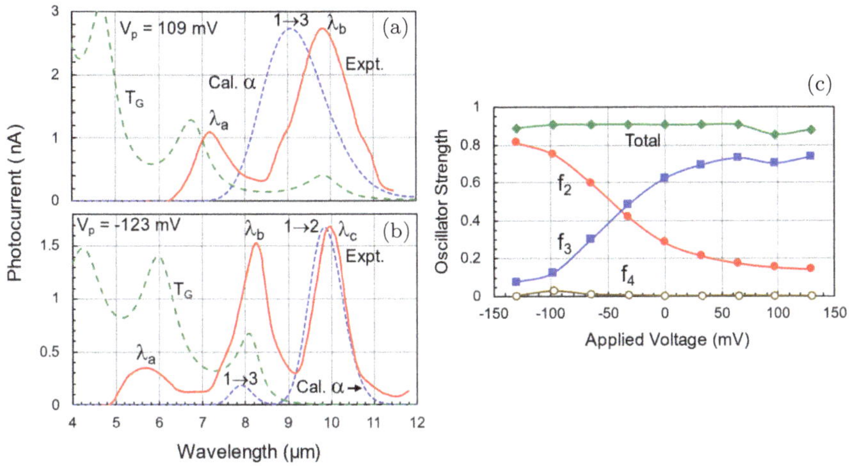

Figure 2.24. The experimental I_p spectrum and calculated α and T_G at (a) forward bias and (b) reverse bias. α and T_G are in arbitrary units. (c) The calculated oscillator strengths of the first three excited states and their total value under different biases.

oscillator strength, and the corresponding transition shifts to a longer peak value λ_{2p}. The main transition is then from E_1 to E_3 under positive bias. On the other hand, when reverse bias is applied, e_2' moves toward e_3' at first and gains oscillator strength, resulting in a closely spaced double peak. When e_2' moves past e_3', the coupled E_2 state is mainly localized in W_1 and gains a large f. The main transition is then from E_1 to E_2 under negative bias.

However, the photocurrent spectrum shown in Fig. 2.24 shows more peaks than expected. For example, the calculated α curve shows that under 109 mV in Fig. 2.24(a), there should only be one absorption peak at $\lambda_3 = 9.1\,\mu$m from $1 \rightarrow 3$ transition. Instead, two photocurrent peaks at $\lambda_a = 7.2\,\mu$m and $\lambda_b = 9.8\,\mu$m are observed. Meanwhile, under -123 mV in Fig. 2.24(b), one major α peak from $1 \rightarrow 2$ transition and 1 minor peak from $1 \rightarrow 3$ transition are predicted. However, three I_p peaks at $\lambda_a = 5.7\,\mu$m, $\lambda_b = 8.3\,\mu$m, and $\lambda_c = 9.9\,\mu$m are observed. Although the two longer wavelength peaks λ_b and λ_c match quite well with the α prediction, the magnitude of the λ_b peak is much larger than that predicted from f_3. Note that the two λ_a peaks under both biases cannot represent the f_4 peak because λ_4 is at 5.2 μm under forward bias and 4.6 μm under reverse bias. They are much shorter than $\lambda_a = 7.2\,\mu$m and 5.7 μm, respectively. In addition, the calculated f_4 are 6×10^{-4} and 1×10^{-4}, respectively, which are insignificant compared to f_2 and f_3 in Fig. 2.24(c). As in the usual B-B detectors, nearly all the oscillator strength goes to the bound states, which yield a constant sum of $\sim$0.90 under all bias in this example. Therefore, the λ_a peaks cannot represent absorption peaks.

In the detection of light, a detector must generate free conducting electrons after absorption in order to produce photocurrent. In a B-B detector, the production of photocurrent can be seen as a two-step process: an optical transition from the QW ground state to a localized excited state and a tunneling transfer of the localized state to a conducting barrier state. With an absorption QE η and a tunneling probability γ, the probability of producing a free electron for each incident photon is $\eta_d = \eta\gamma$. The detection QE η_d is defined analogously to the external QE of a photodiode, which is the probability of producing an electron that passes through the p/n junction for each incident photon. Together with a finite gain g of a QWIP, the conversion efficiency CE is then equal to $\eta_d g = \eta\gamma g$, which is the probability of producing a physical electron in the external circuit. The value of γ can be obtained from $T_G(E)$ in Eq. (2.19).

At large bias, W_1 and W_2 are largely decoupled, and the photoelectrons mainly reside in the thicker well W_1. Under reverse -123 mV bias, the electrons must tunnel through the 154 Å barrier in order to be free. The calculated transmission coefficient T_G of this barrier assuming linear voltage profile across the structure is shown in Fig. 2.24(b). Using E_1 as the energy reference in the wavelength conversion, T_r rises exponentially with decreasing λ and exhibits oscillations at short λ when the tunneling is into one of the resonant states in the tilted barrier. The large resonant tunneling near λ_b will then produce a large photocurrent peak despite having a small f_3 at this wavelength. A small peak at λ_a can then be attributed to the even larger T_G peak near 6 μm despite the very weak absorption produced by f_3 and f_4 at that wavelength.

Under forward 109 mV bias, the photoelectrons in W_1 must tunnel through the double-barrier structure formed by the thin and thick barriers displayed in Fig. 2.22. The double-barrier structure allows resonant tunneling through e_2' of W_2 and thus attains a larger T_G at $\lambda = 6.7\,\mu\text{m}$ and $9.9\,\mu\text{m}$ as seen in Fig. 2.24(a). These T_G peaks modify the absorption spectrum and produce two photocurrent peaks at λ_a and λ_b, respectively. Through double-barrier resonant tunneling, the magnitude of the photocurrent at λ_b is also twice that under reverse bias. Therefore, quantum tunneling plays an important role in determining the photocurrent spectrum in this coupled excited state detector. Even for a regular MQW structure, the oscillatory quantum mechanical transmission coefficient near the top of the barrier may play a role in determining the observed detection QE $\eta_d(\lambda)$ spectrum.

2.13 Superlattice Detector

By joining quantum wells closely together, a superlattice (SL) structure is formed. Its energy structure forms minibands in the growth direction. If all the well widths are the same, the miniband wavevector q is conserved in the transitions, which are represented by the vertical transitions in the E versus q dispersion diagram. On the other hand, when the well widths are different, the eigenstates have mixed symmetry and transitions among different miniband eigenstates are allowed, which can be used to broaden the detection spectrum. For example, in order to cover the entire 8–14 μm LWIR window, a binary SL shown in Fig. 2.25(a) is considered.

The structure contains 7 SL units, and these units are separated by thick 600 Å $Al_{0.17}Ga_{0.83}As$ barriers. Each SL unit contains four 70 Å and four 75 Å wells alternately placed together. The QWs are separated by 30 Å barriers. The doping is $3.8 \times 10^{17}\,\text{cm}^{-3}$ in the wells and the barriers are undoped. With this N_D, all the eigenstates in the first miniband M_1 are populated. Since the 8 M_1 states have different energies, their electron populations are also different. In this example, the first state E_1 in M_1 is at 39.4 meV and the last state E_8 is at 55.7 meV, while the self-consistent common Fermi

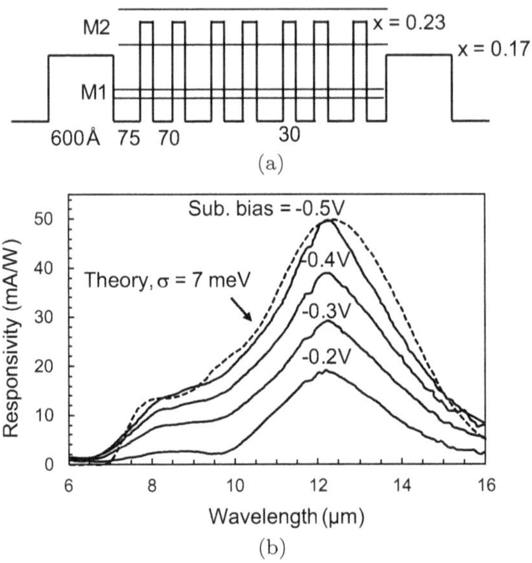

Figure 2.25. (a) The energy diagram of a SL detector. (b) The measured responsivity (solid curves) at different substrate bias voltages. The fitted spectral lineshape of $\lambda\alpha(\lambda)$ with $\sigma = 7\,\text{meV}$ is shown as the dashed curve.

level is at 57.8 meV. As a result, the electron population in E_1 is 8.8 times larger than that of E_8, resulting in significant modification of the absorption lineshape. Calculating the absorption coefficient in Eq. (2.7) with different effective 3D doping densities (N_{sn}/L) for different E_n, the spectral responsivity lineshape of the SL is obtained in Fig. 2.25(b) as the dashed curve. It agrees with the measured spectrum using a constant σ of 7 meV and β_{AlGaAs} of 0.75. The spectral responsivity is seen to extend from 7.5 to 16 μm, covering the entire LWIR atmospheric transmission window.

Using a SL with different combinations of well widths, a QWIP with a wide range of cut-on and cutoff wavelengths can be designed. To expand further from the binary width design, Fig. 2.26(a) shows a SL that contains four different well widths ranging from 50 to 80 Å. With this well width combination, the theoretical α spectrum can be extended from 10 μm to 30 μm in Fig. 2.26(b).

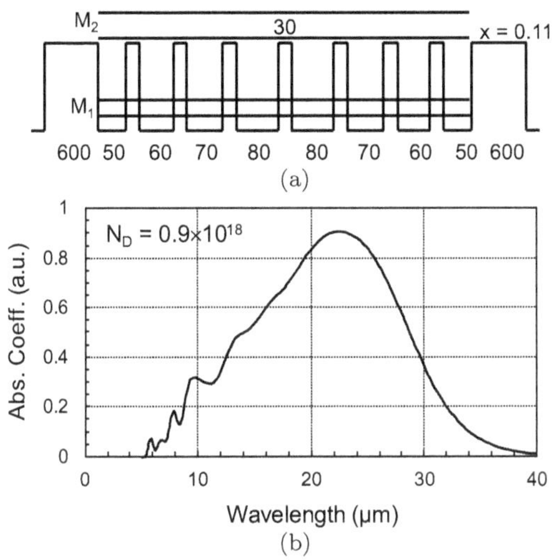

Figure 2.26. (a) A SL design and (b) the calculated absorption coefficient. Adapted with permission from *Opt. Eng.*, 50(6), p. 061005-061005-6 (2011).

2.14 Mid-Wavelength Detector

The $In_yGa_{1-y}As/Al_xGa_{1-x}As$ material system can be used for 3–5 μm MWIR detection. In Fig. 2.27, three typical MQW structures are shown, in which detector A is a B-QB(−) detector, detector B is a B-QB(+) detector, and detector C is a B-QB detector. Det. A consists of 23.5 periods of 300 Å $Al_{0.4}Ga_{0.6}As$ barriers and 5 Å GaAs/30 Å $In_{0.2}Ga_{0.8}As$/5Å GaAs wells. The doping density N_D is $2 \times 10\,\text{cm}^{18}$ in the wells. Det. B consists of 20.5 periods of 300 Å $Al_{0.4}Ga_{0.6}As$ barriers and 5 Å GaAs/25 Å $In_{0.28}Ga_{0.72}As$/5Å GaAs wells doped to $2.5 \times 10\,\text{cm}^{18}$. Det. C consists of 20.5 periods of 200 Å $Al_{0.33}Ga_{0.67}As$ barriers and 5 Å GaAs/30 Å $In_{0.3}Ga_{0.7}As$/5 Å GaAs wells doped to $3 \times 10\,\text{cm}^{18}$.

Using TMM and using $\beta_{\text{AlGaAs}} = 0.60$ and $\beta_{\text{InGaAs}} = 0.63$ in Eq. (2.22), the calculated energy levels of SQW structures are shown in Fig. 2.27. Their E_2 state energy relative to the barrier height H defines their detector type. For Det. A, since $E_2 = 308\,\text{meV}$ is below $H = 320\,\text{meV}$, the E_2 transition dominates the absorption with

$f_2 = 0.99$. Other oscillator strengths are much lower at $f_3 = 0.043$, $f_4 = 0.065$, $f_5 = 0.054$, and $f_6 = 0.036$. The sum of f is 1.19. For Det. B, its E_2 state is slightly above H and therefore the f variation is more gradual at $f_2 = 0.26$, $f_3 = 0.29$, $f_4 = 0.24$, $f_5 = 0.17$, $f_6 = 0.11$, and the sum is 1.13. For Det. C, its $E_2 = 258.5\,\text{meV}$ is nearly aligned with $H = 259.0\,\text{meV}$. The f variation is in between the above two detectors with the sum of 1.16.

To fit the experimental responsivity spectral lineshapes in Fig. 2.27, the calculated $\lambda\alpha(\lambda)$ is shifted in wavelength by $0.165\,\mu\text{m}$, $0.12\,\mu\text{m}$, and $-0.19\,\mu\text{m}$ for Det. A, B, and C, respectively. The individual line broadening σ_n for all E_n is 10.5 meV for Det. A. For Det. B and C, σ_2's increase from 10 meV and 8.5 meV to 15 meV and 23.5 meV, respectively, to smooth out the predicted spectra and match the experimental observation. The increased σ can be explained by the stronger electron intervalley scattering at higher energies. Because of the large doping density and thinner barriers

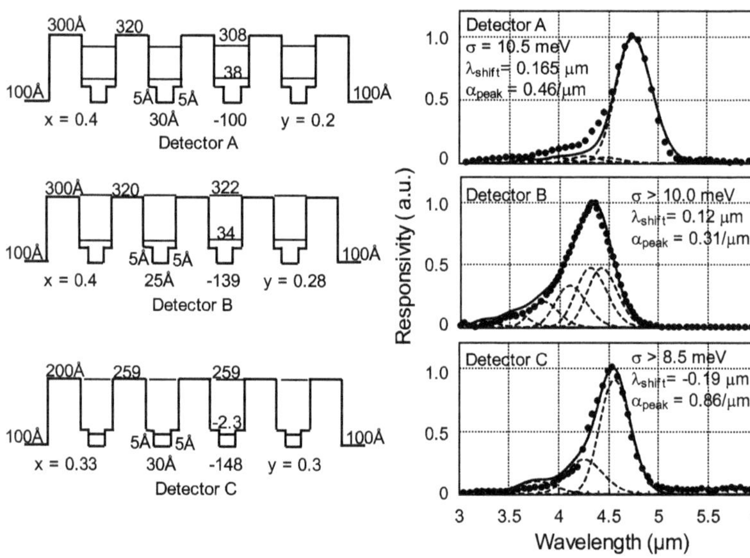

Figure 2.27. The band structures and responsivity spectra of three MWIR MQW detectors. Each represents a 4-QW structure truncated from the actual detector. The numerical valves without units are the energies in the unit of meV. Adapted with permission from [26].

in these detectors, the volume density N_s/L is generally larger than that of the typical LWIR designs. As a result, the calculated peak absorption coefficient α_{peak} of Det. A, B, and C are large at 0.46, 0.31, and 0.86 μm^{-1}, respectively.

Although the above MQW structures provide strong absorption, their spectral widths Γ are about 0.5 μm. The smaller Γ compared to that of the LWIR detectors is due to Γ being scaled as λ^2 for the same line broadening in energy. This narrow bandwidth may not be adequate for many applications. To widen its spectral coverage, SL structures are again considered. The first SL detector, labeled as Det. D, contains eight SL units, each separated by a 300 Å $Al_{0.33}Ga_{0.67}As$ blocking barrier. Each SL unit has five wells and four barriers. The wells are made of 5 Å GaAs/24 Å $In_{0.3}Ga_{0.7}As$/5Å GaAs layers, and the barriers are made of 30 Å $Al_{0.33}Ga_{0.67}As$ layers. The doping density is 2×10^{18} cm^{-3}. The band diagram of a SL unit is shown in Fig. 2.28.

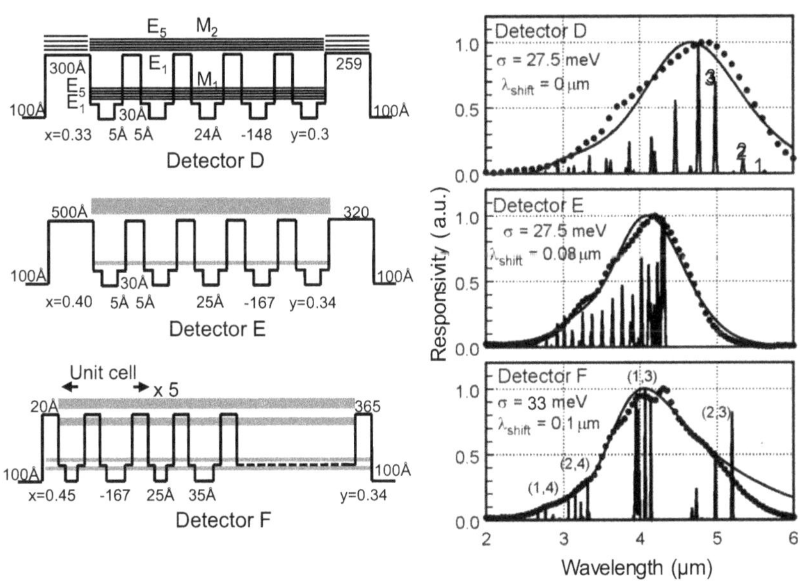

Figure 2.28. The band structure and the responsivity spectra of three MWIR SL detectors. The measured lineshapes are shown as circles while the solid curves are fittings.

If the blocking barriers of Det. D in Fig. 2.28 are replaced by the same 30 Å barriers in the SL, there are only five optical transitions between the first miniband M_1 and the second miniband M_2, and the miniband wavevector q is conserved. The transitions are $(1,5), (2,4), (3,3), (4,2)$, and $(5,1)$, where (i,j) denotes the transition between the ith state in M_1 and the jth state in M_2. With the 300 Å blocking barriers at both ends, these layers introduce their own quantization. The hybridization of the SL states and the barrier states creates three new levels below and one new level above the M_2 miniband. With the hybridized energy levels, the q selection rule is relaxed and there are weak transitions to other miniband states. The number of transition lines, revealed using a small $\sigma = 0.5\,\text{meV}$ in Fig. 2.28, thus increases. The additional transitions to the three new energy states, labeled by 1, 2, and 3, also help to increase the cutoff wavelength and enable better fitting with experiment. Therefore, the blocking barriers have tangible effects on the spectral lineshape. The fitted line broadening σ is 27.5 meV, and there is no adjustment in the peak wavelength. This SL detector has a bandwidth of 1.8 μm, and it is about 3 times wider than the MQW detectors.

Since Det. D is peaked near 5 μm, Det. E is designed to shorten the peak wavelength to ~4 μm. It consists of eight SL units separated by 500 Å $Al_{0.4}Ga_{0.6}As$ blocking barriers. Each unit consists of five wells and four barriers. The wells are made of 5 Å GaAs/25 Å $In_{0.34}Ga_{0.66}As$/5Å GaAs layers, and the barriers are made of 30 Å $Al_{0.4}Ga_{0.6}As$ layers. The doping density is $2 \times 10^{18}\,\text{cm}^{-3}$ in the wells and thin barriers. Since the blocking barrier is thicker than that of Det. D, the number of levels in M_2 is greatly increased from 5 to 24 after level hybridization. In fact, the excited states of Det. E form a quasi-continuum above H with no clear distinction among different minibands. Each SL unit can be regarded as a SQW structure embedded in the barrier material as discussed in Section 2.6. With 5 states in M_1 and 24 states in M_2 between 320 and 500 meV, there are 120 possible transitions. Figure 2.28 shows that out of all these possible transitions, 22 of them have appreciable oscillator strength. Since these transitions are derived from the 5 unperturbed transitions, the oscillator strength follows a similar rising trend as Det. D. With the same 27.5 meV for σ and a shift of the calculated spectrum by $\lambda_{\text{shift}} = 0.08\,\mu$m, the theory matches well with experiment. The peak wavelength is centered near the targeted 4.0 μm, and its Γ is 1.15 μm. The smaller Γ by a factor of 0.64 compared to Det. D is consistent with the expected value of 0.70 from λ^2 scaling.

To further increase the bandwidth, binary SL is adopted. Detector F consists of eight SL units separated by 500 Å $Al_{0.35}Ga_{0.65}As$ blocking barriers. Each SL consists of five subunits. The material composition of the subunit is 20 Å $Al_{0.45}Ga_{0.55}As$/5Å GaAs/35 Å $In_{0.34}Ga_{0.66}As$/5 Å GaAs/20 Å $Al_{0.45}Ga_{0.55}As$/5 Å GaAs/25 Å $In_{0.34}Ga_{0.66}As$/5 Å GaAs layers. The doping density is 2×10^{18} cm^{-3} in the wells and thin barriers. As in the coupled ground state detectors in Section 2.11, the spectral width of the SL detector depends on the electron population in the miniband ground states. With the present doping level, the calculated Fermi level is at 50.0 meV above the GaAs conduction band. The weighted oscillator strengths, $f'_n \equiv N_{sn}f/N_s$, depend on the Fermi energy of a particular ground state n. The lack of parity symmetry in this SL allows optical transitions from M_1 and M_2 to all states in the two upper minibands M_3 and M_4. The four main group of transitions are shown in Fig. 2.28 with a small σ. To fit the experimental spectrum, $\sigma = 33$ meV is obtained. The peak wavelength is at 4.0 μm, and the bandwidth 1.5 μm. Although this design does not completely cover the 2 μm spectral width of the MWIR window, it is 30% wider than that of Det. E. Furthermore, the transitions among minibands create a non-Lorentzian lineshape, which substantially increases the responsivity near its tail. The lower experimental responsivity near the cutoff wavelength can be attributed to the decreasing tunneling probability of the lower-energy photoelectrons.

2.15 Photoconductive Gain and Noise Gain

Since QWIPs are photoconductors, both signal and noise are affected by the photoconductive gain g. In Section 1.4, we derived g to be τ/τ_{tr} for a detector made of homogeneous material, where τ is the recombination lifetime and $\tau_{\rm tr}$ is the transit time across the active thickness. On the other hand, QWIPs contain discrete layer structures, so their effects on g needs to be examined. For example, when using a thicker barrier B where recombination does not occur, τ will increase. However, a thicker B will also increase $\tau_{\rm tr}$, yielding the same g. On the other hand, for the same B, increasing the number of QW periods N will increase $\tau_{\rm tr}$ without affecting τ. Therefore, g should reduce. To give a more descriptive picture of the gain in QWIPs, a new parameter, the capture probability p_c, is introduced [33]. The quantity p_c is the fraction of photoelectrons that will fall back to the ground state of a well when the photoelectrons pass through the well. If τ_i is recombination lifetime of the photoelectrons in the well region and t_w is the dwell time in the well (*i.e.*, the time of a photoelectron staying in the well layer), p_c will be [34, 35]

$$p_c = 1 - e^{-t_w/t_i}, \tag{2.37}$$

which assumes the exponential time decay. Although t_w is governed by the quantum transition process from a well state to a barrier state, the photoelectron transport must obey current continuity in which the number of photoelectrons flowing into a well $Anvt_w$ must be equal to AnW leaving the well, where A is the detector area, n is the photoelectron density, v is the drift velocity in the barrier, and W is the well width. The current continuity thus yields $t_w = W/v$. The fact that the dwell time in a well depends on the drift velocity in the barrier can be explained by the availability of the final states in the barrier for the electrons in the well to transition to. The exponential time decay continues after the remaining photoelectrons pass to the next well at the transit time per period, t_p. Using p_c as the independent variable, one can express g as a function of p_c. This can be achieved by differentiating the time decay function to obtain the probability $P(t)$ of a photoelectron being captured at time t after its creation, which is

$$\begin{aligned} P(t) &= \frac{1}{\tau_i}\exp\left[-\frac{t - nt_p + nt_w}{\tau_i}\right] && \text{for } nt_p < t < nt_p + t_w \\ &= 0 && \text{for } nt_p + t_w < t < (n+1)t_p \end{aligned} \tag{2.38}$$

where n is the well index starting at 0. The recombination lifetime τ of the MQW structure will then be

$$\begin{aligned}\tau &= \int_0^\infty tP(t)dt \\ &= \tau_i + \frac{t_p - t_w}{e^{t_w/\tau_i} - 1}.\end{aligned} \tag{2.39}$$

With the new τ, $g = \tau/\tau_{\text{tr}} = \tau/(Nt_p)$. Substituting $t_w/\tau_i = -\ln(1 - p_c)$ into Eq. (2.39) and taking the limit of $t_w/t_p = W/(W+B) = W/L \ll 1$ in thick barrier QWIPs, the photoconductive gain g is

$$g = \frac{1}{N}\frac{(1-p_c)}{p_c}. \tag{2.40}$$

In Eq. (2.40), g depends on N and p_c in an intuitive manner as $g \propto 1/N$ and when $p_c = 1$, $g = 0$. When p_c is small, $g \propto 1/p_c$.

In homogeneous materials, the photoconductive gain g is the same g determining the generation recombination (g-r) noise. However, the two can be slightly different in MQW structures. Recall that the nonuniform generation of photoelectrons due to Poisson distribution of incoming photons causes generation noise while the fluctuation of lifetime and thus gain cause recombination noise. To account for the fluctuation of τ given by Eq. (2.39) and in realizing a longer duration of current fluctuation gives a larger g-r noise current, i_n [36],

$$\begin{aligned}i_n^2 &= 2egIB\int_0^\infty \left(\frac{t}{\tau}\right)^2 P(t)dt \\ &= 4egIB\frac{1-p_c/2}{1-p_c} \\ &= 4eg_nIB.\end{aligned} \tag{2.41}$$

In Eq. (2.41), we differentiate the photoconductive gain g measured in the photocurrent measurement from the noise gain g_n measured in noise measurement. Equation (2.41) is also applicable only in the limit of $W/L \ll 1$. With g given in Eq. (2.40), the noise gain g_n is

$$g_n = \frac{1}{N}\frac{(1-p_c/2)}{p_c} = \frac{1}{N}\frac{(1-p_c)}{p_c} + \frac{1}{2N}. \tag{2.42}$$

Therefore, g_n of a QWIP is larger than its photoconductive gain g. This difference can be traced back to the assumption that the dwell time t_w in the original well of excitation is the same as that of the subsequent wells. In QWIPs where the electron transport in and out of the QW is governed by quantum transitions, the dwell time t_w is expected to be the same in the original well and the subsequent wells. In a classical detector, the dwell time in the original well is $t_w/2$ since the photoelectrons are created on average at the center of the well, and the recombination is less in the original well. The classical photoconductive gain would then be larger [35]. Therefore, there is a reduction of photoconductive gain in QWIPs due to quantum transport. Assuming half the dwell time in the original well would yield $g = (1-p_c/2)^{1/2}/(Np_c)$, and it will be the same as g_n expressed in Eq. (2.42). Therefore, the classical g in the present formulism is much closer to g_n, and g_n of a QWIP is unchanged from the classical value. When p_c is small, all the gains approach each other and are equal to $1/(Np_c)$.

The photoconductive gain of a QWIP is usually deduced from the noise measurement. It can be performed in a cold chamber without infrared radiation exposure, and the detector is kept at liquid nitrogen temperature where a stable temperature can be maintained. The gain g_n measured is therefore the gain of the thermal current at $T \sim 77\,\text{K}$. It can also be measured at a low $T \sim 10\,\text{K}$ under 300 K background radiation exposure. In this case, the measured g_n is the noise gain of the background photocurrent. The two g_n can be different if the energy distribution of the carriers are substantially different. In the measurement, the detector is biased through a low noise current amplifier. The output voltage of the amplifier is monitored by a spectrum analyzer. The output signal below a few hundred Hertz is usually dominated by the $1/f$ noise of the instrument. Above these frequencies, the signal is flat and dominated by the detector g-r noise. The noise spectrum is averaged over a range of frequencies up to 10 kHz and over several hundred measurements. The value of g_n can then be evaluated by knowing the amplifier gain and the noise bandwidth.

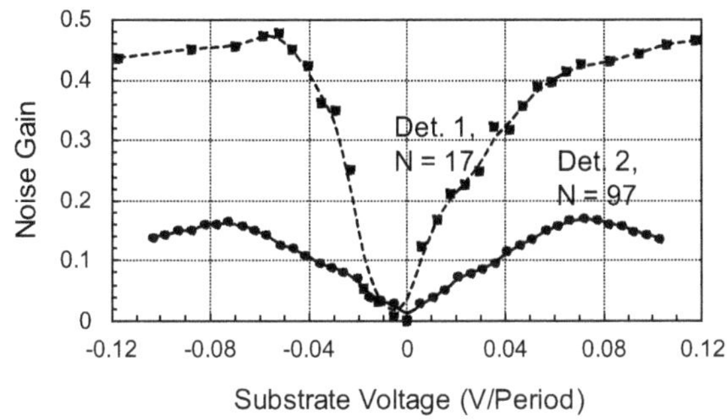

Figure 2.29. The measured noise gain of two detectors having different structural parameters.

In Fig. 2.29, g_n measured from two typical LWIR MQW structures are shown. Det. 1 has the following MQW parameters: thickness $t = 0.94\,\mu$m, number of QW periods $N = 17$, well width $W = 48$ Å, barrier width $B = 500$ Å, Al molar ratio $x = 0.24$, In molar ratio $y = 0$, and doping density $N_D = 1.0$ in $10^{18}\,\text{cm}^{-3}$. Det. 2 has the following parameters: $t = 7.27$, $N = 97$, $W = 49$, $B = 700$, $x = 0.19$, $y = 0.1$, and $N_D = 0.4$ in the same units. Therefore, Det. 1 has a thinner active layer, less QWs, thinner barriers, larger x, and larger doping. The applied bias V is the voltage per QW period on the substrate contact relative to the top contact. Since the two detectors have different period lengths of 548 Å for Det. 1 and 749 Å for Det. 2, the electric fields on the detectors are different. For example, the E field is 1825 kV/m for Det. 1 and 1335 kV/m for Det. 2 at $V = 0.1$ V/period. Figure 2.29 shows that g_n rises almost linearly with V at low V and becomes saturated at large V, after which it can decrease. This trend is mainly caused by the change in the drift velocity even though τ can also be increased by gaining energy from the field. The gain is not completely symmetric with respect to V but the maximum g_n attained are similar in both polarities where $g_n = 0.48$ for Det. 1 and 0.168 for Det. 2. The 2.9 times larger gain in Det. 1 is due to the smaller N. However, it is less than the N ratio of 5.7 because p_c is different. The polarity asymmetry is attributed

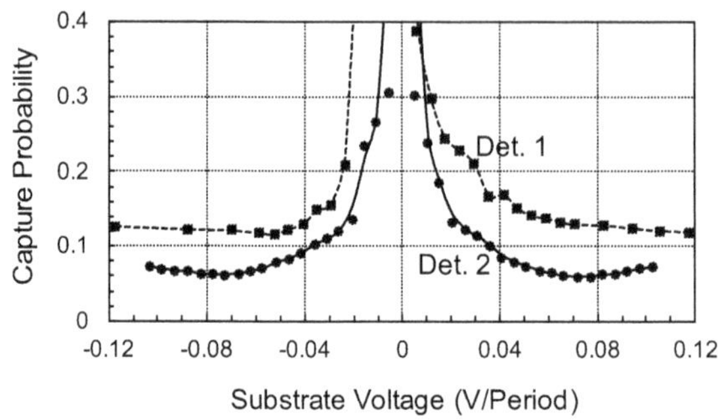

Figure 2.30. The deduced p_c of the two detectors from their noise gain.

to dopant migration during crystal growth, which creates asymmetry in the band bending in the two sides of the QW barrier. It can be minimized by shifting or sharpening the doping profile.

With the measured g_n, the capture probability p_c can be deduced from Eq. (2.42), and it is shown in Fig. 2.30. The value of p_c decreases rapidly with V and settles at $p_c \sim 0.12$ for Det. 1 and $\sim$0.059 for Det. 2. Therefore, Det. 1 has larger carrier recombination in the QW.

The maximum gain and the minimum p_c are set by the saturation velocity v_{sat} in the AlGaAs barrier. The value of v_{sat} is known to be around 0.7×10^7 cm/s but can vary in a limited range depending on the electron populations in the X and L satellite valleys under an applied V and their respective mobilities [37]. Therefore, v_{sat} depends on the barrier composition that sets the energy of the satellite valleys as well as intrinsic and extrinsic scattering effects in the valleys. In order to estimate the order of magnitude of τ of a QWIP, which is given by $\tau = g_n\tau_{\mathrm{tr}} = g_n t/v_{\mathrm{sat}}$ at large V, we assume a common v_{sat} of 0.7×10^7 cm/s in 16 different LWIR QWIPs. The result is shown in Fig. 2.31.

In this set of samples, t ranges from 0.34 to 7.9 μm, N ranges from 6 to 106, W ranges from 46 to 62 Å, B ranges from 500 to 700 Å, x ranges from 0.12 to 0.27, y is either 0 or 0.1, and N_D ranges from 0.2 to 1.5×10^{18} cm^{-3}. With these detector parameters, the shortest τ is 4.4 ps exhibited in a detector with $t = 0.3386$, $N = 6$, $W = 48$, $B = 500$, $x = 0.243$, $y = 0$, and $N_D = 1.0$ in the usual units.

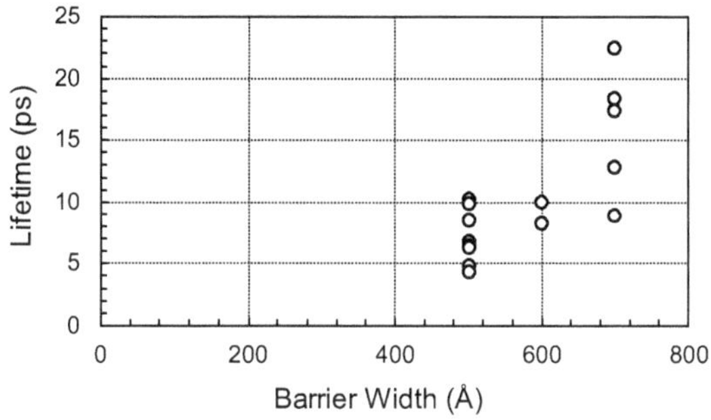

Figure 2.31. The deduced τ at the velocity saturation regime as a function of B.

The largest τ is 22.5 ps from a detector with $t = 7.2653$, $N = 97$, $W = 49$, $B = 700$, $x = 0.21$, $y = 0.1$, and $N_D = 1.0$. The average τ of all samples is about 10 ps. It is much longer than the phonon scattering time τ_{ph} of 0.2 ps at electron energy of ~200 meV above the band edge [38]. This ~50 times longer τ than τ_{ph} is consistent with the presence of the barrier with the B:W ratios of 10–14, where electrons cannot recombine, and the lengthening of τ_i under an applied field.

From Fig. 2.31, τ tends to be longer when B is thicker as expected from $(t_p - t_w) = B/v_{\text{sat}}$ in Eq. (2.39). However, since τ of each detector is not precisely deduced because of the uncertainty of v_{sat}, the B dependence cannot be quantitatively established. By using a dedicated set of detectors having the same structure except their barrier thickness, Rosencher *et al.* confirmed the linear relationship of τ and B [39].

The deduced τ in the order of ~10 ps for QWIPs is extremely short in comparison with the $\tau \sim 10\,\mu$s in a LWIR HgCdTe detector. If QWIPs rely on the same detection mechanism of a photodiode that requires the photoelectrons to travel to the p/n junction to be detected, the photocurrent would be extremely small because of carrier recombination. Fortunately, a photoelectron in QWIPs only needs to survive from carrier recombination in the QW of creation. Even if it is captured in the next well, a finite external photocurrent would have been generated. Since the survival probability $= 1 - p_c \sim 0.9$ under a relatively small field, the short lifetime poses little

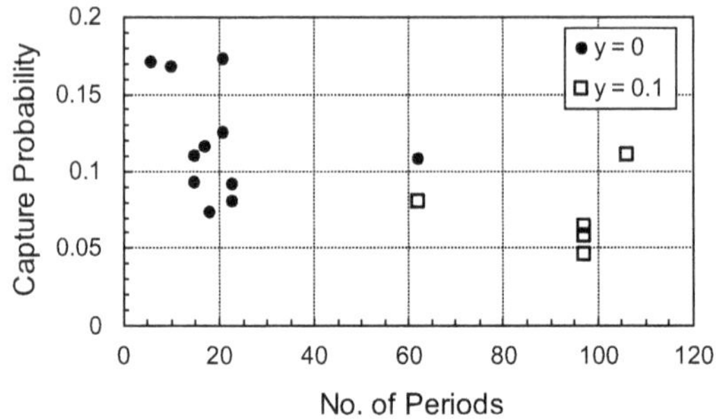

Figure 2.32. The capture probability at the velocity saturation regime with different detector parameters.

adverse effects on the QWIP photocurrent. In fact, the short lifetime can be an advantage in certain respects such as reducing g-r noise, reducing readout charge capacity requirement, and increasing the intrinsic speed of photoresponse.

The adoption of N and p_c in the gain expressions is to separate the interdependence of τ and $\tau_{\rm tr}$. The degree that N and p_c are independent is checked by plotting p_c against N in Fig. 2.32 based on the same set of samples. In this plot, we differentiate those with $y = 0$, which usually have larger x from those with $y = 0.1$, which usually have a lower x. Since lower x gives lower intervalley scattering, samples with $y = 0.1$ tend to have a lower p_c. Accounting for this sampling bias, the average p_c remains relatively constant at ~0.1 from $N = 6$ to 106. At the smallest N, p_c may have a larger value. It can be caused by the lack of photoelectrons in the contact regions, which leads the out-diffusion of the nearby QW photoelectrons and reduces τ_i in the neighboring wells. Therefore, except for the very small N, p_c and N are indeed independent variables. The measured p_c in Fig. 2.32 varies from 0.05 to 0.18, since $p_c = 1 - \exp(-t_w/\tau_i) = 1 - \exp(-W/v_{\rm sat}\tau_i)$. This variation comes from different QW designs that give different W, $v_{\rm sat}$, and τ_i. The plots of p_c against B and N_D do not show obvious trends. Therefore, g does not depend on either B or N_D in the present set of structural parameters.

2.16 Quantum Efficiency Under Edge Coupling

Since the second term in Eq. (2.42) is much smaller than the first term under typical operating conditions where p_c is small, g and g_n are approximately the same. Therefore, we will not differentiate g and g_n in discussing detector performance unless the difference is significant. With $g \approx g_n$, one can deduce the detection QE $\eta_d = \eta\gamma$ from the measured responsivity $R(\lambda, V)$:

$$\eta_d(\lambda, V) = \eta(\lambda)\gamma(\lambda, V) = \frac{hc}{e\lambda}\frac{R(\lambda, V)}{g}. \tag{2.43}$$

In Eq. (2.43), we assume $\eta(\lambda)$ is independent of V in a typical symmetrical QW structure. On the other hand, γ is V dependent, especially for B-B detectors. Therefore, we present four detector types shown in Fig. 2.33. These detectors have different x and W combinations such that $\lambda_H = hc/(H - E_1)$ is at $\lambda = 9.2\,\mu$m at zero volts. λ_H is the absorption wavelength needed to promote an electron

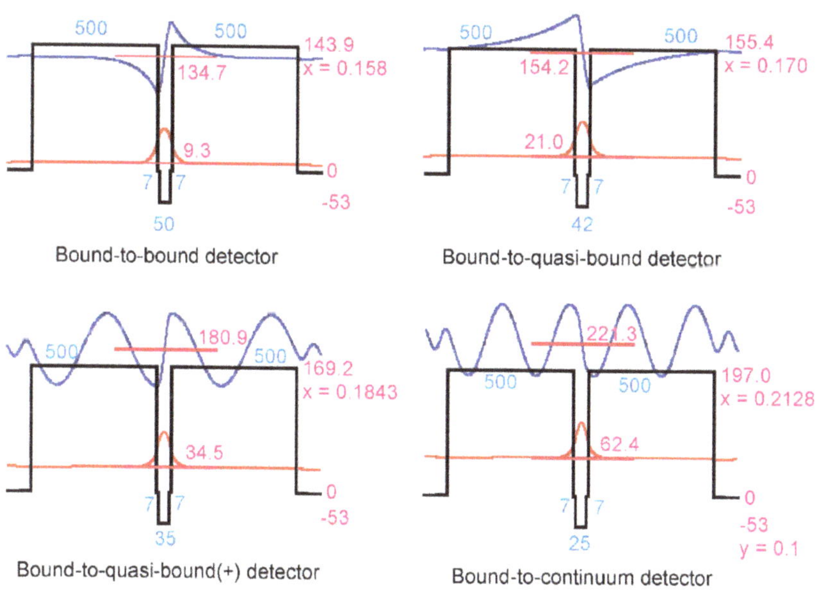

Figure 2.33. Four detector structures with a constant λ_H. The numerals are either layer thickness in Å or energy in meV.

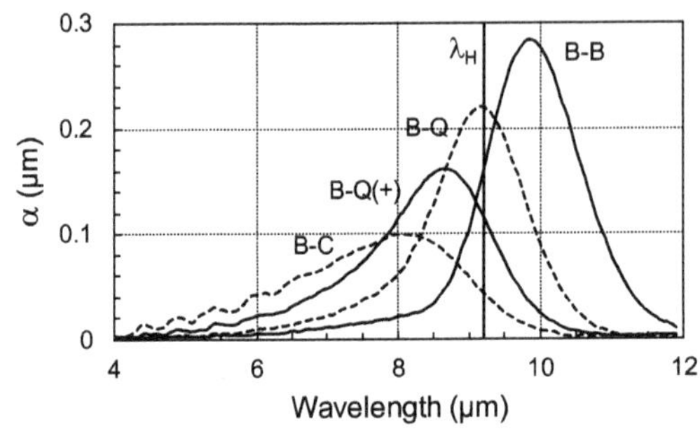

Figure 2.34. The absorption coefficients of the four detectors with the same areal electron density of $6.4 \times 10^{11}\,\mathrm{cm}^{-2}$ per period.

from E_1 to the top of the barriers. It usually represents the measured cutoff wavelength of the detector. Figure 2.33 also shows E_n and ψ_n at which f_n is the maximum.

With the displayed detector structures, $\alpha(\lambda)$ are calculated in Fig. 2.34. The areal doping density is $6.4\times10^{11}\,\mathrm{cm}^{-2}$ for all detectors. For $W = 64\,\text{Å}$, it corresponds to a volume doping density N_D of $1 \times 10^{18}\,\mathrm{cm}^{-3}$. It is seen that the B-B detector has the largest peak α while the B-C detector has the broadest absorption spectrum. The sum of f_n are the same for all detectors and are equal to 1.08. The integrated oscillator strength IAS $= \int \alpha(\lambda)d\lambda$ are therefore very similar. However, if we assume $\gamma(\lambda) = 0$ for $\lambda > 9.2\,\mu\mathrm{m}$ and $\gamma(\lambda) = 1$ when $\lambda < 9.2\,\mu\mathrm{m}$, then IAS $= 0.090$, 0.239, 0.293 and 0.269, respectively, for B-B, B-QB, B-QB(+) and B-C detectors. This example shows that the latter three detectors generate similar amounts of photocurrent above H.

Since the calculated α is for parallel propagating light, η will depend on the light incident configuration and detector geometry. Traditionally, the material properties of QWIPs are characterized using edge coupling geometry shown in Fig. 2.35. In this geometry, the light is incident at a 45° through a polished facet at the edge of a wafer substrate. Since the transition rate in Eq. (2.3) is directly proportional to $|\boldsymbol{A} \cdot \boldsymbol{P}|^2$, $\boldsymbol{A}$ is parallel to $\boldsymbol{E}$ in electromagnetic waves, and $\boldsymbol{P}$ is parallel to the material growth axis, only the transverse magnetic (TM) polarization is able to initiate the transition.

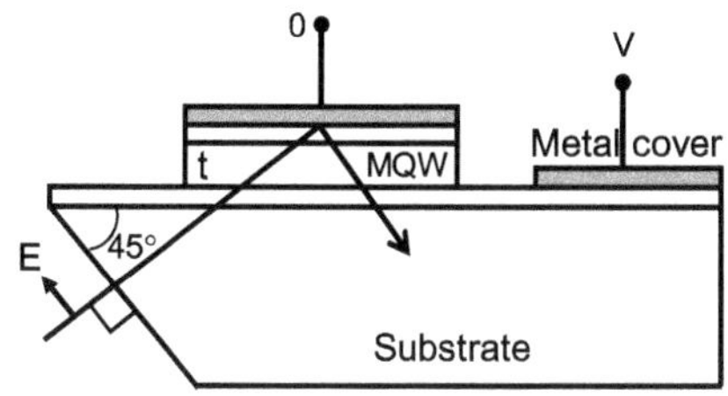

Figure 2.35. The light incident geometry under 45° edge coupling. The detector has an active layer thickness of t, which is sandwiched between two n-doped GaAs contact layers. Au/Ge/Ni/Au is used to make ohmic contacts to these contact layers.

The other polarization, the transverse electric (TE) component, is not absorbed by the detector. Therefore, the maximum QE is 50% under this coupling scheme.

For the TM polarization, part of the light is reflected at the substrate-air interface and part of it transmits into the substrate with transmission coefficient T_s of $4n/(1+n)^2$, where $n = (14.662 - 0.466\lambda)^{1/2}$ is the refractive index of GaAs, and λ is in microns. After entering into the substrate, the light propagates with an incident angle of 45°. It is reflected by the top metal contact and makes another pass in the MQW layer with thickness t. The total optical pathlength is $2t/\cos(45°) = 2\sqrt{2}t$. Since the E field along z is reduced by a factor of $\sin(45°)$, the absorption coefficient α_{45} is reduced by half. Owing to a smaller projected surface, the incident intensity I_s is reduced to $I_s \cos(45°)$ at the detector surface area. Putting all these factors together, the absorption QE η_{45} for unpolarized light under 45° edge coupling is

$$\eta_{45} = \frac{4n}{(1+n)^2} \frac{1}{2\sqrt{2}} [1 - \exp(-\sqrt{2}\alpha t)]. \tag{2.44}$$

In Eq. (2.44), we have ignored the finite absorption in the ohmic metal cover on top and the free carrier absorption in the MQW layer and contact layers. The absorption in the metal and contact layers can reduce η slightly.

The calculated η_{45} with α from Fig. 2.34 is shown in Fig. 2.36. For $N = 62$, the peak η_{45} are 19.2, 16.7, 13.5 and 9.3% for the B-B, B-QB, B-QB(+) and B-C detectors, respectively. The responsivity of

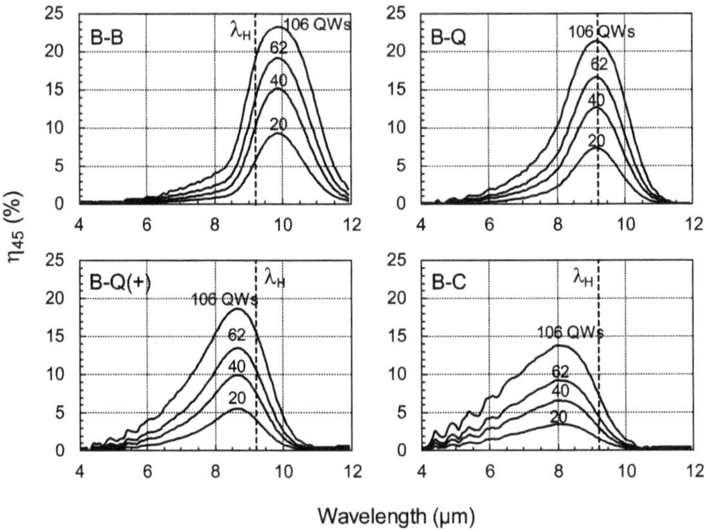

Figure 2.36. The calculated absorption quantum efficiency of different QWIPs under 45° edge coupling.

four detectors was characterized. Two of the detectors have 106 QWs, and the other two have 62 QWs. The QW parameters (x, y, W, B) of one detector are (0.21, 0.1, 50 Å, 700 Å), and it is a B-B detector. The nominal parameters of the other three are the same with (0.19, 0.1, 49 Å, 700 Å). All doping is $1 \times 10^{18}\,\text{cm}^{-3}$. Due to variations in the material growth, the first excited states of the last three wafer materials are either slightly above or below the barrier, leading to three different detector types, B-QB, B-QB(+), and B-C. Note that αt in Eq. (2.44) is independent of B as the period length in α and t cancels out. Therefore, one can compare the calculated η for $B =$ 500 Å and the measured η for $B = 700$ Å under the present coupling geometry.

The responsivity is measured at temperature $T = 10$ K. Figure 2.37 shows the measured $R(\lambda)$. The maximum R except that of the B-B detector has similar magnitude at the highest applied voltage. Under a bias, the barriers are lowered at one side of the QWs. Both B-QB and B-QB(+) detectors evolve toward a B-C detector where the absorption spectra are more exposed above the barriers. Their α

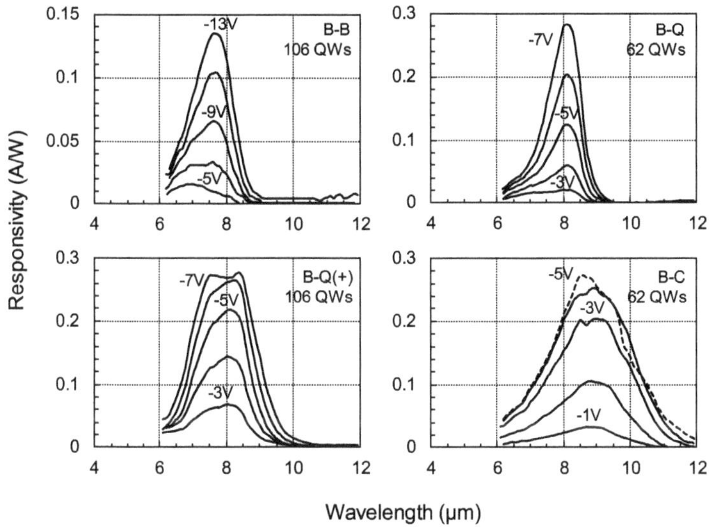

Figure 2.37. The measured spectral R from four detectors under 45° edge coupling.

and R spectra will then become similar at high V. Nevertheless, the B-QB R has a sharper cutoff than the B-QB(+) and B-C detectors. For the B-C detector, it has a similar peak R even though its α is smaller. This is because it has larger γ and g due to the larger photoelectron energy relative to the top of the barriers.

On the contrary, both γ and g are generally small in the B-B detector, which gives it a small R even though it has the largest peak α. From Fig. 2.37, it is seen that R does not change significantly with N. Since η is directly proportional to N while g is inversely proportional to N, these two factors cancel and lead to a relatively constant R.

To obtain $\eta_d = \eta_{45}\gamma$, the noise gains of the detectors are measured, from which the deduced η_d of the detectors are shown in Fig. 2.38. For the B-B detector, the peak $\eta_d = 23.5\%$ at $V = -13\,\text{V}$ and $7.6\,\mu\text{m}$, which is very close to 23.3% predicted for η_{45} at $N = 106$. Therefore, γ is near unity at this bias, and the lower R is due to a small $g = 0.079$ at this V. For the B-QB detector, the measured η_d is 9.2% at $-7\,\text{V}$, which is much smaller than η_{45} of 16.7% at $N = 62$.

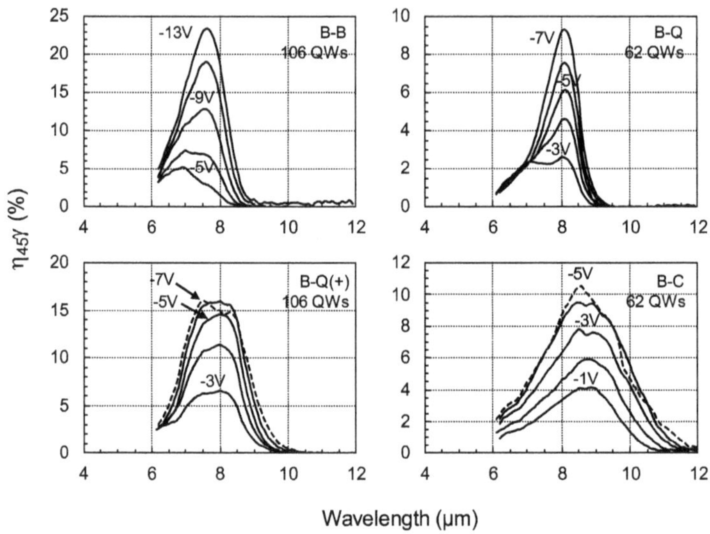

Figure 2.38. The deduced $\eta_d = \eta_{45}\gamma$ from the spectral responsivity and noise measurements under 45° edge coupling.

The value of γ is thus still much less than 1 at this bias. The value of η_d is hence expected to further increase beyond -7 V but it has not been verified by experiment.

For the B-QB(+), the measured η_d saturates at $V = -6$ V, at which $\eta_d = 15.9\%$, which matches quite well with η_{45} of 18.6% at $N = 106$. For the B-C, the measured η_d saturates around $V = -4$ V, at which $\eta_d = 9.5\%$, which matches well with η_{45} of 9.3% at $N = 62$. From Fig. 2.38, γ increases with V even when the photoelectrons are above H although the increase is less than that below H. Therefore, γ is more appropriately interpreted as the electron transfer probability. Being an unbound state is a necessary but not sufficient condition for $\gamma = 1$. It needs an electric field to break the reflection symmetry of the MQW so that the transfer of photoelectrons is primarily to one side of the barrier. The value of γ is 1 only when all the electrons move in the direction of the applied bias. From the η_d saturation voltage of ~ -4 V for the 62 QW B-C detector and ~ -6 V for the 106 QW B-QB(+) detector, the applied field needs to be ≥ 60 mV/period to yield $\eta_d = \eta\gamma = \eta$. In general, the photocurrent of a QWIP depends on two voltage-dependent variables γ and g, and I_p in Eq. (1.21) should be expressed as

$$I_p(\lambda, V) = A_d \frac{\eta_d \Phi \tau}{A_d t} ev = \eta_d \Phi e \frac{\tau}{\tau_{\mathrm{tr}}} = e\eta(\lambda)\Phi(\lambda)\gamma(\lambda, V)g(V). \quad (2.45)$$

Note that the capture probability p_c also attains its minimum at $V \sim 60\,\mathrm{mV/period}$ in Fig. 2.30. Therefore, QWIPs are usually operated at this V range. Although the peak value of the η_{45} calculation agrees with that of the η_d measurement at large V, η_d generally shows a narrower linewidth except in the B-C detector because of the decreasing $\gamma(\lambda)$ at the longer λ.

2.17 Detector Dark Current

The intrinsic dark current of a QWIP can be classified into three main transport mechanisms, which are direct tunneling between adjacent ground states, thermally assisted tunneling (TAT) near the top of the barrier, and thermionic emission (TE) over the barrier. These processes are indicated in Fig. 2.39. Each of these processes can be expressed by the usual formula, $J = nev_d$, where J is the current density, n is the electron volume density, and v_d is the electron drift velocity. In the usual QWIP design, one can employ very thick barriers to reduce the direct tunneling current to an arbitrarily low level. Therefore, this current component is not important to the usual QWIP operation.

The TAT current originates from the thermal excitation of the ground state electrons to yield a large in-plane energy, E_{in}. Upon

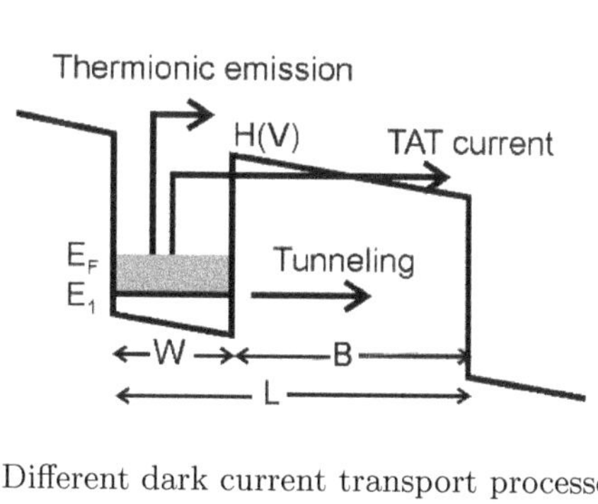

Figure 2.39. Different dark current transport processes in a QWIP.

a scattering process, this large E_{in} transforms into a large E_z, with which a larger transfer probability γ results. The value of γ thus depends on the total energy $E = E_{\text{in}} + E_z$ under the TAT process. The effective 3D electron density for the TAT process is given by

$$\begin{aligned} n_{\text{TAT}}(E,T,V) &= g_{3D} f(E,T)\gamma(E,V) \\ &= \frac{m^*}{\pi\hbar^2 L}\frac{\gamma(E,V)}{1+\exp\left(\frac{E-E_f-E_1}{kT}\right)}, \end{aligned} \tag{2.46}$$

where E is the total energy, T is the detector temperature, and V is the applied voltage per period. $g_{3D} = g_{2D}/L = m^*/(\pi\hbar^2 L)$ is the effective 3D density of states for the ground state electrons, and $f(E,T)$ is the Fermi distribution function. L is the period length, and m^* is the effective mass of the well. The energy reference is at the center of the well.

Under the Wentzel–Kramers–Brillouin (WKB) approximation,

$$\begin{aligned} \gamma(E,V) &= \exp\left\{-\frac{4B}{3e\hbar V}(2m_b^*)^{1/2}[(H(V)-E)^{3/2} \right. \\ &\qquad \left. -(H(V)-eV-E)^{3/2}]\right\} \quad \text{for } E < H(V)-eV \\ &= \exp\left[-\frac{4B}{3e\hbar V}(2m_b^*)^{1/2}(H(V)-E)^{3/2}\right] \\ &\qquad \text{for } H(V) > E > H(V)-eV \\ &= 1 \quad \text{for } E > H(V), \end{aligned} \tag{2.47}$$

where $m_b^* = m^*(1+1.24x)$ is the effective mass of the barrier, and $H(V)$ is the barrier height under V. The drift velocity v_d under TAT is taken to be the usual electron drift velocity in the barrier region, which can be expressed in the form of

$$v_d(V) = \frac{\mu V/L}{\sqrt{1+\left(\frac{\mu V}{L v_{\text{sat}}}\right)^2}}, \tag{2.48}$$

where μ is the low field mobility and v_{sat} is the saturation velocity, which is about 1×10^7 cm/s. Combining n_{TAT} and v_d, the TAT current density is given by

$$\begin{aligned} J_{\mathrm{TAT}}(V,T) &= \int_{E_1}^{\infty} J(E,V,T)dE \\ &= \int_{E_1}^{\infty} en_{\mathrm{TAT}}(E,V,T)v_d(V)dE \\ &= \int_{E_1}^{\infty} \frac{m^*}{\pi\hbar^2 L}\frac{\gamma(E,V)}{1+\exp\left(\frac{E-E_f-E_1}{kT}\right)}\frac{\mu V/L}{\sqrt{1+\left(\frac{\mu V}{Lv_{\mathrm{sat}}}\right)^2}}dE, \end{aligned} \quad (2.49)$$

where $J(E,V,T)$ is the current density per meV. The parameters μ, v_{sat} and $H(V)$ are treated as fitting parameters to the experimental data.

The TE current is the direct thermal excitations to the E_n eigenstates of the SQW. The thermal current originated from E_n is similarly given by

$$J_{TE}(V,T) = \sum_{n=2}\int_{E_n}^{\infty} \frac{m^*}{\pi\hbar^2 L}\frac{\gamma(E,V)}{1+\exp\left(\frac{E-E_f-E_n}{kT}\right)}\frac{\mu V/L}{\sqrt{1+\left(\frac{\mu V}{Lv_{\mathrm{sat}}}\right)^2}}dE, \quad (2.50)$$

where $\gamma(E,V)=1$ in WKB approximation for $E_n > H$. However, as seen in the photocurrent in B-C detectors, $\gamma < 1$ and it increases with V even if $E > H$. Therefore, WKB approximation overestimates both TAT and TE currents with $E > H$. For simplicity, one can regard the $E > H$ component of the TAT current in the WKB approximation of Eq. (2.47) as the TE current. The parameter μ can then be used as the overall fitting parameter to explain the dark current.

We use two detector structures labeled A and B as examples. Detector A has the nominal QW parameters $(x, y, W, B, N_D) = (0.12, 0.1, 50\,\text{Å}, 700\,\text{Å}, 0.9\times10^{18}\,\mathrm{cm}^{-3})$ and B has (0.21, 0.1, 50 Å, 700 Å, $0.9\times10^{18}\,\mathrm{cm}^{-3}$). In fitting the spectral responsivity of A, β_{AlGaAs} in Eq. (2.22) is 0.91, with which $H_A(0) = 134.4\,\mathrm{meV}$. The calculated E_n are 25.6, 136.4, 143.0, and 154.6 meV. At zero volts, A is a B-QB(+) detector. For detector B, $\beta_{\mathrm{AlGaAs}} = 0.75$,

$H_B(0) = 199\,\text{meV}$, and $E_n = 34.1$, 196.4, 206.9, and 224.2 meV. Since E_2 is lower than H at zero volts, B is a B-B detector.

In order to obtain a qualitative understanding of the current-voltage (I-V) characteristics, the functions $f(E)$ and WKB $\gamma(E)$ are plotted at $T = 77\,\text{K}$ and $V = 80\,\text{mV}$ in Fig. 2.40. Their product has a maximum just below $H(V)$. The TAT current is therefore conducting just below the tip of the barrier. It turns out that the peak position does not change substantially with T. For a very thick barrier, the decrease of γ is much steeper than the rise of $f(E)$ as E decreases as shown in Fig. 2.40.

With an almost impenetrable barrier, the majority of tunneling is always close to the tip of the barrier. This situation leads to a well-defined activation energy for the TAT process at each V, and the magnitude of the TAT current thus follows an exponential law.

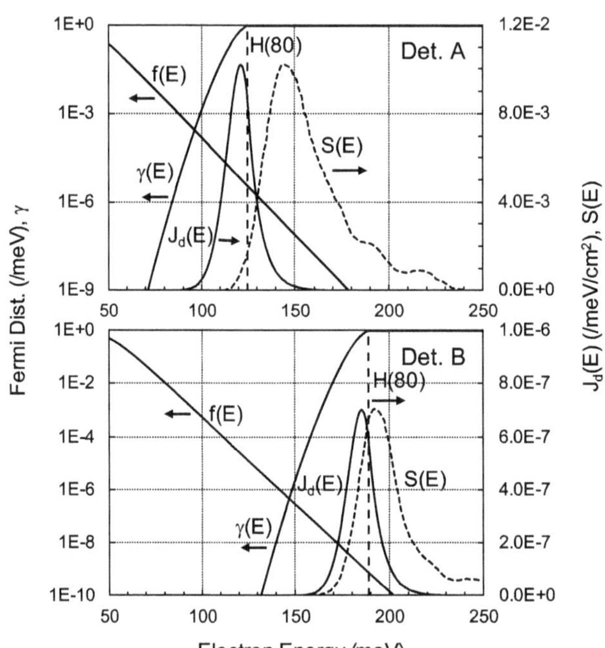

Figure 2.40. The curves of $f(E)$, $\gamma(E)$, and $J(E)$ of a B-QB(+) detector (A) and a B-B detector (B) at 77 K and 80 mV/period. The spectral responsivity $S(E)$ in arbitrary units and $H(80\,\text{mV})$ are also shown. $J_d(E)$ adopts μ and v_{sat} as deduced in Fig. 2.41.

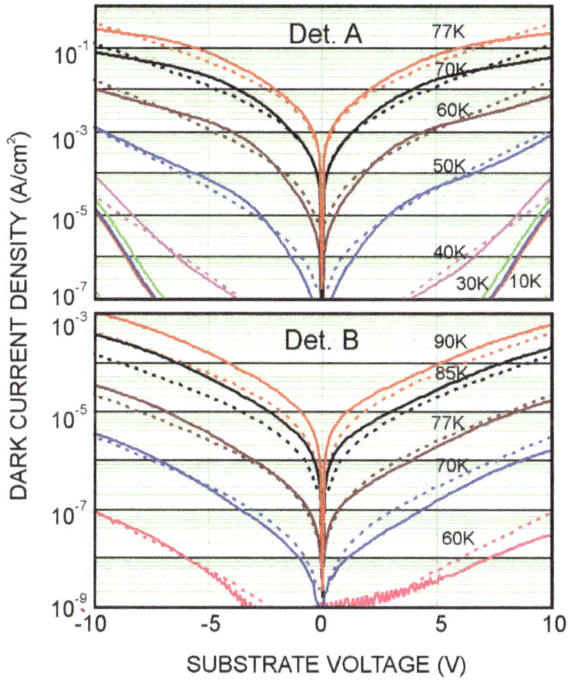

Figure 2.41. The measured (solid curves) and the fitted (dashed curves) dark currents under different biases and temperatures.

Figure 2.40 also shows the spectral responsivity lineshape $S(\lambda) \equiv \Sigma\lambda f_n \rho_n(\lambda)$ of the respective detectors calculated from Eqs. (2.9) and (2.34). At $V = 80\,\text{mV}$, the spectral peak and the TAT current peak are well separated for the B-QB(+) detector. For the B-B detector, $J_d(E)$ and $S(E)$ have a larger overlap.

Figure 2.41 shows the measured dark current-voltage characteristics in solid curves at different T. Both detectors contain 106 QWs, with which $V = V_a/106$ in Eq. (2.49) and V_a is the applied voltage on the bottom contact. To fit the data with Eq. (2.49), we assume $\mu = 400\,\text{cm}^2/\text{Vs}$ for A and $120\,\text{cm}^2/\text{Vs}$ for B, $H_{A,B}(V) = H_{A,B}(0) - 0.12 \cdot V$, and $v_{\text{sat}} = 1 \times 10^7$ cm/s. With the three adjustable parameters, μ, v_{sat} and $H(V)$, the I–V characteristics of both detectors can be fitted simultaneously in a wide range of V and T, which shows the validity of the dark current model.

At $T < 40\,\text{K}$, the measured current in A is higher than the calculated TAT current (not shown), which indicates the presence of extrinsic impurity-assisted tunneling current. Similarly, at $T \gg 77\,\text{K}$, the measured current in B is larger than the TAT current, which shows the increasing importance of the TE current at high temperatures. The current saturation behavior of both detectors is consistent with the expected saturation velocity in AlGaAs. The 3.3-times-lower μ in B is attributed to the larger intervalley scattering, stronger alloy scattering in the barrier, and stronger quantum reflections at the taller barrier boundaries. The lower μ and thus a longer transit time are also consistent with the lower noise gain of 0.06 of B compared to 0.3 of A at the same V_a of $-5\,\text{V}$. The lowering of the barrier height H by the field is nearly twice that expected from a linear potential profile. This indicates that the tilted barrier is increasingly more transparent to tunneling, and the effective barrier height is lowered faster than the tip of the barrier. Det. B also shows large current at negative substrate bias, which is from dopant migration during the material growth.

Overall, the present TAT current model provides an adequate description of the QWIP dark current under the usual operating conditions. Within this model, the dark current peak is always located just below $H(V)$ in a wide range of T. Therefore, there is a well-defined activation energy $E_a(V)$ for each V, and it is $H(V) - E_1 - E_f$. The detector dark current can then be written as

$$\begin{aligned} J_d(V,T) &= J_0(V)\exp\left(-\frac{E_a}{kT}\right) \\ &= J_0(V)\exp\left(-\frac{H(V)-E_1}{kT}\right)\exp\left(\frac{E_f}{kT}\right) \\ &= J_0(V)\exp\left(-\frac{hc}{\lambda_c kT}\right)\exp\left(\frac{N_D W}{g_{2D} kT}\right), \end{aligned} \tag{2.51}$$

where $J_0(V)$ is a prefactor, λ_c is the cutoff wavelength of the detector under V, N_D is volumetric doping density, W is the well width, and g_{2D} is the 2D density of states. In Eq. (2.51), we have assumed the detector to be operated in the B-QB(+) mode, in which λ_c is aligned with $H(V)$, or more precisely, aligned with the dark current peak. Under the usual detector operating voltage where the drift velocity is close to v_{sat}, $J_0(V)$ is insensitive to the low field mobility. As a result, the dark current of a detector with different λ_c and N_D under the usual operating voltage can be estimated from Eq. (2.51) once $J_0(V)$ is extracted from the experimental result of any B-QB(+) detector. For example, from a detector with $\lambda_c = 9.2\,\mu$m at $V = 80\,$mV, $N_D = 1 \times 10^{18}\,$cm^{-3} and $W = 49\,$Å, $J_0(80\,$mV) is deduced to be $1.79 \times 10^4\,$A/cm^2 from J_d measured at 77 K. With this value of J_0, $J_d(V, T)$ of a B-QB(+) detector with any λ_c, N_D, and W can be estimated from Eq. (2.51). The calculated result for a constant $W = 50\,$Å is shown in Fig. 2.42.

In Fig. 2.42, $J_d(T)$ is shown for detectors with different λ_c and N_D. The applied field is assumed to be 80 mV/period required for nearly saturated photocurrent. Figure 2.42 shows that a detector with a large λ_c and a large N_D will need to operate at a low T to achieve

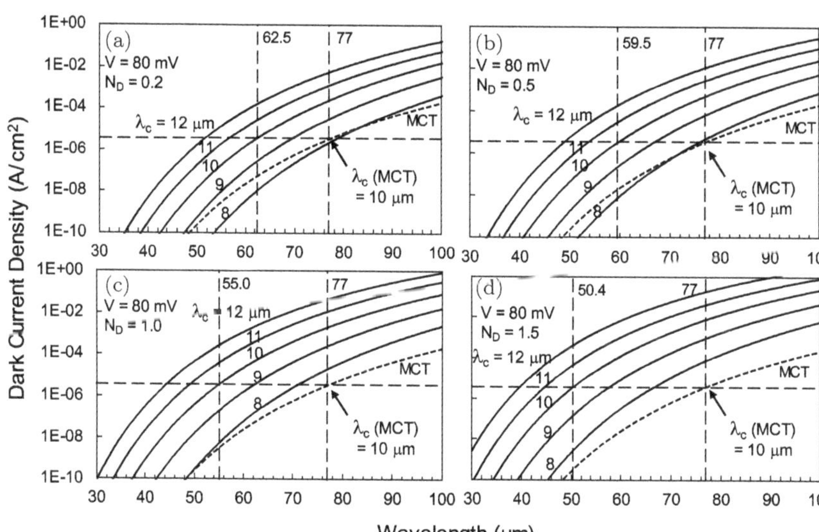

Figure 2.42. The projected dark current density of a B-QB(+) detector under an applied field of 80 meV/period as functions of cutoff wavelength and temperature having doping density = (a) 0.2, (b) 0.5, (c) 1.0, and (d) 1.5 in the unit of $10^{18}\,$cm^{-3}.

the same dark current. For example, to match the dark current of a 10 μm cutoff HgCdTe detector following "Rule 07" operated at 77 K [40], $N_d = 0.2$ (in the unit of $10^{18}\,\text{cm}^{-3}$) detector in Fig. 2.42(a) will need to operate at a lower T of 62.5 K. Likewise, $N_d = 0.5$, 1.0, and 1.5 detectors will need to operate at 59.5, 55.0, and 50.4 K, respectively. For QWIPs, the typical N_d is from 0.5 to 1.0, depending on the application priorities. Therefore, QWIPs need to operate ~20 K colder to yield the same dark current.

In Fig. 2.42(b) with $N_D = 0.5$, at $T = 77$ K, J_d of a 10 μm QWIP is $5.31 \times 10^{-4}\,\text{A/cm}^2$, while that of a HgCdTe detector is $3.10 \times 10^{-6}\,\text{A/cm}^2$. Therefore, the dark current of a QWIP is 171 times higher. The larger J_d is usually attributed to its short lifetime of ~10 ps compared to the ~10 μs of a HgCdTe detector. For HgCdTe detectors, the dominant dark current is the diffusion current J_{diff} in the neutral region and generation-recombination current J_{GR} in the depletion region. In addition to these major currents, there are also smaller currents due to band-to-band tunneling and trap-assisted tunneling. The two dominant currents both depend on the appropriate lifetimes explicitly [41], where

$$J_{\text{diff}} = \frac{en_i^2 t_{\text{diff}}}{N_D}\left(\frac{1}{\varphi\tau_R} + \frac{1}{\tau_A} + \frac{1}{\tau_{\text{SRH}}}\right), \tag{2.52}$$

and

$$J_{\text{GR}} = \frac{en_i W}{\tau_{\text{SRH}}}, \tag{2.53}$$

where n_i is the intrinsic carrier concentration, t_{diff} is the diffusion region thickness, N_D is the majority carrier concentration, τ_R is the radiative recombination lifetime, φ is the photon recycling factor, τ_A is the Auger recombination liftetime, τ_{SRH} is the Shockley-Read-Hall recombination lifetime through impurity states in the bandgap, and W is the width of depletion region. J_{diff} depends on lifetime explicitly because the magnitude of the diffusion current depends on the carrier concentration gradient, which is in turn proportional to $1/\tau$. J_{GR} depends on τ_{SRH} because the injection of carriers into the depletion region creates an excess of carrier concentration in this region and thus a higher SRH recombination rate than that in thermal equilibrium. The transport of the carriers via SRH process is thus higher with larger $1/\tau_{\text{SRH}}$. On the other hand, the dark electron density n_{TAT} of a QWIP in Eq. (2.49) is that in thermal equilibrium, in which the stronger thermal excitation rate is balanced by the equally strong thermal relaxation rate. Following the principle of detailed balance, n_{TAT} and J_d are independent of τ. Since J_d of HgCdTe explicitly depends on τ while J_d of QWIP does not, their difference cannot be related to τ.

Instead, their difference can be better understood using

$$J_d(T) = 17940 \exp\left(-\frac{hc}{\lambda_c kT}\right) \exp\left(\frac{N_D W}{g_{2D} kT}\right) \frac{A}{\text{cm}^2}, \tag{2.54}$$

for QWIPs from Eq. (2.51) at $V = 80\,\text{mV/period}$, and

$$J_d(T) = 8367 \exp\left(-C\frac{hc}{\lambda_c(T)kT}\right) \frac{A}{\text{cm}^2}, \tag{2.55}$$

at saturated reverse bias for HgCdTe photodiodes that follow "Rule 07" [42]. In Eq. (2.55), the inverse diode ideality factor C is found to be 1.16239. Since the HgCdTe bandgap $E_g(T)$ is a function of T for a fixed Cd mole fraction x, the cutoff wavelength

$$\lambda_c(T) = \frac{hc}{E_g(T)} \tag{2.56}$$

varies with

$$E_g(T, x) = E_g(0, x) + 5.35 \times 10^{-4}(1 - 2x)T\,eV, \tag{2.57}$$

when T changes. Inserting Eqs. (2.56) and (2.57) into Eq. (2.55) for $x = 0.22539$, one obtains the dashed curves in Fig. 2.42 where $\lambda_c(77\,\text{K}) = 10.00\,\mu\text{m}$.

Remarkedly, the prefactor in Eq. (2.54) for QWIPs is only $\sim$2 times larger than that in Eq. (2.55) for HgCdTe. Since these prefactors account for voltage-dependent transport properties such as carrier mobility and density gradient, etc., the fact that QWIPs are operated near velocity saturation regime at high voltage surprisingly does not cause this prefactor to be much larger than that of HgCdTe in the low-voltage diffusion regime. Therefore, high-voltage operation of QWIPs is not the cause of the large dark current difference. The first exponential terms in Eqs. (2.54) and (2.55) differ by an inverse ideality factor C. The presence of $C > 1$ for HgCdTe essentially raises its dark current activation energy E_a from $E_g(T) = hc/\lambda_c(T)$ to $C \cdot hc/\lambda_c(T)$. When $C = 1.16239$, this 16.239% increase in E_a significantly reduces its dark current relative to QWIPs with $C = 1$. For example, the ratio of the first exponential terms at $T = 77\,\text{K}$ and $\lambda_c = 10\,\mu\text{m}$ is 20.77. The second exponential term in Eq. (2.54) for QWIPs further increases its J_d by 3.85 times when $N_D = 0.5 \times 10^{18}\,\text{cm}^{-3}$, $W = 50\,\text{Å}$, and $T = 77\,\text{K}$. Different from a semiconductor that operates on interband absorption, a QWIP needs external doping to populate the ground state for intersubband absorption. The doping increases the Fermi energy and thus reduces $E_a = H(V) - E_1 - E_f$. The need for doping will always render a smaller E_a for QWIPs than other interband detectors. For $W = 50\,\text{Å}$ and $N_D = 0.5 \times 10^{18}\,\text{cm}^{-3}$, $E_f = 8.94\,\text{meV}$. The finite E_f has a larger effect on longer wavelength detectors since they operate at lower T where the second exponential term will be larger for the same E_f. In short, relative to HgCdTe photodiodes, QWIP dark current is larger because (1) the high-field operation of QWIPs increases J_d by 2.14 times, (2) the smaller C increases J_d by 20.77 times, and (3) the need for doping increases J_d by 3.85 times. Putting all together, J_d of a QWIP is 171 times higher than HgCdTe detector at 77 K and $\lambda_c = 10\,\mu\text{m}$. To reach the same level of J_d, the QWIP will have to operate 17 K colder. At a higher N_D of $1 \times 10^{18}\,\text{cm}^{-3}$, the QWIP will have to operate 22 K colder. Therefore, the major cause for the larger dark current in QWIPs is their smaller inverse ideality factor rather than the shorter recombination lifetime.

2.18 Hot-Electron Transistor

In Section 2.17, it was shown that the majority of dark current of a B-QB(+) QWIP is conducted via TAT current and it is well separated from the photoelectron distribution. The increase of TAT current below the tip of the barrier accelerates the decrease of the apparent H(V) in explaining J_d(V). Therefore, by placing a quantum barrier next to a QWIP and aligning its barrier height between the two current distributions, a larger H(V) and a smaller I_d can be obtained after the current passing through the quantum barrier. In this structure, the added barrier acts as an electron energy high-pass filter.

To estimate the potential advantage, the theoretical dark current J_d fitted to detector A in Section 2.17 is decomposed into two components. The low-energy TAT component transports below the barrier and the high-energy TE current J_H transports above $H(V)$ as shown in Fig. 2.40. If we assume only J_H is able to go through the filter barrier, the dark current transfer ratio through the filter will be $\alpha_d \equiv J_H/J_d$. Figure 2.43(a) shows the calculated J_d and J_H, with which α_d is plotted in Fig. 2.43(b). From Fig. 2.43, the filter is more effective in reducing the dark current at low temperatures and

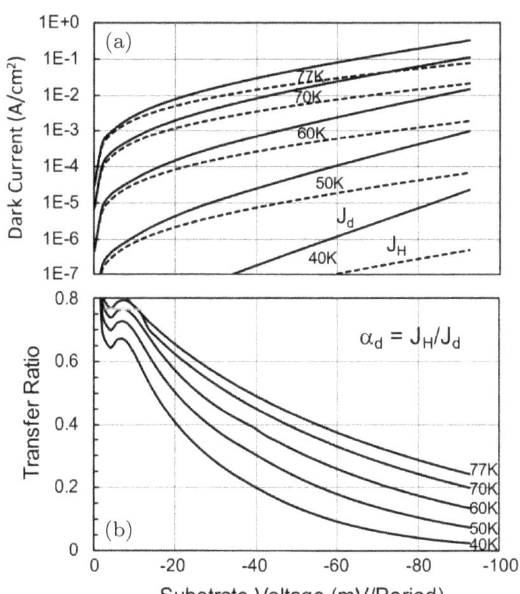

Figure 2.43. (a) The solid curves show the fitted total dark current J_d of a detector with $\lambda_c = 11.6\,\mu$m. The dashed curves show the current component J_H above the barrier height. (b) The dark current transfer ratio α_d through a high-pass filter.

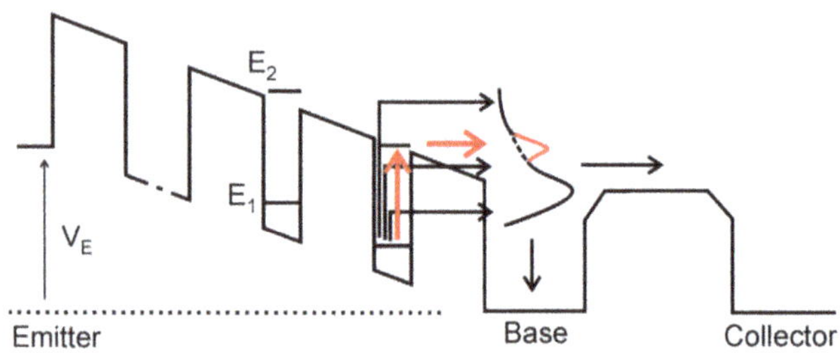

Figure 2.44. The band diagram of a typical IHET. The arrows show the dark current and the photocurrent paths.

high bias where TAT current below $H(V)$ dominates. Since Det. A is cutoff at $11.6\,\mu$m, it is expected to operate at T around 50 K. At 80 mV/period bias, α_d is around 0.10.

To implement current filtering, a pixel needs to have two independent contacts besides the common emitter contact. This detector structure is known as the infrared hot-electron transistor (IHET) [43]. In this device, the QWIP is located between the emitter and the base, and the filter barrier is located between the base and the collector. The band diagram of an IHET is shown in Fig. 2.44. It is usually connected in the common base configuration. The emitter voltage V_E is used to control the QWIP bias and the collector voltage V_C is to control the filter bias. By changing V_C, the fraction of QWIP current flowing into the collector can be regulated. In focal plane array (FPA) implementation, the emitter layer and the base layer can be common to all pixels. The output signal from each pixel is connected to the collector. In this FPA architecture, the common base will replace the usual substrate contact and an extra common emitter contact is biased externally. Measurement on small format arrays show no electrical crosstalk among different pixels in this common base configuration, showing the viability of the approach. An advantage of this scheme is the small bias required between the base and the collector. The small bias is more compatible with the readout circuits designed for photodiodes. Other advantages include the elimination of $1/f$ noise from impurity tunneling and the reduction of dielectric relaxation time of a QWIP at low background by using thinner QWIP barriers [44,45].

According to the energy diagram in Fig. 2.44, when $|V_E|$ increases, the hot-electron distribution will shift upward relative to the filter. Similarly, when V_C is increased positively, the filter barrier height will reduce relative to the distribution. Applying either bias will increase the transfer of electrons into the collector. By monitoring α_d as a function of V_E or V_C, the energy distribution can be deduced. For high-pass filters, the hot-electron energy distribution $\rho(E)$ is directly proportional to either $d\alpha/dV_E$ at a constant V_C or dJ_C/dV_C at a constant V_E, where α is J_C/J_E, and J_C and J_E are the current densities measured at the collector and emitter, respectively. For narrow band-pass filters, which are made of double barrier structures, $\rho(E)$ is directly proportional to α.

Figure 2.45 shows the hot-electron spectroscopy of an IHET at $T = 10\,\text{K}$ with and without a $\lambda = 10\,\mu\text{m}$ laser light source. The IHET consists of 50 periods of GaAs/$Al_{0.27}Ga_{0.73}As$ QWs with $W = 50\,\text{Å}$, $B = 150\,\text{Å}$, and $N_D = 2 \times 10^{18}\,\text{cm}^{-3}$. The base is made of 1500 Å GaAs doped with $0.2 \times 10^{18}\,\text{cm}^{-3}$. The filter is a 700 Å and $x = 0.27$ barrier. In this thin barrier detector at $T = 10\,\text{K}$, the dark current is conducted via direct tunneling in the ground states. Therefore, a $\rho(E)$ peak is observed at a large $V_E = -4.3\,\text{V}$, at which the low-energy ground state electrons can pass through the filter. With an infrared source, an additional $\rho(E)$ peak appears

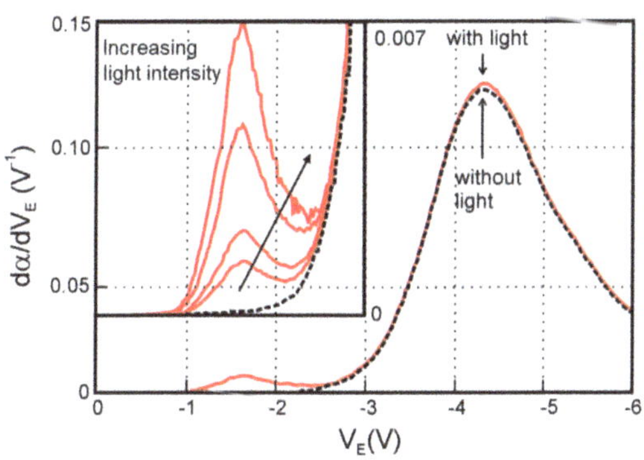

Figure 2.45. Hot-electron distribution observed in an IHET with and without radiation at $T = 10\,\text{K}$. The insert shows the distributions at different levels of radiation. Adapted with permission from *Phys. Rev. B*, 42(14), p. 9166 (1990).

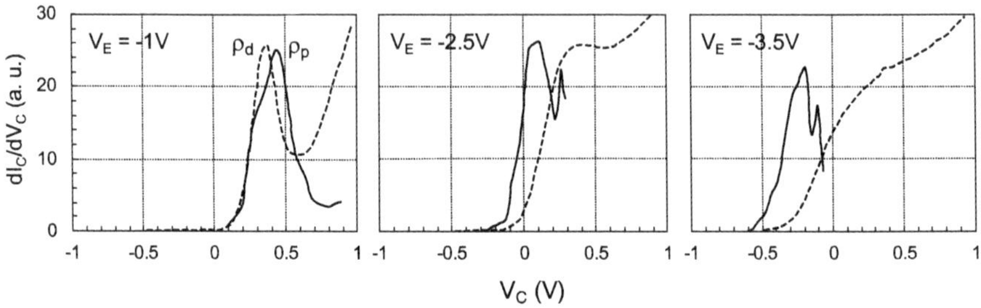

Figure 2.46. Hot-electron distribution observed in an IHET with and without radiation at $T = 77\,\mathrm{K}$. The dark current energy distribution is obtained from DC measurements, while the photocurrent distribution is obtained from AC measurements.

at a lower $V_E = -1.6\,\mathrm{V}$. The magnitude of this photocurrent peak varies with the light intensity as shown in the inset of Fig. 2.45. Interestingly, there are no additional peaks appearing at the larger V_E, and the dark current distribution is unchanged. This result shows the photoelectron distribution does not mix with the dark electron distribution in the base region and the photoelectrons can be referentially admitted into the collector at a low V such as at $-2.2\,\mathrm{V}$, with which I_p/I_d ratio can be increased.

Figure 2.46 shows the hot-electron spectroscopy of another IHET having 50 QWs with $x = 0.25$, $W = 40\,\text{Å}$, $B = 200\,\text{Å}$, and $N_D = 1.2 \times 10^{18}\,\mathrm{cm}^{-3}$. The base is made of 300 Å GaAs doped with $1.2 \times 10^{18}\,\mathrm{cm}^{-3}$. The filter is a 2000 Å and $x = 0.25$ barrier. The hot-electron spectroscopy measurement is performed at $T = 77\,\mathrm{K}$. In this measurement, the DC collector dark current $I_C(V_C)$ is first measured as a function of V_C at a fixed V_E, with which the dark current distribution ρ_d can be deduced using $\rho_d \propto dI_C/dV_C$. The collector photocurrent is then measured under a 9.25 μm laser source using AC lock-in techniques. By measuring the AC signal, the DC dark current at 77 K will not contribute to the photocurrent distribution ρ_p, and ρ_p is again proportional to the AC dI_C/dV_C. At $V_E = -1\,\mathrm{V}$ in Fig. 2.46, a distinct ρ_d peak is measured at $V_C = 0.4\,\mathrm{V}$ in the DC measurement. Since this peak coincides with the AC photocurrent peak, it can be identified to be the TE current from the E_2 resonant state of the QWs. The rising background is the TAT

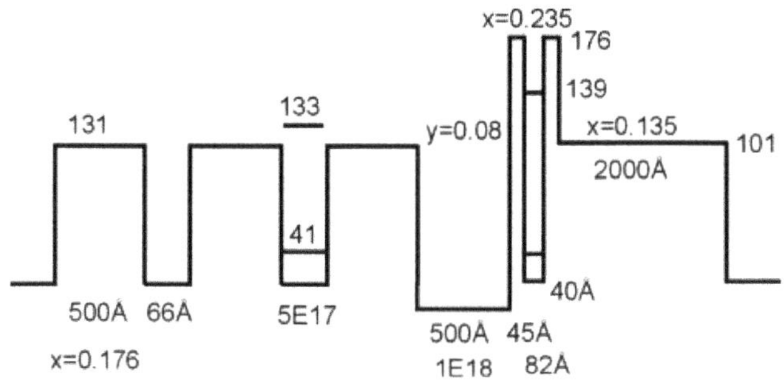

Figure 2.47. The band structure of an IHET with a band-pass filter. The numerals without units are the energies in meV.

current conducted below the barrier for this thin-barrier QWIP. At higher V_E, the TE component rapidly diminishes relative to the TAT component and becomes negligible at $-3.5\,\text{V}$ while the photocurrent peak remains. These two examples show that the IHET structure can be used to study QWIP transport properties in greater detail.

Since the photocurrent distribution is rather sharp even after passing through the base region, one can design a narrow band-pass filter to selectively collect the photoelectrons and reject the wider distributed dark electrons. One example of the IHET structures is shown in Fig. 2.47. The QWIP consists of 30 QW periods. It has a responsivity peak at $13.2\,\mu\text{m}$ and a cutoff at $14.0\,\mu\text{m}$. The transmission of the band-pass filter is at 139 meV with a bandwidth of 10 meV.

Figures 2.48(a) and (b) show the emitter and collector current-voltage characteristics at $V_C = 0\,\text{V}$, respectively. The solid curves are the dark current J_d at different T, while the dashed curves are the 300 K background photocurrent J_p under 45° edge coupling. Both emitter and collector photocurrents are measured at 10 K with $f/1.5$ optics. Notice the occurrence of the plateau regions in Fig. 2.48(b) at $T \sim 50\,\text{K}$ and $V_E \sim -0.7\,\text{V}$ for the collector dark current J_{cd} and photocurrent J_{cp} even though the corresponding emitter current rises rapidly in this bias regime. These characteristics indicate the filter is blocking electrons in a wide range of energies. At $V_E = -0.5\,\text{V}$ and $T = 60\,\text{K}$, α_p is 10 times larger than α_d, with which J_p is

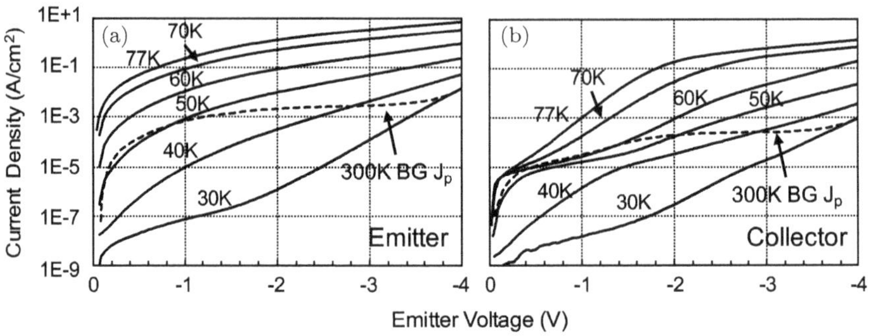

Figure 2.48. The measured detector I-V characteristics at (a) the emitter and (b) the collector of an IHET with a band-pass filter at $V_C = 0\,\text{V}$. Adapted with permission from *Appl. Phys. Lett.*, 66(1), pp. 90–92 (1995).

larger than J_d at 60 K rather than at 50 K. This example shows that the IHET structures can be used to improve the QWIP operating temperature.

References

1. Easki, L. and Tsu, R. (1970). Superlattice and negative differential conductivity in semiconductors, *IBM J. Res. Develop.*, 14, pp. 61–65.
2. Chiu, L. C., Smith, J. S., Margalit, S. and Yariv, Y. (1983). Internal photoemission from quantum well heterojunction superlattices by phononless free-carrier absorption, *Appl. Phys. Lett.*, 43, pp. 331–332.
3. Smith, J. S., Chiu, L. C., Margalit, S., Yariv, A. and Cho, A. Y. (1983). A new infrared detector using electron emission from multiple quantum wells, *J. Vac. Sci. Technol.*, B1, pp. 376–378.
4. Coon, D. D. and Karunasiri, R. P. G. (1984). New mode of IR detection using quantum wells, *Appl. Phys. Lett.*, 45, pp. 649–651.
5. Coon, D. D., Karunasiri, R. P. G. and Liu, L. C. (1985). Narrow band infrared detection in multiquantum well structures, *Appl. Phys. Lett.*, 47, pp. 289–291.
6. West, L. C. and Eglash, S. J. (1985). First observation of an extremely large-dipole infrared transition within the conduction band of a GaAs quantum well, *Appl. Phys. Lett.*, 46, pp. 1156–1158.
7. Levine, B. F., Malik, R. J., Walker, J., Choi, K. K., Bethea, C. G., Kleinman, D. A. and Vandenberg, J. M. (1987). Strong 8.2 μm infrared intersubband absorption in doped GaAs/AlAs quantum well waveguides, *Appl. Phys. Lett.*, 50, pp. 273–275.

8. Levine, B. F., Choi, K. K., Bethea, C. G., Walker, J. and Malik, R. J. (1987). Quantum well avalanche multiplication initiated by $10\,\mu$m intersubband absorption and photoexcited tunneling, *Appl. Phys. Lett.*, 51, pp. 934–936.
9. Levine, B. F., Choi, K. K., Bethea, C. G., Walker, J. and Malik, R. J. (1987). New $10\,\mu$m infrared detector using intersubband absorption in resonant tunneling GaAlAs superlattices, *Appl. Phys. Lett.*, 50, pp. 1092–1094.
10. Choi, K. K., Levine, B. F., Malik, R. J., Walker, J. and Bethea, C. G. (1987). Periodic negative conductance by sequential resonant tunneling through an expanding high-field superlattice domain, *Phys. Rev. B*, 35, pp. 4172–4175.
11. Choi, K. K., Levine, B. F., Bethea, C. G., Walker, J. and Malik, R. J. (1987). Multiple quantum well $10\,\mu$m GaAs/Al_xGa_{1-x}As infrared detector with improved responsivity, *Appl. Phys. Lett.*, 50, pp. 1814–1816.
12. Choi, K. K., Levine, B. F., Bethea, C. G., Walker, J. and Malik, R. J. (1987). Photoexcited coherent tunneling in a double-barrier superlattice, *Phys. Rev. Lett.*, 59, pp. 2459–2462.
13. Levine, B. F., Bethea, C. G., Hasnain, G., Walker, J. and Malik, R. J. (1988). High-detectivity $D^* = 1.0 \times 1010\,\text{cm}\sqrt{\text{Hz}}/\text{W}$ GaAs/AlGaAs multiquantum well wavelength $= 8.3\,\mu$m infrared detector, *Appl. Phys. Lett.*, 53, pp. 296–298.
14. Levine, B. F., Bethea, C. G., Choi, K. K., Walker, J. and Malik, R. J. (1988). Bound-to-extended state absorption GaAs superlattice transport infrared detectors, *J. Appl. Phys.*, 64, pp. 1591–1593.
15. Levine, B. F., Hasnain, G., Bethea, C. G. and Chand, N. (1989). Broadband 8–$12\,\mu$m high-sensitivity GaAs quantum well infrared photodetector, *Appl. Phys. Lett.*, 54, pp. 2704–2706.
16. Levine, B. F., Bethea, C. G., Hasnain, G., Shen, V. O., Pelve, E., Abbott, R. R. and Hsieh, S. J. (1990). High sensitivity low dark current $10\,\mu$m GaAs quantum well infrared photodetectors, *Appl. Phys. Lett.*, 56, pp. 851–863.
17. Choi, K. K. (1993). Detection wavelength of quantum-well infrared photodetectors, *J. Appl. Phys.*, 73, pp. 5230–5236.
18. Gunapala, S. D., Liu, J. K., Park, J. S., Lin, T. L. and Sundaram, M. (2001). Infrared radiation detecting device, US Patent No. 6,211,529.
19. Kastalsky, A., Duffield, T., Allen, S. J. and Harbison, J. (1988). Photovoltaic detection of infrared light in a GaAs/AlGaAs superlattice, *Appl. Phys. Lett.*, 52, pp. 1320–1322.
20. Goossen, K. W., Lyon, S. A. and Alavi, K. (1988). Photovoltaic quantum well infrared detector, *Appl. Phys. Lett.*, 52, pp. 1701–1703.

21. Giorgetta, F. R., Baumann, E., Graf, M. *et al.*, (2009). Quantum cascade detectors, *IEEE J. Quant. Electr.*, 45, pp. 1029–1042.
22. Choi, K. K., Dutta, M., Newman, P. G., Saunders, M.-L. and Iafrate, G. J. (1990). 10 μm infrared hot-electron transistors, *Appl. Phys. Lett.*, 57, 1348–1350.
23. Kamgar, A., Kneschaurek, P., Dorda, G. and Koch J. F. (1974). Resonance spectroscopy of electronic levels in a surface accumulation layer, *Phys. Rev. Lett.*, 32, pp. 1251–1254.
24. Kneschaurek, P., Kamgar, A. and Koch, J. F. (1976) Electronic levels in surface space charge layers on Si(100), *Phys. Rev. B*, 14, pp. 1610–1622.
25. Wheeler, R. G. and Goldberg, H. S. (1975). A novel voltage tunable infrared spectrometer-detector, *IEEE Trans. Elect. Dev.*, ED-22, pp. 1001–1009.
26. Choi, K. K., Bandara, S. V., Gunapala, S. D., Liu, W. K. and Fastenau, J. M. (2002). Detection wavelength of InGaAs/AlGaAs quantum wells and superlattices, *J. Appl. Phys.*, 91, pp. 551–564.
27. Bastard, G. and Brum, J. A. (1986). Electronic states in semiconductor heterostructures, *IEEE J. Quantum Electr.*, QE-22, pp. 1625–1644.
28. Choi, K. K. (1997). *The Physics of Quantum Well Infrared Photodetectors*, World Scientific.
29. Shi, J.-J. and Goldys, E. M. (1999). Intersubband optical absorption in strained double barrier quantum well infrared photodetectors, *IEEE Trans. Elect. Dev.*, 46, pp. 83–88.
30. See for example: Mahan, G. D. (1981). *Many-Particle Physics*, Plenum Press, p. 256.
31. Duke, C. B. (1969). *Tunneling in Solids*, Academic Press, p. 211.
32. Osamura, K. and Murakami, Y. (1972). Free carrier absorption in *n*-GaAs, *Jpn. J. Appl. Phys.*, 11, pp. 365–371.
33. Liu, H. C. (1992). Photoconductive gain mechanism of quantum-well intersubband infrared detectors, *Appl. Phys. Lett.*, 60, pp. 1507–1509.
34. Choi, K. K. (1994). Generalized relationship between gain, noise, and capture probability of quantum well infrared photodetectors, *Appl. Phys. Lett.*, 65, pp. 1266–1268.
35. Choi, K. K. (1996). Reduction of photoconductive gain in quantum well infrared photodetectors, *J. Appl. Phys.*, 80, pp. 1257–1259.
36. Kingston, R. H. and MacAdam, D. L. (1978). *Detection of Optical and Infrared Radiation*, Springer, pp. 56–74.
37. Arabshahi, H., Khalvati, M. R. and Rezaee Rokn-Abadi, M. (2008). Temperature and doping dependencies of electron mobility in InAs, AlAs and AlGaAs at high electric field application, *Bra. J. Phys.*, 38, pp. 293–296.

38. Bernardi, M., Vigil-Fowler, D., Ong, C. S., Neaton, J. B. and Louie, S. G. (2015). Ab initio study of hot electrons in GaAs, *Proc. Natl. Acad. Sci.*, 112, pp. 5291–5296.
39. Rosencher, E., Luc, F., Bois, P., Nagle, J. and Cordier, Y. (1993). Capture time versus barrier thickness in quantum-well structures measured by infrared photoconductive gain, *Appl. Phys. Lett.*, 63, pp. 3312–3315.
40. Tennant, W. E. (2010). "Rule 07" revisited: still a good heuristic predictor of p/n HgCdTe photodiode performance?, *J. Electron. Mater.*, 39, pp. 1030–1035.
41. Kopytko, M., Sobieski, J., Gawron, W. and Martyniuk, P. (2022). Study of HgCdTe (100) and HgCdTe (111)B heterostructures grown by MOCVD and their potential application to APDs operating in the IR range up to 8 μm, *Sensors*, 22, p. 924.
42. Schuster, J., DeWames, R. E. and Wijewarnasuriya, P. S. (2017). Dark currents in a fully-depleted LWIR HgCdTe p-on-n heterojunction: analytical and numerical simulations, *J. Electron. Mater.*, 46, pp. 6295–6305.
43. Choi, K. K., Dutta, M., Newman, P. G., Saunders, M.-L. and Iafrate, G. J. (1990). 10 μm infrared hot-electron transistors, *Appl. Phys. Lett.*, 57, pp. 1348–1350.
44. Arrington, D. C., Hubbs J. E., Gramer M. E. and Dole G. A. (2000). Nonlinear response of QWIP detectors: summary of data from four manufacturers, *Proc. SPIE*, 4028, pp. 288–299.
45. Choi, K. K., Kennerly, S. W., Yao, J. and Tsui, D. C. (2001). Characteristics of QWIPs at low background, *Inf. Phys. & Tech.*, 42, pp. 221–235.

Chapter 3

Electromagnetics

3.1 Introduction

In Chapter 2, we have discussed the general properties of quantum well infrared photodetector (QWIP) materials. However, the invention of this detector material only accomplishes half of what is needed for infrared imaging. To absorb light, a QWIP must sense a finite optical electric field E_z perpendicular to its layers. In a two-dimensional (2D) focal plane array, the detector elements are facing the scene at the perpendicular angle as depicted in Fig. 1.7. The optical electric field $\boldsymbol{E}$ thus polarizes parallel to the QW layers, which cannot induce an absorption. Although the 45° edge coupling geometry depicted in Fig. 2.35 is convenient for detector characterization, it is not feasible for 2D imaging. A light coupling scheme suitable for normal incidence is thus required. To be able to design a useful coupling structure, some knowledge on light propagation in a multi-layer, structured medium will be needed. In this chapter, we will give a brief account on some basic physics of electromagnetism and mathematical tools relevant to the light coupling design of QWIPs. These designs can also be useful for other detectors with isotropic absorption properties.

3.2 Optical Properties of Dielectrics and Metals

A QWIP usually consists of a doped multiple quantum well (MQW)-absorbing layer sandwiched by contact layers at the top and bottom, and it is covered by a metal contact layer depicted in Fig. 2.35.

Besides the infrared absorption in the MQW layer discussed in Chapter 2, the doped contact layers and the metal layer can also absorb some of the radiation and reduce the absorption in the active layer. Although this parasitic absorption is usually negligible, it can be more significant when the optical electric field in these layers are intensified in some of the coupling designs and at very long wavelengths. To better understand the absorption properties of these materials, we start with the common Drude-Lorentz oscillator model for the material permittivity ε. Derived from the Maxwell equations, the wave equation in a material can be expressed as [1]

$$\nabla^2 \boldsymbol{E} = \frac{1}{c^2}\frac{\partial^2 \boldsymbol{E}}{\partial t^2} + \mu_0 \frac{\partial \boldsymbol{J}}{\partial t} + \mu_0 \frac{\partial^2 \boldsymbol{P}}{\partial t^2}, \tag{3.1}$$

where $\boldsymbol{J}$ is the current density of free electrons and $\boldsymbol{P}$ is the electric polarization of bound electrons. Using a classical model for harmonic motion for both free and bound electrons, the complex ε is

$$\varepsilon = 1 - \frac{\omega_p^2}{\omega^2 + i\frac{\omega}{\tau}} + \frac{N_b e^2}{m\varepsilon_0}\sum_j \frac{f_j}{\omega_j^2 - \omega^2 - i\gamma_j\omega}, \tag{3.2}$$

where

$$\omega_p = \sqrt{\frac{N_f e^2}{m^* \varepsilon_0}} \tag{3.3}$$

is the plasma frequency, N_f is the free electron density, m^* is the effective electron mass, and τ is the current relaxation time. N_b is the density of bound electrons, m is the bare electron mass, and f_j and γ_j are the oscillator strength and damping coefficient, respectively, at the jth resonant oscillating frequency ω_j. ω_p is the oscillation frequency of the free electrons when they are collectively displaced. The real part of Eq. (3.2) contributes to the overall permittivity of the material. From Eq. (3.2), the free carriers reduce ε while the bound carriers generally increase it. The sign of contribution is related to the phase of the electron motion relative to the driving force.

In general, $\varepsilon = \aleph^2$ where $\aleph = (n + i\kappa)$ is the complex refractive index, n is the refractive index, and κ is the extinction coefficient. The values of n and κ are contributed from both the free charges and bound charges. From the definition of $\aleph$, $\text{Re}(\varepsilon) = n^2 - \kappa^2$, and $\text{Im}(\varepsilon) = 2n\kappa$. For an electromagnetic (EM) wave with the incident wavevector $k = 2\pi/\lambda$, the wavevector in the material is $(n + i\kappa)k = nk + i/\delta$, where $1/\delta = 2\pi\kappa/\lambda$ is the inverse E field decay length in the material.

In the doped contact layers, the free carriers come from doping, therefore N_f in Eq. (3.3) is the doping density, N_D. The optical absorption of the free carriers can be obtained from its contribution to $\text{Im}(\varepsilon)$. The free carrier absorption coefficient α_f is then given by

$$\begin{aligned}\alpha_f = \frac{2}{\delta_f} &= \frac{4\pi}{\lambda}\frac{1}{2n} Im\left(\frac{\omega_p^2}{\omega^2 + i\frac{\omega}{\tau}}\right)\\ &= \frac{1}{4\pi^2}\frac{e^2}{n\varepsilon_0 m^* c^3 \tau} N_D \lambda^2. \end{aligned} \tag{3.4}$$

Equation (3.4) shows that α_f is directly proportional to N_D, λ^2, and $1/\tau$. If the contacts are heavily doped, α_f will be large in the long wavelengths. Figure 3.1 shows α_f interpolated from the experimental data conducted by Osamura and Murakami [2] at 90 K. The value of α_f is not sensitive to temperature and it has a stronger wavelength dependence than λ^2 predicted in Eq. (3.4). The stronger λ dependence can come from the λ dependence of τ. At $\lambda > 10\,\mu$m, α_f at high doping is comparable to the typical α of a QWIP material

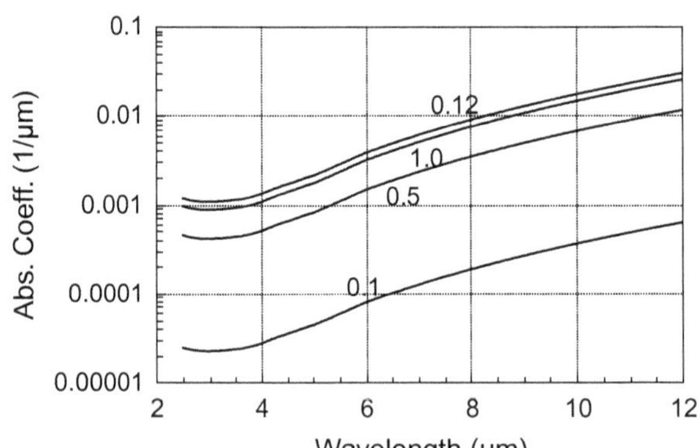

Figure 3.1. The free carrier absorption coefficient at different N_D (in $10^{18}\,\text{cm}^{-3}$).

shown in Fig. 2.20. Therefore, the contact absorption should be part of the consideration in the detector design.

The top of a QWIP is covered with a composite metal layer to provide ohmic contact to the detector. A suggestion for the layer combinations is Pd (50 Å)/Ge (200 Å)/Au (300 Å)/Pd (50 Å)/Au (3000 Å). After e-beam deposition, the layer is annealed at 350°C for 25 min. During the annealing, Ge/Au forms a eutectic solution and diffuses into the QWIP top contact layer, thus creating a heavily n-doped layer (~100 Å thick) for ohmic contact formation. The first Pd layer is to prevent spikes in the Ge/Au diffusion and the second Pd layer is used as a barrier to prevent intermixing of gold. The thick gold layer is used to reduce the contact resistance with the indium bump contact to the Si readout circuit and as an optical reflector for the incident light.

In a metal, its free electron density N_f is close to the atomic density and far larger than the contact doping density. Therefore, its contribution to the metal ε is also much stronger. The value of ε_m from the conducting electrons is

$$\begin{aligned}\varepsilon_m &= 1 - \frac{\omega_p^2}{\omega^2 + i\frac{\omega}{\tau}},\\ &= 1 - \frac{\omega_p^2\tau^2}{(1+\omega^2\tau^2)} + i\frac{\omega_p^2\tau^2}{\omega\tau(1+\omega^2\tau^2)},\\ &= 1 - \frac{\sigma\tau}{\varepsilon_0(1+\omega^2\tau^2)} + i\frac{\sigma}{\varepsilon_0\omega(1+\omega^2\tau^2)},\end{aligned} \tag{3.5}$$

where

$$\omega_p^2 = \frac{Ne^2}{m^*\varepsilon_0} = \left(\frac{Ne^2}{m^*}\tau\right)\frac{1}{\varepsilon_0\tau} = \frac{\sigma}{\varepsilon_0\tau}, \tag{3.6}$$

N is the conducting electron density, and σ is the electrical conductivity. For gold, $\omega_p = 1.38 \times 10^{16}$ rad/s, while for Pd, $\omega_p = 1.83 \times 10^{16}$ rad/s, which are in the ultraviolet regime.

When $\omega\tau \ll 1$, $\text{Im}(\varepsilon_m) = \sigma/(\varepsilon_0\omega)$, and the large σ of the conducting electrons dominates over the dielectric polarizability of the valence electrons in determining n and κ. In this case, $n = \kappa = (1+i)\sqrt{(\mu_0\sigma c^2/2\omega)}$. The E field decay length into the metal δ, which is also known as the skin depth, is

$$\delta = \frac{1}{\kappa k} = \sqrt{\frac{2}{\mu_0 \sigma \omega}}. \tag{3.7}$$

However, at the infrared wavelengths, $\omega\tau$ is in the order of unity, and Eq. (3.7) cannot be used. For example, at $\lambda = 10\,\mu$m, $\omega = 1.9 \times 10^{14}$ rad/s. With the typical τ in the 10^{-14} s range, $\omega\tau \approx 2$. Therefore, δ at the infrared frequencies should be deduced from the theoretical or experimental κ at these frequencies. Using the experimental κ in literature [3, 4], Fig. 3.2(a) shows the skin depths $\delta = 1/\kappa k$ of gold and palladium used in the QWIP contact material. Gold has a smaller skin depth δ_{Au} of about 200 Å, and Pd has a larger δ_{Pd} of ~380 Å. The penetration depth, which is the decay length of light intensity in the material, is half of the skin depth.

When the metal layer thickness is much larger than δ, the incident light from the air cannot transmit through the layer. Part of it is absorbed in the layer and part of it is reflected back to the air. The light reflectance in this case can be obtained by the usual Fresnel's equations [1]. For normal incidence, the absorptance A of the thick metal films is given by

$$A = 1 - R = 1 - \frac{(1-n)^2 + \kappa^2}{(1+n)^2 + \kappa^2}, \tag{3.8}$$

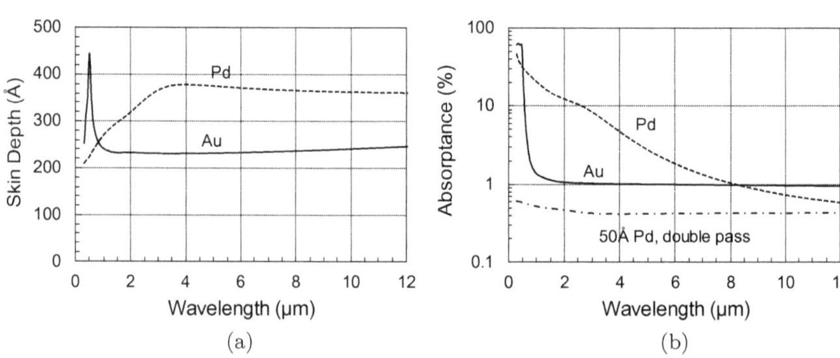

Figure 3.2. (a) The skin depths of Au and Pd in the infrared wavelengths. (b) The absorptance of thick and thin films of the materials.

where R is the reflectance. For thick Au films, the total absorption is about 1% at $\lambda > 2\,\mu$m as shown in Fig. 3.2(b). For Pd, the absorption can be high at 10% at $\lambda = 2\,\mu$m and it decreases exponentially toward the longer wavelengths. Since both Pd layer thicknesses t are 50 Å in the metal contact, the absorption in each layer is estimated to be less than 0.3% using $\eta = 1 - \exp[-2t/(\delta_{\mathrm{Pd}}/2)]$ for double pass of light in each layer. In considering further reduction of Pd thickness upon thermal annealing, the infrared absorption in the Pd layers can be neglected. The thick gold layer in the metal contact thus defines the EM properties of the layer.

For a typical detector with isotropic absorption, the light reaching the top gold contact is substantially reduced after passing through the absorbing layer. Therefore, the gold absorption will be much less than 1%. On the other hand, the QWIP material does not absorb light when the light makes the first vertical pass into the material. The full intensity of light hits the gold surface, resulting in a more substantial absorption. In addition, the light coupling structures in QWIPs may further increase the optical electric field in the gold layer and therefore create stronger absorption. The gold absorption generally cannot be ignored.

3.3 Surface Plasmon-Polaritons

In general, the absorption in the metal layer depends on the EM modes established at the metal/top GaAs contact interfacc. If there are no scatterings in the metal, that is if $\tau \to \infty$, ε in Eq. (3.5) is real and

$$\varepsilon_m = 1 - \frac{\omega_p^2}{\omega^2}. \tag{3.9}$$

With $\mathrm{Im}(\varepsilon_m) = 0$, $(n\omega)^2 = \varepsilon_m\omega^2 = (ck)^2$. The dispersion relation of the EM wave in a metal, which refers to as the bulk plasma wave, is

$$\omega^2 = \omega_p^2 + c^2k^2, \tag{3.10}$$

and it is a free propagating wave with a group velocity v_g equal to

$$v_g = \frac{d\omega}{dk} = \frac{c^2k}{\sqrt{\omega_p^2 + c^2k^2}} = c^2\frac{k}{\omega} = \frac{c^2}{v_p}, \tag{3.11}$$

where $v_p = \omega/k$ is the phase velocity. Since the frequency of this bulk plasma wave is above ω_p, it cannot be excited by the lower-frequency infrared radiation. However, there is another branch of EM wave that exists at a lower ω, and it resides at the metal/semiconductor or dielectric interface. This EM wave is known as the surface plasmon-polariton (SPP) wave. A polariton wave is an EM wave coupled to the elementary excitations of a material medium. SPP wave is one of the examples when the in-plane electric polarization (*i.e.*, the *p*-polarization or the transverse magnetic mode) of an EM wave couples to the surface plasmons of the metal. SSP has a different dispersion relation than that of light in the dielectric and therefore cannot be excited in a planar surface. The light absorption in a planar gold contact is unaffected. However, when the gold layer is nonplanar as in some of the QWIP structural designs, SPP can be excited by light and increase gold absorption.

To better understand the regular SPP on a planar surface, Fig. 3.3 shows an EM wave configuration in the vicinity of a dielectric/metal interface. Their electric fields are in the x-z plane. These two waves can be expressed as

$$\boldsymbol{E}_d = (E_{xd}, 0, E_{zd}) \exp[i(k_{xd}x - \omega t)] \exp(ik_{zd}z), \tag{3.12}$$

and

$$\boldsymbol{E}_m = (E_{xm}, 0, E_{zm}) \exp[i(k_{xm}x - \omega t)] \exp(ik_{zm}z). \tag{3.13}$$

Both waves satisfy Maxwell wave equations in the respective medium as well as the boundary condition at the interface. In each medium,

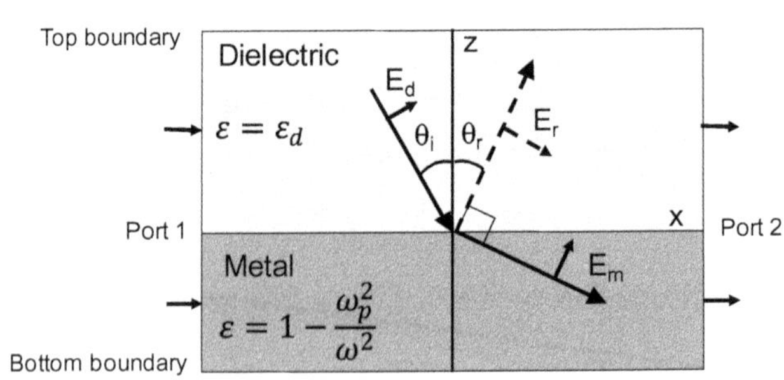

Figure 3.3. The Brewster mode E in the x-z plane. E_r is a field of zero amplitude.

the wave equations are

$$\nabla^2 \boldsymbol{E}_{d,m} = \mu_0 \varepsilon_{d,m} \frac{\partial^2 \boldsymbol{E}_{d,m}}{\partial t^2}, \tag{3.14}$$

where ε_d is the dielectric constant of the dielectric material and ε_m is that of the metal. Without considering electron scattering, ε_m is given by Eq. (3.9). Differentiating Eq. (3.12) and Eq. (3.13) yields

$$k_{zd}^2 = \frac{\omega^2}{c^2}\varepsilon_d - k_{xd}^2, \tag{3.15}$$

and

$$k_{zm}^2 = \frac{\omega^2}{c^2}\varepsilon_m - k_{xm}^2. \tag{3.16}$$

At the interface where $z = 0$, $k_{xd} = k_{xm} \equiv k_x$ from phase matching. Since there is no free charge in each medium, $\nabla \cdot D_{d,m} = 0$, where D is the electric displacement, which gives

$$E_{xd}k_x + E_{zd}k_{zd} = 0, \tag{3.17}$$

$$E_{xm}k_x + E_{zm}k_{zm} = 0, \tag{3.18}$$

and

$$\frac{k_{zd}}{k_{zm}} = \frac{E_{xd}}{E_{xm}}\frac{E_{zm}}{E_{zd}}. \tag{3.19}$$

At the interface, $E_{xd} = E_{xm}$, and $D_{nd} = D_{nm}$, where D_n is the normal component of D. The latter implies $\varepsilon_d E_{zd} = \varepsilon_m E_{zm}$ and

$$\frac{k_{zd}}{k_{zm}} = \frac{\varepsilon_d}{\varepsilon_m}. \tag{3.20}$$

Multiplying Eq. (3.15) with ε_m^2 and Eq. (3.16) with ε_d^2, we arrive at

$$k_x = \frac{\omega}{c}\sqrt{\frac{\varepsilon_d \varepsilon_m}{\varepsilon_d + \varepsilon_m}}. \tag{3.21}$$

The EM modes near the metal/dielectric interface thus acquire dispersion relations described by Eq. (3.21). Substituting ε_m from

Eq. (3.9), two branches of EM modes are obtained with the dispersion relations

$$\omega = ck\left(\frac{\varepsilon_d+1}{2\varepsilon_d}\right)^{1/2}\left\{1+\frac{\varepsilon_d}{\varepsilon_d+1}\left(\frac{\omega_p}{ck}\right)^2 \pm\left[1+\frac{2\varepsilon_d(\varepsilon_d-1)}{(\varepsilon_d+1)^2}\left(\frac{\omega_p}{ck}\right)^2+\left(\frac{\varepsilon_d}{\varepsilon_d+1}\right)^2\left(\frac{\omega_p}{ck}\right)^4\right]^{1/2}\right\}^{1/2}. \tag{3.22}$$

The branch with the positive sign in the second line is the Brewster mode and that with the negative sign is the Fano mode. The two modes are displayed in Fig. 3.4 for gold as the metal and GaAs ($n_d = 3.1$) as the dielectric.

For the Brewster mode, ω is above ω_p, in which ε_m is positive and less than 1, and k_x is real in Eq. (3.21). Since k_x is smaller than the wavevector of light $k_d = \omega\sqrt{\varepsilon_d}/c$ in the dielectric, the Brewster mode can be excited by an incident light with an appropriate angle θ_i to match k_x along the x-axis. Substituting k_x in Eq. (3.21) into Eq. (3.15) and Eq. (3.16) shows that k_{zd}^2 and k_{zm}^2 are both positive.

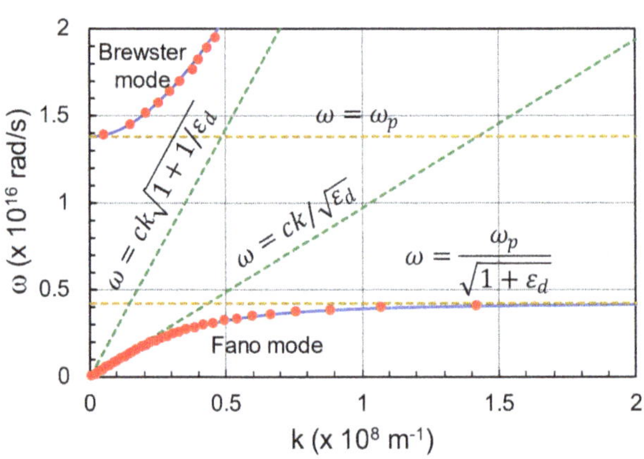

Figure 3.4. Two branches of EM solutions (solid curves) in an infinite space containing a planar Au/GaAs interface. The solid curves are from Eq. (3.22) while the circles are from FEM modeling discussed in Section 3.6. The dashed lines are the asymptotic solutions.

Therefore, k_{zd} and k_{zm} are real. Eq. (3.20) shows that they have the same sign and therefore both waves travel at similar forward directions as depicted in Fig. 3.3. The Brewster mode thus represents a propagating EM solution near the interface. When k_x is very large, ω asymptotically approaches $ck\sqrt{(1+1/\varepsilon_d)}$. This mode is known as the Brewster mode because with k_{zd} in Eq. (3.15), k_{zm} in Eq. (3.16), and k_x in Eq. (3.21), they imply $\tan(\theta_i) = k_x/k_{zd} = n_m/n_d$ and $\tan(\theta_t) = k_x/k_{zm} = n_d/n_m$. Therefore, $\theta_t = 90° - \theta_r$, and the angle θ_i is therefore the usual Brewster angle. At this angle, the incident light propagates into the metal without reflection. The reflected amplitude would be represented by the dashed arrow in Fig. 3.2 if it has a finite amplitude. The absence of reflection is caused by the fact that the induced electric dipole oscillations in the metal are parallel to the reflected direction. Since dipole oscillation does not radiate in the parallel direction, there will be no reflected radiation. With $E_r = 0$, the fields E_d and E_m in Eq. (3.12) and Eq. (3.13) represent a solution to Maxwell equations in this material layer configuration. The Brewster mode has nearly the same dispersion relation as that of bulk plasmon in Eq. (3.10) when k is small but they differ by a constant factor of $\sqrt{(1+1/\varepsilon_d)}$ when k is large.

When ω decreases from ω_p, $\varepsilon_m < 0$ and k_x is a complex number. At sufficiently small ω, when $|\varepsilon_m| > \varepsilon_d$, k_x becomes real again. Since k_x is now larger than $k_d = \omega\sqrt{\varepsilon_d}/c$ in the dielectric, both k_{zd} and k_{zm} are imaginary and have opposite signs in Eq. (3.20), which means both waves decay away from the interface with a decay constant $\delta_d = 1/k_{zd}$ in the dielectric and a decay constant $\delta_m = 1/k_{zm}$ in the metal. When $|\varepsilon_m| \gg \varepsilon_d$ at small k_x, $\delta_m \ll \delta_d$. This EM mode, which is known as the Fano mode, is therefore a bound state solution to Maxwell equations in this material layer configuration. It is also commonly referred to as the SPP wave. Using GaAs/Au interface as an example for the lower branch in Eq. (3.22) yields the dispersion relation for the Fano mode in Fig. 3.4. Alternatively, by performing boundary mode analysis (BMA) in EM modeling on the GaAs/Au interface in Fig. 3.3 with the same ε_m and ε_d, one can obtain the same dispersion relations of the plasmon modes numerically, which are shown as circles in Fig. 3.4. BMA within the finite element method (FEM) will be discussed in Section 3.6. At very small k_x, $\omega \approx ck_x/\sqrt{\varepsilon_d}$, which is close to the light dispersion in the dielectric. Therefore, in the small k_x limit, the presence of the interface does not alter Maxwell solution significantly in the dielectric. On the other hand, at large k_x, $\omega \to \omega_p/\sqrt{(1+\varepsilon_d)}$, which is independent of k_x. The decay constants δ_d and δ_m can be obtained from Eq. (3.15), Eq. (3.16), and Eq. (3.22), and they are

$$\delta_d = \frac{\lambda}{2\pi}\left|\sqrt{\frac{\varepsilon_m+\varepsilon_d}{\varepsilon_d^2}}\right| \tag{3.23}$$

and

$$\delta_m = \frac{\lambda}{2\pi}\left|\sqrt{\frac{\varepsilon_m+\varepsilon_d}{\varepsilon_m^2}}\right|, \tag{3.24}$$

where $\lambda = 2\pi c/\omega$ is the wavelength in air. Using the Drude model of Eq. (3.9) for ε_m with $\omega_p = 1.83 \times 10^{16}$ rad/s for gold and $\varepsilon_d = (3.1)^2 = 9.61$ for GaAs, the decay constants in the vertical direction are shown as solid curves in Fig. 3.5. The decay constant in gold δ_m is relatively constant from $\lambda = 0.5$ to $12\,\mu$m and is equal to ~ 217 Å. It is nearly the same as the skin depth of ~ 230 Å in Fig. 3.2 deduced from the experimental κ of gold with air interface. This similarity

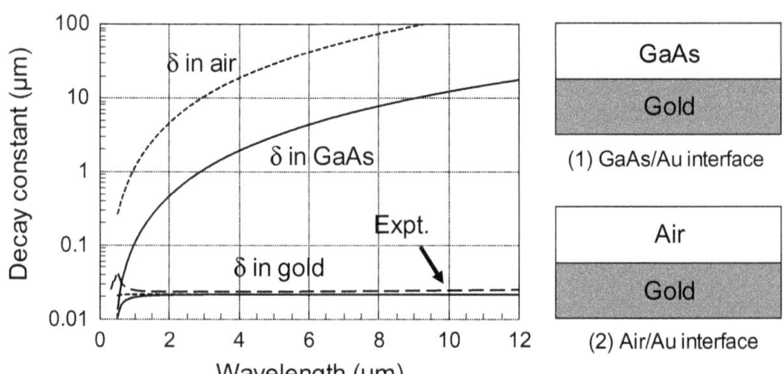

Figure 3.5. The vertical decay constants of the materials in two types of interfaces (1) and (2) as a function of free space wavelength $\lambda = 2\pi c/\omega$. The analysis on δ_{Au} (solid curve) agrees with that in Fig. 3.2(a) (dashed curve) deduced from the experimental κ of Au.

stems from the fact that in the infrared wavelengths, $|\varepsilon_m| \gg \varepsilon_d$, and δ_m is independent of ε_d in Eq. (3.24). Therefore, the skin depth of a metal is independent of the dielectric cover.

The decay constants in air and GaAs in Fig. 3.5, on the other hand, are different and are many orders larger than that of gold. The surface waves in the dielectric side are therefore much less confined than in the metal. The vertical confinement also decreases with increasing λ. In addition, at large λ, their wave dispersions in air or in GaAs are very close to that of the freely propagating waves without the interface. Therefore, the plasmonic effects in a planar dielectric surface are negligible in the infrared wavelengths.

By using Eq. (3.9), we have ignored the damping effects of electron scattering in metal. Therefore, there is no energy loss from the EM fields in metal, and the two EM modes propagate unattenuated. Nevertheless, if the scattering effects are included by adopting the complex ε_m in Eq. (3.5), SPP waves will have a finite propagating length L_{SPP}. Expressing ε_m in Eq. (3.5) as $\varepsilon_m = \varepsilon_R + i\varepsilon_I$ and under the condition $|\varepsilon_R| \gg |\varepsilon_I|$, k_x acquires an imaginary part,

$$\begin{aligned} k_x &= \text{Re}(k_x) + i\,\text{Im}(k_x) \equiv k_{\text{SPP}} + i\,\text{Im}(k_x) \\ &= \frac{\omega}{c}\left(\frac{\varepsilon_d\varepsilon_R}{\varepsilon_d + \varepsilon_R}\right)^{1/2} + i\frac{\omega}{c}\left(\frac{\varepsilon_d\varepsilon_R}{\varepsilon_d + \varepsilon_R}\right)^{3/2}\frac{\varepsilon_I}{2\varepsilon_R^2}. \end{aligned} \tag{3.25}$$

With $\text{Im}(k_x)$ in Eq. (3.25), L_{SPP} of SPP intensity is

$$L_{\text{SPP}} = \frac{1}{2Im(k_x)}. \tag{3.26}$$

Figure 3.6 shows the calculated L_{SPP} curves for both types of interfaces using $\omega_p = 1.38 \times 10^{16}$ rad/s and $\tau = 1.5 \times 10^{-14}$ s for gold. The value of τ is chosen to fit the experimental data of the air/Au interface in the telecommunication wavelengths. Similar to the shorter decay length in the vertical direction, the SPP wave also has a smaller propagation length along the interface when it is covered with a dielectric material. Notice that L_{SPP} is macroscopic at the infrared wavelengths in the order of millimeters at the GaAs/Au interface and in the order of centimeters at the air/Au interface. BMA not only yields the dispersion relation in Fig. 3.4, it also gives

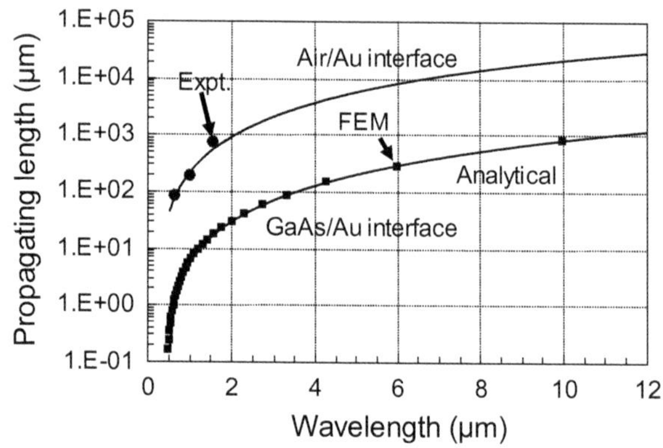

Figure 3.6. The solid curves are the calculated SPP propagating lengths in two types of interfaces based on the analytical model. A current relaxation time τ of 15 fs is used to fit the air/Au data in literature. The squares are from boundary mode analysis within FEM.

$\mathrm{Im}(k_x)$, with which L_{SPP} can also be obtained from EM modeling. It is shown as squares in Fig. 3.6 and agrees well with the analytical analysis.

From EM modeling, one can also obtain the electric field profile, which can be used to visualize the SPP wave. At different k_{SPP} values in Fig. 3.4, the vertical field E_z and the $|E|$ directions can be computed, and two examples are shown in Fig. 3.7. At the smaller k_{SPP} in Fig. 3.7(a), E_z is weaker and spreads wider into the GaAs layer than that at the larger k_{SPP} in Fig. 3.7(b).

Since $\varepsilon_R < 0$, $k_{\mathrm{SPP}} \equiv \mathrm{Re}(k_x)$ is always larger than $k = \omega\sqrt{\varepsilon_d}/c$ in the dielectric for the same ω. Therefore, SPP cannot be directly excited by incident light. It can, however, be excited under certain optical configurations such as those displayed in Fig. 3.8. The optical arrangement shown in Fig. 3.8(a) is known as the Otto configuration and that in Fig. 3.8(b) is the Kretschmann-Raether configuration [5]. They both employ a prism with a dielectric constant ε_p larger than ε_d of the dielectric film. At an incident angle θ, the incident light $k_x(\theta_i) = k\sqrt{\varepsilon_p}\sin(\theta_i)$. If θ_i is larger than the critical angle for total internal reflection at the prism/film interface, the optical field is evanescent across the dielectric in the Otto configuration or

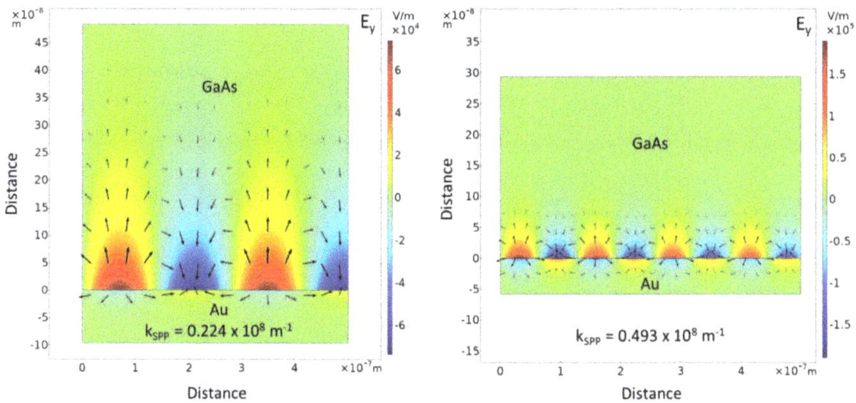

(a) (b)

Figure 3.7. The contour and the intensity map of the vertical field E_z and the $|E|$ direction near the GaAs/Au interface at $k_{\mathrm{SPP}} = k_x =$ (a) 0.224×10^8 rad/m and (b) 0.493×10^8 rad/m.

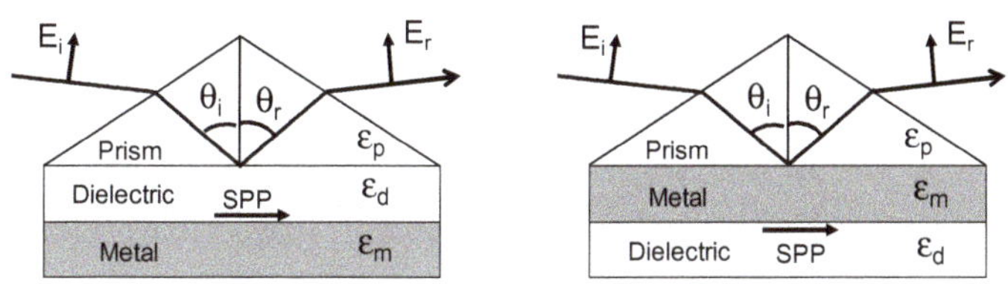

(a) (b)

Figure 3.8. The excitation of SPP waves under the (a) Otto configuration and (b) Kretschmann-Raether configuration.

across the metal in the Kretschmann-Raether configuration. When the layers are optically thin (*i.e.*, $< \lambda/n_d$ or λ/n_m), the incident light is still able to create a field with $k_x = k\sqrt{\varepsilon_p}\sin(\theta_i)$ at the dielectric/metal interface. If $k_x(\theta_i)$ matches k_{SPP} in Eq. (3.25), SPP waves will be excited at the interface. At this matching angle, the incident energy converts into SPP energy and the internal reflection is reduced.

Alternatively, SPP can also be excited on the surface of a grating shown in Fig. 3.9(a). With a grating, the grating k vector,

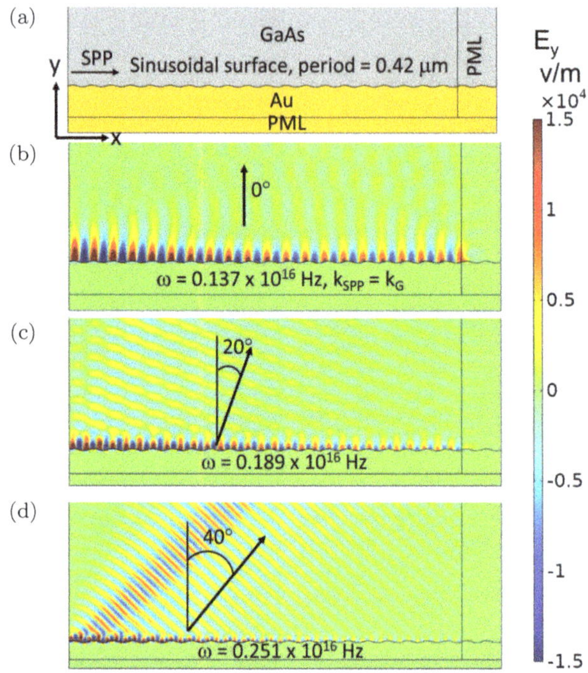

Figure 3.9. (a) The modeled periodic structure. (b), (c), and (d) are the modeled vertical electric field E_y at $\omega = 1.37\,\text{PHz}$, $1.89\,\text{PHz}$, and $2.51\,\text{PHz}$, respectively.

$k_G = 2\pi N/p$, can be subtracted from k_{SPP} in the phase-matching condition, where N is an integer and p is the grating period. For $N = 1$, the phase-matching condition will be $n_d k_0 \sin(\theta) = k_{\text{SPP}} - k_G$, with which

$$\theta = \sin^{-1}\left(\frac{k_{\text{SPP}} - k_G}{n_d k_0}\right), \tag{3.27}$$

where $n_d = \sqrt{\varepsilon_d}$ is the refractive index of the dielectric layer, and $k_0 = \omega/c = 2\pi/\lambda$ is the free space optical k vector. To show an example using the same layer structure in Fig. 3.7, a sinusoidal surface with $p = 0.42\,\mu\text{m}$ and modulation amplitude of $\pm 0.02\,\mu\text{m}$ is considered in Fig. 3.9(a). The same BMA is performed in this layer structure and found that the ω-k dispersion remains the same.

The SPP wave is simply riding on the corrugated surface as seen in Fig. 3.9 at different k_{SPP}. The k_{SPP} is not physically reduced by k_G.

In BMA, the inverse process is analyzed, in which a SPP wave is launched from the left in Fig. 3.9(a), and its optical radiating properties are calculated. In the presence of the grating, the SPP wave is no longer completely confined to the interface as in a planar surface. Instead, part of its energy is radiated at a specific angle θ when the SPP wave travels along the interface. For the present $p = 0.42\,\mu\text{m}$, $k_G = 2\pi/p = 1.496 \times 10^7$ rad/m ($N = 1$). When $\omega = 0.137 \times 10^{16}$ Hz, $k_{\text{SPP}} = 1.487 \times 10^7$ rad/m, which is nearly the same as k_G. Equation (3.27) thus yields $\theta = 0°$ at this ω, and the surface emits EM wave at the normal direction when the SPP wave is launched. Note that at a propagating angle θ, the total electric field $\boldsymbol{E} = \boldsymbol{E}_x + \boldsymbol{E}_y = E\cos\theta\boldsymbol{x} + E\sin\theta\boldsymbol{y}$, where $\boldsymbol{x}$ and $\boldsymbol{y}$ are unit vectors. The ratio of E_y/E_x is therefore tanθ. When $\theta = 0°$, $E_y = 0$, which explains the small E_y when the wave propagates away from the interface in Fig. 3.9(b). If E_x is plotted, a strong modulation wave will be seen. Due to radiation loss, the SPP amplitude is seen to diminish from left to right. (In COMSOL, the y-axis is designated to be the vertical axis in 2D models.)

At the higher ω, k_{SPP} increases much faster than $n_d k_0$ while k_G remains constant, thus θ increases rapidly. At $\omega = 0.189 \times 10^{16}$ Hz, $n_d k_0 = 1.949 \times 10^7$ rad/m while $k_{\text{SPP}} = 2.143 \times 10^7$ rad/m, Eq. (3.27) yields $\theta = 19.39°$. The modeled vertical field E_y profile is shown in Fig. 3.9(c), where a stronger E_y wave propagating at 20° is observed. Similarly, at $\omega = 0.251 \times 10^{16}$ Hz, $n_d k_0 = 2.599 \times 10^7$ rad/m while $k_{\text{SPP}} = 3.154 \times 10^7$ rad/m, the calculated θ is 39.6°, which agrees with the modeled E_y profile in Fig. 3.9(d). SPP waves can be generated by thermal excitation. Indeed, highly coherent and directional thermal radiation originated from surface waves has been observed in SiC grating structures [6], which make use of surface phonon-polaritons to achieve negative permittivity, and the emission angle obeys the grating formula of Eq. (3.27). The inverse process of surface plasmon-polaritons creation by incident light with gold grating has also been demonstrated by Dan'ko *et al.* [7], and the predicted SPP dispersion is observed.

3.4 Optical Transfer Matrix Method

The optical properties of a multilayer composite thin film are usually calculated using the optical transfer matrix method (TMM), similar to that used in solving the Schrodinger equation in MQW structures in Section 2.4. Optical TMM is relevant to QWIPs when the reflection at the detector-slanted sidewalls is used for optical coupling. Since TMM is used only to calculate the coefficients of reflection R, transmission T, and the absorption A, one only needs to know their respective ratios relative to the incident beam and certain simplifications can be made. One simplification is to relate these ratios with the electric and magnetic field components parallel to the film surface rather than the full E and H vectors. The ratios of the components are equal to that of the full vectors for isotropic materials, and it is easier to monitor the parallel components as they are continuous across the layer interfaces from Maxwell equations. The second is the choice of a more convenient unit, the free space unit, as explained by Macloed [8]. Using Macloed's treatment, the magnitudes of $\boldsymbol{E}$ and $\boldsymbol{H}$ of a harmonic wave in a nonmagnetic medium are related by

$$H = \sqrt{\frac{\varepsilon}{\mu_0}} E = yE = \sqrt{\frac{\varepsilon_r \varepsilon_0}{\mu_0}} E = n y_0 E, \tag{3.28}$$

where ε is the permittivity, y is the optical admittance, n is the refractive index, and $y_0 = \sqrt{(\varepsilon_0/\mu_0)}$ is the free space admittance. The wave irradiance $\mathcal{J}$ in W/m^2 can be expressed as

$$\mathcal{J} = \frac{1}{2} Re(\boldsymbol{E} \times \boldsymbol{H}^*) = \frac{1}{2} n y_0 E^2, \tag{3.29}$$

where $\boldsymbol{E}$ and $\boldsymbol{H}$ are in the phaser notation. If we choose the units of E and H such that y_0 is numerically equal to 1, then the optical admittance y will be numerically equal to the refractive index n, and $H = nE$. Since we will see later that both the numerator and denominator of R and T, as well as the absorption A, depend

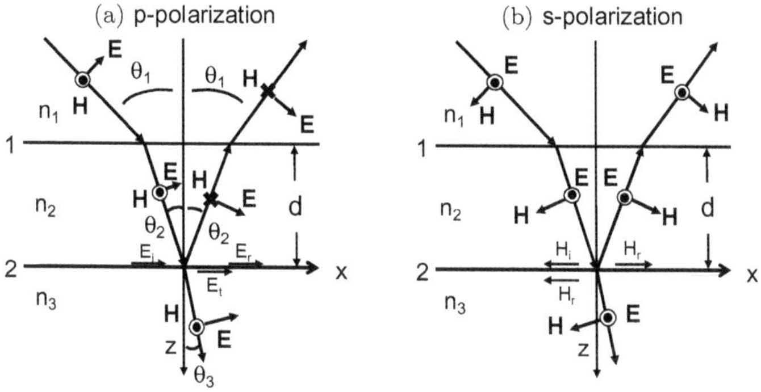

Figure 3.10. The field configurations of (a) p-polarized incident light and (b) s-polarized light onto a thin film with thickness d.

linearly on y_0, the numerical value of y_0 will be factored out in these quantities. Therefore, we can treat n to be the optical admittance in determining the ratio.

Following the approach in [8], a transfer matrix M of a thin film with refractive index n_2 sandwiched between a cover medium with n_1 and a substrate medium with n_3 can be obtained. The layer structure is shown in Fig. 3.10. Referring to Fig. 3.10(a) for p-polarization or TM mode and denoting E and H to be the tangential components, since there is no upward propagating wave from the substrate, the continuity of E and H across interface 2 yields

$$\begin{aligned} E_i + E_r &= E_t, \\ y_2 \frac{E_i}{\cos(\theta_2)} - y_2 \frac{E_r}{\cos(\theta_2)} &= y_3 \frac{E_t}{\cos(\theta_3)}. \end{aligned} \tag{3.30}$$

Solving Eq. (3.30) gives

$$R = \left(\frac{E_r}{E_i}\right)^2 = \left(\frac{\eta_2 - \eta_3}{\eta_2 + \eta_3}\right)^2, \tag{3.31}$$

and

$$T = \frac{\eta_3}{\eta_2}\left(\frac{E_t}{E_i}\right)^2 = \frac{\eta_3}{\eta_2}\frac{(2\eta_2)^2}{(\eta_2 + \eta_3)^2}. \tag{3.32}$$

where $\eta \equiv y/\cos(\theta)$ is the layer admittance for p-polarized light under oblique incidence. The additional prefactor η_3/η_2 in T is related to $\mathcal{J}$ being proportional to the layer admittance in Eq. (3.29). The results in Eq. (3.31) and Eq. (3.32) show that R and T at an interface (interface 2) is known if its E and H across the interface are calculated.

For the entire layer structure in Fig. 3.10(a), one can imagine the two interfaces 1 and 2 being collapsed into one single interface and relate E_1 and H_1 in the cover to E_3 and H_3 in the substrate. To do so, one needs to relate E_1 and H_1 to E_2 and H_2. Different from the substrate, the thin film contains both the incident wave and the reflected wave. We need to distinguish E^+ and H^+ from the incident beam and the E^- and H^- from the reflected beam. In terms of these new quantities, the left-hand side of Eq. (3.30) at the interface 2 is

$$\begin{aligned} E_2 &= E_2^+ + E_2^-, \\ H_2 &= \eta_2 E_2^+ - \eta_2 E_2^-. \end{aligned} \tag{3.33}$$

From Eq. (3.33), one can express E_2^+ and E_2^- in terms of E_2 and H_2. H_2^+ and H_2^- can also be obtained with $H_2^+ = \eta_2\, E_2^+$ and $H_2^- = \eta_2 E_2^-$. With the incident wave phase convention given by $\exp[i(\omega t - kr)]$, E_1^+ and H_1^+ at interface 1 are related to those in interface 2 by $E_1^+ = E_2^+ \exp(i\delta)$ and $H_1^+ = H_2^+ \exp(i\delta)$, where

$$\delta = \frac{2\pi n_2 d \cos\theta_2}{\lambda}, \tag{3.34}$$

and d is the film thickness. Similarly, the E^- and H^- in the reflected beam are given by $E_1^- = E_2^+ \exp(-i\delta)$ and $H_1^- = H_2^- \exp(-i\delta)$. Regrouping the incident and reflected E and H at interface 1,

$$\begin{aligned} E_1 &= E_1^+ + E_1^- = E_2 \cos\delta + H_2 i \sin\delta/\eta_2 \\ H_1 &= H_1^+ + H_1^- = E_2 i\eta_2 \sin\delta + H_2 \cos\delta. \end{aligned} \tag{3.35}$$

With Eq. (3.35), the transfer matrix $M_{1,2}$ can then be set up, whence

$$\begin{bmatrix} E_1 \\ H_1 \end{bmatrix} = M_{1,2} \begin{bmatrix} E_2 \\ H_2 \end{bmatrix} = \begin{bmatrix} \cos\delta & i\sin\delta/\eta_2 \\ i\eta_2 \sin\delta & \cos\delta \end{bmatrix} \begin{bmatrix} E_2 \\ H_2 \end{bmatrix}. \tag{3.36}$$

Since $E_2 = E_3$ and $H_2 = H_3$ at the interface 2, and with $\eta_3 = H_3/E_3$,

$$\begin{bmatrix} B \\ C \end{bmatrix} \equiv \begin{bmatrix} E_1/E_3 \\ H_1/E_3 \end{bmatrix} = \begin{bmatrix} \cos\delta & i\sin\delta/\eta_2 \\ i\eta_2 \sin\delta & \cos\delta \end{bmatrix} \begin{bmatrix} 1 \\ \eta_3 \end{bmatrix}. \tag{3.37}$$

When there are q number of thin films stacked between a semi-infinite incident layer (0th layer) and a semi-infinite substrate layer ($q+1^{\text{th}}$ layer),

$$\begin{bmatrix} B \\ C \end{bmatrix} \equiv \begin{bmatrix} E_1/E_{\text{sub}} \\ H_1/E_{\text{sub}} \end{bmatrix} = \left\{ \prod_{r=1}^{q} \begin{bmatrix} \cos\delta_r & i\sin\delta_r/\eta_r \\ i\eta_r \sin\delta_r & \cos\delta_r \end{bmatrix} \right\} \begin{bmatrix} 1 \\ \eta_{\text{sub}} \end{bmatrix}, \tag{3.38}$$

where E_{sub} is the tangential E field at the substrate interface. The ratio $C/B = H_1/E_1 = \eta$ is the combined composite layer admittance presented to the incident beam. So far, the derivation is for p-polarization. The transfer matrix M turns out to be the same for the s-polarization (TE mode) provided that $\eta_r = y_r \cos(\theta_r)$. Therefore, for s- and p-polarized light,

$$\begin{aligned} \eta_{sr} &= y_r \cos\theta_r = N_r y_0 \cos\theta_r \\ \eta_{pr} &= \frac{y_r}{\cos\theta_r} = \frac{N_r y_0}{\cos\theta_r}. \end{aligned} \tag{3.39}$$

In Eq. (3.39), we used the complex refractive index N_r for the $\sqrt{\varepsilon_r}$ in Eq. (3.28), with which Eq. (3.38) is also applicable to isotropic absorbing layers. The angles θ_r of individual layers are connected by Snell's law, $N_r \sin(\theta_r) = N_0 \sin(\theta_0)$ for all r. Due to fundamental reasons [8], N_0 of the cover medium must be real. Working in the free space units, where $y_0 = 1$, and with Snell's law, η_{sr} and η_{pr} can be expressed as

$$\begin{aligned} \eta_{sr} &= N_r \cos\theta_r = \sqrt{N_r^2 - N_0^2 \sin^2\theta_0} \\ \eta_{pr} &= \frac{N_r}{\cos\theta_r} = \frac{N_r^2}{N_r \cos\theta_r} = \frac{N_r^2}{\sqrt{N_r^2 - N_0^2 \sin^2\theta_0}}. \end{aligned} \tag{3.40}$$

The resulting η of the composite film is $\eta = C/B$. Having found the composite layer η_s for s-polarization and η_p for p-polarization, the respective reflectance R in each polarization, without p and s subscripts in η, B, and C, is

$$R = \left(\frac{\eta_0 - \eta}{\eta_0 + \eta}\right)^2 = \left(\frac{\eta_0 B - C}{\eta_0 B + C}\right)\left(\frac{\eta_0 B - C}{\eta_0 B + C}\right)^*, \tag{3.41}$$

and

$$T = \frac{\eta}{\eta_0}\frac{(2\eta_0)^2}{(\eta_0+\eta)^2} = \frac{C}{B}\frac{4\eta_0 BB^*}{(\eta_0 B + C)^2}$$
$$= \frac{4\eta_0 CB^*}{(\eta_0 B + C)(\eta_0 B + C)^*} = \frac{4\eta_0\,\mathrm{Re}(\eta_{\mathrm{sub}})}{(\eta_0 B + C)(\eta_0 B + C)^*}. \quad (3.42)$$

The last step in Eq. (3.40) has made use of the unimodular properties of the individual $M_{r,r+1}$ so that the overall M is also unimodular and hence $\mathrm{Re}(CB^*) = \mathrm{Re}(C^*B) = \mathrm{Re}(1 \cdot \eta_{\mathrm{sub}})$. Finally, the total absorptance A of the composite thin film excluding the substrate is $A = 1 - R - T$, which gives

$$A = \frac{4\eta_0\,\mathrm{Re}(BC^* - \eta_{\mathrm{sub}})}{(\eta_0 B + C)(\eta_0 B + C)^*}. \quad (3.43)$$

To illustrate the use of TMM, we consider a simple example, in which a thin layer of silicon of 1 μm thick is suspended in air. The light is incident normally from the top as illustrated in the inset of Fig. 3.11(b), and the (n, κ) values of Si are shown in the inset of Fig. 3.11(c).

After multiple bounces of light inside the layer, the unabsorbed light emerges from the bottom surface. At normal incidence, θ_0 is 0° and there is no difference between s- and p-polarizations. The calculated R, T, and A are shown in Fig. 3.11(a)–(c). All these quantities oscillate as functions of λ because of optical interference between the top and bottom surfaces. These oscillations are commonly known as Fabry-Perot oscillations. When λ is small, κ is large, and the light is either reflected or absorbed. There is little light transmitting through the layer. At the longer λ, κ and thus A are smaller, resulting in larger oscillations in R and T.

To verify the TMM calculation, we compare the results with two other theoretical approaches. One is optical ray tracing and the other is FEM. In ray tracing, one assumes the light decaying exponentially along the optical path. Without taking optical interference into

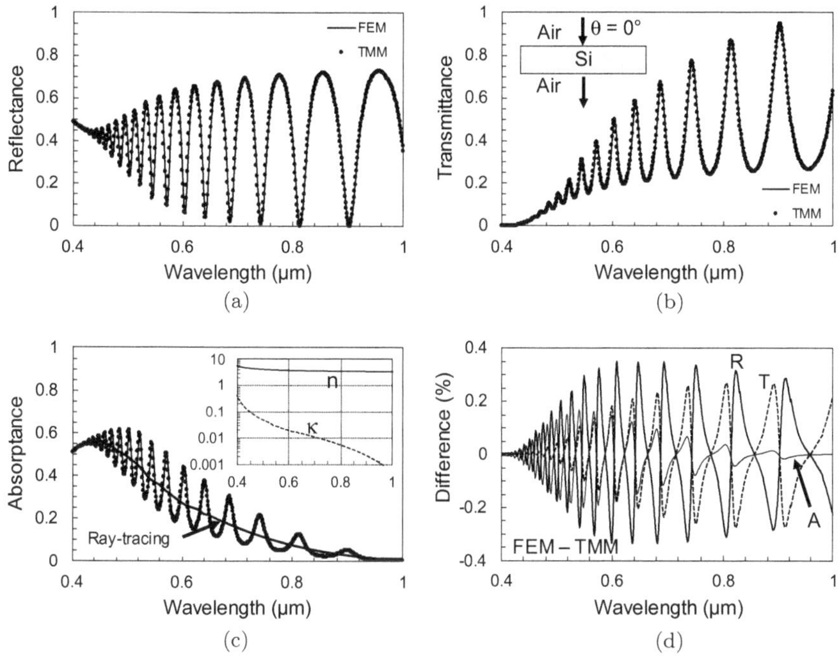

Figure 3.11. (a), (b), and (c) show the calculated R, T, and A, respectively, using TMM, FEM, and ray tracing methods. (d) shows the difference in R, T, and A calculated by FEM relative to TMM.

account, the absorptance $A(\lambda, t)$ is given by

$$\begin{aligned} A(\lambda, t) &= \frac{4n}{(1+n)^2}(1-e^{-\alpha t})\sum_{s=0}^{\infty}\left(\frac{1-n}{1+n}\right)^{2m} e^{-s\alpha t}, \\ &= \frac{4n}{(1+n)^2}\frac{(1-e^{-\alpha t})}{1-\left(\frac{1-n}{1+n}\right)^2 e^{-\alpha t}}, \end{aligned} \tag{3.44}$$

where t is the layer thickness and the summation is over each pass of light upon internal reflection at one of its surfaces. $S = 0$ is for the first pass of light into the material. The factor $[(1-n)/(1+n)]^2$ is the internal reflectance at the Si/air interface. Since Eq. (3.44) does not consider optical interference, it represents the classical value. The ray tracing A is plotted in Fig. 3.11(c). It runs through near the midpoints of the Fabry-Perot oscillations and thus the TMM solution is in qualitative agreement with the classical solution.

A more quantitative comparison can be made with FEM. FEM is a numerical method, in which the field quantities are obtained locally in each meshed element in the model space. The modeled quantities, R, T, and A, from TMM agree with those from FEM in Fig. 3.11, supporting the validity of both methods. Since FEM is a numerical method, its accuracy depends on the density of the adopted mesh. In Fig. 3.11(d), the linear size of the meshed element is set to be less than $1/8^{\text{th}}$ of the wavelength of light in the silicon layer. With this mesh size, their differences in R, T, and A are no more than 0.4% in the entire spectrum. If the mesh size is enlarged to $1/5^{\text{th}}$ of the internal wavelength, the difference will be as large as 2%.

When θ_0 is different from 0°, the optical properties are different for different polarizations. In Fig. 3.12, the R and T are shown separately for TE mode and TM mode. For a fixed wavelength, the reflection and transmission properties can also be plotted as a function of

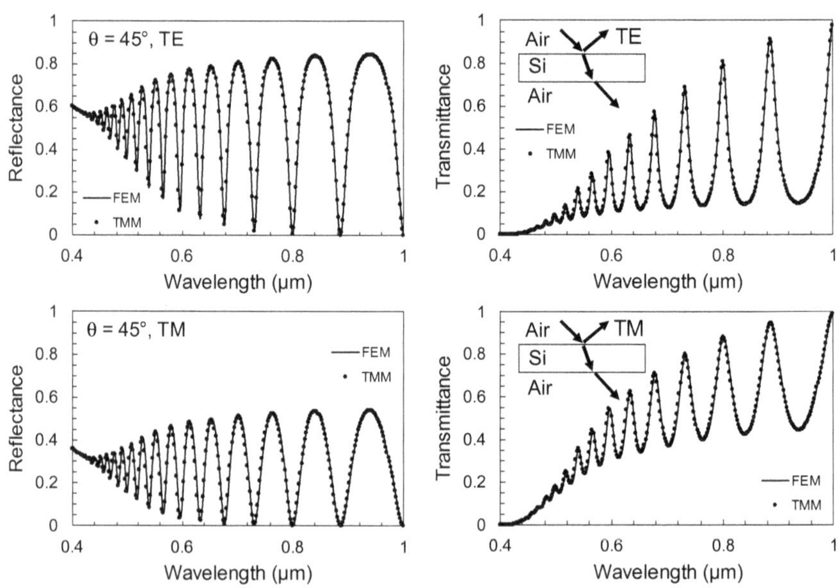

Figure 3.12. The calculated R and T based on TMM and FEM at 45° angle of incidence for TE mode (above) and TM mode (below).

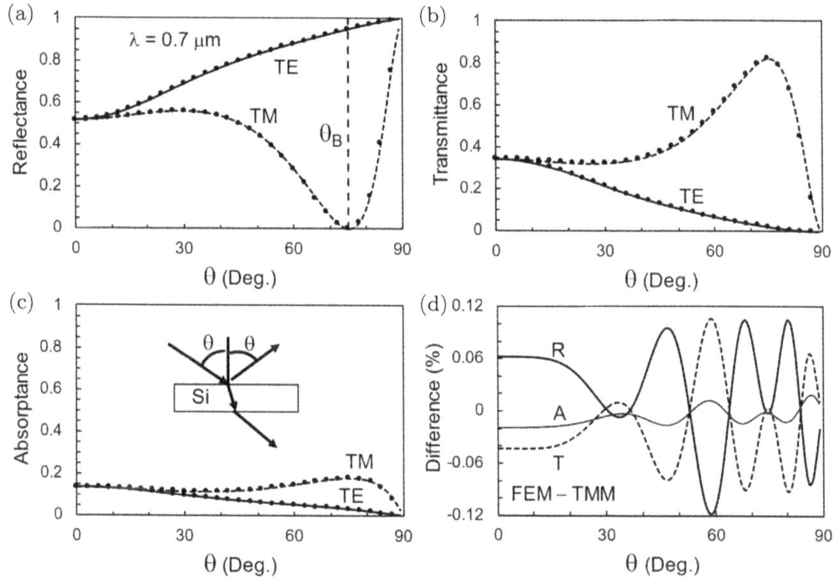

Figure 3.13. (a), (b), and (c) The calculated R, T, and A, respectively, based on TMM (circles) and FEM (solid curves) at $\lambda = 0.7\,\mu$m. (d) The differences in R, T, and A of the two methods for p-polarization. The differences for s-polarization are similar but not shown.

incident angle θ. For example, at $\lambda = 0.7\,\mu$m, (n, κ) of the Si layer is $(3.765, 1.062 \times 10^{-2})$. Applying Eq. (3.41) through Eq. (3.43) yields the optical characteristics shown in Fig. 3.13. Due to the large n of Si, R is usually large unless θ is equal to the Brewster angle θ_B given by $\tan^{-1}(n)$ of 75.1° in Fig. 3.13(a). TMM and FEM in this case differ by less than 0.12% at mesh size of $1/8^{\text{th}}$ of the internal wavelength.

Having established the TMM validity, one can apply it to calculate the reflectivity R of a composite layer structure used in QWIPs. This layer structure is used in the corrugated-QWIP (C-QWIP) structure to be discussed in Chapter 4. A unit cell of the corrugation is schematically shown in Fig. 3.14(a). On top of the MQW sidewalls, there is a composite layer usually made of MgF_2 and Au. It is used to isolate the MQW from the surrounding epoxy material while keeping a large R at the sidewalls. Since the inclined angle θ of the sidewalls is about 48° after chemical etching, the normal incident light will hit

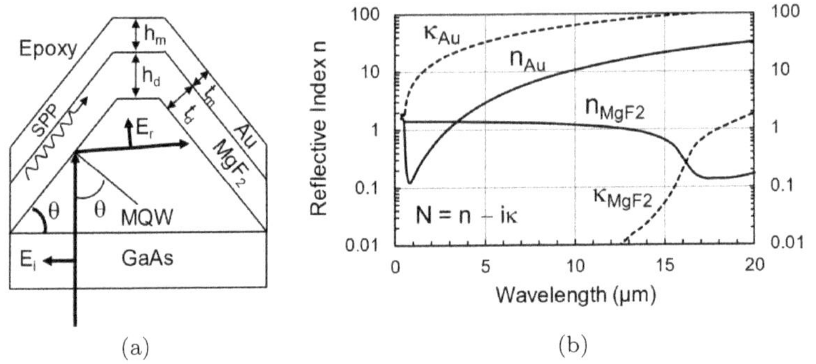

Figure 3.14. (a) The corrugated-QWIP geometry having $\theta = 48^\circ$. The reflected electric field E_r is polarized vertically even though E_i is polarized horizontally. (b) The (n, κ) of MgF_2 and Au films covering the corrugations.

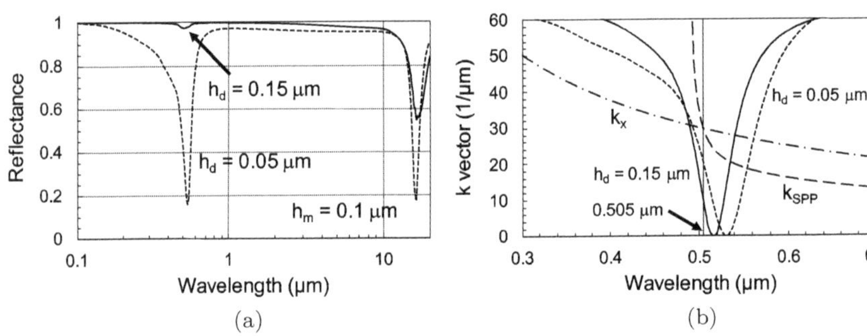

Figure 3.15. (a) The calculated R of the composite film for two MgF_2 thicknesses h_d. (b) The k vectors and R lineshape at a narrow range of λ. $k_x = k_{SPP}$ at the incident wavelength of $0.505\,\mu$m.

this layer surface at the same θ. Upon reflection, the light changes its direction and travels nearly parallel to the MQW layer. The reflected TM mode thus acquires the needed vertical polarization to induce intersubband transitions, even though the TE mode remains uncoupled.

From TMM, one can calculate R of the composite layer efficiently with different MgF_2 and Au thicknesses. Using the complex refractive indices of MgF_2 and Au shown in Fig. 3.14(b), R of two MgF_2 layer heights $h_d = 0.05$ and $0.15\,\mu$m is shown in Fig. 3.15(a). In Fig. 3.15, the layer height h is the layer thickness deposited on top of a planar surface, which is monitored by a thickness sensor during layer deposition. Since the films are deposited on an inclined surface, the layer thickness t perpendicular to the surface is $t = h \cdot \cos\theta$. The gold layer height is assumed to be the same at $h_m = 0.1\,\mu$m. The result shown in Fig. 3.15(a) indicates that R is not very sensitive to h_d. With a relative thick $h_d = 0.15\,\mu$m, R is large in a wide spectral range from 0.1 to 12 μm with $R = 0.967$ at $\lambda = 10\,\mu$m. Therefore, this composite layer structure is suitable for optical coupling.

The anomalous dispersion of the refractive index n_{Au} below 0.5 μm in Fig. 3.14(b) is due to the excitation of surface as well as bulk plasmons in gold while that of $n_{\mathrm{MgF_2}}$ near 16 μm is due to the excitation of MgF_2 optical phonons. Since $n_{\mathrm{MgF_2}} \approx 1.38$ near 0.5 μm, which is less than n_{GaAs} of 3.24 in this calculation, the composite layer structure forms the Otto configuration in Fig. 3.8(a), and SPP is excited at the MgF_2/Au interface when the phase-matching condition, $k_x = n_{\mathrm{GaAs}} 2\pi \sin(48°)/\lambda = k_{\mathrm{SPP}}$, is satisfied. According to $n_{\mathrm{MgF_2}}$ and n_{Au} in Fig. 3.14(b), k_{SPP} and k_x as a function of λ is shown in Fig. 3.15(b). At $\lambda = 0.505\,\mu$m, k_{SPP} is equal to k_x and SPP is excited. The creation of SPP explains the small R at this wavelength indicated in Fig. 3.15. It is interesting that the R minimum is closer to the matching λ when the exciting surface is farther away from the gold surface. For the R minimum at $\lambda = 16.3\,\mu$m, it is caused by the optical phonon absorption in the MgF_2 layer.

The reflectance curves in Fig. 3.15(a) can also be obtained by FEM, which yields nearly identical results. In addition to the R curves, FEM can also give the E field distribution in the layer structure. For example, Fig. 3.16 shows the vertical E_y distribution and $|E|$ direction when 1 W of optical power in the TM mode is incident onto MgF_2/Au with $h_d = 0.05\,\mu$m and $h_m = 0.1\,\mu$m at 48° incident angle.

At $\lambda = 0.53\,\mu$m where R is at the minimum, a strong SPP wave can be seen at the MgF_2/Au interface and substantial E field is located inside the gold layer. In addition, a weaker SPP wave also exists at the Au/epoxy interface, which corresponds to the Kretschmann-Raether configuration in Fig. 3.7(b). Since $\kappa_{\mathrm{MgF_2}}$ at this λ is extremely small and is less than 10^{-9}, all the SPP absorption

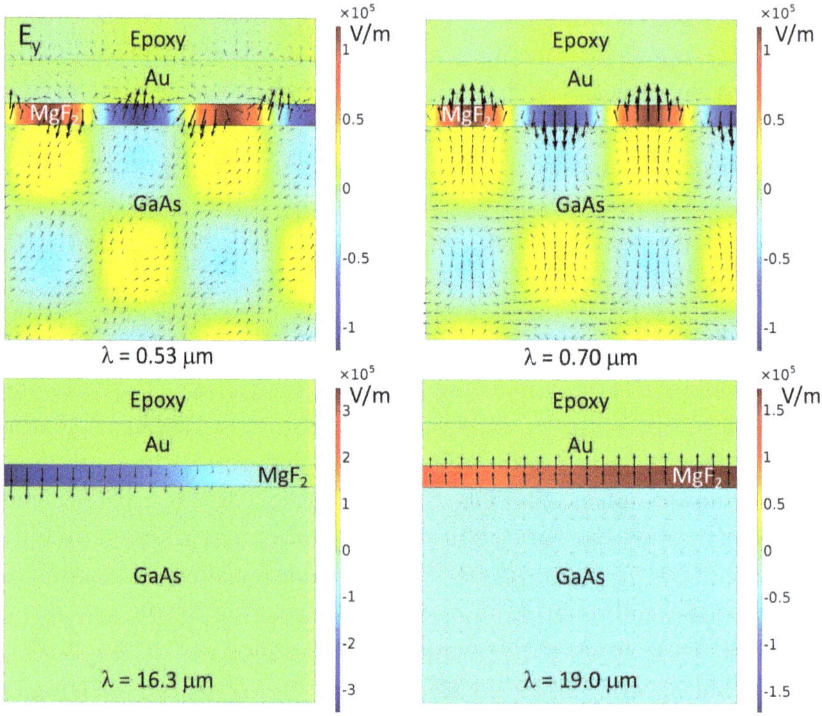

Figure 3.16. The modeled E_y of a MgF_2/Au composite layer sandwiched between a GaAs medium and an epoxy medium at different incident wavelengths. The angle of incidence is 48°.

occurs in the gold layer. Away from the plasmon resonance, such as at $\lambda = 0.7\,\mu$m, the composite layer regains its reflectance as the E field in Au layer and its absorption are greatly reduced. The E field in the GaAs region thus forms substantial interference patterns between the incident beam and the reflected beam. At $\lambda > 1\,\mu$m, Au is nearly a perfect conductor and the plasmonic effects are weak. The E field at interfaces becomes more uniform as for $\lambda = 16.3\,\mu$m and $19.0\,\mu$m shown in Fig. 3.16. The composite layer thus behaves as an optical mirror with high reflectivity except when the optical absorption in the MgF_2 layer is strong.

3.5 Modal Transmission-Line Model

The optical TMM is efficient in calculating the optical properties of multi-layer optical thin films. It can be further expanded to films containing lateral periodic geometries. This method is known as the modal transmission-line (MTL) model, and it has been applied to calculate the absorption of QWIPs having 2D geometries such as C-QWIPs and linear gratings. The detailed formulation can be found in Refs. [9–11]. Here, we briefly outline this approach. Following the coordinates of Fig. 3.17, we consider only TM modes having transverse field components $H_m \equiv -H_{ym}$ and $E_m \equiv E_{zm}$ while $H_{xm} = H_{zm} = E_{ym} = 0$, where m is the index of individual normal modes satisfying Maxwell's equations.

The remaining component E_{xm}, which is responsible for infrared absorption, is obtained from the field equation

$$i\omega\varepsilon_0\varepsilon_x(z)E_{xm} = \frac{\partial H_{ym}}{\partial z}, \tag{3.45}$$

after H_{ym} is known. In Eq. (3.45), $\varepsilon_x(z)$ is the x component of the diagonal dielectric tensor of the uniaxial QWIP material embedded in the cover medium and is periodic in the z direction. In the MTL approach, the detector structure is divided along x into many sublayers identified by an index j. Each sublayer is taken thin enough such that the structural variation along x is negligible within each layer. Similar to the TMM formulism, only the tangential

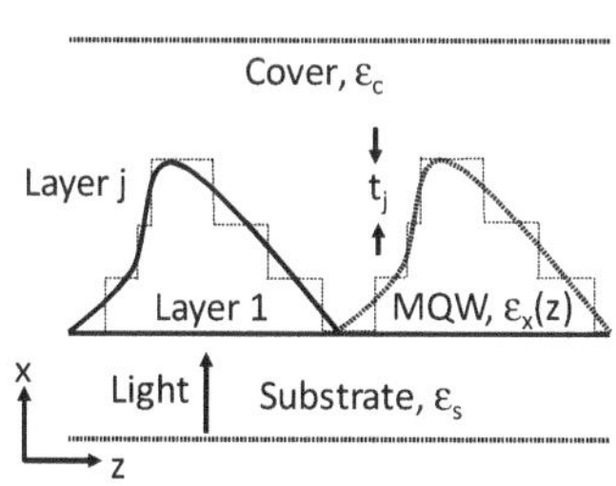

Figure 3.17. The corrugation structure and the coordinate system considered in the modal transmission-line model. The periodic structure is subdivided into thin layers shown by the dotted lines where the structural variation in the vertical direction is negligible.

components are considered, and these components (pointing in the z direction) can be written in the MTL form

$$E_j(x,z) = \sum_m v_{jm}(x) e_{jm}(z),$$
$$H_j(x,z) = \sum_m i_{jm}(x) h_{jm}(z), \tag{3.46}$$

with

$$v_{jm}(x_j) = f_{jm} \exp[-ik_{jm}(t_j - x_j)] + g_{jm} \exp[ik_{jm}(t_j - x_j)],$$
$$i_{jm}(x_j) = Y_{jm}\{f_{jm} \exp[-ik_{jm}(t_j - x_j)] - g_{jm} \exp[ik_{jm}(t_j - x_j)]\}, \tag{3.47}$$

where e_{jm} and h_{jm} are the known m^{th} E and H modes in the jth layer satisfying Maxwell equations, f_{jm} and g_{jm} are coefficients to be determined, k_{jm} is the wavevector of the EM wave in the x direction, t_j is the layer thickness, $0 < x_j < t_j$ is the local spatial variable measured from the bottom of the layer j, $Y_{jm} = \omega\varepsilon_0 / k_{jm}\gamma_{j0}$ is the admittance of the layer, and γ_{j0} is the arithmetic average of $1/\varepsilon_j(z)$. In Eq. (3.46), if the layer j is uniform (*i.e.*, $\varepsilon_j(z)$ is a constant), each mode is a diffracted field of order n, with $e_{jm}(z) = h_{jm}(z) = \delta_{nm} \exp(ik_{zn}z)$, where δ_{nm} is the Kronecker's delta function,

$$k_{zn} = k_0 \left(\sqrt{\varepsilon_j} \sin\theta + \frac{n\lambda}{p} \right), \quad n = 0, \pm 1, \pm 2, \pm K \tag{3.48}$$

where $k_0 = 2\pi/\lambda$, θ is the incident angle, p is the period, λ is the vacuum wavelength, and K is the maximum order included in the analysis. If the layer j is a periodic layer, the expressions for k_{jm}, $e_{jm}(z)$ and $h_{jm}(z)$ are more complicated, but they can be obtained by solving Maxwell's equations inside each thin layer region. Each modal function can then be expressed as a Fourier series with known coefficients $a_{nm}^{(j)}$ and $b_{nm}^{(j)}$ in the form

$$e_{jm}(z) = \sum_n a_{nm}^{(j)} \exp(ik_{zn}z),$$
$$h_{jm}(z) = \sum_n b_{nm}^{(j)} \exp(ik_{zn}z). \tag{3.49}$$

Inserting Eq. (3.49) into Eq. (3.46) and interchanging the summations in Eq. (3.46) yields

$$\begin{aligned} E_j(x,z) &= \sum_n V_{jn}(x)\exp(ik_{zn}z), \\ H_j(x,z) &= \sum_n I_{jn}(x)\exp(ik_{zn}z), \end{aligned} \tag{3.50}$$

where

$$\begin{aligned} V_{jn}(z) &= \sum_m a_{nm}^{(j)} v_{jm}, \\ I_{jn}(z) &= \sum_m b_{nm}^{(j)} i_{jm}. \end{aligned} \tag{3.51}$$

Noting that $a_{nm}^{(j)} = b_{nm}^{(j)} = \delta_{nm}$ in the uniform layers, the field components $E_j(x,z)$ and $H_j(x,z)$ in all layers can then be expressed as a Fourier series of spectral orders n. The continuity of tangential electric and magnetic fields across the boundary requires that $E_j = E_{j+1}$ and $H_j = H_{j+1}$ at each interface, and since the Fourier components are orthogonal, it follows that $V_{jn} = V_{j+1,n}$ and $I_{jn} = I_{j+1,n}$ for all n. These boundary conditions, together with the given incidence conditions with the known E_z in the substrate, completely specify f_{jm} and g_{jm} and thus yield the EM field distribution throughout the entire structure.

After the distribution of $|E_{\text{vert}}|^2 \equiv |E_x|^2$ in the active volume V_r is obtained from Eq. (3.45), the theoretical external unpolarized quantum efficiency η is given by the ratio of the optical absorbed power P_{abs} over the unpolarized incident power P_{inc} in air,

$$\begin{aligned} P_{\text{abs}} &= \int_{AL} dI(\boldsymbol{r}) = \int_{AL} \alpha(\lambda) I(\boldsymbol{r},\lambda) d^3r \\ &= \frac{\alpha(\lambda)n(\lambda)\varepsilon_0 cL}{2} \int_A |E_{\text{vert}}|^2 d^2r, \end{aligned} \tag{3.52}$$

where A is the cross-sectional area of a single corrugation, L is the arbitrary corrugation length in the y-direction, $I(r)$ is the optical intensity of the TM mode, and $\alpha(\lambda)$ is the absorption coefficient of QWIPs for the vertical E component. For normal incidence, the TM mode power P_{sub} inside a thick substrate layer and the unpolarized P_{inc} are related by

$$P_{\text{sub}} = pL\frac{n_{\text{sub}}\varepsilon_0 c}{2}|E_{\text{sub}}|^2 = pL\frac{4n_{\text{sub}}}{(1+n_{\text{sub}})^2}\frac{P_{\text{inc}}}{2}, \tag{3.53}$$

where p is the corrugation period, and $T_{\text{sub}} = 4n/(1+n)^2$ is the transmission coefficient of normal incident light from air into a thick substrate. Under normal incidence, the total E field in P_{sub} and the tangential component are equal and is denoted by E_{sub}. Using the MTL formulation, the fractions of power $\eta = P_a/P_{\text{inc}}$ absorbed in the periodic region, $R = P_r/P_{\text{inc}}$ reflected from the periodic region, and $T = P_t/P_{\text{inc}}$ transmitted through the periodic region can all be calculated.

In Fig. 3.18(b), the quantities η, R, and T as a function of p are plotted for a given C-QWIP geometry and incident wavelength λ of 8.5 μm shown in Fig. 3.18(a). This geometry is characterized by a quantity known as the corrugation fill factor S, which is the total sidewall projection length divided by the period. Namely, $S = 2t_a\cot(\theta)/p$, where t_a is the active MQW layer thickness, and it is equal to the fraction of incident light being reflected by the sidewalls.

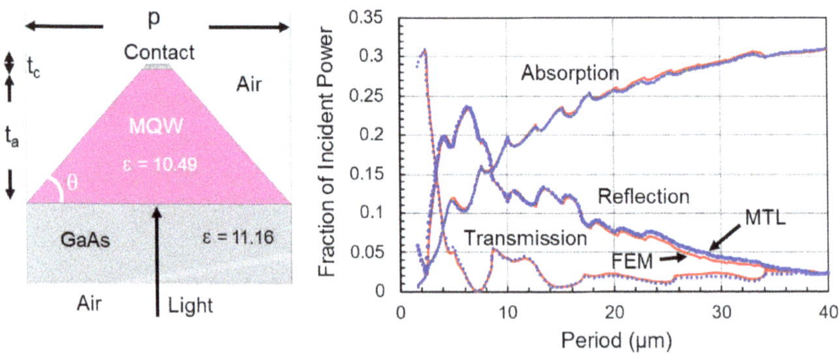

(a) (b)

Figure 3.18. (a) The basic detector geometry of a C-QWIP structure, which consists of a thick GaAs substrate that serves as a common ground contact, a MQW-absorbing layer, and a GaAs top contact layer. (b) Fractions of absorbed, reflected and transmitted powers vs. period, as determined by the MTL model (dots) and FEM (solid curves).

In this example, S is kept at a constant of 0.884 for all p. For example, when $p = 5.5\,\mu\text{m}$, the active layer t_a is $2.7\,\mu\text{m}$ for $\theta = 48°$. The GaAs top contact layer thickness t_c is set to be $0.1\,\mu\text{m}$ for this p, and the ratio t_c/t_a is kept constant for other p. The value of α is chosen to be $0.22\,\mu\text{m}^{-1}$. The present C-QWIP is not covered by a material so that its sidewalls are in direct contact with air.

In Fig. 3.18(b), when p increases, the detector area also increases, which leads to a larger fraction of light being absorbed. However, the absorption η is limited to ~31% for very large corrugations with $p > 40\,\mu\text{m}$. This limit is due to the fact that the TE mode is not coupled and only a fraction of incident light, $T_{\text{sub}} = 0.709$, is transmitted into the substrate. With an anti-reflection coating, a maximum η of 44% can be theoretically reached with this S. More discussions on the C-QWIP characteristics and performance will be given in Chapter 4.

3.6 Finite Element Method

In the above discussions, we have been using Finite Element Method (FEM) to compare and substantiate different calculation methods. Based on advanced computational algorithms, FEM is versatile in obtaining accurate EM solutions to devices with arbitrary geometry and material properties. With the increasingly developed commercial software packages in EM modeling, the EM properties of many optical devices can be modeled accurately. Different device designs described in this book are based on a particular FEM package offered by COMSOL, Inc. [12]. In the following, we will only give a brief outline on its working principle. Interested readers should consult dedicated monographs [13–15].

Starting with the basic Maxwell equations and assuming a harmonic wave with a propagating phase $e^{j\omega t}$,

$$\nabla \cdot \boldsymbol{D} = \varepsilon_r \varepsilon_0 \nabla \cdot \boldsymbol{E} = \rho, \tag{3.54}$$

$$\nabla \cdot \boldsymbol{B} = \mu_r \mu_0 \nabla \cdot \boldsymbol{H} = 0, \tag{3.55}$$

$$\nabla \times \boldsymbol{E} = -\frac{\partial \boldsymbol{B}}{\partial t} = -j\omega \mu_r \mu_0 \boldsymbol{H}, \tag{3.56}$$

$$\nabla \times \boldsymbol{H} = \boldsymbol{J}_v + \frac{\partial \boldsymbol{D}}{\partial t} + \boldsymbol{J}_c = \sigma \boldsymbol{E} + j\omega \varepsilon_r \varepsilon_0 \boldsymbol{E} + \boldsymbol{J}_c, \tag{3.57}$$

where $\boldsymbol{E}$, $\boldsymbol{H}$, $\boldsymbol{D}$, and $\boldsymbol{B}$ are the electric field intensity, magnetic field intensity, electric flux density, and magnetic flux density, respectively. The conduction electric current density $\boldsymbol{J}_v$ accounts for induced current flow in the medium, while the electric current density $\boldsymbol{J}_c$ represents the excitation current source. The electric charge density ρ is the free electric charge in the medium.

Taking the curl of Eq. (3.56) and substituting Eq. (3.57) into the curl equation gives

$$\nabla \times \left(\frac{1}{\mu_r}\nabla \times \boldsymbol{E}\right) - k_0^2\left(\varepsilon_r - \frac{j\sigma}{\varepsilon_0\omega}\right)\boldsymbol{E} = \boldsymbol{J}_c, \tag{3.58}$$

where $k_0 = \omega\sqrt{(\mu_0\varepsilon_0)}$ is the wave vector in free space. In Eq. (3.58), the medium is assumed to obey Ohm's law such that $\boldsymbol{J}_v = \sigma\boldsymbol{E}$. Equation (3.58) is the general Maxwell wave equation solved by COMSOL.

Treating the differential as an operator, Eq. (3.58) is in the form of

$$\left[\nabla \times \left(\frac{1}{\mu_r}\nabla \times\right) - k_0^2\left(\varepsilon_r - \frac{j\sigma}{\varepsilon_0\omega}\right)\right]\boldsymbol{E} \equiv L\boldsymbol{E} = \boldsymbol{g}, \tag{3.59}$$

where $\boldsymbol{g}$ is the short hand of $\boldsymbol{J}_c$. Equation (3.59) stands for the so-called strong form of the formulation, in which the solution $\boldsymbol{E}$ satisfies the equation at every point in the considered domain Ω. This statement is equivalent to

$$\int_\Omega \delta(r)(L\boldsymbol{E})d^3r = \int_\Omega \delta(r)\boldsymbol{g}d^3r, \tag{3.60}$$

where $\delta(r)$ is the Dirac delta function. In the FEM approach, the strict requirement of Eq. (3.60) is relaxed, and the solution $\boldsymbol{E}$ only needs to satisfy

$$\int_\Omega w(r)(L\boldsymbol{E})d^3r = \int_\Omega w(r)\boldsymbol{g}d^3r, \tag{3.61}$$

where $w(\boldsymbol{r})$ is a distribution function over a small sampling volume anywhere within Ω. Equation (3.61) then represents the weak form of the formulation. Obviously, the solution is increasingly accurate if $w(\boldsymbol{r})$ is finite over a smaller volume. To yield a numerical solution, the domain Ω is divided into a mesh of finite elements in the form of tetrahedral or quadrilateral, etc. An approximate solution, $\boldsymbol{E}_a(\boldsymbol{r})$, is sought at the vertices of the elements. Inside an element, $\boldsymbol{E}_a(x, y, z)$ is obtained from interpolation among the end points. If the element is of the first order, the interpolation will be linear, while it will be quadratic if the element is of the second order. Higher orders can also be implemented. The interpolation function is known as the shape function α of the element. For example, for a tetrahedral element m with 4 vertices $i = 1, 2, 3, 4$,

$$\boldsymbol{E}_a(x, y, z) = \sum_{1}^{4} \alpha_i(x, y, z)\boldsymbol{E}_{mi}, \tag{3.62}$$

where $\boldsymbol{E}_{mi}$ is the value to be determined at the vertex i at (x_i, y_i, z_i). For linear shape functions, α_1 is 1 at (x_1, y_1, z_1). It decreases linearly to 0 at other vertices. It decreases quadratically for second order, etc. The same is true for other α_i.

Substituting Eq. (3.62) into Eq. (3.61),

$$\sum_{m=1}^{N} \int_{\Omega} w(\boldsymbol{r})[L\boldsymbol{E}_m(\boldsymbol{r})]d^3r = \int_{\Omega} w(\boldsymbol{r})\boldsymbol{g}d^3r, \tag{3.63}$$

where N is the number of elements in Ω. While the sampling function $w(\boldsymbol{r})$ can be any localized function, it can be chosen to be the same as the shape function of the element, which is known as the Galerkin's method. For example, if α_1 is chosen to be the sampling function, a linear expression

$$\begin{aligned} &\boldsymbol{E}_1\langle\alpha_1, L\alpha_1\rangle + \boldsymbol{E}_2\langle\alpha_1, L\alpha_2\rangle + \boldsymbol{E}_3\langle\alpha_1, L\alpha_3\rangle + \boldsymbol{E}_4\langle\alpha_1, L\alpha_4\rangle \\ &\quad \equiv \sum_{j=1}^{4} \boldsymbol{E}_j C_{1j} = \langle\alpha_1, \boldsymbol{g}\rangle, \end{aligned} \tag{3.64}$$

can be derived to relate the vertex $\boldsymbol{E}_i$ to the excitation $\boldsymbol{g}$ within the element, where $\langle u, Lv\rangle$ stands for $\int uLvd^3r$, and $C_{ij} = \langle\alpha_i, L\alpha_j\rangle$ are known coefficients from the α definitions. Repeating sampling with α_2 to α_4, a total of 4 equations can be established for 4 unknowns $\{E_i\}$. However, a vertex may be shared with a number of neighboring elements. The excitation in each neighboring element contributes to that vertex. Therefore, for a particular $\boldsymbol{E}_j$ in element m,

$$\sum_{nn}\sum_{j=1}^{4}\boldsymbol{E}_j^{(m)}C_{ij}^{(m)}=\sum_{j=1}^{4}\boldsymbol{E}_j^{(m)}\sum_{nn}C_{ij}^{(m)}=\sum_{nn}\langle\alpha_i^{(m)},\boldsymbol{g}^{(m)}\rangle, \quad (3.65)$$

where nn stands for nearest neighbor elements that share the same vertex j with element m. If we give each vertex a global label k, the vertex j in element m will have a unique identification, with which

$$\sum_{k\epsilon S}\boldsymbol{E}_k\sum_{nn}C_{ik}\equiv\sum_{k\epsilon S}\boldsymbol{E}_kD_{ik}=\sum_{nn}\langle\alpha_i^{(m)},\boldsymbol{g}^{(m)}\rangle, \quad (3.66)$$

where S is a set of vertices belonging to element m, and D_{ik} is a known quantity from the C_{ij} coefficients. Equation (3.66) forms a set of coupled linear equations that relate $\boldsymbol{E}_k$ at each vertex k to the excitation $\boldsymbol{g}$ in all elements containing k. This set of equations, with proper indexing, can be converted into a matrix form and be solved by numerical methods. After the approximate electric field distribution $\boldsymbol{E}_a(\boldsymbol{r})$ is obtained, $\boldsymbol{E}_a(x, y, z)$ at any location can be found by interpolation. Looking further into steps in setting up this set of coupled linear equations into matrix form is beyond the scope of this text. Instead, we will turn to using FEM as a mathematical tool in optical designs.

The EM modeling in this text is performed using COMSOL, version 5.5. The detailed operating procedure can be found in COMSOL Multi-Physics Reference Manual. The techniques can also be learned in online training courses and tutorials. Here, we will briefly outline the modeling procedures using several examples.

3.6.1 *Surface Plasmon-Polariton Modeling*

To obtain the lower branch of the plasmon dispersion curve in Fig. 3.4, which is the Fano mode, we open the software with the Model Wizard, choose the *2D Space Dimension*, add the *Electromagnetic Waves, Frequency Domain interface*, and add the *Boundary Mode Analysis* preset study. BMA is used to compute the propagation constants (k_x in 2D) and propagating mode shapes for a given frequency at a port boundary. In the *Global Definitions* section, *parameters* node, we define $f0$ as a frequency variable (and input a value of interest), free space wavelength (λ = c_const/$f0$), material parameters for gold (plasma angular frequency $\omega_p = 1.38 \times 10^{16}$ rad/s and damping $= 1/\tau = 6.667 \times 10^{-13}$ s^{-1}) to define $\varepsilon_{\mathrm{Au}}$ in the Drude model, material parameters for GaAs ($\varepsilon_d = n_d^2 = 3.1^2$), GaAs layer thickness ($a_1 \cdot \lambda$), Au layer thickness ($a_2 \cdot \lambda$), and the distance between the two port boundaries in Fig. 3.3 (*e.g.*, 0.5 μm). The constants a_1 and a_2 are set so that the GaAs and gold thicknesses are larger than δ_d and δ_m in Eq. (3.23) and Eq. (3.24), respectively, at any calculated $f0$.

After setting parameters in the *Component* 1 section, we construct the geometry and input materials' relative permeabilities (= 1) and conductivities (= 0). In the *Electromagnetic Waves, Frequency Domain (emw)* section, we input *Wave Equation, Electric* for Au using the Drude-Lorentz model by specifying ω_p, γ, oscillator strength = 1, and resonant frequency = 0. In the same section, we input *Wave Equation, Electric* for GaAs by specifying ε_d. The *emw* section is populated with several default settings such as solving for in-plane E field component, full field simulation, and quadratic shape functions.

After setting the material geometry and properties, we specify boundary conditions in the *emw* section. The top and bottom boundaries in Fig. 3.3 are set to *Scattering Boundary Condition, no incident field, Plane wave type*, and *Second Order*. Scattering boundary condition allows the boundary to be transparent to the impinging wave. The transparency varies with the incident angle. Employing second order rather than first order improves the transparency. The two left boundaries in Fig. 3.3 are set to *Port* 1, *Numeric type, Wave Excitation on,* $\theta_{\mathrm{in}} = 0$, *and* $d_{\mathrm{offset}} = 0$. Specifying numeric port triggers a BMA node in Study 1 as Step 1. One can add BMA at any time from the dropdown panel of Study 1. We add another port 2 and include the right boundaries in this port. Port 2 entries are *Numeric type, Wave Excitation off,* $\theta_{\mathrm{in}} = 0$, *and* $d_{\mathrm{offset}} = 0$. This action triggers another Step 2 BMA node. In Step 3, *Frequency Domain*, we enter *Frequency Unit: Hz and Frequencies:* $f0$, and this completes the structure specifications.

The next step is to specify the density of the finite elements in *Component* 1 section. A denser mesh will yield a more accurate solution but the computation time is also longer and more random-access memory is required. In 2D modeling, one can usually afford the *Extremely fine* setting in the *Physics-controlled mesh* in the *Sequence type* specification, which is chosen for the present example. The general rule of thumb for the minimum mesh size is about $1/5^{\text{th}}$ of the wavelength in the medium for quadratic shape elements. Quadratic shape element is the default setting in *emw* modeling, and the mesh size can be specified using the *User-controlled Sequence* type. The required size depends on the error tolerance, the optical interference effects, and the physical quantity of interest. For example, if the electric field at a particular spot is of interest and this spot is near a sharp metal corner, a dense mesh will be required in this region. In this case, it is also useful to define a specific point in the geometry so that a mesh node is created at this spot in the meshing operation. The E field can then be evaluated at the exact location without interpolation of the shape functions. On the other hand, if the quantity of interest is an integrated value such as the total absorption in the volume, the exact value at any particular point is not as important, and the mesh size can be larger. The usual step for checking the adequacy of mesh density is to repeat the modeling with a denser mesh.

After creating the mesh, we initiate a *parametric sweep* in *Study* 1. From Fig. 3.4, the upper ω limit of the Fano branch is 4.13×10^{15} rad/s, which gives the upper f_0 limit of $\omega/2\pi = 6.6\times10^{14}$ Hz. For each f_0, *Study* 1 contains three steps. Step 1 is to perform BMA on port 1. In this node setting, we use *Effective mode (refractive) index* n_{eff}, *Mode analysis frequency* $= f_0$, *manual mode search method, Desired number of searched modes* $= 1$, *Search for modes* (n_{eff}) *around* $= 12$, and *Mode search method around shift* $=$ *Closest in absolute value*. The effective mode index n_{eff} is the real part of $k_x c/\omega$, where k_x is the complex propagation constant of the mode to be solved. For the Fano branch, the analytical solution shows k_{SPP} and thus n_{eff} to be larger than any other nearly degenerate optical modes propagating in GaAs, which have $n_d = 3.1$, so the choice of a large

n_{eff} (*e.g.*, 12) ensures the SPP mode is selected. We sweep f_0 from 1×10^{13} Hz to 6.6×10^{14} Hz in 2×10^{13} Hz step size. The random-access memory used during computation is about 2 GB, and the total runtime is about 3 minutes on a personal computer. This resource represents the lower limit in EM modeling. A larger 3D problem requires orders of magnitude larger in both random-access memory and runtime.

The upper branch of the plasmon dispersion in Fig. 3.4, the Brewster mode, is modeled by setting the BMA similarly except *Search for modes* (n_{eff}) *around* = 0.8 and *Mode search method around shift* = *Smaller real part.* This setting is again guided by the analytical solution that the n_{eff} is less than 1 for the Brewster mode, and its n_{eff} is smaller than any other propagating modes in GaAs. The E field of the chosen mode is found to be strong only in the metal side, and the wave travels away from the interface at an angle when the wave is launched at the left boundaries. The E field in the dielectric is nearly zero. This mode thus resides exclusively in gold and acquires its refractive index. The visualization of the field profile thus has helped to understand the physical nature of this mode. The Brewster mode is less important in device application.

After the parametric sweep, the *Results* section contains the *Study* 1/*Parametric solutions* 1 in the *Dataset* node. In the *Derived Values* node, we initiate *Global Evaluation* for this dataset and select *emw.beta_1* in the *Expression* entry, which is to evaluate k_x for each f_0. A table, *Table 1*, is generated, in which the frequency setting f_0, the returned frequency f, and the propagation constant k_x are presented. Plotting $\omega = 2\pi f$ against $\text{Re}(k_x)$ yields the dispersion curves in Fig. 3.4, while plotting $|1/(2\text{Im}(k_x))|$ yields the propagating length in Fig. 3.6. This example shows that although EM modeling can accurately reproduce the analytical result, it needs to be heavily guided by prior knowledge on the expected solution in order to arrive at the correct answer. In an open system, many different propagating k_x modes can be nearly degenerate, and mode analysis will return a number of k_x values for a given f_0. Additional physical insights are required to single out the relevant mode of interest.

Nevertheless, once the model is established, the model can be applied to more complicated structures that do not have simple analytical solutions. To be cautious, each new solution should not be fully accepted as real unless it is validated by other means. For example, the scattering amplitudes of SPP by shallow gratings may not be obtained easily by analytical means. It can, however, be modeled by the same FEM procedures without much additional efforts other than constructing a grating profile in the *Geometry* node. Since the scattering angle of the modeled wavefront obeys the grating formula, the modeled solution is expected to be valid. However, even if it is valid, the solution is still an approximation, and its accuracy can always be improved with finer mesh and better treatment of boundary conditions.

The wave amplitudes in Fig. 3.9 are more accurately presented by adding a Perfectly Matching Layer (PML) to the right of Port 2. PML is an artificial layer in which a complex scaled coordinate is implemented to achieve a matching impedance to the physical layer while absorbing all outgoing energy, thereby realizing an open system. It exhibits lower reflection than the second-order scattering boundary condition when the incident angle is large. To add the PML layer, we first construct a layer to the right of the GaAs layer and define it to have the same material properties. We construct similar PML layers for gold to the right and underneath the gold layer. After constructing the layers, we designate them to be *Artificial Domains* in the *Definitions* under *Component* 1 section. In the *Port 2* node, we activate the *slit condition on interior port* with *Domain-backed Slit-type* and set the correct *Power Flow Direction.* All the exterior boundary conditions except Port 1 are set to second-order scattering boundary condition to increase the transparency of the boundaries. The plots of E_x and E_y (the vertical axis is designated as the y-axis in COMSOL 2D models) can be obtained in the *Results* section, *Electric Field (emw)* node, *Surface* plot by specifying *emw.Ex* or *emw.Ey* in the *Expression* panel. To view the dynamic propagation of E_y in time, we perform an animation in COMSOL on *emw.Ey* in the *Electric field, Surface plot* node by setting in the *Animation, Dynamic data extraction, Full harmonic, 36 Frames.* When the phase of the wave increases in time, the waves will move in the direction of propagation. One may need to further adjust the PML properties to reduce reflection if it is of interest.

3.6.2 *Modeling of Optical Properties of a Si Slab*

In Figs. 3.11–3.13, the optical properties of a 1 μm Si slab are modeled in the wavelength range between 0.4 and 1 μm. In the *parameter* section of the model, we enter the wavelength (0.4 μm), the thickness of the Si layer (1 μm), the thicknesses of the bottom and top air layers (both are 1 μm), the length of all layers (1 μm), and the incident angle (angle = 0 or 45 [deg]). In the *Global definitions*, we define *interpolation functions* of the refractive index $\aleph = (n, \kappa)$ of Si shown in Fig. 3.11 by importing the data files. An *analytic function* of the absorption coefficient of Si

$$\alpha(wavelength) = \frac{4\pi\kappa(wavelength)}{wavelength}, \tag{3.67}$$

is then defined. After constructing the geometry and specifying the air material properties for the top and bottom layers, in the *emw* section, we add a *Wave Equation, Electric* node for Si. In this node, we specify *Refractive index* and use *User defined real and imaginary parts* from the $\aleph$ data files. We specify top air boundary as *Port* 1, bottom air boundary as *Port* 2, and other boundaries as *periodic boundaries*. In *Port* 1, we specify the port as *periodic*, *Wave excitation:* on, P_{in}: 1 W, *Input quantity: Electric Field*, *Electric mode field amplitude*: E_z = 1 V/m (the out-of-plane component for TE mode), *Angle of incidence* α: angle, *Include in automatic diffraction order Calculation*: checked, and *Refractive index, real part*: 1 (*i.e.*, n_{air}). For TM mode calculation, we select *Input quantity: Magnetic field* and H_z = 1 A/m. The specified P_{in} overrides the E_z and H_z magnitudes and they are for direction specifications only. *Port* 2 is similarly set except the *wave excitation* is off. In *Periodic Condition 1* node, we specify *Floquet periodicity* type and *k-vector for Floquet periodicity* to be *From periodic port.* The Floquet periodic condition enforces the phases φ on two parallel boundaries to be related by $\varphi(x + d_x, y + d_y) = \varphi(x, y)\exp(ik_x d_x + ik_y d_y)$, where $(k_x, k_y) = (k\cos\alpha, k\sin\alpha)$, and d_x and d_y are the distances between the two parallel surfaces in x and y directions, respectively. In the *mesh* section, we adopt mesh size $= 1/8^{\text{th}}$ of the wavelength inside the Si slab in all layers. This mesh size was found to produce errors less than 0.4% under normal incidence in Fig. 3.11 and 0.12% under 45° incidence. In the mesh operation, we first mesh the left boundaries and copy the mesh to the right boundaries before meshing the interior areas. The periodic condition requires the mesh in the periodic surfaces to be exactly the same.

In *Study* 1 section, we add *parameter sweep* in wavelength between 0.4 and 1.0 μm. In step 1 under Study 1, we specify *Frequencies* = c_const/wavelength, and hit *compute* to start the simulation. After the simulation, we add a *1D plot group* in the *Results* section, in which a *Global* evaluation is initiated. In this *Global* panel, we specify in *Dataset*: Solution 1/Parametric solutions 1 and in *Expression:* abs(emw.S11)^2, which computes the reflection coefficient R. Similarly, we compute the transmission coefficient T by inputting in *Expression*: abs(emw.S21)^2. The absorption coefficient A can be obtained by $1 - R - A$. However, this approach only calculates the total absorption in all the layers within the structure. Although it is applicable in the present example where there is only one absorbing layer, it is more flexible by calculating the specific A of Si $= P_{\text{abs}}/P_{in}$, where P_{in} has been set to be 1 W, and

$$\begin{aligned} P_{\text{abs}} &= \int_A dI(\boldsymbol{r}) = \int_A \alpha(\lambda) I(\boldsymbol{r}, \lambda) d^2 r \\ &= \frac{\alpha(\lambda) n(\lambda) \varepsilon_0 c}{2} \int_A |E|^2 d^2 r, \end{aligned} \tag{3.68}$$

where $\alpha(\lambda)$ is given by Eq. (3.67), and $n(\lambda)$ is the inputted Si Re($\aleph$). To perform the calculation, in the *Results* section, *Derived Values* node, we add the *Surface Integration 1*, select the Si domain, input (emw.normE)^2*c_const*epsilon0_const*n(wavelength)*alpha(wavelength)/2 in the *Expression*, and compute. These simulations yield the R, T, and A curves as a function of λ in Fig. 3.11 and Fig. 3.12. Similarly, to yield the angular dependence at a particular λ as in Fig. 3.13 for $\lambda = 0.7\,\mu$m, we need to specify this wavelength

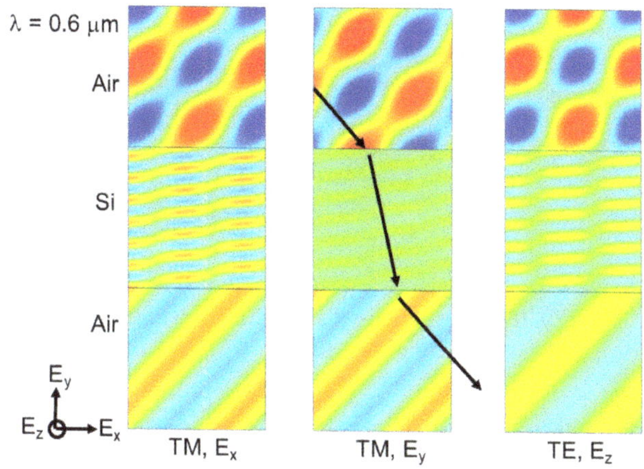

Figure 3.19. The E field component profiles of waves incident at 45° with $\lambda = 0.6\,\mu$m onto a $1\,\mu$m Si slab suspended in air.

value in the *parameter* section, and add *Study* 2 in the model. In the parameter sweep node, we specify the *parameter name* to be angle and range(0[deg],1[deg],89[deg]) to perform the angular sweep. Figure 3.19 shows an example of the E field plot for $\theta = 45°$ and $\lambda = 0.6\,\mu$m under both TM and TE polarizations. At the top air layer, there is interference between the incident and reflected waves, while the bottom air layer only contains transmitted wave. The transmission coefficients are different among the three waves but the angle of refraction is the same. Based on Eq. (3.68), the absorption of a C-QWIP can also be modeled by FEM and its result in Fig. 3.18(b) agrees with the MTL calculation.

3.7 45° Edge Coupling

QWIP materials possess uniaxial properties in that they only absorb vertically polarized light. To overcome this polarization requirement, the detectors are often characterized under 45° illumination at a polished substrate facet. Based on the detector geometry shown in Fig. 2.35 of Chapter 2, the absorption quantum efficiency η is given by Eq. (2.44) using classical ray tracing technique. In general, if the facet is polished at an angle θ shown in Fig. 3.20, the classical $\eta_c(\theta)$

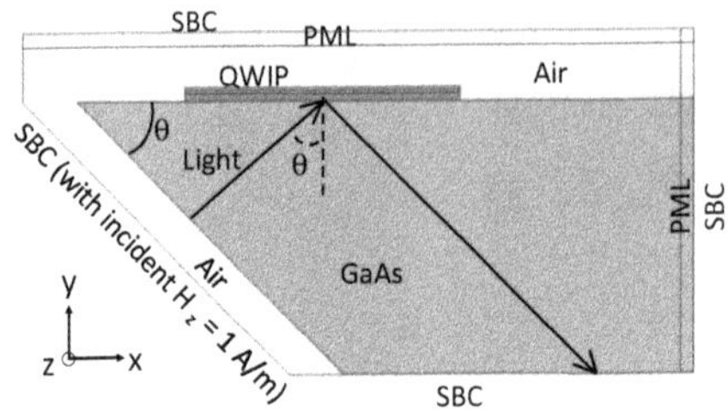

Figure 3.20. The QWIP geometry used in FEM modeling. Scattering boundary condition is set at all exterior boundaries.

for unpolarized light will be

$$\eta_c(\theta) = \frac{4n}{(1+n)^2}\frac{\cos\theta}{2}\left[1 - \exp\left(-\frac{\alpha\sin^2\theta{\cdot}2t}{\cos\theta}\right)\right], \tag{3.69}$$

where α is the absorption coefficient for vertical E field. In Eq. (3.69), $4n/(1+n)^2$ is the substrate transmission for normal incidence onto the facet. The first $\cos(\theta)$ accounts for the dilution of the optical flux on the detector area. The 1/2 term accounts for the inactivity of the TE mode. The $\sin^2\theta$ accounts for the smaller E_y compared to the incident E. The $2t/\cos(\theta)$ term is the optical pathlength in the active layer with thickness t for two passes of light. If we further assume that both $n = 3.239$ and $\alpha = 0.15\,\mu\text{m}^{-1}$ are wavelength-independent, $\eta_c(\theta)$ in the ray tracing model will be independent of λ, *e.g.*, with $t = 3\,\mu$m, η_c is 3.59%, 12.0%, and 6.22% at $\theta = 20°$, 45°, and 80°, respectively. The maximum η_c occurs at 56.5°, where $\eta_c = 13.5\%$.

Although Eq. (3.69) contains the main geometrical effects, it nevertheless ignores optical interference and any optical effects related to the finite mesa structure. Since 45° edge coupling is commonly used in characterizing the intrinsic absorption of the QWIP material and in single element detection, it is important to understand this detector geometry more thoroughly.

With the detector geometry in Fig. 3.20, the z direction is invariant, and we can use 2D modeling and consider only the TM mode, in which the incident H is pointing in the z direction. For a given incident H, the incident power in air is

$$P_{\text{inc}} = dL\frac{n_{\text{air}}\varepsilon_0 c}{2}E_{\text{inc}}^2 = dL\frac{n_{\text{air}}\varepsilon_0 c}{2}\left(\frac{Z_0 H}{n_{\text{air}}}\right)^2 = dL\frac{\varepsilon_0 c Z_0^2}{2n_{\text{air}}}H^2, \tag{3.70}$$

where d is the detector dimension along x, L is the arbitrary dimension along z, and $Z_0 \approx 377\Omega$ is the vacuum impedance. With the absorbed power given by Eq. (3.52), η for unpolarized light is

$$\begin{aligned}\eta &= \frac{1}{2}\frac{\alpha(\lambda)n(\lambda)\varepsilon_0 cL}{2dL\frac{\varepsilon_0 c Z_0^2}{2n_{\text{air}}}H^2}\int_A |E_y|^2 d^2r \\ &= \frac{\alpha(\lambda)n(\lambda)n_{\text{air}}}{2dZ_0^2H^2}\int_A |E_y|^2 d^2r. \end{aligned}\tag{3.71}$$

By modeling E_y in the detector volume, the unpolarized η can be readily obtained and compared with the classical model. When the incident medium is not air, η can be evaluated by replacing n_{air} with that of the medium in Eq. (3.71).

In COMSOL, we input constants α, n, and the length of the detector (100 μm) in the *parameters* section. The incident H_z is set to be 1 A/m in the *Scattering boundary condition* node. We define an analytical Im(ε) function, ei(wavelength) = n*α*wavelength/ 2π, with which the diagonal relative permittivity (n^2, n^2-j*ei(wavelength), 1) of the active layer can be defined in the *Wave Equation, Electric* node. Actually, Im(ε_{zz}) is ignored in the 2D model and can be any number. In the mesh, we use $1/5^{\text{th}}$ of the internal wavelength in each layer. In the PML layer, we use a mapped mesh. To increase the accuracy, we input 2 in the *Scale Geometry, y-direction scale* to increase the mesh density by a factor of 2 in the vertical direction. For the angular parametric sweep, we set $\lambda = 10\,\mu$m in the parameter section. For the wavelength sweep, we set three different representative angles at 20°, 45°, and 80°. After the computation, we perform a surface integration to yield η in percentage in the *Results, Derived Values* section, by entering the expression, conj(emw.Ey)*emw.Ey*alpha*n/2/d/Z0^2*100 in the *Dataset*.

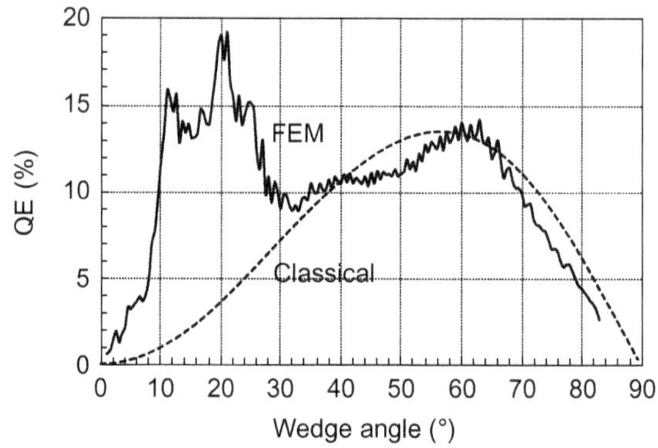

Figure 3.21. The quantum efficiency η of an edge-coupled QWIP with different wedge angles using classical ray tracing and finite element method. $\lambda = 10\,\mu$m.

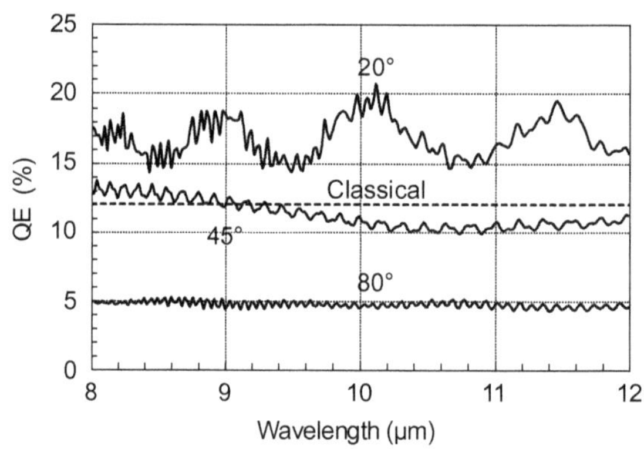

Figure 3.22. η_{EM} as a function of λ for three different wedge angles, 20°, 45°, and 80°. The ray tracing η_c at 45° is shown as the dashed line.

The results of the modeled η_{EM} are shown in Fig. 3.21 and Fig. 3.22. η_{EM} can be very different from η_c when θ is smaller than 30°. At the larger angles, they become similar. In particular, at 45°, η_{EM} in Fig. 3.22 is relatively wavelength-independent and is very close to η_c. Therefore, this common characterization technique is relatively accurate in extracting the intrinsic lineshape of intersubband absorption as seen in Fig. 2.20, although it may introduce small distortions. To understand the displayed quantum efficiency

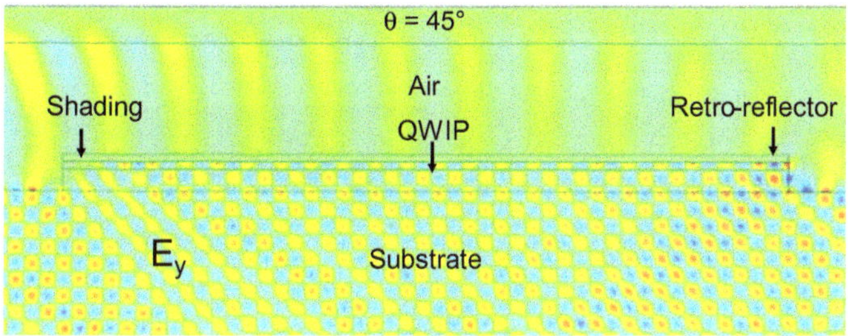

Figure 3.23. The vertical E_y field distribution at $\lambda = 10\,\mu$m under 45° edge coupling, which shows two classical effects, shading and retro-reflection, that can modify η_c.

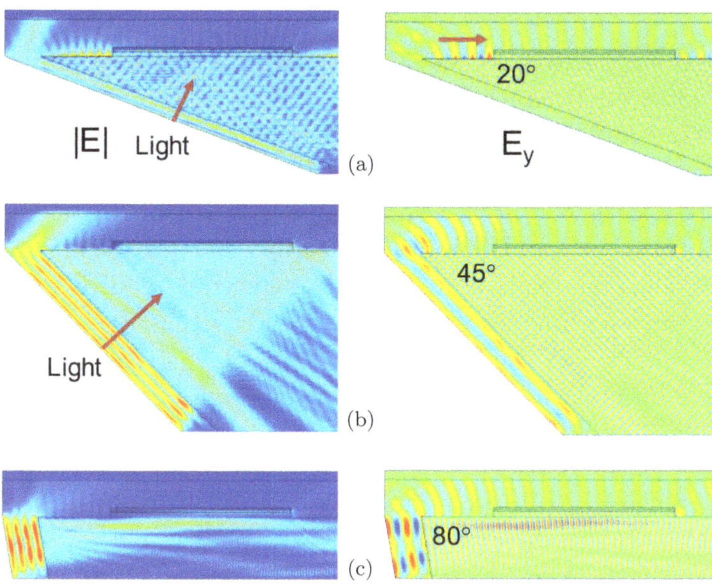

Figure 3.24. The total E field and the vertical E_y component at $\lambda = 10\,\mu$m with $\theta =$ (a) 20°, (b) 45°, and (c) 80°. A strong vertically polarized wave is generated with $\theta = 20°$.

variations in greater detail, the total $|E|$ field and the vertical E_y component are plotted in Figs. 3.23–3.25.

Figure 3.23 shows that the incident and reflected beams produce an interference pattern under 45° incidence with constant amplitude

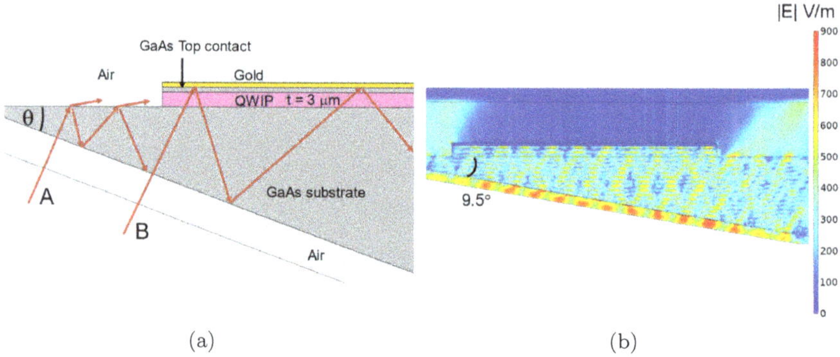

Figure 3.25. (a) The possible light path when θ is smaller than 45°. (b) Multiple passes of light are seen inside the detector at $\theta = 9.5°$.

along the x direction. When λ increases, this interference pattern shifts vertically, producing a slow oscillating quantum efficiency in Fig. 3.22. The higher frequency oscillations are due to Fabry-Perot oscillations from the right-hand mesa corner shown in Fig. 3.23, which acts as a retro-reflector in forming a resonant etalon with the substrate surface. Its frequency can be changed by varying the detector location relative to the substrate edge, which, in effect, changes the length of the etalon.

Figure 3.23 also shows another effect, which is shading at the left-hand corner, where the incident light cannot reach this corner classically. This shading effect consistently lowers η_{EM} at $\theta > 60°$ in Fig. 3.21. Nevertheless, shading and retro-reflection effects at these corners do not significantly change the classical prediction. The more unexpected change is the large η_{EM} at the small θ around 20°. The small vertical E_y at these angles is not expected to produce a large absorption. It can be understood from Fig. 3.24(a) that at 20°, there is a strong surface wave induced at the air/semiconductor interface. This wave is primarily polarized vertically with a large E_y and advances from left to right. It enters into the active layer through the mesa sidewall and produces a large absorption. This dynamical phenomenon can be observed by performing an animation on emw.Ey in the *Electric field, Surface plot* node. The occurrence of a surface wave can be explained by the optical refraction at the substrate top surface outside the detector shown in Fig. 3.25(a) where the beam A is refracted at a large angle and travels nearly parallel to the surface. Near the critical angle for total internal reflection of about 18.0° for $n = 3.239$, the refracted beam is nearly parallel to the surface. With multi-beam interference, the strongest surface wave is found to occur near 20° for the 10 μm light. Therefore, strong vertical field can be produced adjacent to the detector even with a small wedge angle. This multi-beam interference is sensitive to λ, leading to relatively large variations in the spectral response in Fig. 3.22.

Above 20°, total internal reflection at the GaAs/air interface occurs and the surface wave disappear. Nevertheless, η_{EM} is still higher than η_c when $\theta < 30°$. It is due to another phenomenon where the number of optical passes in the detector is larger than two assumed in classical ray tracing. As shown in Fig. 3.25(b), with sharp wedges, the light reflected from the metal surface can hit the polished facet and reflect back to the detector, leading to multiple passes. Only when θ increases to 45°, the reflected beam from the metal travels parallel to the facet and will not be reflected by it, thus fulfilling the assumption in the classical formulation. The number of passes for a given θ depends on the length d of the detector. For the present $d = 100\,\mu$m, the multiple bounce effect is seen to drop to zero when θ is larger than 32° in Fig. 3.21.

In summary, rigorous EM modeling has verified the relatively constant coupling efficiency of the 45° edge coupling geometry and thus this geometry is useful for characterizing the absorption lineshape of the QWIP material. It also happens that η_{EM} and η_c are nearly the same in the modeled λ range at this particular angle. Therefore, one can use the classical expression to calculate α conveniently when η_{45} is known from experiment. This confirmation has helped QWIP production. At the same time, EM modeling also reveals unexpected optical effects, such as surface refraction and wave guiding that can significantly change the classical prediction at small coupling angles. It also quantifies the expected optical effects on shading and retro-reflection. Since $\eta_{EM} = 19.2\%$ at $\theta = 21°$ and it is 11.2% at 45° (for $\alpha = 0.15\,\mu\text{m}^{-1}$), the wedge angle of 21° is more effective in light coupling for single detector detection. However, the detector must be packaged properly so that it will not affect the surface wave at the leading edge.

References

1. Fowles, G. R. (1975). *Introduction to Modern Optics*, Dover Publications.
2. Osamura, K. and Murakami, Y. (1972). Free carrier absorption in n-GaAs, *Jpn. J. Appl. Phys.*, 11, pp. 365–371.
3. Olmon, R. L., Slovick, B., Johnson, T. W., Shelton, D., Oh, S.-H., Boreman, G. D. and Raschke, M. B. (2012). Optical dielectric function of gold, *Phys. Rev. B*, 86, p. 235147.
4. Rakić, A. D., Djurišic, A. B., Elazar, J. M. and Majewski, M. L. (1998). Optical properties of metallic films for vertical-cavity optoelectronic devices, *Appl. Opt.*, 37, pp. 5271–5283.
5. Raether, H. (1988). *Surface Plasmons on Smooth and Rough Surfaces and on Gratings*, Springer Tracts in Modern Physics, Vol. 111, Springer, p. 11.
6. Greffet, J. J., Carminati, R., Joulain, K. *et al.* (2002). Coherent emission of light by thermal sources, *Nature*, 416, pp. 61–64.
7. Dan'ko, V., Dmitruk, M., Indutnyi, I. *et al.* (2017). Au gratings fabricated by interference lithography for experimental study of localized and propagating surface plasmons, *Nano. Res. Lett.*, 12, p. 190.
8. Macleod, H. A. (2018). *Thin-Film Optical Filters*, 5th Ed., CRC Press.
9. Tamir, T. and Zhang, S. (1996). Modal transmission-line theory of multilayered grating structures, *J. Lightwave Technol.*, 14, pp. 914–927.
10. Yan, L., Jiang, M., Tamir, T. and Choi, K. K. (1999). Electromagnetic modeling of quantum-well photodetectors containing diffractive elements, *IEEE J. Quantum Electron.*, 35, pp. 1870–1877.
11. Jiang, M., Tamir, T. and Zhang, S. (2000). Modal theory of diffraction by multilayered gratings containing dielectric and metallic components, *J. Opt. Soc. Amer. A*, 18, pp. 807–820.
12. COMSOL, Inc., 100 District Avenue Burlington, MA 01803 USA.
13. Jin, J. (2014). *The Finite Element Method in Electromagnetics*, 3rd Ed., John Wiley & Sons.
14. Sadiku, M. N. O. (1992). *Numerical Techniques in Electromagnetics*, CRC Press.
15. Zhu, Y. and Cangellaris, A. C. (2006). *Multigrid Finite Element Methods for Electromagnetic Field Modeling*, John Wiley & Sons.

Chapter 4

Corrugated-Quantum Well Infrared Photodetectors

4.1 Introduction

In Chapter 3, we briefly described the use of corrugated structure for optical coupling. In this chapter, we will present this optical scheme in more detail. In two-dimensional (2D) infrared imaging, the detector array is facing the scene perpendicularly. The optical electric field E is thus polarized parallel to the quantum well layers and produces no absorption. Although 45° edge coupling is convenient for detector characterization, it is incapable of 2D imaging. A light coupling scheme for normal incident light is needed. In literature, there are many approaches. For example, Goossen *et al.* [1, 2] and Hasnain *et al.* [3] used diffraction gratings fabricated on top of the detectors to redirect light propagation. Andersson *et al.* [4–6] and Lundqvist *et al.* [7] made detailed grating analysis to determine the optimum structure. The result forms the basis of the current quantum well infrared photodetector (QWIP) focal plane array (FPA) technology. Sarusi *et al.* [8] used random reflectors to scatter light. Rokhinson *et al.* [9] and Choi *et al.* [10] used QWIP grid lines as individual optical antennas for light coupling. Schimert *et al.* [11] and Chen *et al.* [12] used collective resonance in optical cavities to create vertical polarization. Kalchmai *et al.* [13] used photonic crystal slab. Choi *et al.* designed individual pixels as resonators for polarization conversion and field enhancement [14, 15]. All of these coupling schemes make use of optical diffraction and interference to convert the incident parallel E field into vertical E field. These coupling schemes will be covered in the next chapter. In this chapter, we will continue to discuss corrugated QWIPs (C-QWIPs). It utilizes an intuitive detector geometry, in which the detector mesa is etched into prism- or pyramid-shaped structures. The sidewall angle is tuned to be about 45°. The incident light hitting the sidewalls will then be reflected into nearly parallel propagation. This coupling scheme is expected to be less dependent on the wavelength than those relying on optical diffraction and interference. The C-QWIP was first demonstrated by Chen *et al.* [16–18].

Although a C-QWIP with bare sidewalls shown in Fig. 3.18 is able to reflect light internally at the multiple quantum well (MQW)/air interface, the reflection is subject to the absorption properties of the backfill material external to the sidewall surface. Therefore, a more complete structure contains a non-absorbing dielectric material deposited on the sidewall and possibly a metal layer on top of the dielectric. In FPA fabrication, epoxy resin material is often used to bond the detector array to the Si readout electronics to strengthen its mechanical properties. This material replaces air as the filling material on top of the corrugation. However, the epoxy ring in the material contains C–O–C bonds and they exhibit strong infrared absorption in the 6–15 μm band [19]. The light reflected at the inclined sidewall exhibits attenuated total internal reflection at a weakened level. Fortunately, the evanescent field extends only a short distance away from the sidewall, and the external absorption can be minimized by placing a spacer layer to keep epoxy away from the surface. A dielectric layer with small (n, κ) such as MgF_2 with its complex refractive index shown in Fig. 3.14(b) will be suitable for this purpose. The detector can be further isolated from epoxy with a metal layer such as gold depositing on the dielectric.

4.2 Optical Coupling by Inclined Sidewall Reflection

The C-QWIP structure is motivated by the notion of inclined sidewall reflection. A C-QWIP may contain one set of inclined sidewalls in a prism-shaped geometry (PR-QWIP) or two sets of inclined sidewalls in a pyramid-shaped geometry (PY-QWIP). Ideally, the sidewalls are perfectly reflecting and its base angle θ is 45°. In this ideal case, the external quantum efficiency (QE) η referenced to unpolarized light can be calculated conveniently from the classical ray tracing model. From the decay of light along the horizontal path after reflection from the two sidewalls, the classical QE η_c of a PR-QWIP is given by [20, 21]

$$\begin{aligned}\eta_c &= \frac{4n_s}{2(1+n_s)^2}\frac{1}{p}\int_0^{t_a}\int_0^{p-2y}\alpha[e^{-\alpha x}+e^{-\alpha(p-2y-x)}]dxdy \\ &= \frac{4n_s}{(1+n_s)^2}\frac{1}{p}\left[t_a+\frac{e^{-\alpha p}}{2\alpha}(1-e^{2\alpha t_a})\right],\end{aligned} \tag{4.1}$$

where n_s is the GaAs substrate refractive index, p is the corrugation period, t_a is the active MQW layer thickness, and y is the vertical coordinate. The 1/2 factor in the first line accounts for the inactivity of the transverse electric mode, and $4n_s/(1+n_s)^2$ is the substrate transmission T_{sub} under normal incidence. There is no contribution from the other two vertical sidewalls in the classical model. The effectiveness of C-QWIP coupling is determined by the sidewall fill factor S, which is the projected area onto the base over the total base area of a corrugation, *i.e.*, $S = 2t_a\cot(\theta)/p = 2t_a/p$ for a 45° prism. A small S represents a shallow trapezoid. C-QWIPs with different S thus fall into different classes.

We can extend the ray tracing model to a PY-QWIP. In this geometry, all sidewalls contribute to light coupling. However, η will not be substantially larger because more active material is removed and each surface still can only couple to one polarization. By considering the absorption in each square slice of area $(p-2y)^2$ at a height y above the base, the ideal η_c of a PY-QWIP is

$$\begin{aligned}\eta_c &= \frac{4n_s}{2(1+n_s)^2}\frac{1}{p}\int_0^{t_a}2\frac{(p-2y)}{p}\int_0^{p-2y}\alpha[e^{-\alpha x}+e^{-\alpha(p-2y-x)}]dxdy \\ &= \frac{4n_s}{(1+n_s)^2}\frac{e^{-\alpha p}}{\alpha^2p^2}\{1+\alpha p+2\alpha^2t_a(p-t)e^{\alpha p} \\ &\quad +[\alpha(2t_a-p)-1]e^{2\alpha t_a}\}.\end{aligned} \tag{4.2}$$

4.3 Prism-Shaped C-QWIP with Thick Substrate

An ideal PR-QWIP with 45° sidewalls is depicted in Fig. 4.1. From Eq. (4.1), its η_c is plotted in Fig. 4.2(a) for different S. In particular, when $p = 25\,\mu$m, $S = 0.8$, and $\alpha = 0.22\,\mu\text{m}^{-1}$, QE is 26.7%.

With a large corrugation, the average distance between opposite sidewalls is large. The optical pathlength is much longer than the decay length $\sim 1/\alpha$, and most light from sidewall reflection is absorbed. QE thus tends to saturate. The saturation value is limited by S and T_{sub}.

In Section 3.5, we showed that the classical QE can be modified by optical interference. Therefore, the PR-QWIP QE is modeled again with the finite element method (FEM). We use a slightly larger θ of 48° usually obtained from wet chemical etching of sidewalls, although its QE does not differ appreciably from that of 45° sidewalls. To maintain the same S, the MQW layer is thicker than the ideal structure. For example, if $p = 25\,\mu$m and $S = 0.8$, t_a will be 11.1 μm

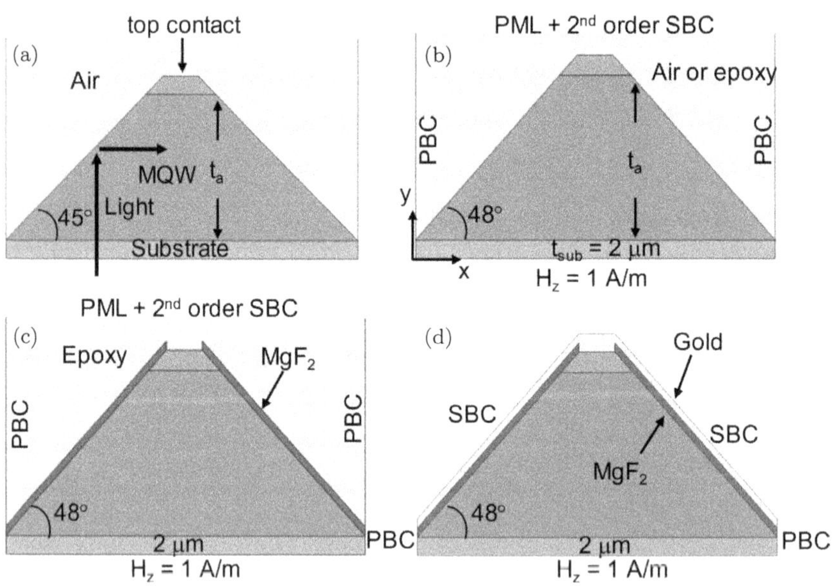

Figure 4.1. (a) The ideal PR-QWIP used in ray tracing. The QWIP in electromagnetic (EM) modeling may be covered by (b) either air or epoxy, (c) MgF_2 and epoxy, or (d) MgF_2 and Au.

instead of 10.0 μm. The thicker layer is balanced out by the slightly off-horizonal direction of the reflected light, which leads to similar QE. The detector geometry in the model is shown in Fig. 4.1(b), and the FEM QE for $\lambda = 8.5\,\mu$m with air cover is shown in Fig. 4.2(b). In this model, the GaAs substrate thickness is assumed to be thick such that the dense Fabry-Perot (FP) oscillations associated with the substrate cavity are not observed and the light is incident into the substrate with the average Fresnel transmission coefficient. The detector QE from Eq. (3.71) is then equal to

$$\eta_{\text{EM}}(p) = \frac{4n_s}{(1+n_s)^2}\frac{1}{2}\frac{n_a n_s \alpha}{pZ_0^2 H_z^2}\int_A |E_y|^2 d^2 r. \tag{4.3}$$

In Eq. (4.3), H_z is the internal out-of-plane H, which can be set at any arbitrary value in the present linear system such as 1 A/m, $n_s = 3.34$, MQW $n_a = 3.24$, and MQW $\alpha = 0.22\,\mu\text{m}^{-1}$. In the model space, the substrate thickness t_{sub} is set to be close to the internal λ, such as $2\,\mu$m in this example. The modeled η_{EM} is shown in Fig. 4.2(b). It is smaller than η_c at small p as expected

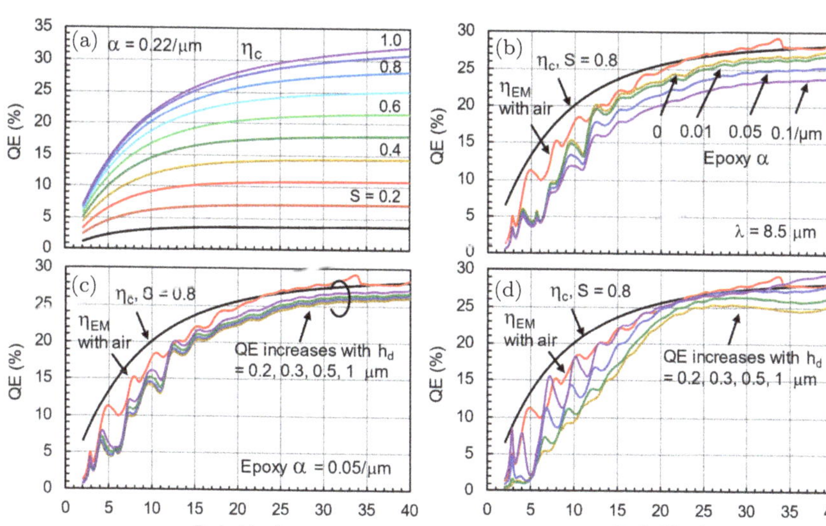

Figure 4.2. External unpolarized QE at $\lambda = 8.5\,\mu$m with thick substrate when the sidewalls are covered by (a) air, (b) epoxy, (c) MgF_2 and epoxy, and (d) MgF_2 and Au.

when the corrugations are more transparent to the radiation due to interference. As p increases to $\sim$25 μm, the modeled QE approaches η_c and eventually exceeds it slightly because of the additional optical coupling from the top contact sidewalls.

The C-QWIP QE is affected by the material covering the sidewalls. If epoxy resin is used, its typical n of 1.5 reduces the critical angle for total internal reflection and thus reduces optical confinement. Even if there is no infrared absorption, an epoxy cover reduces QE as shown by the epoxy $\alpha = 0$ curve in Fig. 4.2(b). When epoxy is more absorptive, η decreases further. This epoxy absorption is caused by the presence of the evanescent E field outside the sidewalls.

Since the evanescent field decays exponentially away from the sidewall surface, the epoxy absorption can be reduced by depositing a non-absorbing spacer layer onto the sidewall. The spacer n should be low to keep the decay length of the field short. The small n and κ can be satisfied by MgF_2 shown in Fig. 3.14(b). With MgF_2 as the spacer in Fig. 4.1(c), the modeled QE is shown in Fig. 4.2(c) for different spacer heights h_d from 0.2 μm to 1 μm. A thicker spacer helps to bring QE closer to η_{EM} with air cover. Specifically, at $p = 25\,\mu$m, when $h_d = 0.2, 0.3, 0.5$, and 1 μm, QE is 24.4, 24.7, 25.1, and 25.6%, respectively, for a fixed $\alpha_{\text{epoxy}} = 0.05/\mu$m. They are increasingly close to $\eta_{\text{EM}} = 27.1\%$. Therefore, with the right choice of epoxy, a relatively thin layer of MgF_2 will be sufficient to minimize the adverse effects of epoxy.

One can also add a layer of gold on MgF_2 to isolate the epoxy completely from the sidewalls. The resulting QE for the Fig. 4.1(d) geometry is shown in Fig. 4.2(d). Surprisingly, the gold layer generally reduces QE at small p and h_d. It turns out that the presence of a metal layer creates a second strongly reflecting surface at the Au/MgF_2 interface that changes the optical interference inside the detector. Figures 4.3(a) and (b) show a comparison without and with gold, respectively, when $p = 10\,\mu$m and $h_d = 0.2\,\mu$m. The maximum vertical E_y with gold is only about half of that without gold. The QE with gold is therefore reduced. This change is brought by the presence of two reflecting beams from each side of the corrugation, one from the MgF_2/MQW interface and another from Au/MgF_2 interface, and

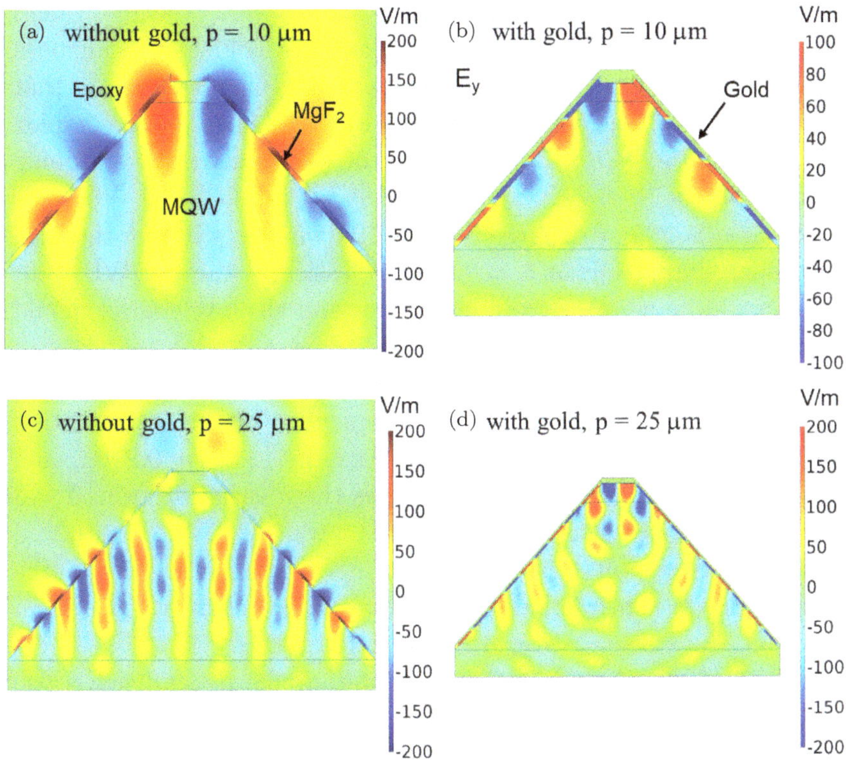

Figure 4.3. The vertical E_y pattern when $H_z = 1$ A/m, $\lambda = 8.5\,\mu$m, and $h_d = 0.2\,\mu$m at (a) $p = 10\,\mu$m without gold, (b) $p = 10\,\mu$m with gold, (c) $p = 25\,\mu$m without gold, and (d) $p = 25\,\mu$m with gold. Note the different y scale in (b).

they interfere with each other according to the relative phase. This reflecting condition is different from that without gold, in which the reflection from the MgF_2/air or MgF_2/epoxy interface is very weak due to the similarities in their n values where $n_{\mathrm{MgF2}} \simeq 1.2$ and $n_{\mathrm{epoxy}} \simeq 1.5$. These small n values also enable the sidewall reflection without gold to be similar to that of the bare sidewall. Even though QE with gold generally reduces, it can nevertheless be large and approaches η_{EM} with air cover in Fig. 4.2(d) at large h_d or at large p. Therefore, one can design a sensitive C-QWIP with gold cover using an appropriate MgF_2 thickness for a given corrugation period and wavelength. For large-period C-QWIPs, the h_d requirement is relaxed. For example, at $p = 25\,\mu$m, QE is 24.9, 25.9, 26.7, and

26.6%, respectively, for $h_d = 0.2$, 0.3, 0.5, and 1 μm. All these values are close to $\eta_{\text{EM}} = 27.1\%$. Overall, having a gold cover can be a general solution for C-QWIPs when α_{epoxy} is uncertain. Note that even at large p, the field pattern in Fig. 4.3(d) is more complex than that in Fig. 4.3(c) in the presence of two effectively overlapping optical cavities.

When p is 20 μm or larger, C-QWIPs with or without the gold cover offer similar QE at $\lambda = 8.5\,\mu$m. Classically, the C-QWIP coupling should be wavelength independent, and all wavelengths should have the same η shown in Fig. 4.2(a). However, this picture can change with optical interference. The intrinsic coupling characteristic can be known from EM modeling using fixed material properties. Using $n_s = 3.34$, $n_a = 3.24$, $\alpha = 0.22\,\mu\text{m}^{-1}$, $t_d = 0.2\,\mu$m, and $\alpha_{\text{epoxy}} = 0.05\,\mu\text{m}^{-1}$, Figure 4.4(a) shows the η variation from $\lambda = 3$ to 15 μm for $p = 25\,\mu$m and $S = 0.8$. Overall, C-QWIPs with and without gold exhibit relatively constant coupling efficiency in the modeled range. Their magnitudes, however, are slightly lower than that from ray tracing classical QE as noted before.

Besides its wavelength dependence, the angular dependence is also of interest. It affects the effective QE using different optics. The angular dependence can be modeled more conveniently using port boundary condition instead of using active scattering boundary condition. Due to the strong optical reflection by the retro-reflecting properties

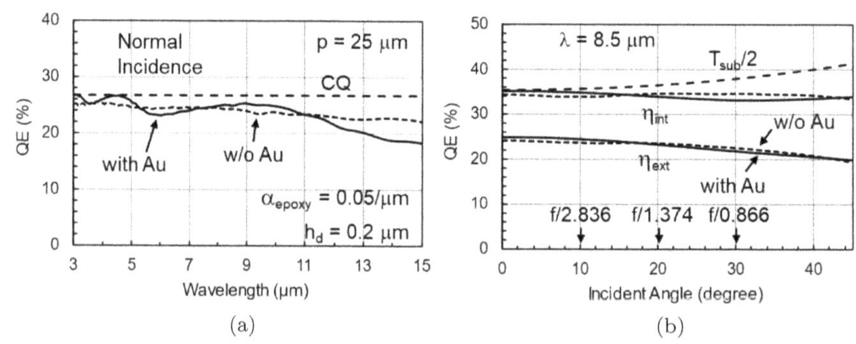

Figure 4.4. (a) The wavelength dependence of C-QWIP coupling with and without gold cover. (b) The angular dependence of C-QWIP coupling with and without gold cover.

of C-QWIPs, we need to use domain-backed, slit interior port boundary condition for wave launching in *Port* 1. The domain at the back is set to be Perfectly Matched Layer to absorb the reflected wave as shown in Fig. 4.5(a).

In using the port boundary condition, we can optionally specify a dummy passive port as *Port* 2. In the *Floquet PBC* panel, we use *From periodic port* in *k-vector* in *Floquet periodicity* tab. In the *Port* 1 panel, we specify *port input power* P_{in} to be 1 W, *Input quantity* to be *Magnetic field*, $H_0 = 1$ A/m (a dummy value overridden by P_{in}), *Angle of incidence* α to be ang, where ang is a sweep parameter in the study, *real part of refractive index* to be n_s, and *automatic diffraction order* is included in the port analysis. (Note that the COMSOL defined angle of incidence α should not be confused with the absorption coefficient α.)

The internal incident angle α and the external incident angle θ is related by Snell's law, where $n_{\text{air}}\sin(\theta) = n_s\sin(\alpha)$. For oblique incidence, Eq. (4.3) is modified to give an external unpolarized $\eta(\theta)$ of

$$\eta_{\text{ext}}(\theta) = \frac{T_{\text{sub}}(\theta)}{2}\cos\theta\left(\frac{c\varepsilon_0\alpha n_a}{2P_{\text{in}}}\int_A |E_y|^2 d^2r\right), \tag{4.4}$$

where $T_{\text{sub}}(\theta)$ is obtained from the Fresnel equation,

$$T_{\text{sub}}(\theta) = 1 - \left|\frac{-n_s^2\cos\theta + \sqrt{n_s^2 - \sin^2\theta}}{n_s^2\cos\theta + \sqrt{n_s^2 - \sin^2\theta}}\right|^2. \tag{4.5}$$

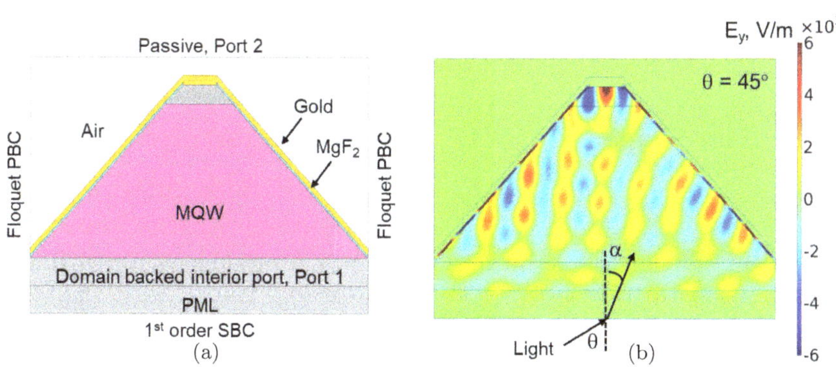

Figure 4.5. (a) The model setup and (b) the modeled E_y at $\theta = 45^\circ$ and $P_{\text{in}} = 1$ W.

The factor of 2 under T_{sub} in Eq. (4.4) accounts for the inactivity of the TE mode, and $\cos(\theta)$ accounts for the dilution of intensity falling onto an inclined surface. The bracketed term in Eq. (4.4) is referred to as the internal QE for transverse magnetic (TM)-polarized light. For unpolarized light, η_{int} is half of the bracketed term. The modeled unpolarized η_{int} for both gold-covered and -uncovered detectors varies only very slightly with θ in Fig. 4.4(b). With $n_s = 3.39$, the refracted angle $\alpha = 12.2°$ is small when the incident angle $\theta = 45°$. The small refracted angle thus makes the angular dependence small. The corresponding field distribution at 45° incidence is shown in Fig. 4.5(b). Figure 4.4(b) also shows $T_{\text{sub}}/2$ given by Eq. (4.5). It increases gradually as θ approaches the Brewster angle of 73.6° for the TM mode. The value of η_{ext} should then increase. However, since the incident light spreads out to a larger area at a larger θ, the detector with a fixed area will receive less light by a factor of $\cos(\theta)$. Including this factor gives a gradually decreasing η_{ext} in Fig. 4.4(b). However, the decrease is negligible with f-number as small as 1.5. Therefore, for the usual optics used in QWIP FPAs, C-QWIPs can be considered as angular independent, in addition to wavelength independent.

To compare η_{ext} obtained by using active scattering boundary condition in Fig. 4.4(a) and active port boundary condition in Fig. 4.4(b), η at $8.5\,\mu$m in Fig. 4.4(a) is 24.41% without gold and 24.91% with gold, while η at 0° in Fig. 4.4(b) is 24.33% without gold and 24.98% with gold. Therefore, the two approaches agree with each other under the same modeling condition.

4.4 Prism-Shaped C-QWIP with Thin Substrate

At large p or h_d, QE of a PR-QWIP with thick substrate can be approximately obtained by classical ray tracing. This feature is convenient in single-detector detection and material characterization. On the other hand, for FPA applications, the substrate is usually thinned either partially or completely. In the latter case, only a thin common ground contact layer remains. Removing most of the substrate is necessary to avoid thermal mismatch between GaAs

FPAs and Si readout integrated circuits (ROICs). With a thin ground contact/substrate layer, which is simply referred to as the substrate, resonant cavity effects will be more apparent in each corrugation. There are two cavity effects. The first is the FP oscillations brought by the retro-reflection property shown in Fig. 4.6(a). The second is light trapping in Fig. 4.6(b) when θ is not exactly equal to 45°.

For the first cavity effect, when the sidewall angle is 45°, two sets of coherent interference, labeled as interference 1 and 2, can take place. In interference 1, the incident beam r_1 partially transmits into the substrate as r_3. The light makes a round trip inside the corrugation and returns as r_4. If r_4 interferes with the reflected beam r_2 destructively, the substrate reflection will be reduced and r_3 will be strengthened, analogous to the usual FP resonances in parallel planes. The analogy is applicable because all the incident light has the same path length of $(2t_g + 2h + w) = (2t_g + p)$ in the corrugation between the two substrate reflections, independent of the incident location, where t_g, h, and w are defined in Fig. 4.6. The retro-reflector resonance, however, differs from a FP etalon in that it possesses another interference set (#2), in which r_2 interferes directly with another beam r_5 incident on the opposite sidewall.

Since there is a phase shift of π upon 45° internal reflection for the TM mode, the optical $\boldsymbol{E}$ vector switches direction at the inclined

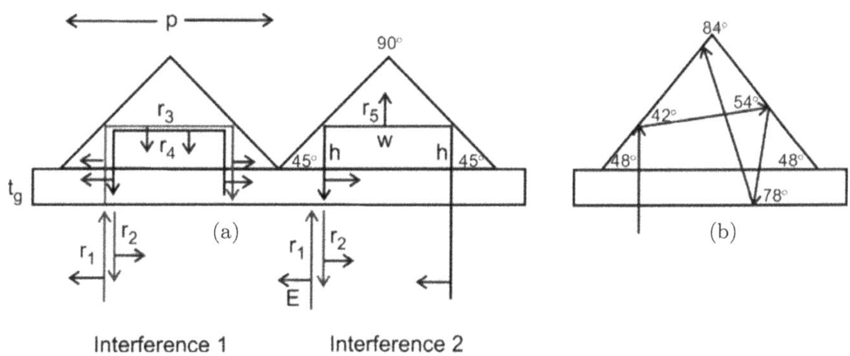

Figure 4.6. Possible optical paths in C-QWIPs with thinned substrate when the sidewall angle is (a) 45° and (b) 48°. Adapted with permission from *Inf. Phys. & Tech.*, 50, pp. 124–135 (2007).

surfaces. Therefore, resonant peaks for interference 1 occur when the sum of the phase shift of r_3 and r_4 is $2N\pi$, where N is an integer. Hence, the peaks occur at

$$2\left(\frac{2\pi n_a p}{\lambda} + 2\frac{2\pi n_s t_g}{\lambda}\right) = 2N\pi$$
$$\Rightarrow \lambda = \frac{2n_a p + 4n_s t_g}{N}, \tag{4.6}$$

where n_a and n_s are the MQW and substrate refractive indices, respectively. Likewise, the rays r_5 and r_2 interfere destructively at

$$\frac{2\pi n_a p}{\lambda} + 2\frac{2\pi n_s t_g}{\lambda} = (2M+1)\pi$$
$$\Rightarrow \lambda = \frac{2n_a p + 4n_s t_g}{2M+1}. \tag{4.7}$$

Since the first series contains all integers N and the second series contains only odd integers, the odd integer series receives twice the enhancements from both interferences 1 and 2. Therefore, the peak value of the odd integer series will always be larger than that of the even integer series. As a result, the detector QE spectrum displays only the odd integer series. To show this cavity effect, an ideal C-QWIP geometry with $\theta = 45°$ is modeled in Fig. 4.7. The classical

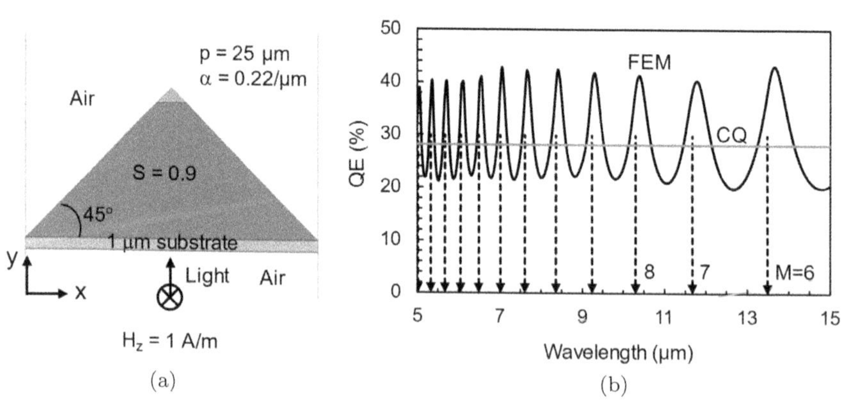

Figure 4.7. (a) The ideal C-QWIP with 45° angle used in FEM modeling and (b) the modeled external QE spectrum. The arrows are given by Eq. (4.7) with M from 6 to 17.

QE for $\alpha = 0.22\,\mu\text{m}^{-1}$ and $S = 0.9$ is 28.2%. The external QE is modeled by FEM with the substrate thickness t_g of $1\,\mu$m. The modeled QE oscillates about the classical value as expected from a FP etalon. The peak wavelengths follow the odd integer series although the simple analysis gives slightly shorter wavelengths. We expect similar oscillations to persist when θ deviates slightly from 45°.

For the light-trapping effect in Fig. 4.6(b), when θ is different from 45°, the partially reflected light from the substrate will not only produce an oscillatory spectrum but also increase the average QE as suggested by ray tracing. To study this effect, the same detector geometries displayed in Fig. 4.3(c) and (d) are adopted, whose thick substrate QE, η_{thick}, is shown in Fig. 4.4(a). For the detector without gold cover, when $t_{\text{sub}} = 1.43\,\mu$m, the modeled η_{thin} oscillates slightly above η_{thick} as shown in Fig. 4.8(a), showing a small light-trapping effect. For the C-QWIP with gold cover, η_{thin} is raised more substantially in Fig. 4.8(b).

When the present t_{sub} is set at $0.96\,\mu$m, η_{thin} shows a broad maximum around 30% centered at $\lambda = 8.5\,\mu$m, and it exceeds the classical QE of 26.69%. Therefore, having a gold cover can be advantageous. Because of the presence of two reflecting interfaces at each sidewall, the FP oscillations are more irregular than those shown in Fig. 4.7 for bare sidewalls.

In Section 2.17, the QWIP dark current is shown to depend on the doping density N_D of the MQW material. N_D should be minimized

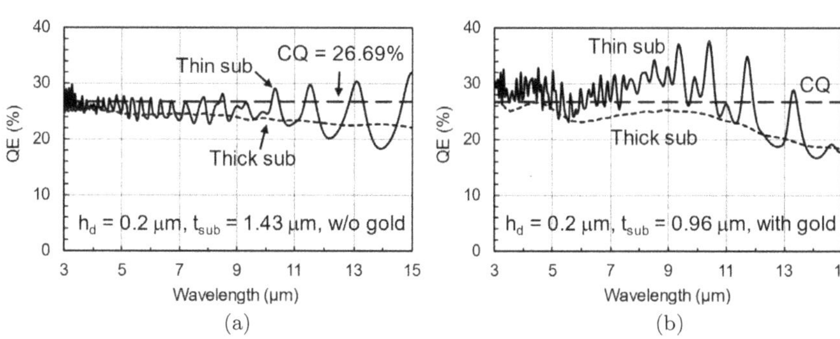

Figure 4.8. The modeled external unpolarized QE for $p = 25\,\mu$m (a) without and (b) with gold cover. Both contain a $0.2\,\mu$m tall MgF_2 layer.

to achieve higher operating temperature. Meanwhile, $\alpha = 0.22\,\mu\text{m}^{-1}$ used in Fig. 4.8(b) is for a bound-to-quasi-bound (B-QB) detector with $\text{N}_D = 1 \times 10^{18}\,\text{cm}^2$ in a 52 Å quantum well with a 500 Å barrier. To know N_D dependence, we modeled QE with different α in Fig. 4.9(a). Since α is linearly proportional to N_D, QE at different N_d can then be known and is plotted in Fig. 4.9(b). It is seen that a C-QWIP only needs a small α of $\sim 0.07\,\mu\text{m}^{-1}$ or a low doping of $\sim 0.35 \times 10^{18}\,\text{cm}^{-3}$ to achieve QE of 30%. These low values are due to the large absorption volume in a $p = 25\,\mu\text{m}$ detector. Higher N_d will increase I_d without increasing I_p significantly.

Another property of interest is the optical crosstalk in an FPA. The incident light into one detector pixel can be scattered into another pixel and causes ghost images. To estimate the extent of this adverse effect, thirteen gold-covered C-QWIP from Fig. 4.8(b) are modeled, in which the light at $\lambda = 8.5\,\mu\text{m}$ is incident into the array through a $3 \times 25\,\mu\text{m}$ aperture placed at $2\,\mu\text{m}$ away from the substrate as shown in Fig. 4.10(a). Under this illumination condition, the center pixel (#7) is uniform illuminated. Due to optical diffraction at the aperture, there is a finite spread of radiation into neighboring pixels #5 and #9. The modeled QE for all pixels in the array is shown in Fig. 4.10(b). While the center pixel #7 has QE of 33.6%, pixel #8 has QE of 33.0% and pixel #9 has QE of 0.35%. The slightly smaller QE of pixel #8 than pixel #7 by 0.6% is attributed to a small amount of incident light being spread to #9, which reduces

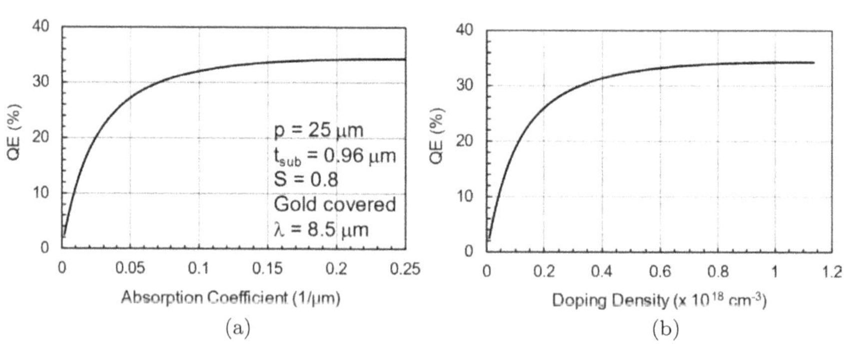

Figure 4.9. The external unpolarized QE at $8.5\,\mu\text{m}$ with different (a) α and (b) N_D.

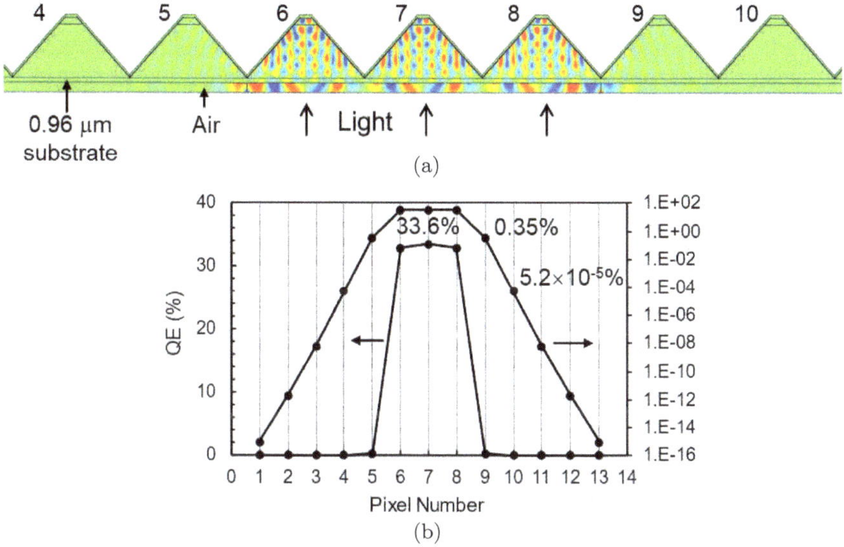

Figure 4.10. (a) The modeling setup for optical crosstalk study and (b) the modeled QE as a function of pixel number in the array in linear scale and in logarithmic scale.

#8's absorption and increases #9's absorption. The actual scattered light from #8 to #9 is much smaller than their apparent QE ratio of $0.35/33.0 = 1.1\%$.

The level of optical crosstalk can be more accurately accessed from QE ratio between #10 and #9, which is $5.2 \times 10^{-5}/0.35 = 1.5 \times 10^{-4}$, and this ratio is maintained to be the same for further pairs. The scattering of light from pixel to pixel is therefore at the 10^{-4} level. This low level of crosstalk is due to several reasons. Firstly, the detector volume is large, so most light with TM polarization entering the detector volume is absorbed. Secondly, due to the retro-reflecting properties of the corrugations, the unabsorbed light is reflected back into free space in nearly normal direction. Thirdly, the substrate is thin at $0.96\,\mu$m. The light can only scatter into the next pixel through a narrow region. These factors allow thinned C-QWIPs to have extremely low optical crosstalk compared to other coupling schemes that deflect light at large angles. The low crosstalk is critical in obtaining high-definition images as witnessed in Section 4.11.

The above modeling is carried out in two dimensions, where the geometry is assumed to be uniform in the third dimension. Using 2D modeling, the structure can be modeled efficiently in high spatial resolutions, with which highly accurate results can be obtained. However, in imaging arrays, the pixels are fitted into square unit cells with structural variation in the third dimension. To maintain a uniform corrugated cross-section, one may use plasma etching to create vertical sidewalls at both ends to form a prism structure. For example, a $2\,\mu$m gap can be etched between pixels to form a $25 \times 23\,\mu\text{m}^2$ area pixel in a $25\,\mu$m pitch array. The additional sidewall surfaces can help light coupling but the smaller pixel volume reduces the material for absorption. It will need 3D modeling to know the actual change.

3D modeling follows the same steps as 2D modeling. However, 3D modeling needs N times more meshing nodes than 2D, where $N \simeq L/d$, L is the linear dimension of the model, and d is the linear size of the mesh. For an approximate cubical modeling space, this difference can be 10 to 1000 times larger. Therefore, there are several iterative solvers in speeding up the process, such as the GMRES (generalized minimum residual) method. COMSOL automatically chooses the optimum solver for the model. Unfortunately, for structures that need periodic boundary condition (PBC), COMSOL often resorts to using the direct solver, which does not offer any speed improvement. For periodic arrays, other boundary conditions can incur errors as seen in the above crosstalk modeling. Due to incident light spreading into other pixels, pixel #8 has a slightly smaller QE than pixel #7. Nevertheless, this example also shows that one can either accept this level of error or enlarge the modeling space so that the illumination is uniform in the pixel of interest. In this approach, one can adopt scattering boundary condition at the edges and partition the model into smaller segments along the planes of symmetry. The solution of the smaller segment can then be obtained faster.

In 3D modeling, there are two independent polarizations in the incident x-y plane while the light travels in the z direction. Instead of performing modeling twice, one can model incident light with diagonal polarization, *i.e.*, with the azimuthal angles of 45° or 135°.

For structures with 4-fold symmetry, there is no difference among different azimuthal angles. For 135° polarization, the input $\boldsymbol{H}$ field (H_x, H_y, H_z) can be set to be (1, 1, 0) A/m. The unpolarized external QE η will then be

$$\eta = \frac{\alpha(\lambda)n(\lambda)n_{\text{air}}}{p^2 Z_0^2(\sqrt{2})^2} \int_V |E_z|^2 d^3r, \tag{4.8}$$

where $\alpha(\lambda)$ is for z-polarized light, $n(\lambda)$ is the refractive index of the absorbing material, p is the pixel period, and $\sqrt{2}$ accounts of the strength of $\boldsymbol{H}$ in the diagonal direction. This value of η assumes incidence from air. If the light is launched inside the substrate, n_{air} is replaced by $n_{\text{sub}}(\lambda)$ and the known substrate transmission coefficient should to be included in Eq. (4.8) to obtain the external η. The integral in Eq. (4.8) is the volumetric integral instead of the areal integral in 2D.

To test modeling in 3D, C-QWIP without gaps is modeled. It is 25 μm in length shown in Fig. 4.11(a). PBC is applied to both vertical surfaces. This setting effectively models an infinitely long corrugation, and 2D and 3D modeling should yield the same QE even though the input $\boldsymbol{H}$ in 3D contains an extra in-plane component. Modeling the structure with the same corrugation cross-section indeed results in the same spectrum shown in Fig. 4.11(b).

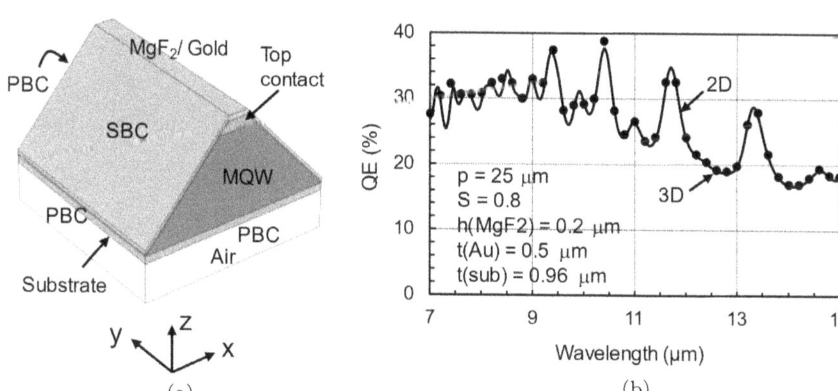

Figure 4.11. (a) The 3D extension of the 2D cross-section. (b) The comparison between the 2D and 3D solutions of the same geometry.

This agreement confirms that the y-polarized light represented by the finite H_x is inactive in the QW absorption. The small differences in Fig. 4.11(b) between the two solutions are attributed to the coarser mesh adopted in the 3D model.

After 3D modeling is verified, a $1\,\mu$m gap is opened at each end of the corrugation as shown in Fig. 4.12(a). This gap can be filled with air or epoxy of different α_{epoxy}. PBC is applied to all connected boundaries. The modeled QE is shown in Fig. 4.12(b). With the additional sidewalls, QE is generally larger irrespective of the covering material in the gaps and despite having less material. The average QE is about 40% between 7 and $11\,\mu$m, which is 25% larger than the 30% without the gaps.

To understand this difference, the E_z profiles of 2D and 3D models are compared in Fig. 4.13. Although the two field patterns are similar in Figs. 4.13(a) and (b), E_z in 3D is generally stronger than in 2D. This arises from the diffraction of E_y by the vertical sidewalls as seen in Fig. 4.13(c). Therefore, both polarizations make contribution to the PR-QWIP absorption, in which the y-polarized light makes up to 25% of the overall QE in the present detector geometry.

The MQW thickness t_a at $p = 25\,\mu$m and $S = 0.8$ is $11.11\,\mu$m. Although this t_a is feasible to grow with the InGaAs/AlGaAs

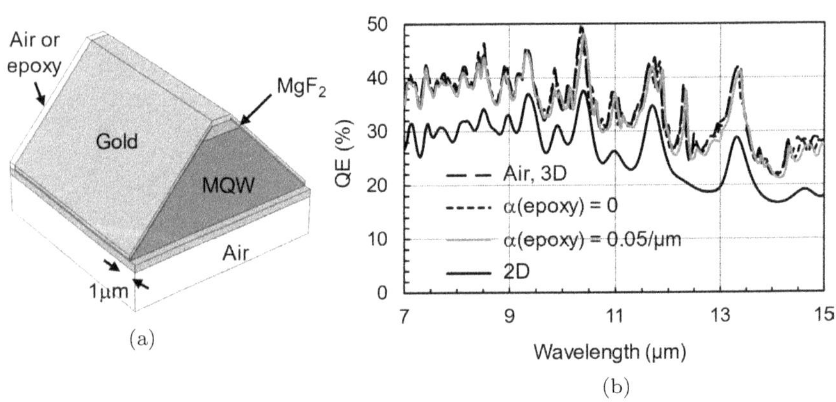

Figure 4.12. (a) The PR-QWIP with a $1\,\mu$m gap at both ends. (b) The modeled external unpolarized QE with different gap-covering materials compared to the 2D model without a gap.

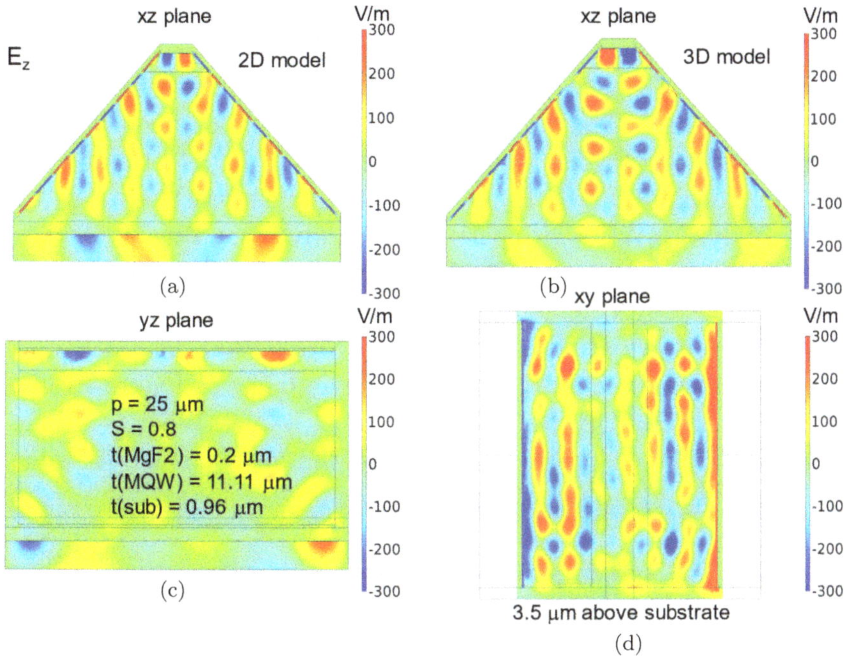

Figure 4.13. (a) The E_z profile in the 2D model. (b), (c), and (d) The E_z profiles in the 3D model at different cut planes.

MQW material, it does have some disadvantages such as the large voltage required to yield a sizable photoconductive gain g and the g saturates at a small value due to the large number of quantum well periods. From Fig. 1.15, g should be larger than 0.1 when there are temporal and spatial noises to yield a satisfactory detection range. There are two approaches in reducing t_a. One is to adopt a thinner MQW layer in a large corrugation and the rest of the volume is filled by contact materials. This structure in effect has thicker top and bottom contact layers and the corrugation is etched deeper into the bottom contact layer. The second approach is to adopt corrugations with smaller period p_c, and a pixel with period p contains several corrugations. For example, a $25\,\mu$m pixel may contain 3 corrugations having base area of $8.33 \times 23\,\mu\text{m}^2$. A $12\,\mu$m pixel may contain 2 corrugations with base area of $6 \times 10\,\mu\text{m}^2$. The individual corrugations can be connected by the gold metal cover or by an indium bump contact across the corrugations. For $p_c = 8.33\,\mu$m, t_a will be $3.70\,\mu$m for $S = 0.8$, and when $p_c = 6\,\mu$m, $t_a = 2.67\,\mu$m. These t_a are 3 to 4 times thinner than the $11.11\,\mu$m in a $25\,\mu$m pixel. The thinner MQW structures are modeled and examples of their E_z field profiles are shown in Fig. 4.14 under the first approach and Fig. 4.15 under the second approach.

To study C-QWIPs with partially filled MQWs, a PR-QWIP with the same detector cross-section in Fig. 4.11 is modeled in 2D. As before, the MQW α is assumed to be $0.22\,\mu\text{m}^{-1}$. The MQW layer is placed arbitrary at the middle of the corrugation, *i.e.*, at half of the total corrugation thickness of $12.49\,\mu$m or $6.50\,\mu$m. Figure 4.16(a) shows the modeled QE spectrum at several t_a values. When t_a is large at $10\,\mu$m, QE is relatively uniform across the spectrum and its value

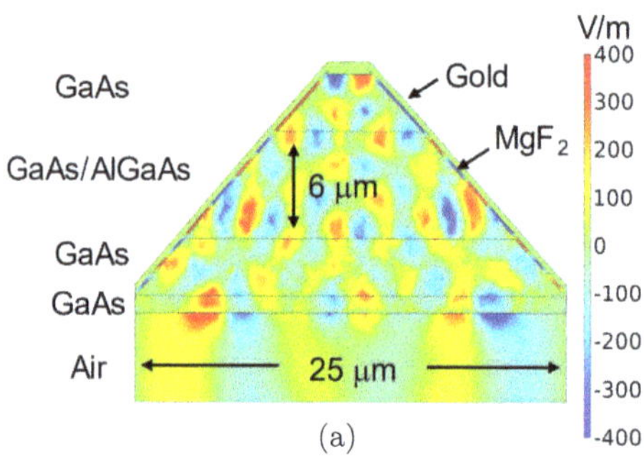

Figure 4.14. 3D model of E_z profiles when the corrugation is partially filled with MQWs with thickness of (a) $6\,\mu$m and (b) $4\,\mu$m.

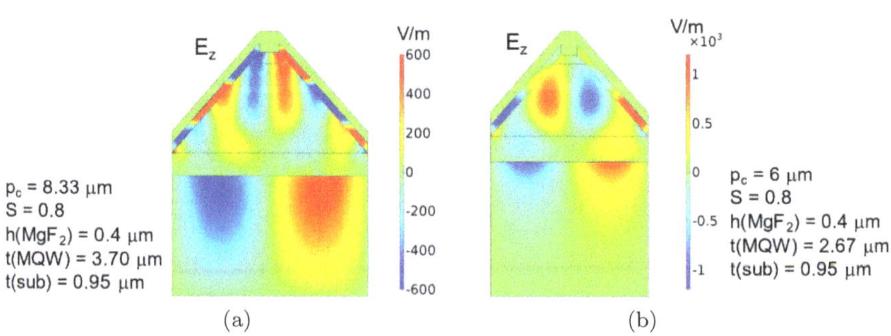

Figure 4.15. 3D model of E_z with corrugation periods of (a) $8.33\,\mu$m and (b) $6\,\mu$m.

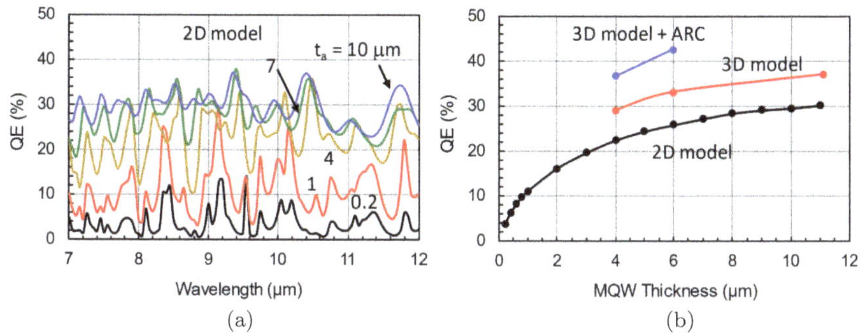

Figure 4.16. (a) 2D external unpolarized QE vs. t_a at different λ. (b) The average QE modeled in 2D, modeled in 3D, and modeled in 3D with ARC.

of ~30% is close to that of the fully occupied corrugation. When t_a is reduced, the average value decreases and the fluctuation amplitude grows because a smaller volume of the E field distribution is being sampled. The averaged spectral QE gradually decreases to 22% at 4 μm in Fig. 4.16(b). Due to the larger QE fluctuation at thinner t_a, this approach will be less useful when t_a is below 4 μm. The E_z profiles in Fig. 4.13(b) and Fig. 4.14 explain the slow QE decrease. When t_a decreases, the internal E_z field increases as seen in the scales of the figures. When the optical damping due to absorption is weaker, the light makes more round trips in the cavity and produces stronger E_z, which in turn lessens the QE decline.

In Fig. 4.16(b), PR-QWIP with additional vertical sidewalls are also shown for different absorber thicknesses. The modeled QE follows the general trend of the 2D model. Specifically, the vertical sidewalls increase QE by a factor of 1.277 from the 2D value at $t_a = 6\,\mu$m.

To further increase QE, one can place an anti-reflection coating (ARC) on the substrate surface to reduce substrate reflection. A simple example is to use a quarter wavelength ARC with $n = \sqrt{n_s} = 1.828$ and thickness $= 8.5/4/n = 1.16\,\mu$m. Figure 4.17 shows the modeled results for two t_a. For $t_a = 6\,\mu$m, the addition of ARC increases the average QE from 33.2% to 42.7%, representing a 28.7% increase. For $t_a = 4\,\mu$m, the average QE increases from 29.1% to 36.8%, a 26.5% increase. Since these increases are close

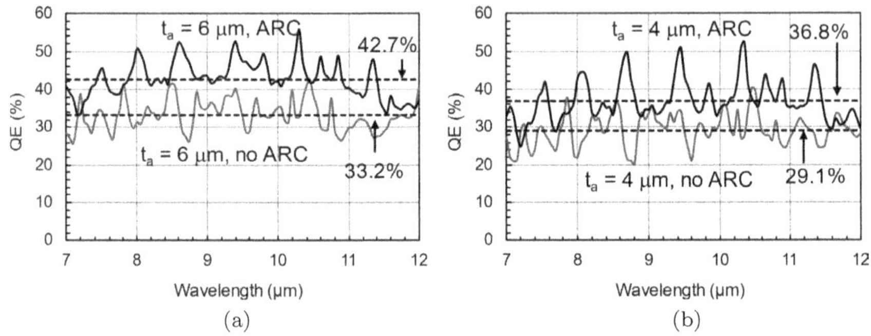

Figure 4.17. The PR-QWIP QE with and without ARC for t_a = (a) 6 μm and (b) 4 μm.

to the substrate transmission loss of 29.1%, the ARC is found to be effective in raising C-QWIP QE. Its t_a dependence is shown in Fig. 4.16(b). Therefore, a thinner MQW layer can be implemented in a 25 μm pixel with performance similar to a fully occupied C-QWIP.

For the second approach of using smaller corrugations, one can adjust the MgF_2 thickness to create strong interference pattern in the corrugations. By choosing t_d and t_{sub} appropriately, it is possible to obtain a large QE at certain λ ranges. For example, by using $t_d = 0.4\,\mu$m and $t_{\text{sub}} = 0.95\,\mu$m, a resonant peak can be realized at $\lambda = 8.5\,\mu$m as shown in Fig. 4.18 for two corrugation p_c periods.

In Fig. 4.18, the structure is first optimized using the 2D model and then a more accurate solution is obtained with 3D modeling. Since the present 3D structures are in the form of small corrugations, its vertical sidewall surface is small and thus it only has a small diffraction effect. The 3D QE is then very close to the 2D QE. Figure 4.18(a) shows that a peak QE of 40% can be obtained with $p_c = 8.33\,\mu$m and $t_a = 3.7\,\mu$m. Similarly, Figure 4.18(b) shows a QE of 50% with $p_c = 6\,\mu$m and $t_a = 2.67\,\mu$m. This level of QE is achieved without ARC and thus offers a better performance than the large corrugation approach for narrow band detection. The large QE is again produced by the large E_z shown in Fig. 4.15 at 8.5 μm.

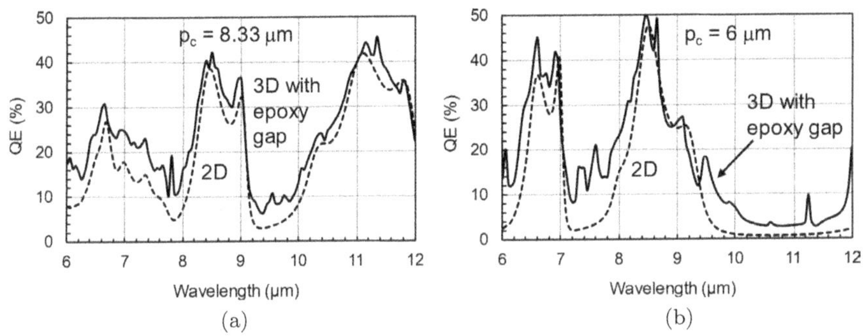

Figure 4.18. The modeled QE under the small corrugation period p_c approach for $p_c =$ (a) 8.33 μm and (b) 6 μm. The length of the corrugation in (a) is 23 μm and in (b) is 10 μm.

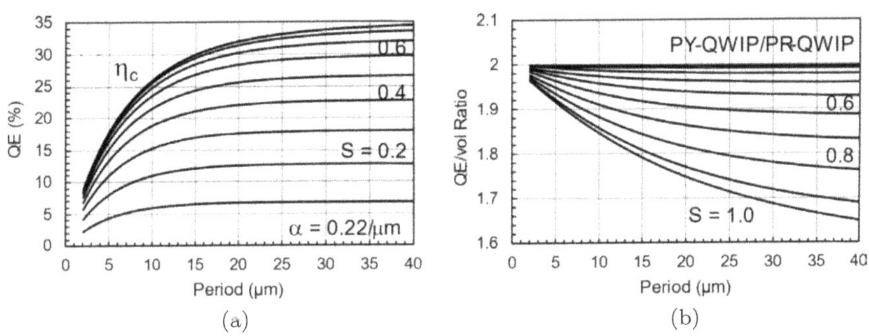

Figure 4.19. (a) The external unpolarized QE of an ideal, bare PY-QWIP with 45° sidewall angle. (b) The (QE/volume) of a PY-QWIP over the (QE/volume) of a PR-QWIP.

4.5 Pyramid-Shaped C-QWIPs

C-QWIP with a truncated pyramid geometry has four inclined sidewalls and therefore is expected to be more efficient in light coupling. It is also polarization-independent because of the square geometry. If the sidewalls are inclined at 45°, QE obtained from the classical model of Eq. (4.2) is shown in Fig. 4.19(a). It rises faster than that of the PR-QWIP shown in Fig. 4.2(a) but saturates only at a slightly larger value of 32.5% compared to the 26.7% for the PR-QWIP at $p = 25$ and $S = 0.8$. Nevertheless, accounting for the smaller dark current in the smaller volume, the PY-QWIP offers a more substantial photocurrent to dark current ratio improvement.

The increase of I_p/I_d is directly proportional to the factor $r \equiv (\eta_{\mathrm{PY}}/V_{\mathrm{PY}})/(\eta_{\mathrm{PR}}/V_{\mathrm{PR}})$, where V is the active volume, and it is shown in Fig. 4.19(b). At very small S, which represents a shallow trapezoid, an extra pair of inclined sidewalls doubles QE but it does not affect I_d appreciably. The value of r is thus 2. At larger S, r is smaller because of the smaller volume for absorption. At $p = 25\,\mu$m and $S = 0.8$, r is 1.8. Therefore, a PY-QWIP is nearly twice as effective as a PR-QWIP in terms of the I_p/I_d ratio. This picture, however, changes when a 3D model of PR-QWIP is considered because of the vertical sidewall contribution. Therefore, the final r is closer to 1.4.

To model a 3D PY-QWIP structure using FEM, we can take advantage of the symmetric geometry and only model a quarter of the structure shown in Fig. 4.20(a). This will reduce the volume by a factor of 4 and increase the mesh density by the same amount for the same number of nodes. However, PBC cannot be applied to the truncated geometry. Applying scattering boundary condition instead will incur a small error in the present structure due to optical diffraction in the air as noted before. For this structure, x-polarized

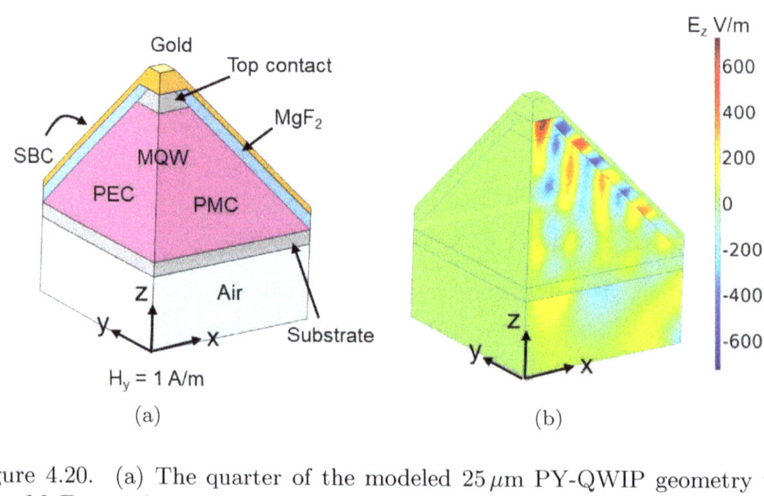

Figure 4.20. (a) The quarter of the modeled $25\,\mu$m PY-QWIP geometry with $1\,\mu$m MgF_2 and Au cover and the boundary condition set up. The sidewall angle is 48°. (b) The modeled E_z at $\lambda = 8.5\,\mu$m. The E_z field on the y-z plane is uniformly zero, while E_z on the x-z plane is antisymmetric in the other half.

light will be launched in the air by inputting $H_y = 1\,\text{A/m}$. Since this structure will not produce E_y, the total $\boldsymbol{E}$ must point perpendicular to the y-z plane. This $\boldsymbol{E}$ symmetry can be enforced by setting the perfect electric conductor boundary condition on the y-z plane and the perfect magnetic conductor boundary condition on the x-z plane as indicated in Fig. 4.19(a). The modeled E_z is shown in Fig. 4.19(b), whose pattern is similar to Fig. 4.13(b).

To be definite, QE of PY-QWIPs having thick substrate is modeled using

$$\eta_{\text{PY}} = 4T_{\text{sub}} \frac{\alpha(\lambda) n(\lambda) n_{\text{sub}}(\lambda)}{p^2 Z_0^2} \int_V |E_z|^2 d^3 r, \tag{4.9}$$

where V is a quarter of the active volume. Figure 4.21(a) shows the modeled QE with different detector configurations. For a bare PY-QWIP with air cover, the modeled QE is slightly larger than the classical value of 32.5% due to the larger sidewall angle. By applying MgF_2/Au cover with a large MgF_2 layer height h_d, η_{PY} approaches classical QE and it is relatively independent of wavelength. When the substrate is thinned to 0.96 μm and light is incident from air, optical interference introduces QE oscillations. The oscillations can be quite large when $h_d = 0.2\,\mu$m without gold or when $h_d = 1\,\mu$m with gold. The peak value exceeds 40%. However, even with $h_d = 1\,\mu$m, the average η_{PY} is about 35%, which is lower than the average QE of 40% of the 3D PR-QWIP in Fig. 4.12. Therefore, the PY-QWIP

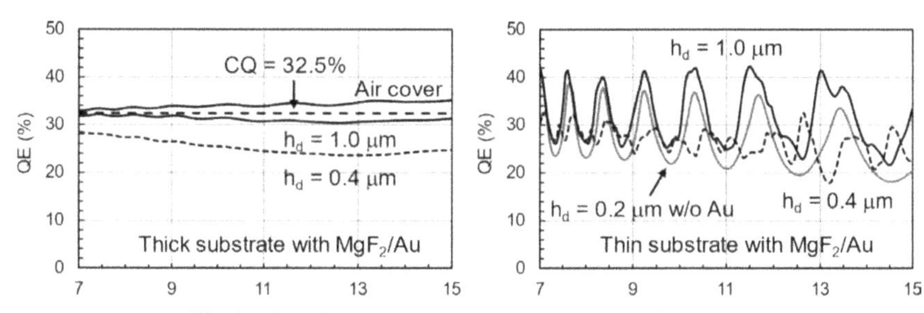

(a) (b)

Figure 4.21. The external unpolarized QE of a PY-QWIP with different covers for (a) thick substrate and (b) thinned substrate. $T_{\text{sub}} = 0.96\,\mu$m.

does not actually offer higher QE in thinned structures. An ARC is expected to further increase η_{PY} similar to a PR-QWIP.

4.6 C-QWIPs for Material Characterization

The C-QWIP structure is also useful for material characterization. In the usual characterization process, the responsivity spectrum $R(\lambda) = e\lambda g\gamma\eta(\lambda)/hc$ is obtained from radiometric measurements, where g is the gain and γ is the tunneling probability of the photoelectrons. Using Eq. (2.44), $\alpha(\lambda)$ of the material can be deduced. This procedure, however, assumes $g\gamma$ to be wavelength-independent, which may not be the case if the photoelectron decay length depends on the initial hot-electron energy. In this section, we describe an experiment in which α and $g\gamma$ are simultaneously obtained as a function of wavelength. In this particular experiment [20], ten PR-QWIPs with different corrugation periods p ranging from 6 to 1200 μm are prepared from the same wafer material. Each detector has the same total area of 700 μm × 1200 μm but different number of corrugations shown in Fig. 4.22(a).

The material contains 40 periods of 50 Å GaAs QWs and 500 Å $Al_{0.27}Ga_{0.73}As$ barriers. The thickness of the active region t_a is

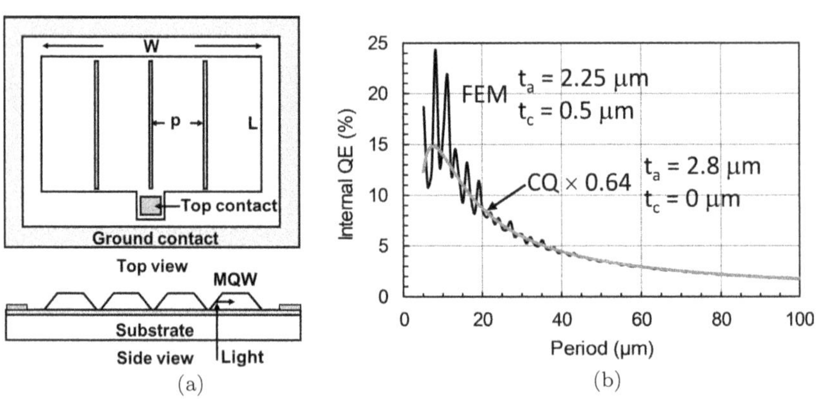

Figure 4.22. (a) The top view and side view of a test detector having 4 corrugations. $L = 700\,\mu$m, and $W = 1200\,\mu$m. Reproduced with permission from [20]. (b) The classical and FEM QE (without vertical sidewall contributions) as a function of period for fixed t_a and t_c. $\lambda = 8.5\,\mu$m.

2.25 μm. The top contact layer t_c is 0.5 μm thick. A total depth of 2.8 μm is etched into the material to expose the active volume. In the bottom of the corrugations, there are a 1.5 μm GaAs contact layer and a 1 μm undoped $Al_{0.8}Ga_{0.2}As$ layer. The doping density in the well is 5×10^{17} cm^{-3}, and the substrate thickness is 200 μm. The responsivity measurement is performed at temperature $T = 10$ K and 77 K. The detector geometry for $p = 300\,\mu$m is shown in Fig. 4.22(a).

The original experiment was analyzed using the classical QE of Eq. (4.1) and assumed the total etched depth of 2.8 μm to be the absorber thickness t_a. The more exact QE is, however, given by the EM solution in Eq. (4.3). To compare the classical η_c with $t_a = 2.8\,\mu$m and FEM η_{EM} with $t_a = 2.25\,\mu$m, the two QEs are calculated in Fig. 4.22(b) with $\alpha = 0.22\,\mu\text{m}^{-1}$ and $\lambda = 8.5\,\mu$m. In the EM model, the sidewall angle is 48°, and the refractive indices of the undoped $Al_{0.8}Ga_{0.2}As$ layer, the GaAs layers, and MQW layer are taken to be 3.00, 3.3407, and 3.2388, respectively. The η_{EM} curve follows the same trend of η_c but oscillates at small p, and η_{EM} is smaller than η_c by a factor of 0.64, *i.e.*, $\eta_{\text{EM}} = 0.64\eta_c$. The smaller η_{EM} is because of the thinner absorber thickness and the semi-transparency of shallow corrugations. Not all light is reflected at the sidewalls. Since the trends of η_c and η_{EM} differ only by a multiplication factor, we can keep the original discussion using $\eta_c(p)$ in the internal responsivity analysis, in which

$$R_{\text{int}}(p) = \frac{e(g\gamma)'\lambda}{hc}\eta_c(p) + R_0, \tag{4.10}$$

where $(g\gamma)'$ is the apparent $g\gamma$ if η_c is employed, and R_0 is the contribution from the vertical sidewalls at both ends and it is assumed to be a constant. If we were to use the more accurate $\eta_{\text{EM}}(p)$ instead,

$$R_{\text{int}}(p) = \frac{eg\gamma\lambda}{hc}\eta_{\text{EM}}(p) + R_0 = \frac{eg\gamma\lambda}{hc}0.64\eta_c(p) + R_0. \tag{4.11}$$

Therefore, the more accurate $g\gamma$ will be $(g\gamma)'/0.64$.

In the original analysis, a quantity *NR* is defined. It is the responsivity normalized by the dark current ratio r_d between a C-QWIP and a planar QWIP without corrugations measured at 77 K.

The adoption of $NR = R_{\text{int}}/r_d$ is to improve data analysis. In fabricating the corrugations, a slight over-etching will reduce the active volume and thus R_{int}, but it also reduces the dark current. By taking the ratio of R_{int}/r_d, the etching nonuniformity can be minimized. Similarly, the larger top contact resistance in the narrower corrugations yields smaller R_{int} and r_d. The change of contact resistance with p can also be corrected using NR.

Consistent with the 45° sidewall angle in deriving the classical QE, r_d is simply equal to $(p - t_a)/p$. The classical NR is

$$NR(p) = \frac{e(g\gamma)'\lambda}{hc}\frac{1}{p - t_a}\left[t + \frac{e^{-\alpha p}}{2\alpha}(1 - e^{2\alpha t_a})\right] + \frac{p}{p - t_a}R_0, \quad (4.12)$$

where $t_a = 2.8\,\mu\text{m}$. In the experiment, R_{int} of the C-QWIPs is measured at different λ, applied voltage V, and temperature T. Figure 4.23 shows R_{int} of a $p = 8\,\mu\text{m}$ detector at $T = 10\,\text{K}$ and 77 K.

Figure 4.24(a) is R_{int} at 2 V and 77 K with different p. Dividing R_{int} with r_d gives the NR curves in Fig. 4.24(b). NR is the largest at $p = 6\,\mu\text{m}$ because it has the largest sidewall fill factor S. From the $NR(\lambda)$ spectra, $NR(p)$ can be extracted at each λ as shown in Fig. 4.25. These data points are then fitted to Eq. (4.12) using α, $(g\gamma)'$, and R_0 as fitting parameters, where α determines the slope of the curve, $(g\gamma)'$ determines the magnitude, and R_0 determines NR offset at large p. The data generally fall onto the fitting curves closely [20].

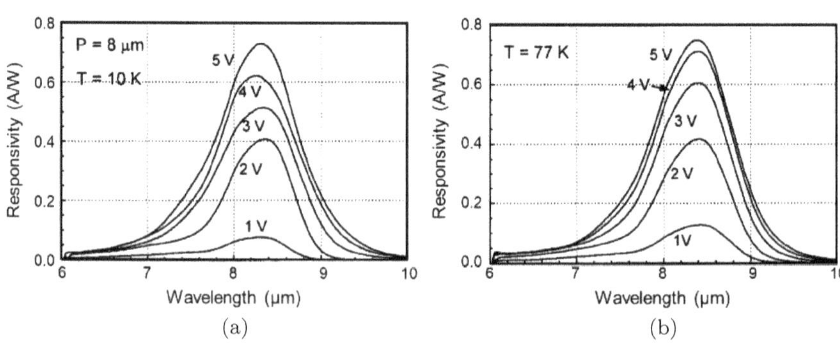

Figure 4.23. The measured R_{int} for the C-QWIP with $p = 8\,\mu\text{m}$ at (a) 10 K and (b) 77 K. Reproduced with permission from [20].

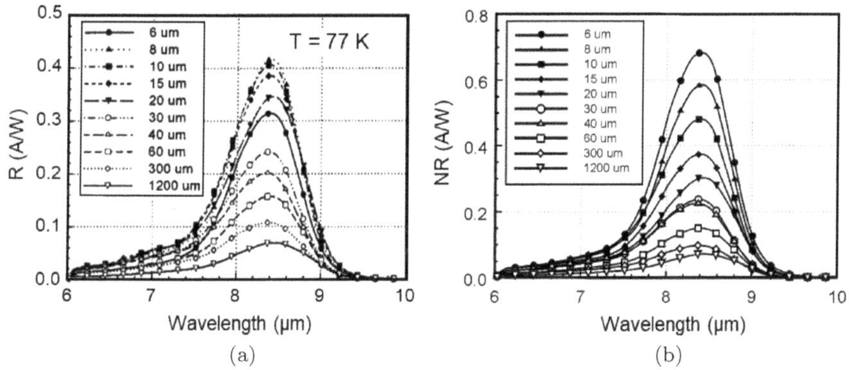

Figure 4.24. The measured (a) responsivity and (b) normalized responsivity at 2 V and 77 K for C-QWIPs with different p. Reproduced with permission from [20].

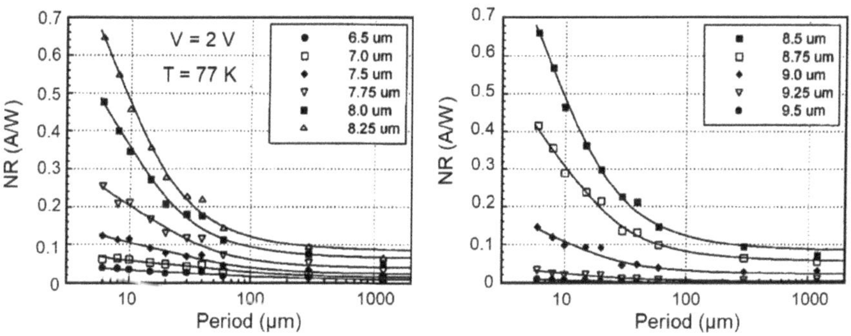

Figure 4.25. The extracted NR vs. p for different λ at 2 V and 77 K. Reproduced with permission from [20].

Figure 4.26 shows the fitted α at two T. In this plot, the result at the longer λ is omitted when the curve slope is too shallow to be extracted. At either T, α is a peaked spectrum as expected from intersubband transition and its value is between 0.15 and 0.24 μm^{-1}, depending on T and V. The intensity decay length is then between 4.2 and 6.7 μm at the absorption peak. The peak wavelength is at 8.4 μm and it nearly coincides with the responsivity peak. So does the half-maximum cut-on wavelength at 7.8 μm. On the other hand, the cut-off wavelength is extrapolated to 9.2 μm, which is longer than the 8.8 μm for R, indicating that the lower-energy photoelectrons are

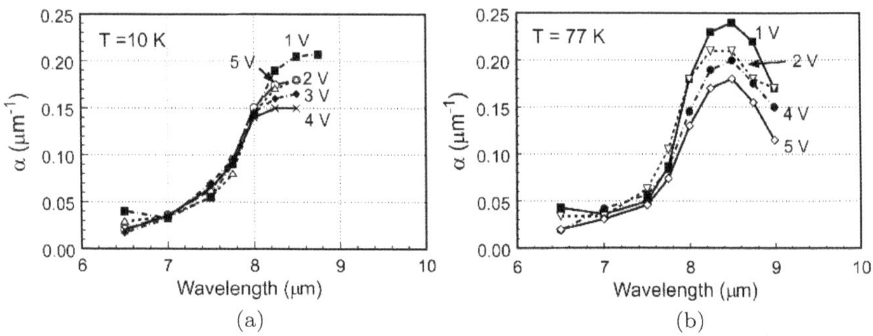

Figure 4.26. The deduced $\alpha(\lambda)$ of the QWIP material at different V and T. Reproduced with permission from [20].

not completely free. In general, α is broader than R. Therefore, R can only be an approximate representation of the α spectrum.

At large V or at low T, α is generally smaller. The former is due to the formation of a depleted space charge region at large V that reduces ground state electrons. The latter is due to the depopulation of ground state electrons back to the donor states at low T. Therefore, α depends on operating conditions. Since η_c shares the same p dependence as η_{EM} in Fig. 4.22(b), the value of α is unaffected using EM analysis. The deduced experimental α is thus accurate and is in reasonable agreement with the theoretical value of $0.22\,\mu\text{m}^{-1}$ from Eq. (2.9).

From the curve fitting, the apparent $(g\gamma)'$ is also obtained, and it is converted to $g\gamma = (g\gamma)'/0.64$ expected using η_{EM} for the analysis. The value of $g\gamma$ is shown in Fig. 4.27. Different from α, $g\gamma$ shows negligible T dependence, indicating that the electron transport is essentially the same at 10 K and 77 K.

At a small bias near 1 V, $g\gamma$ is approximately constant over a wide range of λ. Under a small driving field, the higher-energy photoelectrons thermalize onto the barrier conduction band edge rapidly, making both the recombination lifetime τ, the transit time $\tau_{tr} = t/v_d$, and their ratio g independent of λ. The small $g\gamma(= 0.047)$ at 1 V indicates a large capture probability $p_c = (gN + 1)^{-1} = 0.34$, assuming $\gamma \simeq 1$ at large energy. Near both ends of the spectrum, $g\gamma$ finally decreases but for different reasons. At the short

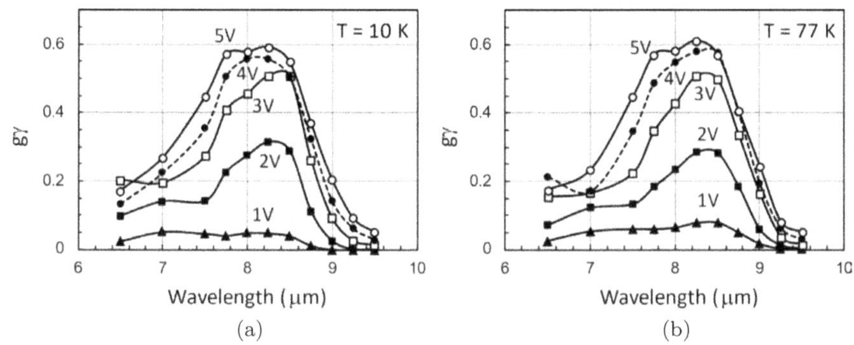

Figure 4.27. The $g\gamma$ spectrum at different temperature and bias. Reproduced with permission from [20].

λ, intervalley scattering into GaAs satellite valleys in the well, which is located ~300 meV above the Γ valley, increases τ_{tr}. The estimated λ threshold for this process is 6 μm for Fermi level of 90 meV, consistent with the decreasing g toward 6 μm. At the other end of the spectrum, the reduced tunneling probability γ reduces gγ.

Figure 4.27 also reveals the bias dependence of g. When V increases, the drift velocity v_d increases, which gives a rising $g\gamma$. However, the acceleration is not uniform. Lower-energy electrons near the top of the barrier, which is far below the L valleys, accelerate faster and acquire a larger v_d. The curve of $g\gamma$ thus develops into a peaked spectrum. At very large V, $g\gamma$ eventually saturates to ~0.6 because of intervalley scattering. The saturation voltage is set by the L valleys in the barrier, which are ~108 meV above the barrier, corresponding to 4.3 V across the entire QWIP. The last parameter extracted is R_0. Its contribution to NR can be extracted at the largest $p = 2000\,\mu$m in Fig. 4.25, and it amounts to 9.5% of NR at $p = 6\,\mu$m.

Since α and $g\gamma$ are distinct physical quantities, their spectra are different. In Fig. 4.28, their normalized spectra are shown at $V = 1$ and 5 V with $T = 77$ K along with the normalized R spectrum. In this plot, the difference of the α and $g\gamma$ spectra are clearly observed. R is bound by α at shorter λ and $g\gamma$ at longer λ. The reduction of R at the longer λ is expected from the small γ of a B-QB detector.

From the noise measurement, the average gains of 77 K I_d and background I_p can be deduced from the generation-recombination

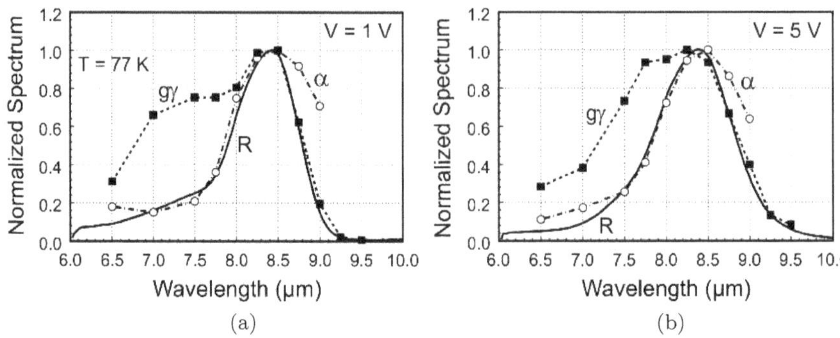

Figure 4.28. The figure compares the spectral shape of R, α, and $g\gamma$ at two voltages. Reproduced with permission from [20].

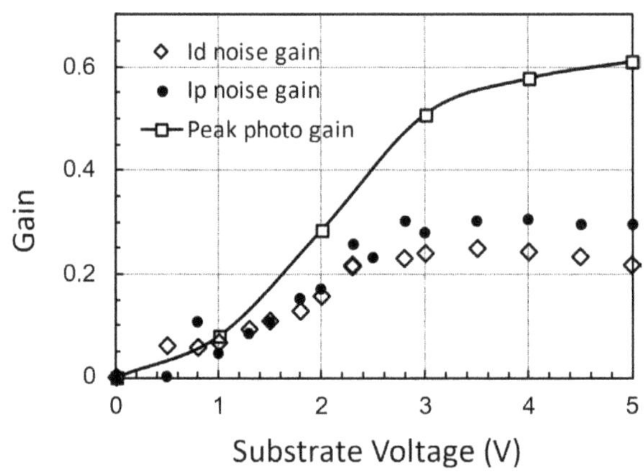

Figure 4.29. The gain values deduced from the background photocurrent noise, the dark current noise, and the peak value of $g\gamma$. Adapted with permission from [20].

noise expression. To compare them to the peak $g\gamma$ values, the three quantities are shown in Fig. 4.29. Below 2.5 V, the dark current g_d and the photocurrent g_p coincide, indicating that they are sampling the same electron distribution. Higher than 2.5 V, g_d and g_p are saturated. The value of g_d, however, saturates at a smaller value and decreases at higher V while g_p remains constant. At large bias, the dark current is increasingly dominated by thermally activated tunneling current, which transports at lower energies and thus has a shorter mean free path than the photoelectrons conducting above the barriers. The peak value of $g\gamma$, on the other hand, samples at a unique photoelectron energy where the mean free path is the longest. Therefore, its value is about double the average values represented by g_p and g_d.

In this section, we have shown a simple and yet informative characterization technique for QWIP materials. By measuring the detector responsivity at different corrugation periods, the $\alpha(\lambda)$ and $g\gamma(\lambda)$ spectra can be extracted simultaneously, which are otherwise difficult to obtain. In this method, α of parallel propagating light and $g\gamma$ are directly measured without invoking the dipole sine rule and the equivalence of photoconductive and noise gains. At large bias, g_p and g_d are substantially lower than the peak $g\gamma$. If we use g_p or g_d to deduce η from $R = e\lambda g\gamma\eta/(hc)$ (assuming $\gamma = 1$), η can be substantially over-estimated. For example, at 5 V, $g_p = 0.295$ and $g_d = 0.22$, the peak R_{int} for $p = 8\,\mu$m in Fig. 4.23(b) is 0.75 A/W. With $\gamma = 1$ at 5 V, η_{int} will be 37.5% using g_p and 50.5% using g_d, which far exceed the modeled η_{EM} ~15–25% in Fig. 4.22(b). On the other hand, if the peak $g\gamma$ of 0.61 is adopted, η_{int} will be 18.1%, which agrees with η_{EM}. Therefore, using the measured average $g\gamma$ to deduce η from R can be inaccurate at large bias for some detectors, depending on the $g\gamma$ spectrum. In addition, the increase of R is usually attributed to the increase of $g\gamma$ entirely. In this analysis, α and thus η can decrease with V. At $\lambda = 8.25\,\mu$m, the 6 μm detector has $R = 0.087$ A/W at 1 V and 0.532 A/W at 5 V, a factor of 6.1 increase. Since α decreases by 14% in this voltage change, $g\gamma$ increases by 7.1 times. Therefore, R and $g\gamma$ can have different bias dependence.

4.7 C-QWIP Focal Plane Array Fabrication

From rigorous EM modeling in Section 4.6, we obtained the general optical properties of C-QWIPs. Its external, unpolarized QE is ~29% without ARC and ~37% with ARC for a 4 μm thick MQW absorbing layer. When t_a increases to 6 μm, QE is 33% and 43%, respectively. This QE can be maintained at $N_D > 0.5 \times 10^{18}$ cm^{-3}

shown in Fig. 4.9. The main advantages of the C-QWIP coupling are its wavelength-independent coupling efficiency and relatively low level of pixel crosstalk. The absence of fine features in the detector geometry adds to the simplicity of array fabrication. The same detector geometry can be applied to a wide range of wavelengths from 3 to 15 μm. Economical FPA batch processing can be commenced at the same time without the need for prior knowledge on the absorption properties of individual wafers and the refractive indices of the constituent layers. The production cost can therefore be reduced while the throughput increases. The broadband coupling offsets its lower peak QE value while the lower N_D requirement reduces dark current. Therefore, C-QWIPs have unique advantages. Some of the fabricated C-QWIPs are shown in Fig. 4.30.

In the C-QWIP array fabrication, linear corrugations are first etched into the material using chemical solutions such as 1 H_3PO_4:1 H_2O:3 H_2O. When the corrugations are etched through 2 μm wide linear slits on a photoresist mask that are parallel to the $(01\bar{1})$ plane, the $(1\bar{1}\bar{1})$ crystallographic plane is exposed by the anisotropic etching rates, and the desired inclined sidewalls are created as shown in Fig. 4.30(g). The second step is the etching of vertical sidewalls using plasma etching, which contains BCl_3 and Ar gases. The third step is to deposit contact metals consisting of 50 Å Pd/215 Å Ge/385 Å Au/50 Å Pd/700 Å Au on top and common contact layers. After rapid thermal annealing, optical reflectors consisting of ~2000–4000 Å MgF_2/100 Å Cr/600 Å Au are deposited on pixel sidewalls. Subsequent steps are indium contact deposition, ROIC hybridization, epoxy backfill, and substrate thinning. Substrate thinning is performed by mechanical lapping or diamond turning to about 30 μm. The substrate can also be completely removed using inductively coupled plasma etching containing BCl_3, SF_6, and Ar. Using $Al_{0.4}Ga_{0.6}As$ as the etch stop layer, the etching selectivity is about 6000:1 between the etch stop layer and GaAs substrate [22], and the resulting surface is mirror-like. ARC is usually deposited on the substrate to increase QE and to reduce FP oscillations caused by the uncoated substrate/air interface.

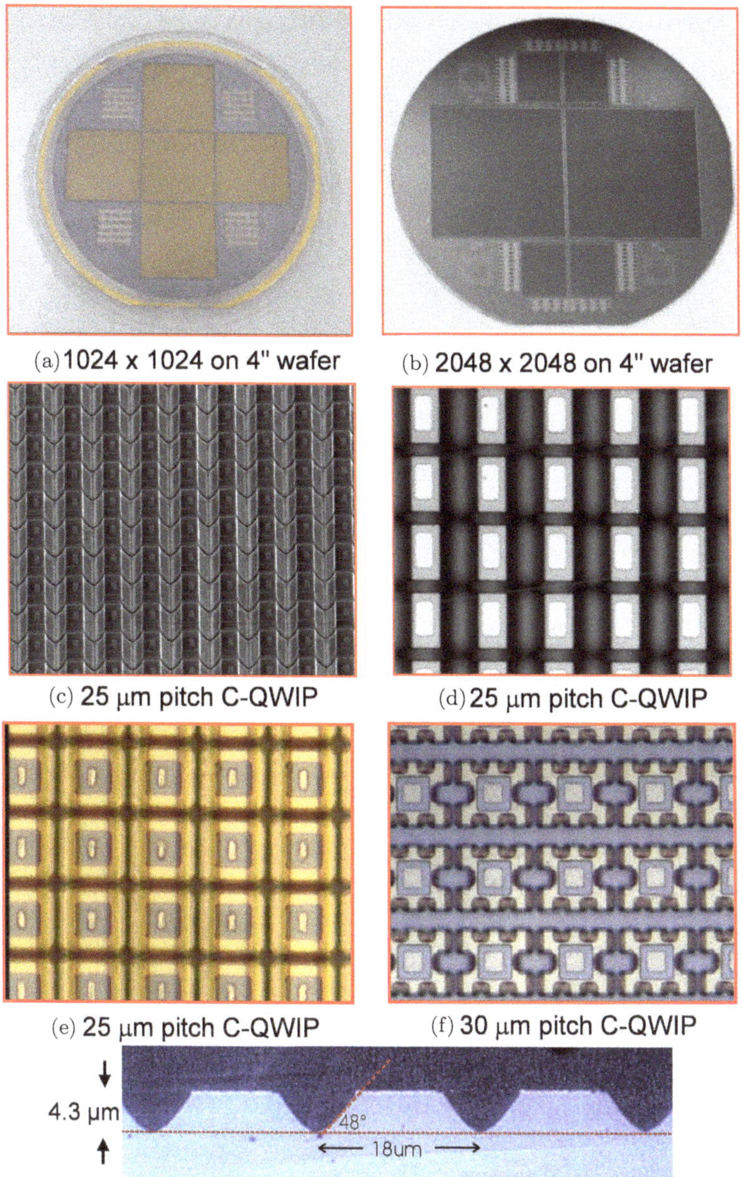

(a) 1024 x 1024 on 4" wafer

(b) 2048 x 2048 on 4" wafer

(c) 25 μm pitch C-QWIP

(d) 25 μm pitch C-QWIP

(e) 25 μm pitch C-QWIP

(f) 30 μm pitch C-QWIP

(g) Cross-section at S = 0.49

Figure 4.30. Processed C-QWIP FPAs.

4.8 Focal Plane Array Readout Integrated Circuits

C-QWIP FPAs are usually hybridized to direct injection readout circuits, which is briefly described here. Figure 4.31 shows the simplified schematic diagram of the electrical circuit. Right after the reset of the storage capacitor C_{int} by switching on the reset field effect transistor (FET) momentarily, the voltage on C_{int} is V_{RST}, and the voltage between the QWIP substrate and the drain is $V_{\text{sub}} - V_{\text{RST}}$. When the injection FET gate is set at V_G, the resistance between source and drain, R_{FET}, is practically infinite when V_S is lower than $V_G - V_{\text{th}}$, where V_{th} is the turn-on threshold voltage, but $R_{\text{FET}} \simeq 0\,\Omega$ when it is higher. The quiescent V_S is therefore very close to $V_G - V_{\text{th}}$, in which $I_D = (V_{\text{sub}} - V_S)/R_{\text{QWIP}}$, where I_D is the drain current and R_{QWIP} is the QWIP resistance under operation. When $V_{\text{RST}} - V_S$ is large, the FET is in saturation mode, in which I_D is independent of the change of $V_{\text{RST}} - V_S$.

When the QWIP is exposed to another level of radiation, R_{QWIP} becomes R'_{QWIP} and the operation point shifts to V'_S. The measured photocurrent i_p is

$$\begin{aligned} i_p &= \frac{V_{\text{sub}} - V_s'}{R'_{\text{QWIP}}} - \frac{V_{\text{sub}} - V_s}{R_{\text{QWIP}}}, \\ &= \frac{V_{\text{sub}} - V_s}{R'_{\text{QWIP}}} - \frac{V_{\text{sub}} - V_s}{R_{\text{QWIP}}} - \frac{\Delta V_s}{R'_{\text{QWIP}}}, \\ &= I_p - \frac{i_p}{g_m R'_{\text{QWIP}}} = \frac{g_m R'_{\text{QWIP}}}{1 + g_m R'_{\text{QWIP}}} I_p, \end{aligned} \tag{4.13}$$

where ΔV_S is the shift of the source voltage, I_p is the intrinsic photocurrent if there was no voltage shift, and $g_m = \Delta I_D/\Delta V_s = i_p/\Delta V_s$ is the transconductance of the FET. In the subthreshold regime, $g_m = eI_D/kT$. The injection efficiency of the FET, η_{inj}, is defined as i_p/I_p. By assuming $R'_{\text{QWIP}} \simeq R_{\text{QWIP}}$ and considering the full detector impedance $R_{\text{QWIP}} + 1/(i\omega C_{\text{QWIP}})$,

$$\eta_{\text{inj}} = \frac{g_m R_{\text{QWIP}}}{1 + g_m R_{\text{QWIP}}} \frac{1}{1 + \frac{i\omega C_{\text{QWIP}} R_{\text{QWIP}}}{1 + g_m R_{\text{QWIP}}}}. \tag{4.14}$$

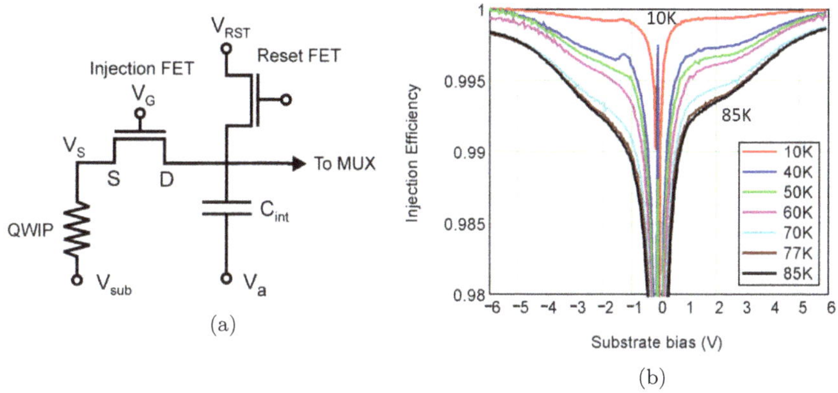

Figure 4.31. (a) A typical direct injection readout circuit with an n-channel injection FET. (b) The calculated injection efficiency of the FET under different bias and temperature of a 10.2 μm cutoff C-QWIP.

When $g_m \gg 1/R_{\text{QWIP}}$, η_{inj} is close to 1 and the measured photocurrent is close to the intrinsic photocurrent. During the signal integration, V_D is constantly changing. However, as long as the FET is in saturation mode (*i.e.*, when $V_D - V_S$ is large), I_D is independent of V_D and a constant i_p can be maintained during the integration cycle. Therefore, the FET serves as a voltage buffer for the detector. Figure 4.31(b) shows the calculated η_{int} of a 25 μm pitch, 10.2 μm cutoff C-QWIP pixel at different T while exposing to a 300 K background under $f/2$ optics. The value of g_m is calculated from the pixel current, C_{QWIP} is estimated to be 6.23 fF, and $\omega = 400\,\pi$. η_{inj} is shown to increase with V_{sub} and with decreasing T. Under the usual operating condition, $\eta_{\text{inj}} > 0.99$ can be achieved.

4.9 Theoretical Photocurrent of C-QWIPs

In Sections 4.4 and 4.5, η of a C-QWIP with a constant α has been calculated accurately using FEM. In this section, we calculate the photocurrent for a given α spectrum. The adopted QW period is displayed in Fig. 4.32. This structural design yields two distinct responsivity spectra under edge coupling in a particular material growth run. They are shown in Figs. 4.32(a) and (b) for wafer

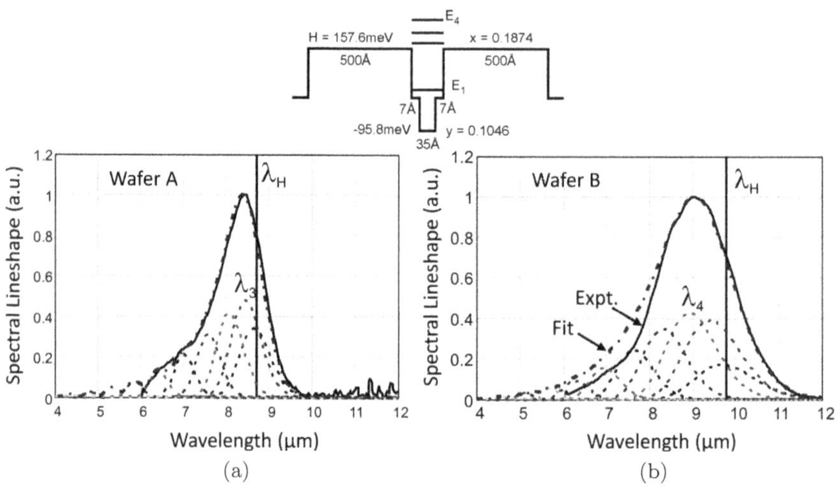

Figure 4.32. The measured responsivity lineshapes and the fitted λf_n curves using the TMM formalism in Section 2.4 based on the same material structure but with different band offset parameters of (a) 0.75 and (b) 0.65. The figure also shows the energy level structure of a QW period.

materials labeled A and B, respectively. Due to the nonuniformity of the molecular beams in the growth reactor, the MQW layer thickness depends on the location of the wafer in the chamber. If the nominal material thickness is assumed, the lineshape of wafer A is consistent with the band offset parameter $\beta_{\mathrm{AlGaAs}} = 0.75$ for $x = 0.1874$ while that of wafer B in Fig. 4.32(b) is consistent with $\beta_{\mathrm{AlGaAs}} = 0.67$. This value of x is measured from a test piece placed at the center of the platen. According to these β values, structure A has an absorption peak 4.4 meV above H and is a B-QB(+) detector. Structure B has an absorption peak 12.2 meV above H and is a bound-to-continuum (B-C) detector. With the fitted spectra in Fig. 4.32, the values of α obtained from Eq. (2.9) are shown in Fig. 4.33(b). The peak α are 0.15 $\mu\mathrm{m}^{-1}$ for A and 0.11 $\mu\mathrm{m}^{-1}$ for B with $N_D = 1 \times 10^{18}\,\mathrm{cm}^{-3}$ and $B = 500$ Å. With the theoretical $\alpha(\hbar\omega)$, the classical η_c of the C-QWIPs assuming unity substrate transmission, $p = 25\,\mu\mathrm{m}$, and $t_a = 11\,\mu\mathrm{m}$ are shown in Figs. 4.33(c) and (d), respectively. Their peak η_c are 36.1% for A and 31.9% for B and their bandwidths are significantly broader than that of α shown in Fig. 4.33(b).

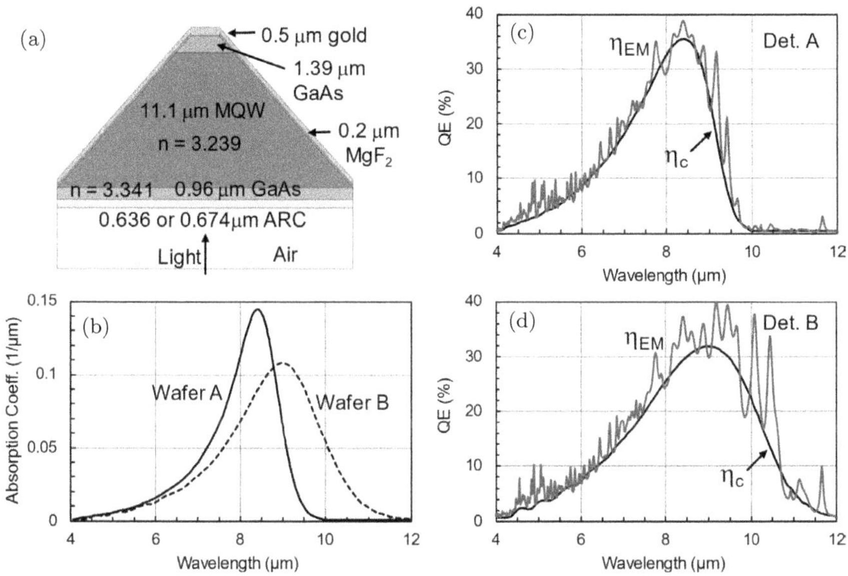

Figure 4.33. (a) The C-QWIP geometry adopted in 2D EM modeling. (b) The calculated α spectrum based on the nominal N_D and the observed lineshape. (c) The modeled external unpolarized η with ARC for Det. A based on the α spectrum for A. (d) The modeled η with ARC for Det. B.

With the C-QWIP structure shown in Fig. 4.33(a), the modeled η_{EM} in Figs. 4.33(c) and (d) give the same QE spectrum as η_c except the appearance of dense oscillations due to optical interference. Therefore, the classical prediction with 100% substrate transmission can be realized with an appropriate ARC. In this analysis, we have assumed fully occupied corrugations but without the contribution from the vertical sidewalls. The similarity between η_c and η_{EM} shows that the classical model can be used to predict and analyze C-QWIP performance.

The photocurrent density J_p generated from a blackbody with $T_B = 300\,\mathrm{K}$ between 7.5 and 12 μm is

$$J_p = \frac{\pi}{4F^2+1}\int_{7.5\,\mu\mathrm{m}}^{12\,\mu\mathrm{m}} eg\gamma\eta(\lambda)L(T_B,\lambda)d\lambda, \tag{4.15}$$

where $F = 2$ is the f-number, and $L(T_B, \lambda)$ is the spectral radiance in Eq. (1.1). If we further assume at certain bias, $\gamma = 1$ and $g = 0.2$, J_p is

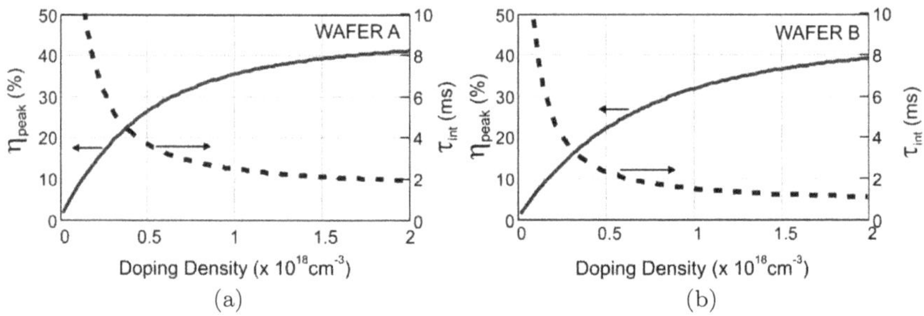

Figure 4.34. The peak η and the corresponding τ_{int} for full well integration as a function of N_D for (a) wafer A and (b) wafer B.

calculated to be $0.133\,\mathrm{mA/cm^2}$ for structure A and $0.222\,\mathrm{mA/cm^2}$ for structure B. A $p = 25\,\mu\mathrm{m}$ C-QWIP pixel will then generate 0.831 nA of photocurrent for A and 1.39 nA of photocurrent for B. If the charge capacity N_C of the readout is 13 Me$^-$, the integration time τ_{int} needed to fill up the charge well by I_p alone is 2.5 ms for A and 1.5 ms for B. For different N_D, τ_{int} varies according to the curves in Fig. 4.34. For example, reducing N_D by half to $0.5 \times 10^{18}\,\mathrm{cm^{-3}}$ increases τ_{int} to 3.5 ms for A and 2.3 ms for B. When the integrated photoelectrons reach N_C without dark current, the temporal noise-limited NEΔT calculated from Eq. (1.43) with thermal contrast of 0.0178/K are 9.2 and 9.7 mK respectively for A and B. NEΔT is unchanged for other N_D if τ_{int} is adjusted to obtain full N_C.

4.10 NEΔT Analysis

The calculated temporal NEΔT of ~10 mK represents the performance limit for the given η, g and N_C in long wave infrared imaging. A more realistic analysis also considers dark current, pixel nonuniformity, and read noise. From the dark current discussion in Section 2.17, J_d can be obtained from Eq. (2.51). The pixel nonuniformity is contributed from both dark current and photocurrent. The former can be due to variation in pixel area and the latter can be due to variation in light coupling efficiency. We consider only photocurrent nonuniformity in this analysis since $I_p > I_d$ in the usual operation. Based on the theoretical J_d and J_p from Eqs. (2.51) and (4.15), NEΔT of FPAs made from materials A and B can be calculated using Eq. (1.43). With I_p nonuniformity σ, Eq. (1.43) becomes

$$\mathrm{NE\Delta T} = \frac{1}{C}\sqrt{\left(\frac{2gN + n_r^2}{N^2}\right)\left(1 + \frac{1}{r}\right)^2 + \sigma^2}, \tag{4.16}$$

where C is the thermal contrast, N is the total collected electrons, n_r is the read noise, and r is J_p/J_d.

With a given QW structure, the only adjustable parameter is the doping density N_D. When N_D increases, J_p initially increases faster than J_d, resulting in a larger r. At certain N_D, r reaches a maximum and decreases. This N_D turns out to be at $0.28\mathrm{N}_0$ for both A and B, where N_0 stands for $1 \times 10^{18}\,\mathrm{cm}^{-3}$ [23]. Soon after, another detector figure of merit, the dark current limited detectivity D^*, defined as $(\eta/2\hbar\omega)(eg/J_d)^{1/2}$, peaks at $0.45N_0$. The maximum D^* gives the largest $J_p/\sqrt{J_d}$ and thus represents the largest signal-to-noise ratio when the dark current noise dominates. To complete the analysis, we also consider $N_D = N_0$ and $2N_0$, where the detectors are operated in large photocurrent and dark current modes. In this analysis, $g = 0.2, n_r = 500e^-, N_C = 13\,Me^-$, $F = 2$, and $\sigma = 0.02\%$ after electronic nonuniformity correction. The calculated NEΔT for the four N_D values is plotted in Fig. 4.35 at different T and τ_{int}.

For structure A with the cutoff wavelength $\lambda_c = 9.2\,\mu\mathrm{m}$, I_d is negligible at 55 K even with a large N_D. At this low T, a large η at a large N_D yields a low NEΔT. In particular, the $2N_0$ detector produces the lowest NEΔT of 14.4 mK in the shortest τ_{int} of 1.8 ms before readout saturation. The same NEΔT, however, will be reached by other N_D detectors at later times when N eventually accumulates to N_C.

At 60 K, I_d of the $2N_0$ detector is already appreciable at $r = 3$. This raises the minimum NEΔT. Nevertheless, if τ_{int} is intended to be less than 1 ms, the $2N_0$ detector with its large η of 41% still provides the highest sensitivity. At 65 K, I_d of the $2N_0$ detector reduces r to 0.6 and severely limits τ_{int} to sub-milliseconds, resulting in a higher NEΔT. However, if there were no N_C limit, this N_D is still a good choice for this temperature. This observation underscores the importance of collecting a large number of photoelectrons during

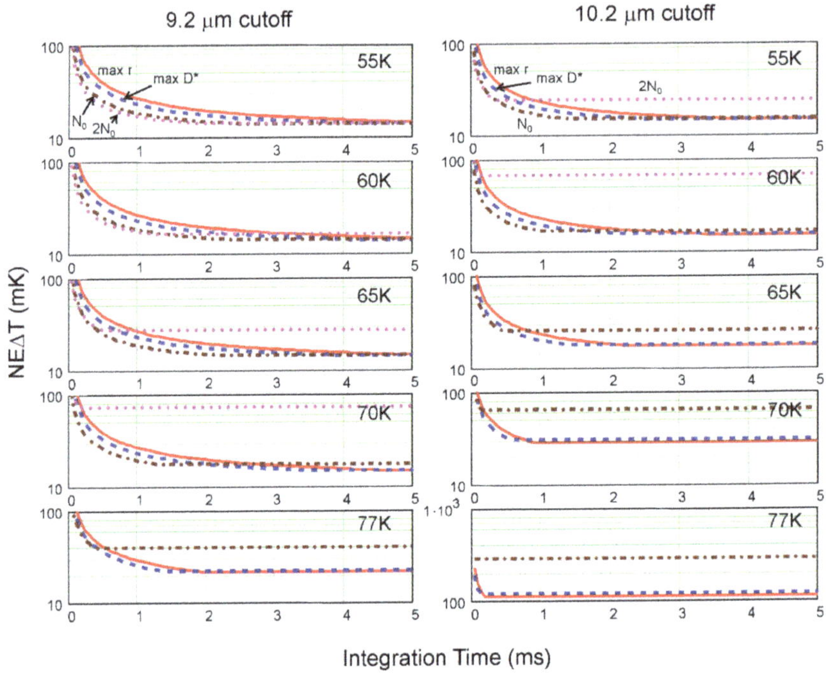

Figure 4.35. The predicted NEΔT of a C-QWIP FPA with (left) wafer A and (right) wafer B at different T and N_D.

the integration time. However, with the limited N_C, the optimum doping density is lowered to N_0.

Towards the higher operating temperatures, a lower I_d lessens the effects of readout saturation. In this case, the $0.46N_0$ detector, at which D^* is maximum, is preferred. However, there is no special significance of being at the D^* maximum other than the fact that it occurs at a lower N_D. Below 77 K, the detection is background limited (BLIP), and the dark current noise and therefore the dark current limited D^* do not play a significant role. In addition, when τ_{int} changes, the noise bandwidth also changes. NEΔT thus depends on $(D^*\tau_{\text{int}})^{-1}$, which in turn depends on r, rather than D^* alone. The situation for material B with $\lambda_c = 10.2\,\mu\text{m}$ is in parallel with the above discussion. From this analysis, a suitable doping density can be selected for different application priorities.

4.11 Experimental C-QWIP Focal Plane Array Performance

In this section, we discuss the experimental results from four C-QWIP FPAs [24]. These FPAs have different cutoff wavelengths, doping densities, active layer thicknesses, and pixel pitch. From these structural variations, one can assess the general performance of C-QWIPs and the validity of the detector model. The labeling of these FPAs is based on an internal tracking scheme. The substrate of the FPAs is mechanically thinned to $\sim$30 μm, and most of them are AR-coated. By comparing those with and without ARC, we found that ARC generally improves the responsivity by 28%, which indicates the substrate transmission is at 90% with ARC. Therefore, for those few FPAs that are not AR-coated, we multiply their η by 1.28 to account for the reflection loss. The ROICs adopted in this work operate on negative substrate voltages (*i.e.*, the common bottom contact is negative relative to the top contact). Due to the dark current asymmetry exhibited in most QWIP materials such as that of Det. B in Fig. 2.41, these FPAs tend to have higher dark current than those operated on positive substrate bias.

In discussing the FPA performance, it is more convenient to label η_d as the detection QE and η_{abs} as the absorption QE such that $\eta_d = \eta_{\text{abs}}\gamma$, where γ is the photoelectron tunneling probability. To determine η_d of an FPA, a test structure, which is from the same wafer as the FPA, is processed into an edge-coupled detector. The material spectral responsivity $R(V, \lambda)$ is measured from this detector at different substrate voltages V. Since the C-QWIP structure preserves the spectral lineshape of the material, this spectrum will be assumed to be the FPA detection spectrum. Also from the same detector, the photoconductive gain $g(V)$ is obtained from measuring the dark current noise at liquid nitrogen temperature. From $R(V, \lambda)$, the normalized absorption spectrum $S(V, \lambda)$ can be obtained from the $R(V, \lambda)/\lambda$ curve as specified by Eq. (4.10). The QE spectrum of a C-QWIP FPA will then be $\eta_d(V)S(V, \lambda)$, where $\eta_d(V)$ is the peak QE.

Next, the FPA pixel current as a function of V is measured in front of a blackbody source through an optical aperture. At each T_B, the I-V characteristics are recorded. To extract $\eta(V)$, $J(V, T_B)$ is fitted to [25]

$$
\begin{aligned}
J(V, T_B) &= J_d(V) + J_p(V, T_B) \\
&= J_d(V) + \frac{\pi}{4F^2 + 1} e\eta_d(V)g(V) \int_{\lambda_1}^{\lambda_2} S(V, \lambda)L(T_B, \lambda)d\lambda,
\end{aligned}
\tag{4.17}
$$

where J is the total current density, J_d is the dark current density, J_p is the photocurrent density, and λ_1 and λ_2 encompass the absorption spectrum. The f-number of the aperture F is 2.2. At each V, the two fitting parameters are $J_d(V)$ and $\eta_d(V)$. Figure 4.36 shows the typical fitting to the data.

4.11.1 *LC1 Focal Plane Array*

The first FPA described is labeled as LC1 with the material structure listed in Fig. 4.37. As usual, the barriers at the two ends on the MQW are 50 Å thick to reduce contact resistance. Figure 4.38(a) shows $R(\lambda)$ at a typical V, from which $S(V, \lambda)$ and hence $CE(V) = \eta_d(V)g(V)$ can be obtained from the blackbody measurement. Meanwhile, the

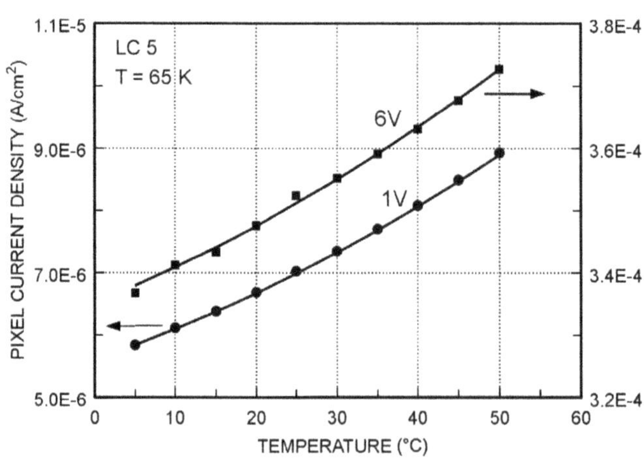

Figure 4.36. The pixel current densities (symbols) of the LC5 FPA at −1 V and −6 V substrate voltages are plotted against T_B. The solid curves are fittings to the data using J_d and η as adjustable parameters. Reproduced with permission from [24].

LC1 wafer material	Material	Doping (E18 cm^{-3})
Top contact	1.5 μm GaAs	0.9
MQW (106 periods) (7.955 μm)	Well: 5Å GaAs / 40Å $In_{0.1}Ga_{0.9}As$ / 5Å GaAs Barrier: 700Å $Al_{0.21}Ga_{0.79}As$	0.9
Bot. contact	2.7 μm GaAs	0.9
Etch stop	500Å $Al_{0.3}Ga_{0.7}As$	undoped
Buffer	2500Å GaAs	undoped
Substrate	GaAs	undoped

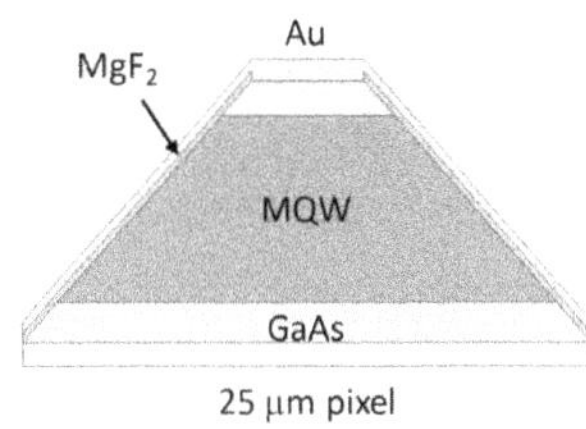

Figure 4.37. The detector material structure and 2D geometry of the LC1 FPA.

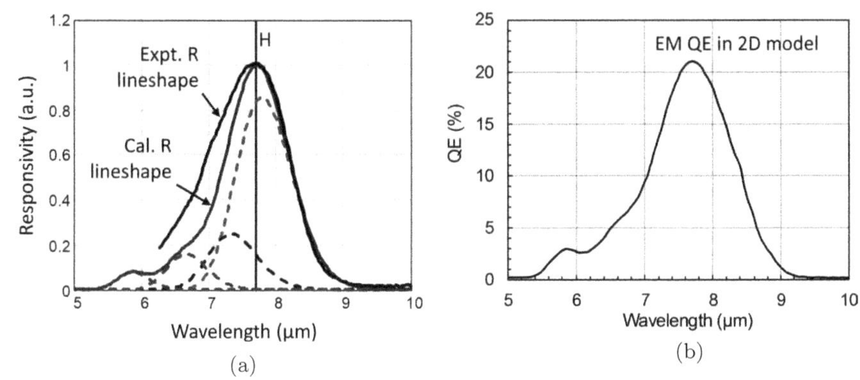

Figure 4.38. (a) The measured R lineshape under edge coupling and the fitted $R \propto \lambda f_n$ lineshape. The dashed curves are for individual excited states. The solid line represents barrier height H. LC1 is a B-B detector since E_1 is below H. Adapted with permission from [24]. (b) The internal unpolarized QE of the C-QWIP in 2D, in which the EM wave is launched inside the GaAs substrate.

theoretical $\eta_{\text{abs}}(\lambda)$ can be calculated based on the material and geometry shown in Fig. 4.37. The input $\alpha(\lambda)$ is calculated from the QW energies and the oscillator strengths obtained from TMM in Section 2.7. The line broadening σ is known by fitting $\lambda\alpha(\lambda)$ to $R(\lambda)$ lineshape as shown in Fig. 4.38(a). With the fitted line broadening, the peak α is calculated to be $0.0864\,\mu\text{m}^{-1}$. Performing 2D EM modeling yields η_{abs} spectrum in Fig. 4.38(b) with a peak value of 21.0%. With vertical sidewall contribution, η_{abs} should increase by a factor of 1.33 to $\eta \sim 27.9\%$.

In addition to the peak value, the spectral width of the η_d spectrum is also important in determining the photocurrent measured from a broadband source. Therefore, a figure of merit, the integrated quantum efficiency IQE, is defined, which is

$$\text{IQE} = \int_{\lambda_1}^{\lambda_2} \eta_d(\lambda) d\lambda, \tag{4.18}$$

where λ_1 and λ_2 (in μm) are the limits of the spectral band of interest. IQE has the unit of μm, and it gives the peak value if the 50% bandwidth is 1 μm. Using the η_{EM} spectrum in Fig. 4.38(b), absorption IQE of LC1 from 2D modeling is 0.329 μm.

The dark current gain of LC1 is shown in Fig. 4.39(a). To yield a sizable g, a large voltage is needed due to the B-B nature of the transition. At $V = -3\,V$, g is 0.042. With the measured CE $= 0.298\%$ at λ_p of 7.7 μm, η_d is deduced to be 7.1%, much less than the η_{abs} of 27.9% from modeling. A larger V is needed to increase γ but it is beyond the biasing limit of the ROIC. A new detector array is then hybridized onto a fan-out circuit without ROIC so that higher bias can be applied. At -11 V, CE is measured to be 2.84%. With $g = 0.077$ from Fig. 4.39, η_d would be 36.9%, which is higher than the 3D η_{abs} of 27.9% by a factor of 1.32. This apparently larger η_d is

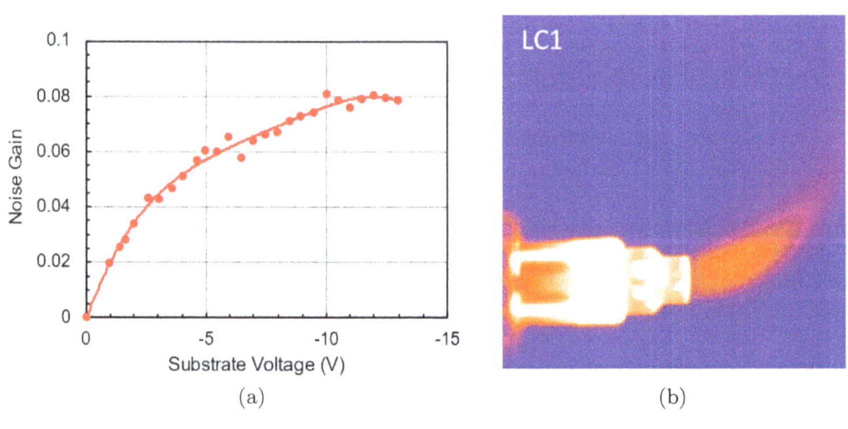

Figure 4.39. (a) The noise gain measured from the 77 K dark current. (b) A 1024×1024 image of a torch taken by the LC1 FPA at $T = 80$ K, $V = -3$ V, and $\tau_{\text{int}} = 16$ ms.

probably due to the actual g of the photocurrent peak being larger than g_d at large bias as described in Section 4.6.

At $T = 70\,\text{K}$, $V = -11\,\text{V}$, and $T_B = 300\,\text{K}$, the fitted J_p is $25.8\,\mu\text{A/cm}^2$ and J_d is $14.6\,\mu\text{A/cm}^2$. With the integration time τ_{int} of 5 ms, N is 7.88 Me^-. Substituting these values and average $g_p \simeq g_d = 0.077$ into Eq. (4.16), the calculated temporal NEΔT is 10.6 mK, where $C = 0.0207$ is used for λ between 7 and $8.4\,\mu$m. At lower bias of -3 V where CE is 9.5 times smaller, a longer τ_{int} is needed for imaging with the present ROIC. Figure 4.39(b) shows an infrared image of a torch using 16 ms integration time.

4.11.2 *LC2 Focal Plane Array*

The LC2 is a B-C detector. The photoelectrons are relatively free and thus the detector requires less bias. This material has the same structure as LC1 except that the barrier Al molar ratio is 0.12 instead of 0.21. The measured responsivity spectrum and the fitted $\lambda\alpha(\lambda)$ are shown in Fig. 4.40(a). Since the spectral width is wide at $\sim 3\,\mu$m, the theoretical α_{peak} is lower at $0.0593\,\mu\text{m}^{-1}$. The 2D η_{EM} is thus small at 15.2% but IQE is substantial at $0.567\,\mu$m. The 3D η_{EM} should be at $1.33 \times 15.2\% = 20.2\%$.

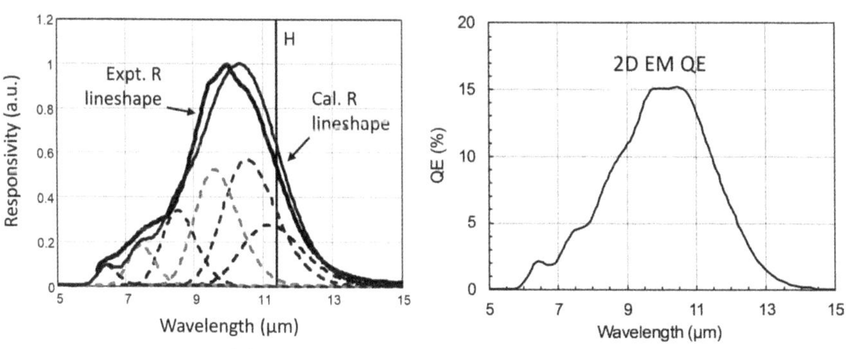

Figure 4.40. (a) The measured responsivity lineshape under edge coupling and the fitted λf_n lineshape of the LC2 material. LC2 is a B-C state detector since the α peak is well above H. Adapted with permission from [24]. (b) The unpolarized internal QE of the C-QWIP pixel in 2D with the integrated $\text{IQE}_{\text{EM}} = 0.567\,\mu$m.

The dark current g is shown in Fig. 4.41(a), and it reaches 0.15 at -5 V, which is nearly twice that of LC1 at -13 V. The larger g is due to the smaller intervalley scattering in the smaller x detectors and a higher photoelectron energy relative to the barrier height. With an ROIC operated at -5 V, CE is measured to be 3.90%, and η_d is deduced to be 25.8% based on g. This value of η_d is ~1.28 times larger than the 3D η_{EM}, similar to the LC1 array and again signaling an underestimate on g_{peak} using g at high V. With a wide spectral width of 3 μm, a large integrated photocurrent of 0.217 mA/cm^2 is observed at -5 V. This level of J_p is comparable to 0.222 mA/cm^2 estimated for Det. B in Section 4.9. At $T = 58$ K, the dark current is still quite high at 0.425 mA/cm^2 due to the longer λ_c, the B-C structure, and the negative substrate bias. Using $\tau_{\text{int}} = 0.5$ ms, N is 12.5 Me$^-$, and NEΔT is 25.9 mK.

Figure 4.41(b) shows CE and $\eta_d = \text{CE}/g$. The rising η_d shows that even for B-C detectors, γ is not unity. The photoelectrons still need an applied electric field to move in a definite direction. Figure 4.41(b) also shows the background photocurrent of a test detector scaled to fit the measured CE. The same V dependence of the directly measured I_p supports the CE measurement. It also indicates a higher CE of 4.9% at -8 V. Figure 4.42 shows infrared images taken by the FPA.

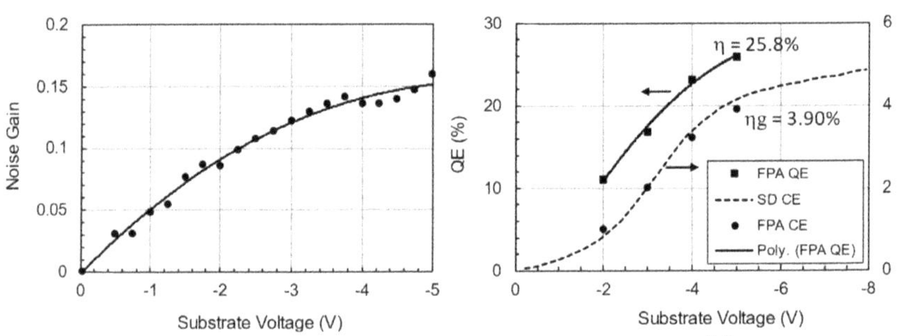

(a) (b)

Figure 4.41. (a) The measured noise gain g_d of the LC2 detector. (b) The measured FPA CE (circles) and $\eta_d = \text{CE}/g$ (squares) as a function of V. The dashed curve shows the background photocurrent I–V characteristics of a test detector in arbitrary units. Adapted with permission from [24].

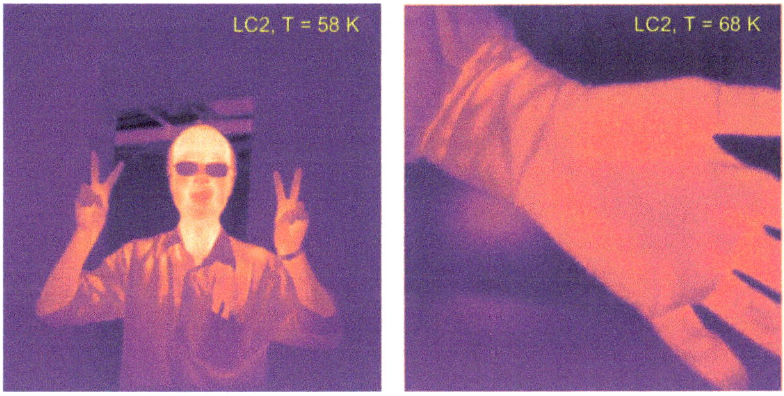

(a) (b)

Figure 4.42. 1024 × 1024 images taken by the LC2 FPA at (a) −3 V, 58 K and 1 ms and (b) −3 V, 68 K, and 0.3 ms.

4.11.3 *LC5 Focal Plane Array*

Although the B-C structure yields a lower bias voltage, it produces large dark current. In the LC5 material, B-QB(+) layer structure is adopted, in which the absorption maximum is ∼$\Gamma/2$ above the barrier height. The layer structure is shown in Fig. 4.43. The material contains 63 QW periods and $t_a = 4.658\,\mu$m. The observed and fitted spectra are shown in Fig. 4.44 and the calculated α_{peak} is $0.0785\,\mu\text{m}^{-1}$. Using the α spectrum and the C-QWIP geometry in Fig. 4.43, the unpolarized internal η_{abs} is modeled to be 12.6% at $\lambda = 9\,\mu$m in Fig. 4.44(b). The 3D η_{abs} is therefore 1.33 times larger at 16.8%.

The gain g is shown in Fig. 4.45(a). At −5 V, $g = 0.19$. The measured CE and the deduced η_d are shown in Fig. 4.45(b). For the FPA with an ROIC, it can be operated only up to −4 V, at which CE = 2.34% and $\eta_d = 13.9\%$. Another array connected directly to a fan-out circuit without an ROIC gives a slightly higher CE of 2.64% and a η of 15.6% at the same V. In general, the two arrays show nearly the same CE in the tested bias range. I_p of a test detector is consistent with the measured CE.

The maximum η from the fan-out is 16.1% at −5 V, which is in good agreement with the 3D η_{EM} of 16.8%, indicating that $g_{\text{peak}} \approx g$

LC5 wafer material	Material	Doping ($E18\ cm^{-3}$)
Top contact	2.0 μm GaAs	1.0
MQW (63 periods) (4.658 μm)	Well: 7Å GaAs / 35Å $In_{0.1}Ga_{0.9}As$ / 7Å GaAs Barrier: 700Å $Al_{0.19}Ga_{0.81}As$	1.0
Bot. contact	5.8 μm GaAs	1.0
Etch stop	500Å $Al_{0.3}Ga_{0.7}As$	undoped
Buffer	2500Å GaAs	undoped
Substrate	GaAs	undoped

(a)

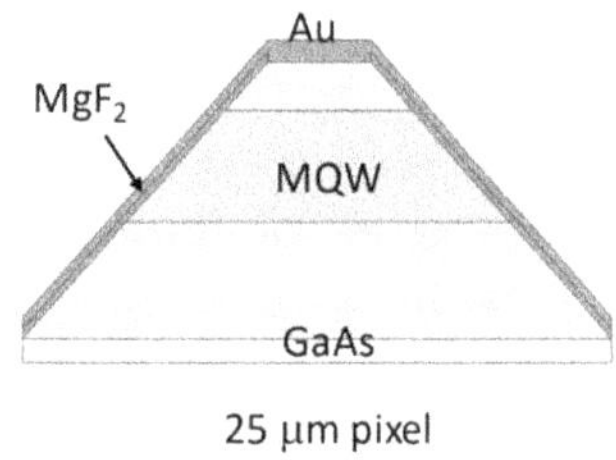

(b)

Figure 4.43. (a) The detector material structure and (b) 2D geometry of the LC5 FPA.

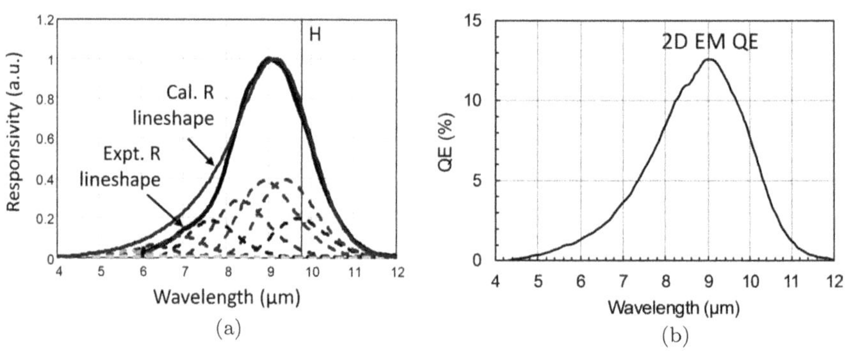

Figure 4.44. (a) The responsivity spectrum under edge coupling and the fitted λf_n lineshape of the LC5 material. H is the position of the barrier height. Adapted with permission from [24]. (b) The unpolarized internal QE of the C-QWIP modeled in 2D with $\mathrm{IQE_{EM}} = 0.352\,\mu$m.

for this detector. The present CE of 3.05% at -5 V is similar to the CE of 2.84% at -11 V for LC1. A thinner active material can thus be used to obtain similar CE as a thicker material under the ROIC bias constraint. At 60 K, J_p is $50.3\,\mu\text{A/cm}^2$ and J_d is $44.8\,\mu\text{A/cm}^2$. With a 2 ms integration time, N is $7.42\,\text{Me}^-$, and NEΔT is 23.0 mK. This NEΔT is consistent with the 20 mK calculated in Fig. 4.35 for the same $10.2\,\mu$m cutoff FPA under the same operating condition.

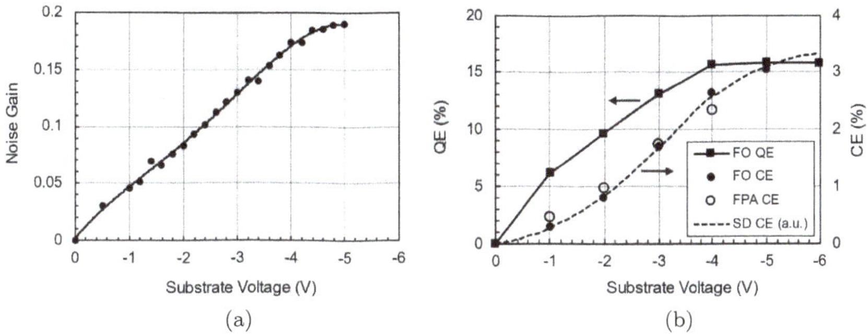

Figure 4.45. (a) g of the LC5 detector. (b) FPA CE (open circles), fan-out circuit CE (closed circles), and $\eta = \mathrm{CE}/g$ (squares) as a function of V. The dashed curve shows the background photocurrent of a test detector in arbitrary units. Adapted with permission from [24].

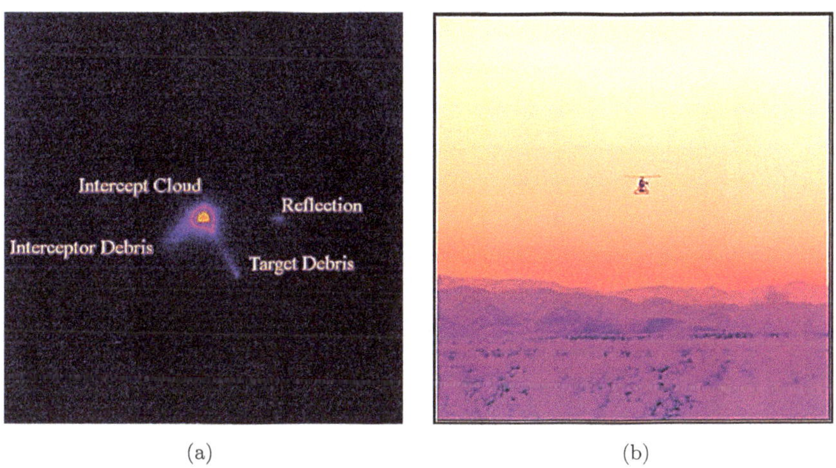

Figure 4.46. Images taken by an LC5 FPA during (a) missile intercept and (b) drone detection.

Figures 4.46 to 4.50 are some of the infrared images taken by the LC5 FPA.

Another wafer material, LC5v2, is very similar to LC5 material. They have the same nominal MQW layer but LC5v2 has a $1\,\mu$m thicker top contact layer and a $1\,\mu$m thinner bottom contact layer so that the MQW layer shifts down by $1\,\mu$m inside the corrugation. The material structure is shown in Fig. 4.51. The QE spectra of

Figure 4.47. A 1024 × 1024 image taken by an LC5 FPA at 55 K, 5 ms, and −2.5 V.

Figure 4.48. An image taken by an LC5 FPA. Adapted with permission from [24].

(a) (b)

Figure 4.49. Images of (a) a building and (b) an engineer taken by an LC5 FPA. Notice that in (b), the person and objects rather than the lighting in the laboratory dominate the reflection from the googles.

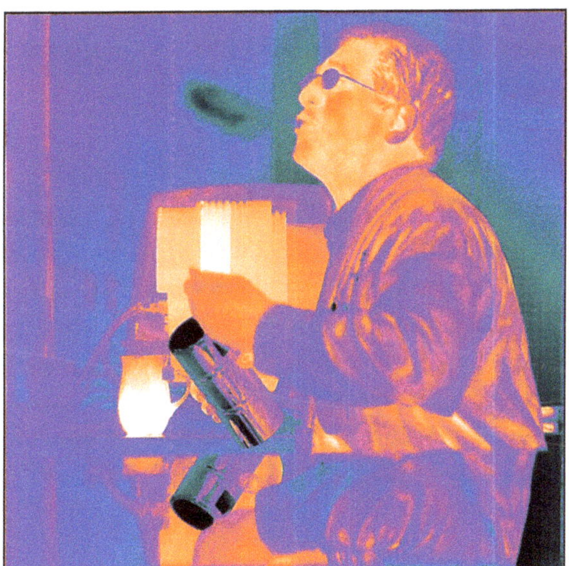

Figure 4.50. An image taken by an LC5 FPA.

the edge-coupled single detector and that of an FPA are measured and compared in Fig. 4.51. The two spectra share near the same spectral lineshape, confirming the wavelength-independence of the C-QWIP coupling. The small wiggles in the FPA spectrum are due

LC5v2 wafer material	Material	Doping (E18 cm^{-3})
Top contact	3.0 μm GaAs	1.0
MQW (63 periods) (4.658 μm)	Well: 7Å GaAs / 35Å $In_{0.1}Ga_{0.9}As$ / 7Å GaAs Barrier: 700Å $Al_{0.19}Ga_{0.81}As$	1.0
Bot. contact	4.8 μm GaAs	1.0
Etch stop	500Å $Al_{0.3}Ga_{0.7}As$	undoped
Buffer	2500Å GaAs	undoped
Substrate	GaAs	undoped

(a)

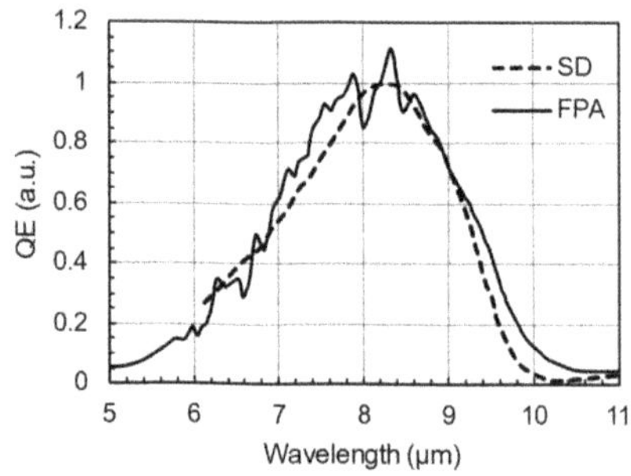

(b)

Figure 4.51. (a) The detector material structure and (b) QE spectral lineshapes measured from LC5v2 edge-coupled detector and FPA.

to optical interference predicted by EM modeling as in Fig. 4.33 when the substrate is removed. The I-V characteristics of the LC5v2 FPA are similar to that of the LC5. Figure 4.52 shows the infrared image of two LC5v2 FPAs operating simultaneously.

4.11.4 *LC6 Focal Plane Array*

The LC6 material has the same QW design as LC5 but with a lower doping of $0.5 \times 10^{18}\,\text{cm}^{-3} = 0.5\,N_0$ in the well and a thicker total t_a thickness ($7.28\,\mu$m vs. $4.66\,\mu$m). The layer structure is shown in Fig. 4.53(a). With the lower doping and the fitted responsivity lineshape in Fig. 4.54(a), the calculated α_{peak} is $0.0392\,\mu\text{m}^{-1}$, half of that of LC5. The pixel pitch of this FPA is $20\,\mu$m instead of $25\,\mu$m. The format of this FPA is 640×512. The MQW layer contains 98 QW periods. With its top contact layer thickness of $0.5\,\mu$m, the etching depth is $7.8\,\mu$m to reach the bottom contact layer. The resulting pixel geometry is shown in Fig. 4.53(b). This detector design thus has a lower intrinsic absorption but a thicker active layer, and it has a smaller sidewall fill factor $S = 0.655$. The 2D EM QE is 9.70% at $8.7\,\mu$m and $\text{IQE}_{\text{EM}} = 0.261\,\mu$m in Fig. 4.54(b). The expected 3D η_{EM} is ~12.9%.

Figure 4.55(a) shows the measured gain g. The ROIC of this FPA can bias up to -8 V. At -6 V, CE = 2.02%. With $g = 0.15$, η is 13.4%, which matches well with η_{EM} of 12.9%. At higher V, η is

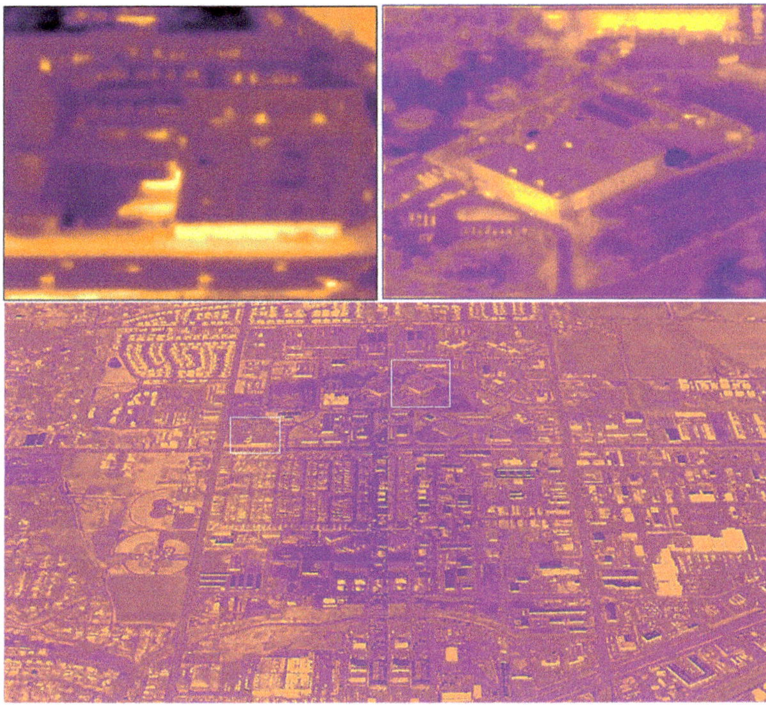

Figure 4.52. An image taken by two 1024 × 1024 LC5v2 FPAs operating simultaneously.

LC6 wafer material	Material	Doping (E18 cm^{-3})
Top contact	0.5 μm GaAs	1.0
MQW (98 periods) (7.280 μm)	Well: 7Å GaAs / 35Å $In_{0.1}Ga_{0.9}As$ / 7Å GaAs Barrier: 700Å $Al_{0.19}Ga_{0.81}As$	0.5
Bot. contact	2.5 μm GaAs	1.0
Etch stop	500Å $Al_{0.3}Ga_{0.7}As$	undoped
Buffer	2500Å GaAs	undoped
Substrate	GaAs	undoped

(a)

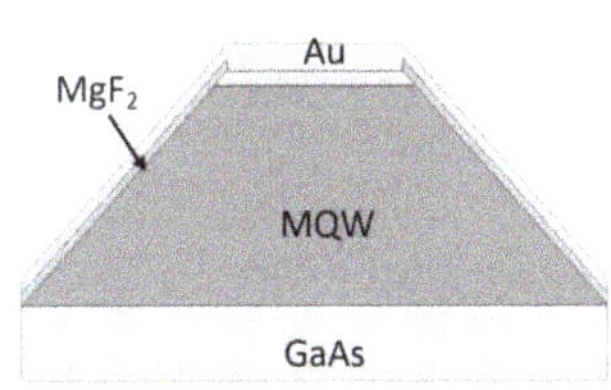

(b)

Figure 4.53. (a) The detector material structure and (b) pixel geometry of the LC6 FPA.

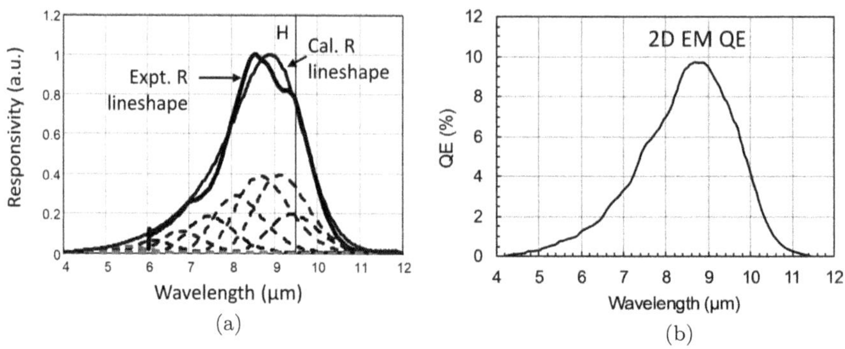

Figure 4.54. (a) The edge-coupled R spectrum and the fitted $\lambda\alpha(\lambda)$ lineshape. Adapted with permission from [24]. (b) The 2D unpolarized internal $\eta_{\rm EM}$ with $\rm IQE_{EM} = 0.261\,\mu m$.

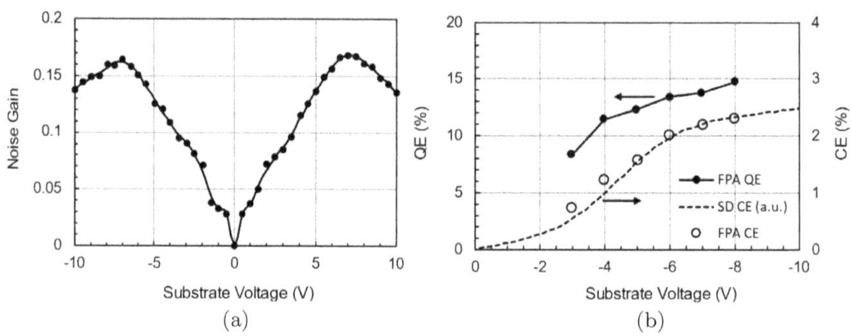

Figure 4.55. (a) g of the LC6 detector. (b) FPA CE (open circles) and $\eta = \mathrm{CE}/g$ (close circles) as a function of V. The dashed curve shows the background photocurrent of an edge-coupled detector in arbitrary units. Adapted with permission from [24].

slightly higher at 15%. However, with the decreasing g, this deduced η is not accurate, and η is expected to be saturated at 13.4%. This example shows that although the corrugation is fully occupied with the active material, the combined effect of having a lower N_D, a smaller corrugation period, and a smaller S yields a smaller η. On the other hand, the LC6 CE of 2.33% and η of 13.4% at $-8\,$V are remarkably close to the LC5 CE of 2.34% and η of 13.9% at $-4\,$V. Therefore, both approaches turn out to be about equal. Figure 4.56 shows an image taken by an LC6 FPA.

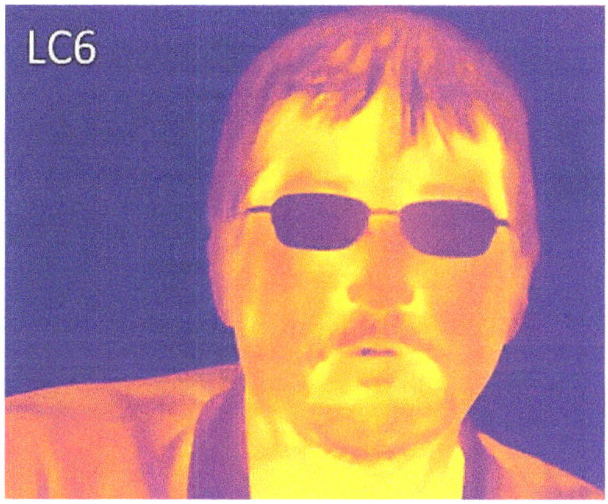

Figure 4.56. An image taken by the LC6 FPA at $T = 68\,\text{K}$, $\tau_{\text{int}} = 1.76\,\text{ms}$, and $V = -7\,\text{V}$.

4.12 Focal Plane Array Characteristics at Different Cutoff Wavelengths

In Section 4.11, the detector modeling is found to be consistent with the experimental finding. It can therefore be used to analyze the C-QWIP performance systematically. In this analysis, the detectors are operated at full bias (~80 mV/period) near QE saturation where $\gamma = 1$. The value of g is taken to be g_1/N_w, where g_1 is the gain for a single QW and N_w is the number of QWs in the detector. G_1 is found experimentally equal to $-54.68x + 26.18$ near full bias, where x is the Al mole fraction of the barrier. The dark current is given by Eq. (2.51) at $V = 80\,\text{mV}$. Four detector designs are considered. The material (x, y) values are (0.19, 0.1), (0.175, 0.1), (0.19, 0), and (0.175, 0), and the barrier thickness is 700 Å. The doping density is set at $1 \times 10^{18}\,\text{cm}^{-3}$. With these detector parameters, the calculated responsivity spectra are shown in Fig. 4.57.

Assuming 25 μm pixel pitch, 90% substrate transmission, and 90% pixel fill factor (*i.e.*, the vertical gap between pixels are 2.5 μm), the calculated classical QE and CE are shown in Fig. 4.58. At $N_w = 60$, the four detectors have $(\lambda_c, \eta, \text{CE}, \Gamma) =$ (8.80 μm, 19.8%, 5.2%, 1.59 μm), (9.42 μm, 19.9%, 5.5%, 1.80 μm), (10.72 μm, 17.8%, 4.7%, 2.68 μm), and (11.65 μm, 18.6%, 5.2%, 3.04 μm), respectively, where Γ is the full width at half maximum

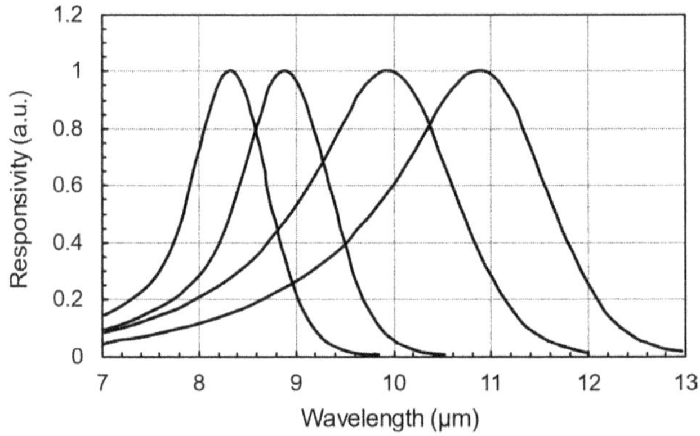

Figure 4.57. The normalized spectral responsivity of four detectors considered in the system analysis. Adapted with permission from [24].

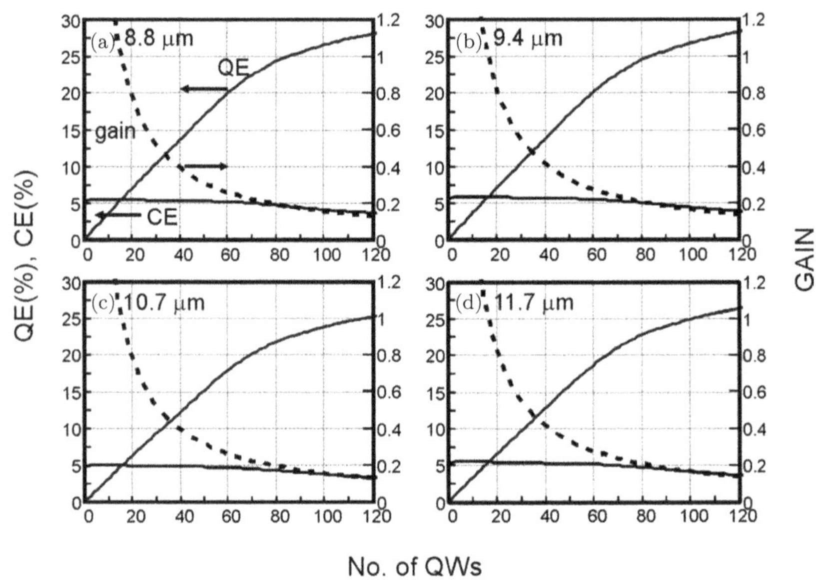

Figure 4.58. Calculated η, fitted g, and CE as a function of N_w for four different detector designs. Reproduced with permission from [24].

of the spectrum. These calculated Γs are generally slightly smaller than experimental values. Figure 4.58 shows η and g do not depend on x or λ_c appreciably. While η increases with N_w, g decreases with it, leading to a relatively constant CE $\approx 5\%$ as a function of N_w.

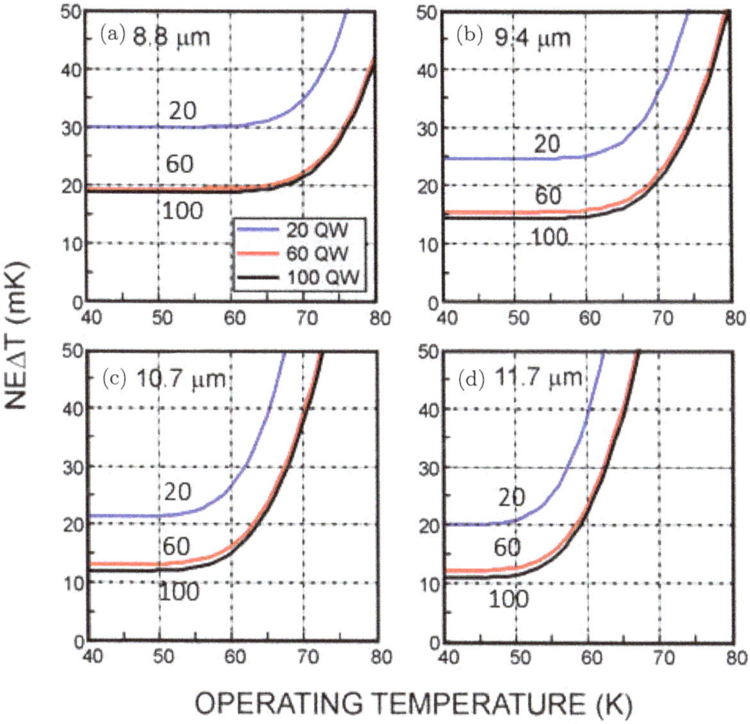

Figure 4.59. Calculated NEΔT of the four designed FPAs as a function of operating temperature assuming $n_r = 900\,e^-$ and $p = 25\,\mu$m. Reproduced with permission from [24].

With the R lineshape in Fig. 4.57 and CE in Fig. 4.58, I_p is calculated from Eq. (4.15) using $F = 2$, cut-on wavelength $= 8\,\mu$m, and $T_B = 300\,$K. After calculating I_p from Eq. (4.15) and I_d from Eq. (2.51), the temporal NEΔT is evaluated using Eq. (4.16) with $\tau_{\text{int}} = 2\,$ms, $n_r = 900$, and $\sigma = 0$. Figure 4.59 shows the respective NEΔT at different T. When $N_w > 60$, all FPAs yield NEΔT below 20 mK between 60 and 70 K, and there are no advantages in using more QW periods. With a finite n_r, the shorter cutoff FPAs have an advantage over the longer cutoff FPAs at high T. To achieve the best sensitivity, the longer cutoff FPAs must operate at a lower T. Since these higher doping detectors are optimized for shorter integration times, their operating T also tends to be lower.

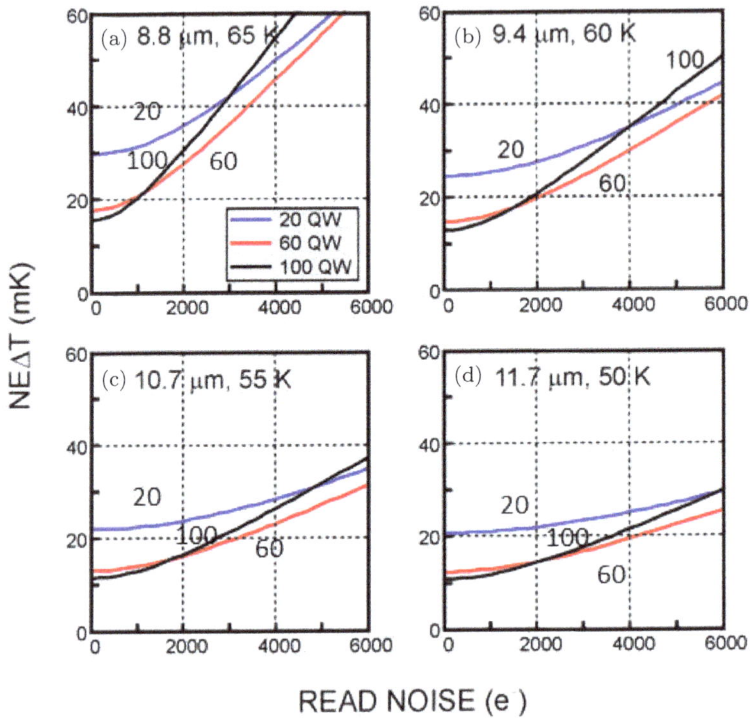

Figure 4.60. Projected NEΔT as a function of n_r for the four FPA designs. Reproduced with permission from [24].

To know the effects of read noise on NEΔT, Figure 4.60 shows that the longer cutoff FPAs are more immune to readout and system noises. Figure 4.60 also shows that when n_r is large, a 100-QW detector will not be as sensitive as a 20-QW detector. When $n_r \gg 2gN$ in Eq. (4.16), the noise is dominated by the readout/system noise. In this case, NEΔT does not depend on the detector noise. The larger CE of the 20-QW detector will then be more sensitive. On the other hand, when n_r is small, the detector signal-to-noise ratio is important, and the lower noise of the 100-QW detector will give a better sensitivity. On balance, a 60-QW detector is suitable for most conditions.

Figure 4.61 shows the number of collected electrons N for $\tau_{\text{int}} = 2\,\text{ms}$. They are between 4 and 20 Me^- at low T, which set the charge

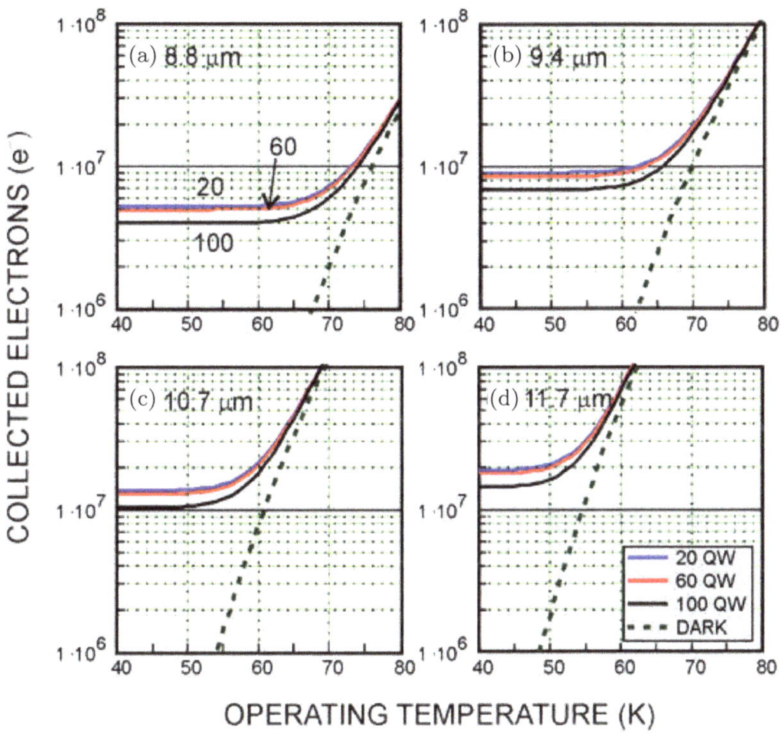

Figure 4.61. Number of collected electrons of the four designed FPAs at different T. Reproduced with permission from [24].

well capacity requirements on the ROICs. From Eq. (4.16), NEΔT is proportional to $1/\sqrt{N}$. When the pixel pitch is smaller, N will be smaller and NEΔT will be larger. When the pixel pitch is 20 μm, NEΔT will be ~5 mK higher than that of the 25 μm pitch as shown in Fig. 4.62.

4.13 C-QWIP Focal Plane Array Spectral Lineshape

In Section 4.12, we assumed all C-QWIP FPAs are with ARC so that there is no substrate reflection. The modeled QE is the internal QE with EM waves launching inside the substrate. Therefore, FP oscillations are not predicted in the modeled spectra. If the FPAs are not coated with an ARC, FP oscillations will be present when the

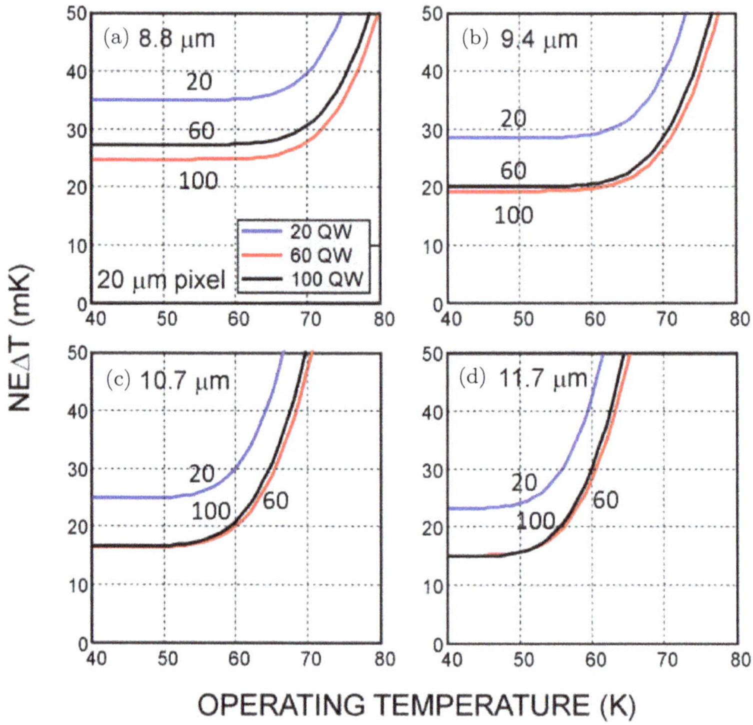

Figure 4.62. Calculated NEΔT of $20\,\mu$m pitch FPAs with the same system parameters. Reproduced with permission from [24].

substrate is either completely or partially removed. In this section, we examine the QE spectrum when the FPAs are uncoated.

4.13.1 *NVL5 Prism-Shaped C-QWIP Focal Plane Array*

The first wafer material is labeled as NVL5. Its layer structure and α spectrum based on edge coupling are shown in Fig. 4.63. It is peaked at $\lambda = 8.7\,\mu$m with a value of $0.145\,\mu\text{m}^{-1}$. The C-QWIP shown in Fig. 4.64 is etched to a depth of ~$11.5\,\mu$m, and the substrate is totally removed. With the designed bottom layer thickness of $4.92\,\mu$m in the material, the exposed bottom contact layer thickness t_b is $4\,\mu$m and the ground contact layer thickness t_g is ~$0.9\,\mu$m. The MgF_2

NVL5 wafer material	Material	Doping (E18 cm^{-3})
Top contact	3 μm GaAs	1.0
MQW (60 periods) (4.50 μm)	Well: 19Å GaAs /24Å GaAs/5Å GaAs Barrier: 700Å $Al_{0.172}Ga_{0.828}As$	2.0 in 24Å GaAs
Bot. contact	4.92 μm GaAs	1.0
Etch stop	1000Å $Al_{0.3}Ga_{0.7}As$	undoped
Buffer	2500Å GaAs	undoped
Substrate	GaAs	undoped

(a)

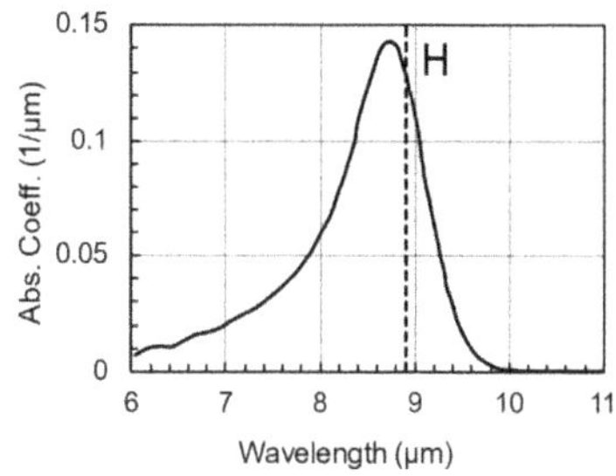

(b)

Figure 4.63. (a) The detector material structure and (b) calculated α of NVL5 PR-QWIP FPA.

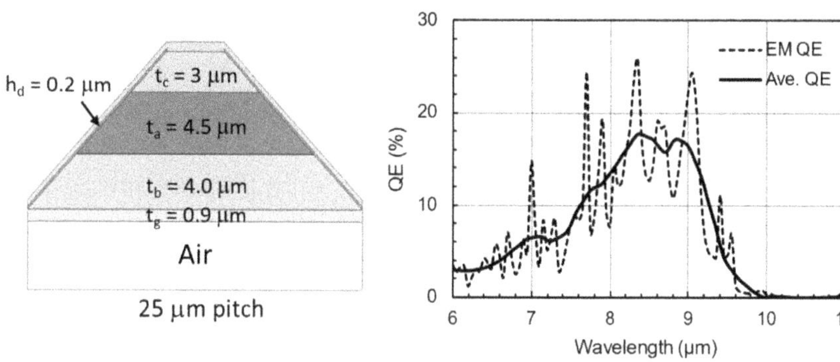

(a) (b)

Figure 4.64. (a) The NVL5 C-QWIP geometry and (b) modeled external unpolarized QE spectra in 2D with $\mathrm{IQE_{EM}} = 0.353\,\mu$m.

thickness h_d is $0.2\,\mu$m. With this α and geometry, the modeled η_{EM} in 2D is shown as the dashed curve in Fig. 4.64. With partially filled corrugation and absence of ARC, strong QE oscillations are present.

The $g(V)$ curve is measured in Fig. 4.65(a). At $V = -2.69$ V, g is 0.092. The detector CE $= \eta_d g = \eta_{\mathrm{abs}}\gamma g$ is measured at the same V with different discrete wavelengths. The converted $\eta_d = \mathrm{CE}/g$ is shown in Fig. 4.65(b). For this array, the observed FP oscillations are substantially weaker and less dense than the modeled η_{EM} in

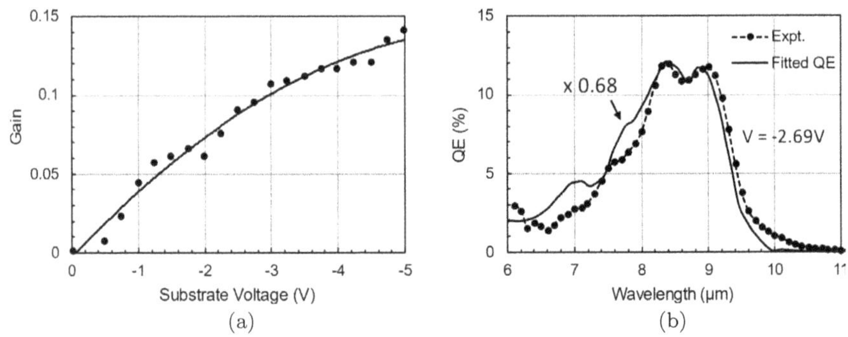

Figure 4.65. The measured (a) noise gain and (b) $\eta_d \equiv \eta_{\text{abs}}\gamma$ (circles). The fitted solid curve is $\eta_{\text{EM}} \times 0.68$.

Fig. 4.64, showing a certain degree of spectral averaging. Indeed, by averaging the oscillations with numerical localized polynomial regression, the oscillations of the average η_{EM} can be fitted to the observed spectrum in Fig. 4.65(b) by adjusting smoothing parameters. The weaker FP oscillations can be due to the finite spectral bandwidth of the measurement and possible fabrication nonuniformity.

At -2.69 V, the deduced η_d is 12.0% at the peak wavelength, which is 0.68 times smaller than the modeled average η_{EM} in Fig. 4.64(b). The smaller η can be explained by γ being less than unity at this V. When the detector is biased to a higher voltage such as -5 V, η is expected to be larger.

4.13.2 *NVL5 Pyramid-Shaped C-QWIP Focal Plane Array*

Another NVL5 wafer is processed into PY-QWIPs. In this array processing, the top contact surface of the wafer is first glued by epoxy to a double-side AR coated silicon wafer. The detector GaAs substrate is then completely removed. After the C-QWIP pixels are fabricated on the detector wafer, the array is hybridized to an ROIC, and the light is incident from the Si substrate side. This C-QWIP array processing does not need the usual MgF_2/Au cover nor the epoxy backfill on top of the corrugations. Therefore, the

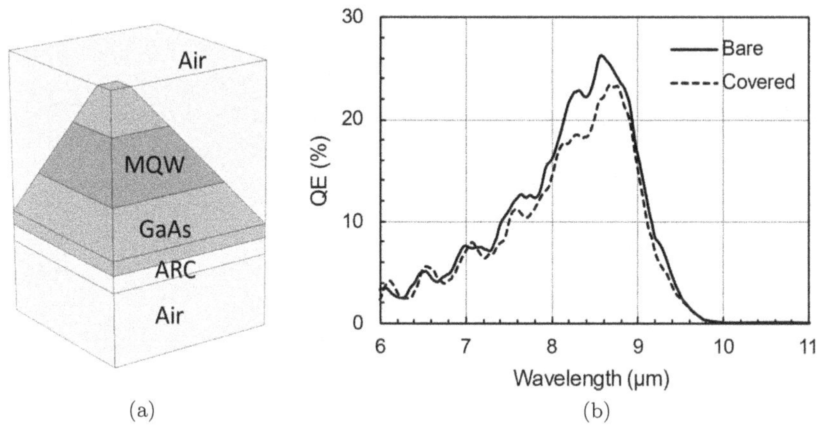

Figure 4.66. (a) The PY-QWIP geometry and the 3D η_{EM} with and without a MgF_2/Au cover. No spectral averaging is applied. (b) The bare $\mathrm{IQE}_{\mathrm{EM}} = 0.378\,\mu$m and the covered $\mathrm{IQE}_{\mathrm{EM}} = 0.347\,\mu$m.

detector pixels contain the bare sidewall surfaces originally envisioned. This is possible because of the mechanical support from the Si wafer.

A quarter of the modeled geometry is shown in Fig. 4.66, in which the coated Si substrate is replaced by a single ARC layer. The modeled 3D η_{EM} with and without the MgF_2/Au cover is shown in Fig. 4.66. The bare sidewall produces a larger QE as expected from Fig. 4.2. With the equivalent ARC, the FP oscillations are much reduced even without spectral averaging.

The measured η_d spectrum is shown in Fig. 4.67(a), and its peak η_d is deduced to be 24% at η_d saturation, which is only slightly less than the modeled bare η_{EM} in Fig. 4.66 by a factor of 0.95, indicating that γ is close to unity.

With ARC, the η_{EM} lineshape reproduces the measured FP oscillations without spectral averaging. The experimental dip at $8.0\,\mu$m in Fig. 4.67(a) can be attributed to the known strong absorption line of the epoxy at this wavelength that is used to bond the FPA to the Si substrate. Figure 4.67(b) shows a 1024×1024 infrared image taken by this PY-QWIP FPA.

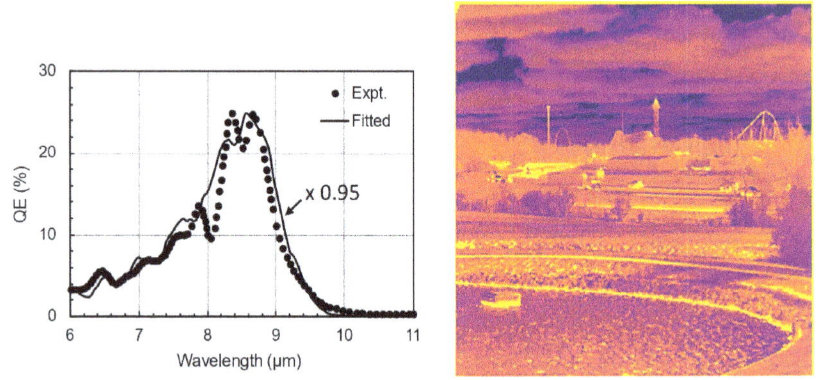

(a) (b)

Figure 4.67. (a) The η_d of a PY-QWIP FPA with bare sidewalls and (b) an infrared image. The experimental minimum at $8\,\mu$m is attributed to epoxy absorption.

4.13.3 *TIRS4 Focal Plane Array with Partially Thinned Substrates*

When the substrate of a C-QWIP FPA is not completely removed but only mechanically thinned to certain values, the FP oscillations may be too dense to be resolved. The observed QE is then averaged over many cycles of oscillations. In this example, the C-QWIP FPAs are mechanically thinned to around $30-50\,\mu$m, and their CE spectra are measured at discrete wavelengths. The wafer material is labeled as TIRS4, and its layer structure is listed in Fig. 4.68(a). It is designed for $p = 25\,\mu$m pitch arrays.

In order to demonstrate a larger format array, which contains 1920×2048 pixels but the ROIC has $20\,\mu$m pixel pitch, a second version of the wafer material, TIRS4v2, is grown [26]. It has a similar MQW structure but has thinner top and bottom contact layers shown in Fig. 4.68(b) so that smaller corrugation FPAs can be fabricated. However, for detector characterization for both wafers, only $25\,\mu$m pitch arrays are used. The test detector geometries are shown in Fig. 4.69(a). With thinner detector layers, the TIRSv2 C-QWIP has a smaller sidewall fill factor S. Based on the edge coupling responsivity spectra, the calculated α spectra for the two

TIRS4 wafer material	Material	Doping (E18 cm^{-3})
Top contact	3.0 μm GaAs	1.0
MQW (60 periods) (4.571 μm)	Well: 60Å GaAs Barrier: 700Å $Al_{0.17}Ga_{0.83}As$	0.6
Bot. contact	5.0 μm GaAs	1.0
Etch stop	1000Å $Al_{0.3}Ga_{0.7}As$	undoped
Buffer	2500Å GaAs	undoped
Substrate	GaAs	undoped

(a)

TIRS4v2 wafer material	Material	Doping (E18 cm^{-3})
Top contact	1.5 μm GaAs	1.0
MQW (60 periods) (4.541 μm)	Well: 55Å GaAs Barrier: 700Å $Al_{0.169}Ga_{0.831}As$	0.62
Bot. contact	3.5 μm GaAs	1.0
Etch stop	1000Å $Al_{0.3}Ga_{0.7}As$	undoped
Buffer	2500Å GaAs	undoped
Substrate	GaAs	undoped

(b)

Figure 4.68. The material structures of (a) TIRS4 and (b) TIRS4v2. TIRS4v2 has thinner top and bottom contact layers and thinner wells.

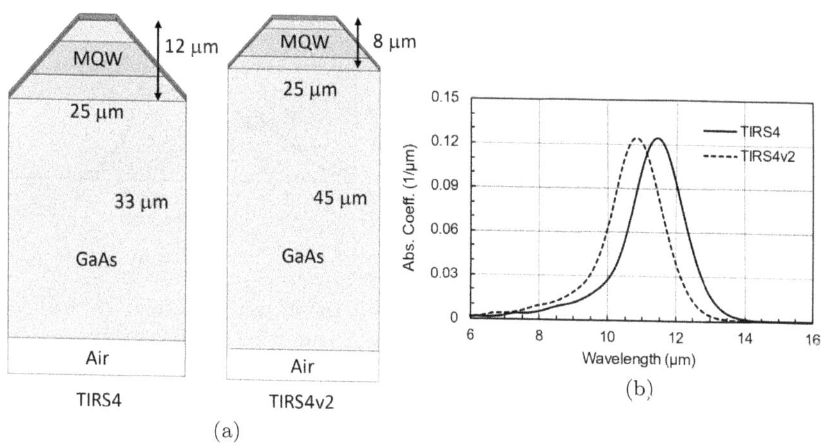

Figure 4.69. (a) The pixel geometries and (b) calculated α of TIRS4 and TIRS4v2 FPAs.

materials are shown in Fig. 4.69(b). With large substrate thicknesses, the modeled QE spectra show dense FP oscillations in Fig. 4.70. Upon spectral averaging, smoother spectra are expected for both detectors.

The measured CE spectra for both 25 μm pitch arrays are shown in Fig. 4.71. The gain curves for these two arrays have not been

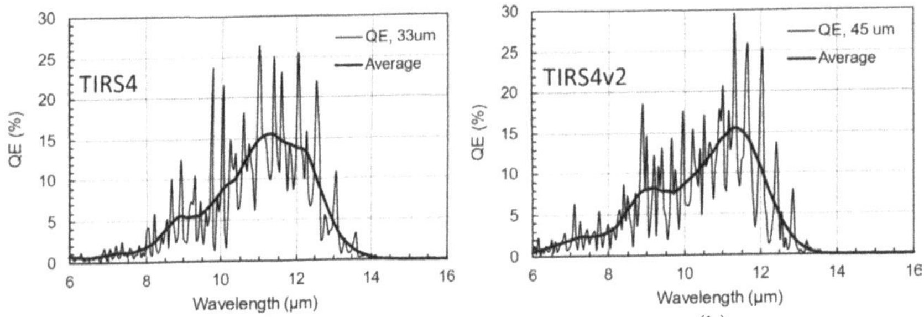

(a) (b)

Figure 4.70. The modeled η_{EM} in 2D (thinner curves) and the averaged values (thick curves) for (a) TIRS4v1 and (b) TIRS4v2 25 μm pitch FPAs. $\mathrm{IQE}_{\mathrm{EM}} = 0.503$ for v1 and 0.485 for v2.

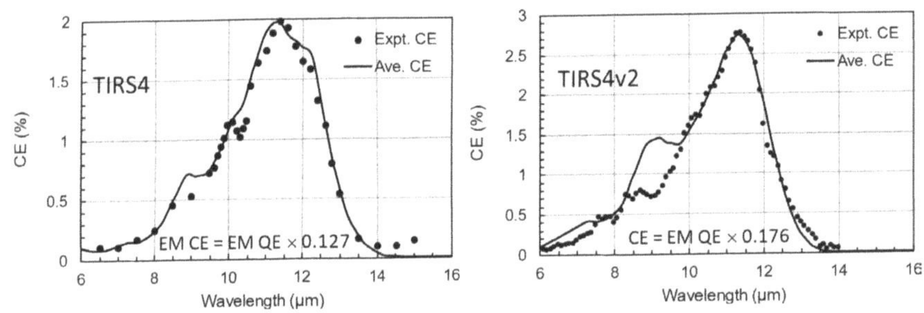

(a) (b)

Figure 4.71. The measured CE (circles) and the fitted CE (solid curves) by scaling η_{EM} with a constant for (a) TIRS4 FPA and (b) TIRS4v2 FPA.

measured. To fit the CE curves, the η_{EM} curve in Fig. 4.70(a) is multiplied by 0.127 and that in Fig. 4.70(b) is multiplied by 0.176. These multiplication factors are typical $g\gamma$ values for 60 QWs. Using $g\gamma$ as the scaling factor, the modeled average CE spectra follow the measured spectra rather closely, thus explaining the small wiggles in the measurement. These arrays are operated at positive substrate voltage, where the dark current is lower. The I-V characteristics of the 25 μm pitch TIRS4v2 C-QWIP pixels are shown in Fig. 4.72(a). The array is background limited (*i.e.*, $I_p = I_d$) at about 46 K. The infrared image of a wire cutting tool taken by the 20 μm pitch TIRS4v2 1920 × 2048 FPA is shown in Fig. 4.72(b).

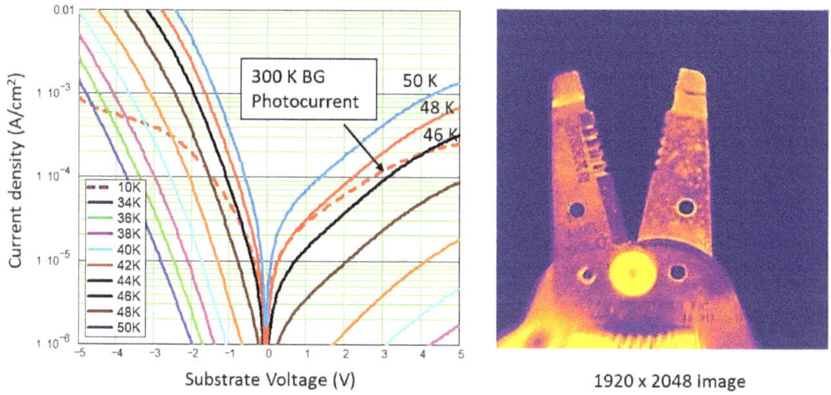

Figure 4.72. (a) The I-V characteristics of TIRS4v2 C-QWIP pixels and (b) the infrared image taken by the FPA, which is in 1920×2048, $20\,\mu$m pitch format.

TIRS5 wafer material	Material	Doping (E18 cm^{-3})
Top contact	3.67 μm GaAs	1.0
MQW (60 periods) (4.571 μm)	Well: 19Å GaAs / 36Å GaAs / 5Å GaAs Barrier: 700Å $Al_{0.172}Ga_{0.828}As$	1.0 in 36Å GaAs
Bot. contact	4.15 μm GaAs	1.0
Etch stop	1000Å $Al_{0.3}Ga_{0.7}As$	undoped
Buffer	2500Å GaAs	undoped
Substrate	GaAs	undoped

(a)

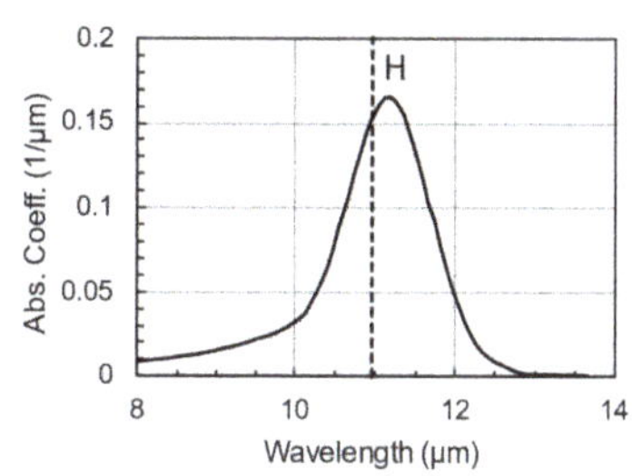

(b)

Figure 4.73. (a) The detector material structure and (b) calculated α of TIRS5 PR-QWIP FPA.

4.13.4 *TIRS5 Focal Plane Array*

TIRS5 wafer material has undergone detailed study [27] so that the C-QWIP performance in the very long wavelength infrared regime can be thoroughly understood. The material layer structure and the corresponding absorption coefficient calculated from Eq. (2.9) are shown in Fig. 4.73.

Although the C-QWIP QE can be accurately modeled by FEM, the observed lineshape on average can be adequately described by the classical model. The partially filled C-QWIPs remains wavelength-independent on average. Therefore, one can simplify the analysis using a corrugation fill factor $\kappa(t_a)$ to account for the effects of a partially filled corrugation. For $S = 0.8$, the total height is $11.11\,\mu$m for $p = 25\,\mu$m. Accordingly, $\kappa(t_a)$ is the fraction of the maximum value when t_a is less than $11\,\mu$m when MQW layer is placed at the center of the corrugation. The function $\kappa(t_a)$ obtained from EM modeling is plotted in Fig. 4.74. Using this κ parameter, the detection QE of a partially filled C-QWIP is given by

$$\eta_c(V,\lambda) = T_{\text{sub}} f_p \kappa \frac{1}{p}\left[t + \frac{e^{-\alpha p}}{2\alpha}\left(1 - e^{2\alpha t}\right)\right]\gamma(V) \tag{4.19}$$

where f_p is the areal fill factor in the presence of a vertically etched gap between two pixels. For the present $t_a = 4.57\,\mu$m, κ is 0.78 from Fig. 4.74.

To know the spectral CE $= \eta_c g$, the photocurrent noise gain g is measured and is shown in Fig. 4.75(a). Assuming $f_p = 0.8$, $g = 0.24$, $\kappa = 0.78$, Eq. (4.19) matches the experimental CE at 5 V using $\gamma = 0.90$ in Fig. 4.75(b). Therefore, η_c with a fitting parameter γ can be used to describe the averaged CE spectrum of a partially thinned C-QWIP. Strictly speaking, γ also contains corrections due

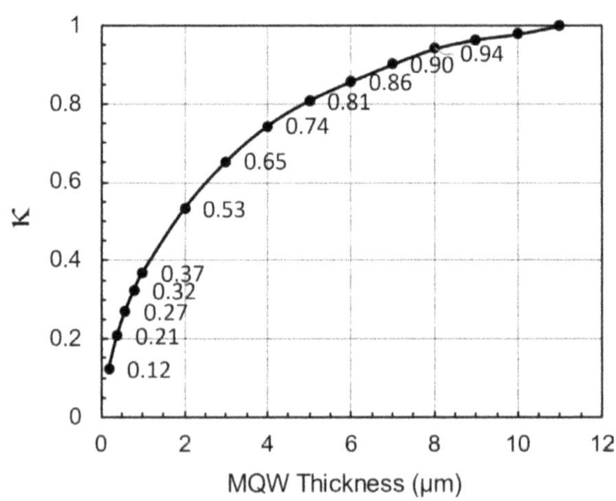

Figure 4.74. The fraction of the maximum QE as a function of MQW thickness placed at the center of the $p = 25\,\mu$m and $S = 0.8$ corrugation.

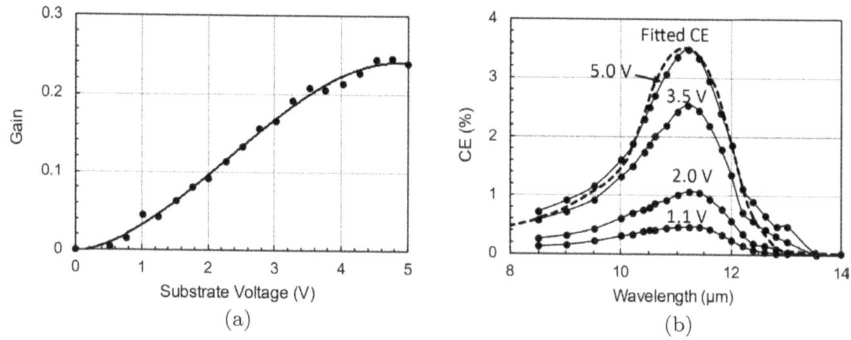

Figure 4.75. (a) The noise gain and (b) conversion efficiency of the TIRS5 FPA. The classical CE model is used to fit the measured spectrum at 5 V. $\mathrm{IQE_{EM}} = 0.477$. Adapted with permission from *Inf. Phys. & Tech.*, 54, pp. 170–176 (2011).

to the presence of vertical sidewalls and metal covers and the use of g instead of the g_{peak} in evaluating CE. With this γ value, the fitted $\eta_c(5\,\text{V}) = 14.7\%$ at λ_p.

In addition to the optical properties, the dark current J_d of this material is also modeled using Eq. (2.49). For convenience, it is rewritten here for discussion:

$$J_d(V,T) = \int_{E_1}^{\infty} \frac{m^*}{\pi\hbar^2 L} \frac{\gamma(E,V)f(E,T)}{\sqrt{1+\left(\frac{\mu V}{L v_{\text{sat}}}\right)^2}} \frac{\mu V}{L} dE, \tag{4.20}$$

where $\gamma(E, V)$ is the tunneling probability in WKB approximation, and $f(E, T)$ is the Fermi-Dirac distribution. At very low T, when $f(E, T)$ is essentially 1 for $E < E_F$ and 0 when $E > E_F$, Eq. (4.20) describes the direct tunneling process. When T increases, the finite $f(E, T)$ above E_F creates the thermally assisted tunneling current. At very high T, more electrons are excited above the barrier with which $\gamma = 1$ and give rise to thermal emission (TE) current. To visualize their relative contributions at different T, the normalized dark current density $J_d(E, V, T)$ inside the integral of Eq. (4.20) is plotted in Fig. 4.76(a) at $V = 83\,\text{mV/period}$ and different T.

Figure 4.76(a) shows that the carrier distribution shifts up in energy rapidly when T increases. At any T except 10 K, the dark carriers mainly come from the thermally activated tunneling process. Only at very high T, the TE current becomes appreciable.

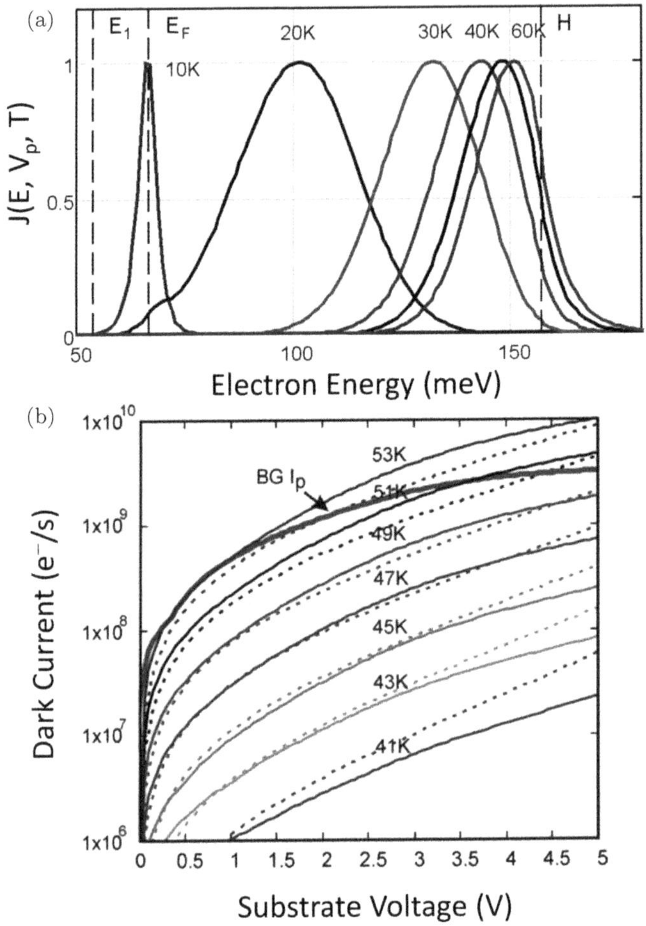

Figure 4.76. (a) The normalized dark current energy distribution at 83 mV/period. (b) The measured dark current density (solid curves) and the fitted dashed curves using Eq. (4.20). The thick solid curve shows the measured background photocurrent. Adapted with permission from *Inf. Phys. & Tech.*, 54, pp. 170–176 (2011).

Figure 4.76(b) shows the measured and fitted dark currents. The fitting parameters used in Eq. (4.20) are $H = 157.1\,\text{meV}$ in the WKB expression, $\mu = 400\,\text{cm}^2/\text{Vs}$ and $v_{\text{sat}} = 1 \times 10^7\,\text{cm/s}$. This value of H is close to the 163.5 meV deduced from the spectral response measurement, and v_{sat} is the same as the known saturation velocity for GaAs. Besides the present TIRS5 material, the fitted values of μ and v_{sat} are found to be the same for other detectors having λ_c ranging from 9 to 13 μm, showing the validity of the dark current model. At shorter wavelengths, such as $\lambda_c = 8.6\,\mu$m, one does need to use a smaller μ of 120 cm^2/Vs for fitting. Figure 4.76(b) also shows the photocurrent J_p generated by a blackbody source with an 8 μm longpass filter under $f/2$ optics. At 5 V, J_p is observed to be 3.24×10^9 e/s/pixel, which is 70% of the calculated 4.62×10^9 e/s/pixel from the spectral measurement. The smaller observed J_p is due to the finite transmission loss of the KRS-5 window in the optical dewar.

Using $J_p(5\text{V}) = 4.62 \times 10^9$ e/s/pixel and $J_d(5\text{V})$ shown in Fig. 4.76(b), the NEΔT of this TIRS5 FPA can be estimated from Eq. (4.16). The input parameters are: $C = 0.0149$, $g = 0.24$, and $n_r = 260$, which is a measured value. To consider only the temporal NEΔT, σ is set to be 0. Figure 4.77 shows the calculated NEΔT at different T and τ_{int}. Using 2 ms integration time, and at $T = 43$ and 51 K, NEΔT is 16 and 22 mK, respectively. Experimentally, operating

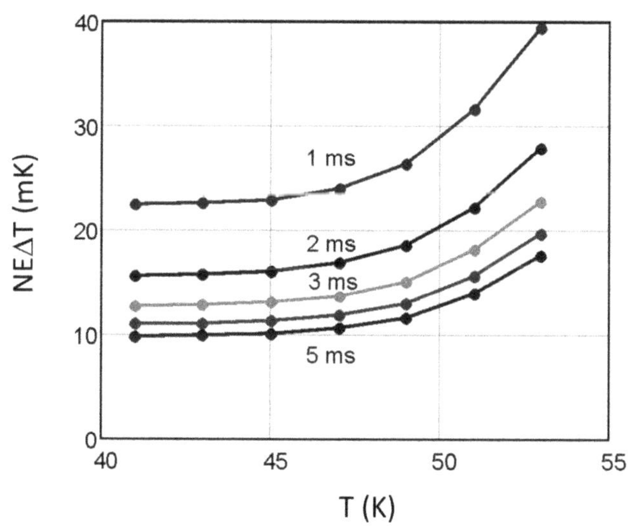

Figure 4.77. The calculated NEΔT of the TIRS5 FPA operated under $F/2$ and 5 V. Adapted with permission from *Inf. Phys. & Tech.*, 54, pp. 170–176 (2011).

the array at 43 K but at a lower voltage of 3.5 V and a longer τ_{int} of 5.5 ms, NEΔT is found to be 17 mK [27], which is in good agreement with the calculated value. We also note that the fabricated FPAs typically have very high operability, being in the range of 99.96%.

4.13.5 *TIRS6 Focal Plane Array*

TIR6 FPAs are processed using a combination of mechanical lapping and selective dry etching to completely remove the substrate. It is used to eliminate the dense FP oscillations in the detection spectrum. However, without an ARC, FP oscillations are still present but with a longer oscillation period. The material layers and detector geometry with 25 μm pixel pitch are shown in Fig. 4.78. The nominal layer structure has been grown twice and the exhibited α spectra are shown in Fig. 4.79(a) as v1 and v2. Their gain curves are measured and shown in Fig. 4.79(b).

Without a substrate, the FP oscillations are less dense. The modeled and averaged η_{EM} in 2D are shown in Fig. 4.80. From the measured CE spectra and g at $V = 2.1$ V for v1 and 2.75 V for v2, the respective $\eta = \text{CE}/g\gamma$ can be deduced assuming $\gamma = 1$. The deduced η and the average η_{EM} are compared in Fig. 4.81.

TIRS6 wafer material	Material	Doping (E18 cm^{-3})
Top contact	3 μm GaAs	1.1
MQW (60 periods) (4.54 μm)	Well: 19Å GaAs / 31Å GaAs / 5Å GaAs Barrier: 700Å $Al_{0.166}Ga_{0.834}As$	1.1 in 31Å GaAs
Bot. contact	5.0 μm GaAs	1.1
Etch stop	1000Å $Al_{0.3}Ga_{0.7}As$	undoped
Buffer	2500Å GaAs	undoped
Substrate	GaAs	undoped

(a)

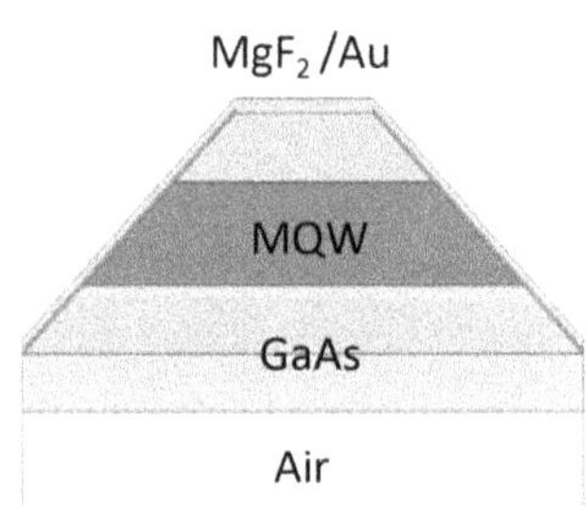

(b)

Figure 4.78. (a) The TIRS6 material and (b) detector geometry.

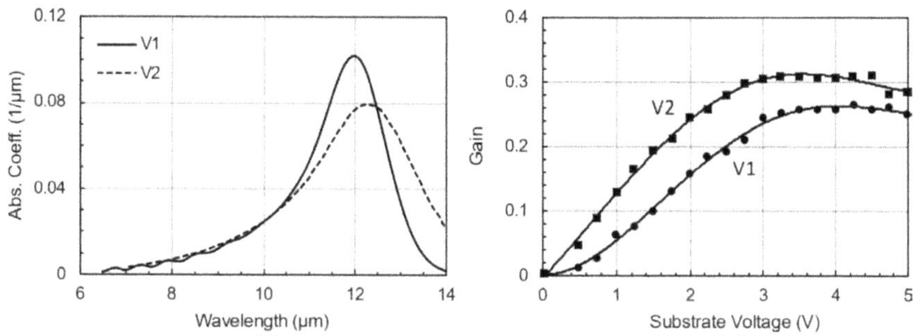

(a) (b)

Figure 4.79. (a) The deduced α spectra from edge coupling and (b) the measured noise gain of TIRS6v1 and TIRS6v2 materials.

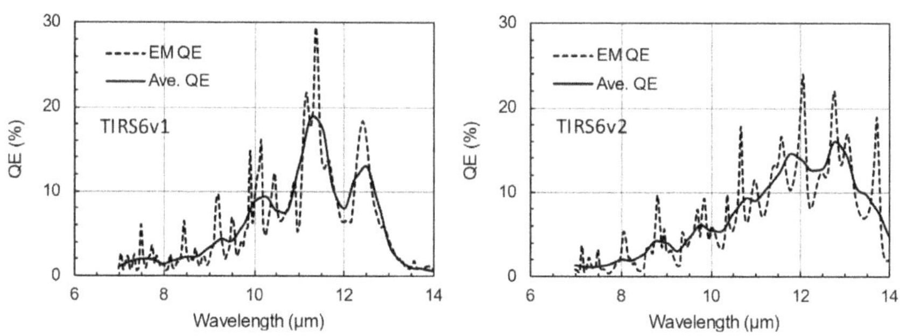

(a) (b)

Figure 4.80. The modeled 2D η_{EM} and the smoothed spectrum for (a) TIRS6v1 FPA and (b) TIRS6v2 FPA. The modeled $\mathrm{IQE}_{\mathrm{EM}}$ for TIR6v1 is 0.637.

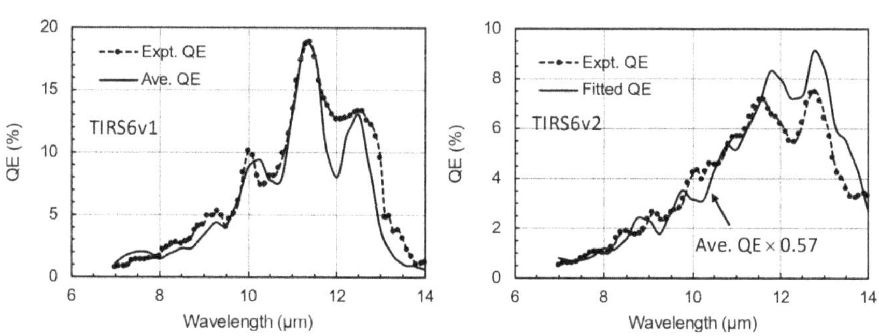

(a) (b)

Figure 4.81. The modeled average η_{EM} (solid curve) and the measured η_d (dashed curve) for (a) TIRS6v1 FPA and (b) TIRS6v2 FPA.

For TIRS6v1, the theory and experiment agree well in both magnitude and lineshape. On the other hand, γ needs to reduce from 1 to 0.57 to fit the data for TIRS6v2. Since the cutoff wavelength of v2 is longer than v1 for the same nominal MQW structure, the actual well width of v2 is probably wider than the nominal value that leads to smaller energy level spacing and the excited states being more confined in the QWs. The lower γ for v2 is thus consistent with the smaller tunneling probability of the photoelectrons, especially for those near the cutoff wavelength.

4.14 Two-Color C-QWIPs

4.14.1 *Two-Color C-QWIPs in Stacking Approach*

QWIPs are narrow band detectors. If two or more QWIP materials are stacked together, they form narrow band multi-color detectors. In order to read out the optical signal simultaneously and separately, several external contacts are needed for each detector. Based on this approach, two-color QWIP cameras have been demonstrated and are commercially available [28, 29]. The typical NEΔT is around 30 to 40 mK and the integration time is 5–10 ms. These FPAs are based on grating coupling, and they will be discussed in Chapter 5. In this section, we will describe using C-QWIPs for two-color detection [30]. One of the applications in two-color imaging is to extract the absolute temperature of an object. Since the thermal radiation of an object depends on the absolute temperature T as well as the emissivity $\varepsilon(\lambda)$, it is difficult to deduce T from the integrated intensity. If, however, $\varepsilon(\lambda)$ happens to be similar in two narrow bands as seen in Fig. 1.3, it is possible to deduce the approximate T from the relative signals from these bands. By taking the ratio P_r of the photocurrent generated in each color, say MW and LW,

$$P_r(T) = \frac{I_p(MW)}{I_p(LW)} = \frac{\int_{MW} L(\lambda, T) R_{MW}(\lambda) d\lambda}{\int_{LW} L(\lambda, T) R_{LW}(\lambda) d\lambda}, \tag{4.21}$$

where $L(\lambda, T)$ is the power spectral radiance in $\mathrm{W/cm^2/\mu m/sr}$, and R is the responsivity, the ε dependence and other geometrical factors can be factored out. The resulting $P_r(T)$ is a montonically increasing function of T, and hence T can be uniquely determined from P_r.

To illustrate this approach, a C-QWIP of geometry in Fig. 4.22 is used. In this case, the material contains two stacks of QWs and there is an extra middle contact. The lower stack is for mid-wavelength detection and the upper stack is for long wavelength detection. The detection peak λ_p is designed to be around $4.5\,\mu$m and $8.8\,\mu$m, respectively. The two stacks are separated by a 5000 Å GaAs layer, which is used for middle contact. The detector area is $1200\,\mu\text{m} \times 700\,\mu$m and the substrate is $150\,\mu$m thick. The corrugation period is $7\,\mu$m. The cross-section of a single period is shown in Fig. 4.82.

The fabricated wafer material also contains edge-coupling detectors. Their responsivity lineshapes at 3 V for the MW and 2 V for the LW are shown in Fig. 4.83. By fitting $\lambda\alpha(\lambda)$ to the R lineshapes, $\alpha(\lambda)$ for both bands can be obtained and they are shown in Fig. 4.84(a). With the fitted absorption bandwidth and the nominal doping density, the peak α is $0.39\,\mu\text{m}^{-1}$ at $4.8\,\mu$m and it is $0.086\,\mu\text{m}^{-1}$ at $8.7\,\mu$m.

Since the detector substrate is unthinned, the external unpolarized η_{abs} of the C-QWIP is modeled by first modeling the 2D geometry in Fig. 4.82 for the internal QE and then multiplying it

2-Color QWIP wafer material	Material	Doping (E18 cm^{-3})
Top contact	0.5 μm GaAs	1.0
LW MQW (23 periods) (1.265 μm)	Well: 50Å GaAs Barrier: 500Å $Al_{0.28}Ga_{0.82}As$	0.5
Middle contact	0.5 μm GaAs	1.0
MW MQW (23 periods) (0.782 μm)	Well: 5Å GaAs/30Å GaAs/5Å GaAs Barrier: 300Å $Al_{0.4}Ga_{0.6}As$	2.0
Bot. contact	0.5 μm GaAs	1.0
Buffer	2500Å GaAs	undoped
Substrate	GaAs	undoped

(a)

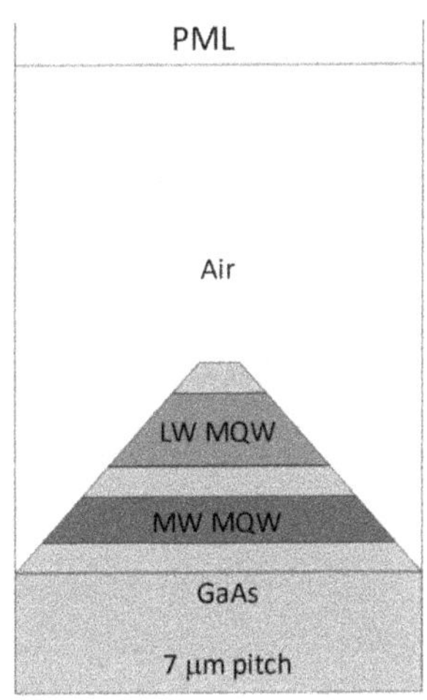

(b)

Figure 4.82. (a) The material structure of a two-color C-QWIP in stacked approach and (b) the adopted C-QWIP geometry.

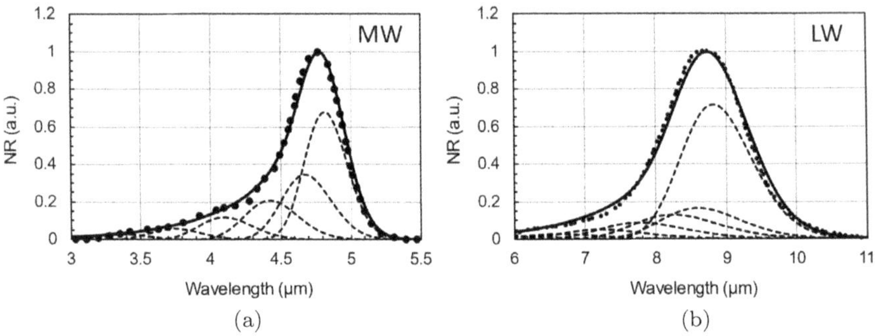

Figure 4.83. The measured responsivity spectrum (circles) and the fitted $\lambda\alpha(\lambda)$ (solid curve) spectrum for (a) the MW stack and (b) the LW stack using edge coupling. The dashed curves are individual contributions from different energy levels.

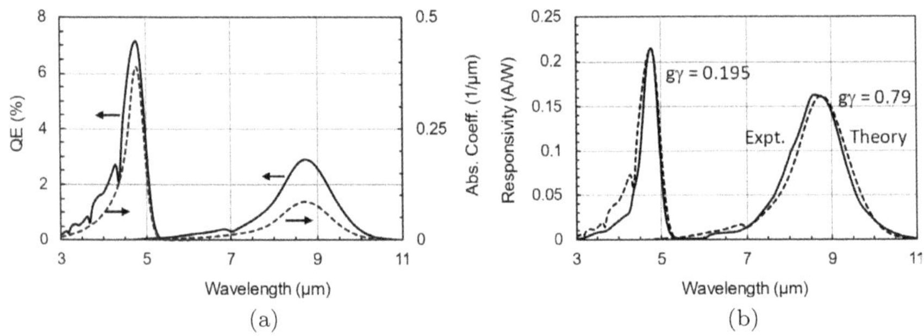

Figure 4.84. (a) The modeled α and external QE of the two-stack $p = 7\,\mu$m C-QWIP. (b) The measured and fitted responsivities of each stack using $g\gamma$ as a fitting parameter.

by T_{sub} of 0.721. The peak η is 7.1% in MW and 2.9% in LW. These small values are due to the adopted small corrugations of $p = 7\,\mu$m, the small individual corrugation fill factors κ, and the low doping of 0.5×10^{18} cm^{-3} in the LW stack. The measured responsivity $R(= \eta_{\text{abs}} e g\gamma\lambda/hc)$ is shown in Fig. 4.84(b). Based on the modeled η_{abs}, the experimental R can be fitted by using $g\gamma = 0.195$ at 3 V for the MW and 0.79 for the LW at 2 V. These values of $g\gamma$ are expected for $N = 23$ QW periods in each stack. The discrepancy of the spectrum towards $\lambda = 3\,\mu$m can be explained by the decreasing $g\gamma$ at the shorter λ as observed in Fig. 4.27.

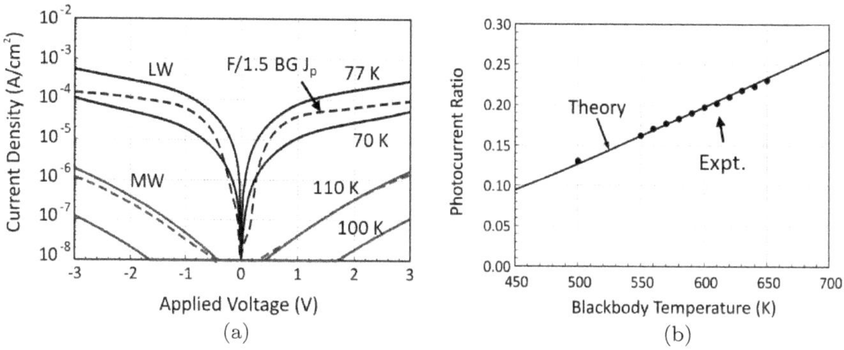

Figure 4.85. (a) The dark current (solid curves) and the 300 K background photocurrent (dashed curves) of the two-stack C-QWIP. (b) The measured photocurrent ratio (circles) and calculated value (solid curve) as functions of the temperature. Reproduced with permission from [30].

The current-voltage characteristics of the C-QWIP are shown in Fig. 4.85(a). The dark current J_d of the C-QWIP is found to be 3.5 times lower than that of the edge-coupled QWIP in the LW stack and 1.25 times lower in the MW stack. These lower dark currents are consistent with the reduction of average widths of the stacks in the corrugation by 2.6 times for the LW and 1.30 times for the MW. The lower J_d helps to maintain $T_{\mathrm{BLIP}} = 75\,\mathrm{K}$ for the LW and 110 K for the MW under $f/1.5$ optics. These values of T_{BLIP} are slightly higher than the 70 K and 107 K for the edge-coupled detector in the LW and MW, respectively [30].

To demonstrate remote temperature sensing, the detector is operated at 10 K without focusing optics. A blackbody source is used as the target. Since the field of view of the detector is wider than the blackbody aperture, an optical chopper is used to modulate the blackbody radiation to isolate it from the ambient radiation. $-2\,\mathrm{V}$ is applied to the top contact and $+3\,\mathrm{V}$ is applied to the bottom contact, while the center common contact is grounded. The two AC photocurrents are measured simultaneously. The blackbody temperature is adjusted from 500 K to 650 K to obtain a contrast from the 300 K chopper. The measured photocurrent ratio $P_r(T)$ is recorded and compared to that calculated from Eq. (4.21) using the

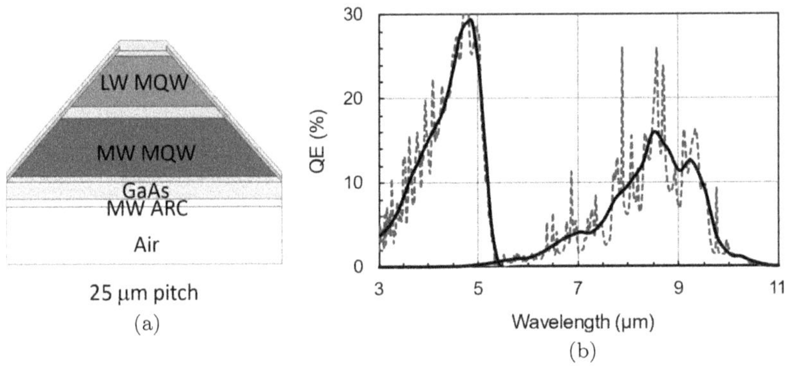

Figure 4.86. (a) A larger corrugation of the two-color C-QWIP with an ARC optimized for 4.8 μm and (b) the corresponding modeled QE with (solid curves) and without (dashed curves) curve smoothing.

experimental R_{MW} and R_{LW} in Fig. 4.84(b). The experimental P_r shown in Fig. 4.85(b) agrees with the calculated value with no fitting parameters. This good agreement shows that there is no unexpected absorption in other wavelengths that can defeat the sensing process. The two-color C-QWIP is thus capable of remote temperature sensing. To increase QE, one can adopt a larger corrugation. For example, by increasing the corrugation period to 25 μm as shown in Fig. 4.86 and employing an MW ARC, the peak QE is modeled to be 29.4% in the MW and 16.0% in the LW.

4.14.2 *Voltage-Tunable Two-Color C-QWIP*

Although stacking multiple narrowband absorbing layers offers a straightforward implementation of multi-color detection, the need for two or more indium contacts for each pixel posts severe limitation on the pixel size, pixel operability, and array pixel counts. A detector with voltage-tunable detection wavelength will eliminate these constraints and be able to adopt the usual detector geometry. However, this approach requires more sophisticated material layer structure and readout electronics. In particular, the ROIC must be able to provide voltage switching functionality that coordinates with signal integration. If a single capacitor is employed for charge integration for all channels, the detection of each color will need to be in sequential manner, and the detection of different colors will not be truly simultaneous [31].

There are several approaches in achieving voltage tunability in QWIPs. They include Stark shift [32–35], multiple transitions in heavily-doped QWs [36], and electron transfer between coupled QWs [37, 38]. All of these approaches have limited wavelength tunability, sensitivity, and operating temperature. A somewhat more practical design is based on the double-superlattice (DSL) [39, 40] structure. The typical band structure of a unit cell is shown in Fig. 4.87. The SL1 QWs, which is for MW detection, are doped at 4×10^{18} cm^{-3}. The same doping is applied to the LW SL2 QWs but only in the middle part (20 Å) of the QWs. Each unit cell is separated by a GaAs buffer layer with thickness of ∼2000 Å and $N_D \sim 1 \times 10^{18}$ cm^{-3}.

The entire structure can contain several DSL unit cells sandwiched between two contact layers. The SLs are joined by a graded barrier. This graded barrier blocks the conduction of the ground minibands but allows the electrons in the second miniband to go through. The detection band of the individual SL can be chosen and optimized independently, which gives the material structure its flexibility. Combining this material with the wavelength-independent C-QWIP coupling will yield a detector that can detect any two infrared spectral bands. The present example is designed to detect

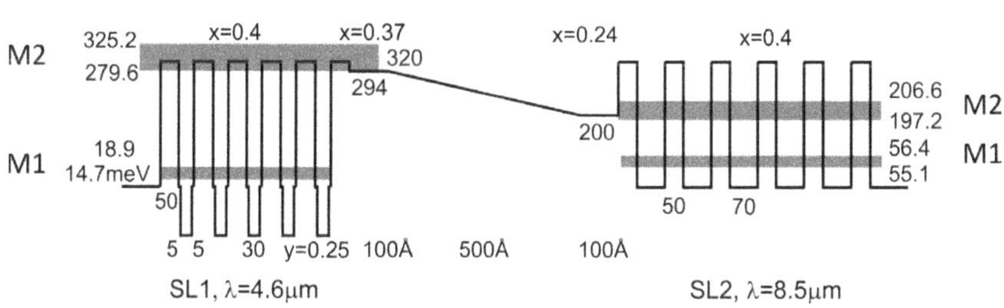

Figure 4.87. Band diagram of a unit cell of a voltage-tunable two-color detector. The numerals without units are either energies in milli-electron volts or thicknesses in angstroms. The Al and In molar ratios are denoted by x and y, respectively. Adapted with permission from *Inf. Phys. & Tech.*, 50, 124–135 (2007).

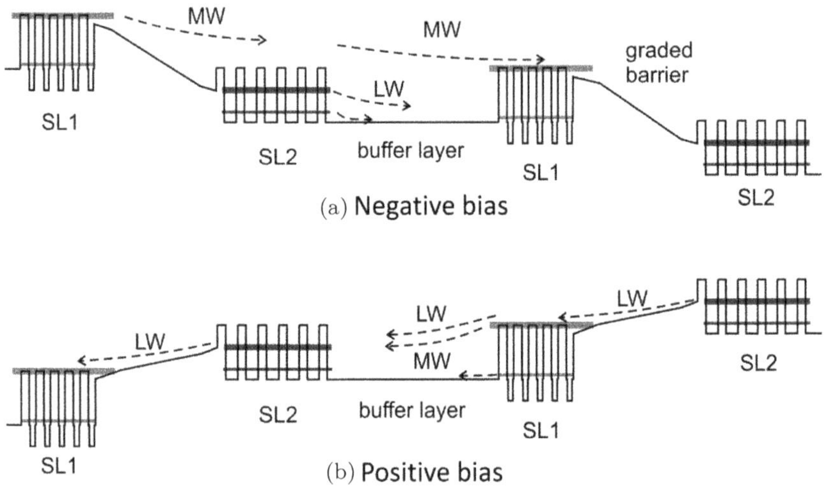

Figure 4.88. The band diagram of the voltage-tunable detector under (a) negative and (b) positive substrate bias. The arrows show the hot-electron trajectories. These hot electrons originate either from thermal excitation or from photoexcitation. Adapted with permission from *Inf. Phys. & Tech.*, 50, 124–135 (2007).

4.6 and 8.5 μm bands. Its material layer parameters are shown in Fig. 4.87.

Figure 4.88 shows two unit cells under different biasing conditions. In each DSL unit cell, the two SLs are highly conducting with and without light because their ground minibands $M1$ allow electrons to conduct freely through the SL barriers. There is no appreciable change in conductance when the $M1$ electrons are excited to $M2$. A standalone SL therefore will not be a photoconductor. When two SLs are separated by a thick barrier, however, the very low $M1$ tunneling probability through the thick barrier will stop the dark current flow. When the detector is exposed to infrared radiation and the DSL is under negative bias as shown in Fig. 4.88(a), the photocurrent can flow from $M2$ of SL1, travel across the thick barrier, and reach the upper minibands of SL2. MW photoconductivity is thus created. The presence of LW radiation can also generate photoelectrons in $M2$ of SL2. However, these photoelectrons will pass into the highly conducting buffer layer, resulting in no impedance change in the DSL.

The connecting buffer layer is designed to attenuate the LW photoelectron energy, so that these electrons cannot change the impedance of the next unit cell and cause spectral crosstalk. Meanwhile, the MW photoelectrons, after gaining energy from the applied electric field across the thick barrier, are able to do so and yield a substantial photoconductive gain. Consequently, the detector detects only in MW under negative bias. When the bias polarity is switched to positive in Fig. 4.88(b), only the SL2 $M2$ photoelectrons pass through the thick barrier and produce a LW photoconductivity. The voltage tunability is thus realized.

Figure 4.89 shows an experimental demonstration of this device concept. The employed C-QWIP geometry is shown in Fig. 4.22 with a corrugation period of 8 μm. The corrugation sidewalls are in direct contact with air. The detector material contains 7 DSL units, giving an absorber thickness of 3.2 μm. The total corrugation height is 3.9 μm, and the substrate thickness is 200 μm. Figure 4.89 shows that the C-QWIP detects only MW under negative bias and only LW under positive bias. There is no spectral crosstalk. Figure 4.90 shows the dark current and background photocurrent under $f/2$ optics. The

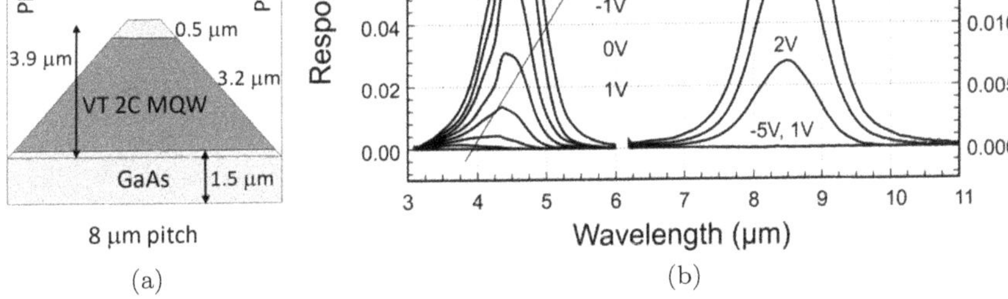

Figure 4.89. (a) The corrugation structure used in modeling and experiment, and (b) the measured responsivity under different voltages. Adapted with permission from *Inf. Phys. & Tech.*, 50, 124–135 (2007).

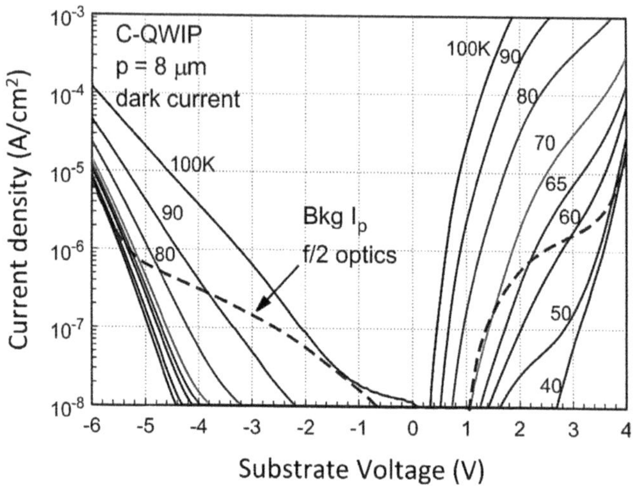

Figure 4.90. I–V characteristics of a voltage-tunable C-QWIP. Adapted with permission from *Inf. Phys. & Tech.*, 50, 124–135 (2007).

detector is BLIP ∼80 K under MW operation and ∼60 K under LW operation, depending on bias.

To model the detector optical properties, each SL is treated as a single QW "molecule". For an SL with 5 QWs and 6 quantum barriers, there are 5 levels in $M1$ and 5 levels in $M2$. Since the $M1$ states are usually occupied by doping electrons, intersubband transitions can be initiated from any $M1$ state to any $M2$ state. Nevertheless, due to the conservation of the miniband wavevectors, only 5 transitions dominate. The 5 $M1$ levels share a common Fermi level E_F, and the 2D electron density n_n occupied in each level E_n is given by $n_n = g_{2D}(E_F - E_n)$, where $g_{2D} = 1.744 \times 10^{36} J^{-1}\text{m}^2$. The total 2D electron density N_s in each SL is the sum of n_n of the M1 states and is equal to the total doping in all 5 QWs in the SL. After solving for the common E_F, the absorption coefficient for each SL with total length $L = 5W + 6B$ is

$$\alpha(\lambda) = \frac{N_s}{L}\frac{\pi e^2 \hbar}{2m^* n \varepsilon_0 c}\sum_{n=1}^{5}\left(\frac{n_n}{N_s} f_n\right)\rho_n(\lambda - \lambda_n) \tag{4.22}$$

following Eq. (2.9). The term $f_n' \equiv n_n f_n / N_s$ can be considered as the effective oscillator strength of the state n accounting for the

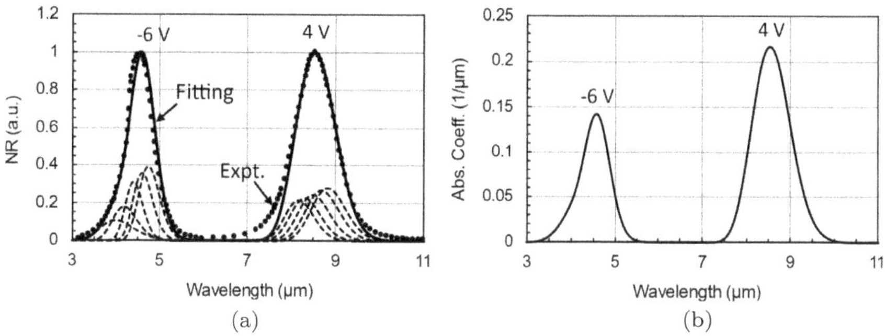

Figure 4.91. (a) The measured normalized responsivity (circles), the fitted $\lambda f_n' \rho_n(\lambda)$ (dashed curves), and the combined $\lambda f_n' \rho_n(\lambda)$ (solid curves). (b) The deduced α at different V.

unequal electron occupation in each state. Since R of a C-QWIP is wavelength-independent, and $R \propto \alpha\lambda \propto \Sigma\lambda f_n' \rho_n(\lambda)$, one can fit the measured R at -6 V and 4 V using the level-broadening parameters σ_n in ρ_n to yield the α lineshape. The fitting is shown in Fig. 4.91(a). To fit the peak positions, the β_{AlGaAs} in Eq. (2.22) is adjusted to 0.55 in MW and 0.65 in LW for $x = 0.4$ in both SLs. After the α lineshape is obtained by knowing ρ_n, α for both SLs can be calculated according to the nominal doping density. However, this α only applies to the specific SL length within the DSL unit cell. Therefore, in EM modeling, we adopt an α averaged over a DSL unit cell, *i.e.*, the individual superlattice length L in Eq. (4.22) is replaced by the total length of the DSL unit cell (3860 Å). The DSL thus acquires a voltage-dependent homogeneous α shown in Fig. 4.91(b).

The peak values of α are $0.14\,\mu\text{m}^{-1}$ and $0.22\,\mu\text{m}^{-1}$ in MW and LW respectively. These values are comparable to a regular QWIP material because each QW in a regular MQW is individually separated by a thick QW barrier, resulting in about the same average doping densities. Specifically, the average N_D of the present DSL is in fact 1.6 times higher than a regular MQW with 700 Å QW barrier doped at $1 \times 10^{18}\,\text{cm}^{-3}$. With a similar absorption bandwidth, α of the present detector is therefore slightly larger. N_D can be reduced if a lower dark current is desired.

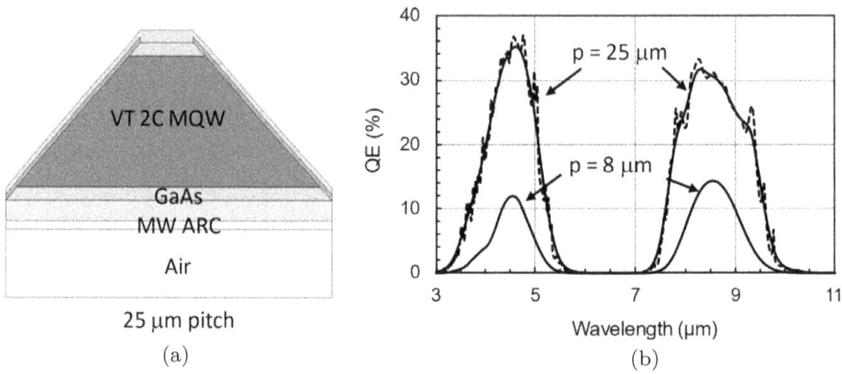

Figure 4.92. (a) A 25 μm C-QWIP with an ARC optimized for 4.8 μm incident light. (b) The modeled QE of an 8 μm pitch corrugation with thick substrate and no ARC, and the modeled (dashed curves) and smoothed (solid curves) QE of a 25 μm pitch corrugation with thinned substrate and ARC.

With the calculated α and the $p = 8\,\mu$m, thick-substrate C-QWIP, the modeled external QE is shown in Fig. 4.92(b). The peak QE is 12.0% and 14.3% in MW and LW, respectively. Figure 4.92(b) also shows that if a larger corrugation with $p = 25\,\mu$m and MW ARC is used as in Fig. 4.92(a), QE can be increased to 33% in both bands. The C-QWIP in this geometry contains 25 DSL unit cells for a total absorber thickness of 9.63 μm.

With the modeled η for the 8 μm C-QWIP, the measured R can be fitted using $g\gamma$ as the fitting parameter. Figure 4.93(a) shows the fitting with $g\gamma = 0.258$ for MW and 0.029 for the LW at the respective voltages. With these values of $g\gamma$, the observed R spectrum is well matched. The values of g are deduced from the noise measurement and they are shown in Fig. 4.93(b). The standard g-r noise expression is still used for this structure, where $i_n^2 = 4egI_d$, i_n is the current noise per $\sqrt{\text{Hz}}$, and I_d is the dark current.

Under a negative bias for MW operation, to yield a sufficient dark current in the measurement, T needs to be 111 K or higher. At $T = 111$ K and 116 K, g increases monotonically with increasing bias and approaches 0.8 at large V. This large g indicates that the M2 electrons from the MW SLs are able to pass through the buffer layers without losing all their kinetic energy. They traverse across most of

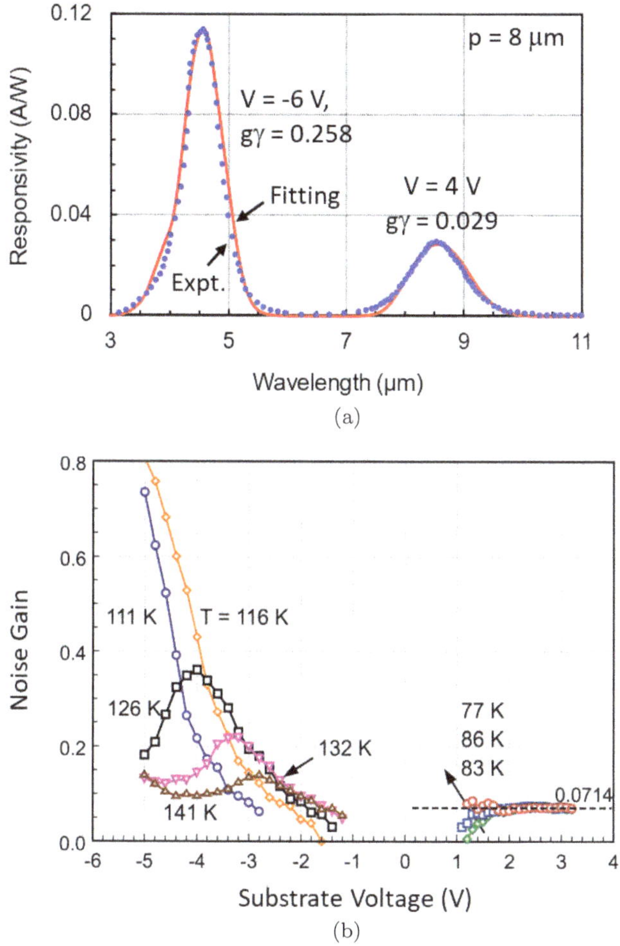

Figure 4.93. (a) The measured (circles) and the fitted (solid curves) responsivity of a $p = 8\,\mu$m C-QWIP at -6 V and 4 V. (b) The deduced gain from noise measurement using the g-r noise expression $i_n^2 = 4egI_d$. Adapted with permission from *Inf. Phys. & Tech.*, 50, 124–135 (2007).

the DSL unit cells as Fig. 4.88(a) illustrates. Under negative bias, the applied field is added to the built-in field of the graded barrier so that the electric field in the barrier layer is large. The MW electrons thus acquire large kinetic energies to overcome the energy attrition in the buffer layer. In contrast, the M2 electrons from the LW SLs under this negative bias do not pass through a high-field region to gain energy. They do not pass through any high-impedance region to produce either photocurrent or dark current.

When a positive substrate voltage is applied, the applied field is mostly canceled by the built-in field of the graded barrier as shown in Fig. 4.88(b). Therefore, the M2 electrons from the LW SLs do not gain as much energy in the graded barriers and thus are not able to pass through as many DSL unit cells. In fact, the measurable g at a finite positive bias in Fig. 4.93(b) is a constant at $\approx 0.0714 = 1/14$. This relatively constant g suggests that the mean free path of the LW thermal electron is a constant in this voltage range. If we assume that the LW electrons can only go through one unit cell and there is no statistical fluctuation in the mean free path, the noise would be shot noise generated in each unit cell where $i_n^2 = 2eI_d$. Since the present detector contains 7 independent unit cells connected in series, the variance of the noise current should be 7 times smaller than that of a single cell. The measured noise current of the detector should then be $i_n^2 = (2eI_d)/7 = 4eI_d/14$. The fact that such constant $g = 1/14$ is observed confirms the assumption that the thermal electrons from a LW SL cannot pass through the second DSL unit cell after traversing through the buffer layer up to 3.2 V.

If g in LW remains 0.0714 at 4 V and $g\gamma = 0.029$ at this bias, γ for the LW SL will then be 0.41. For the MW, the value of photocurrent g cannot be conclusively determined by the dark current noise gain. From Fig. 4.93(b), the noise gain at $T > 126$ K attains a maximum and then decreases at larger V. The decreasing g may indicate that the higher-energy hot electrons can acquire enough energy from the applied field and be scattered into satellite valleys where they lose the forward momentum quickly. Therefore, the photocurrent gain at -6 V depends sensitively on the initial photoelectron-launching energy. If $\gamma \approx 1$ because of the unbound nature of the MW M2 and the large total E field, the measured $g\gamma = 0.258$ at -6 V will yield $g = 0.258$ for the MW photoelectrons. This value of gain implies a mean free path of $0.258 \times 7 = 1.81$ DSL unit length and thus g is a statistical quantity. The MW dark and photo electrons from the M2 miniband are then subject to g-r noise instead of shot noise.

We thus observe two different noise mechanisms under opposite bias in the same detector. Under positive bias, since both η and g are constant, the increase in the LW responsivity in Fig. 4.89 is thus entirely due to the increase of γ in the presence of the applied field.

4.14.3 *Voltage-Tunable Two-Color C-QWIP Focal Plane Array*

The voltage-tunable DSL detector structure has been demonstrated in FPA format [41]. To be compatible with the ROIC biasing limitations, which is about ± 1 V, the detector material contains three DSL units, which has a similar layer structure as that shown in Fig. 4.87. The differences are the barrier width for SL1 is reduced from 50 Å to 40 Å and that for SL2 is reduced from 50 Å to 30 Å. They are to yield broader spectral bands. The two SLs are joined by a graded barrier with the effective Al molar ratio x decreasing from 0.36 to 0.22. Different from the previous detector, in which x is varied by controlling the Al-to-Ga flux ratio, the variation in the present detector is achieved by growing very thin $Al_{0.22}Ga_{0.78}As$/$Al_{0.4}Ga_{0.6}As$ periodic layers having different thickness ratios. The spatially averaged x is designed to decrease gradually from SL1 to SL2.

A detector wafer is fabricated into 45° edge coupling geometry for material characterization. Figure 4.94(a) shows the measured spectral responsivity at $T = 10$ K. It has similar detection peak wavelengths as the previous detector but with broader bandwidths.

Figure 4.94(b) shows the detector total current at different T under 300 K background and $f/2$ optics. Since the thermal dark current is absent at $T = 10$ K, the detector current at this T represents the background photocurrent I_p before the tunneling current sets in at high V. From this measurement, the detection is seen to be background limited around 80 K under MW operation and around 60 K under LW operation depending on the applied voltage.

The FPA format is 256 × 256 FPA with pixel pitch of 30 μm. The mesa height is $t = 2.8$ μm. Figure 4.30(f) shows the top view of the fabricated pixel geometry. Since the present C-QWIP geometry

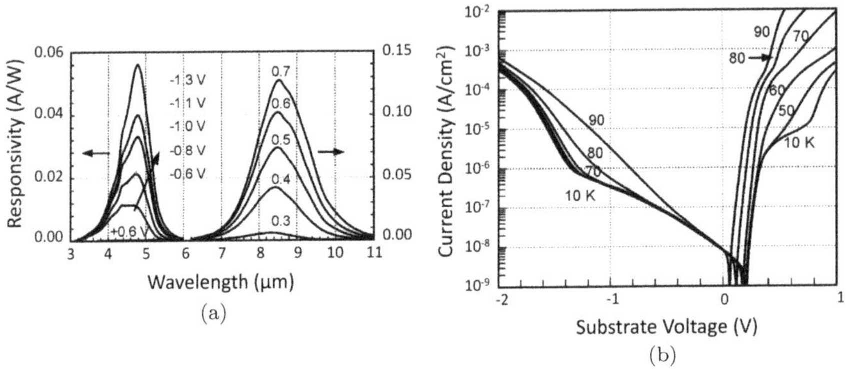

Figure 4.94. (a) The measured spectral responsivity of an edge-coupled QWIP. (b) The measured detector total current, $I_p + I_d$, at different T under a 300 K blackbody radiation. Reproduced with permission from [41].

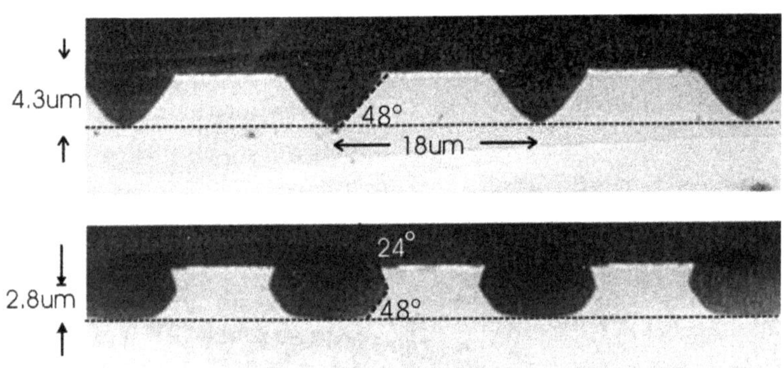

Figure 4.95. The C-QWIP sidewall profile in orthogonal directions at an etching depth of 4.3 μm into GaAs material. The etching mask is an array of squares with 16 μm side length and 18 μm period. Notice the half missing material in the hourglass sidewall direction. Reproduced with permission from *Inf. Phys. & Tech.*, 47, pp. 91–105 (2005).

is fabricated using a single chemical etching process, the detector contains both the regular inclined sidewalls in one direction and the hourglass shaped sidewalls in the other. Figure 4.95 shows the general sidewall profiles when a pixel of square geometry is etched in a single chemical process.

Both of these sidewalls produce nearly the same QE as modeled by the modal transmission-line method [21]. Therefore, all the sidewalls in the present pixel contribute to optical coupling. This single etching process, however, cannot be used to produce large C-QWIPs with $t \simeq 10\,\mu$m since in creating the hourglass sidewalls, large amount of detector material is removed, which reduces the pixel fill factor severely. One of the arrays was hybridized to an ROIC capable of biasing the array in both polarities. The substrate is not thinned nor AR coated. To determine the conversion efficiencies of the FPA, the FPA is placed in front of a blackbody through an $f/2.44$ optical window. The FPA is biased at $V = -1.1$ V for the MW operation and 0.85 V for the LW operation. The operating temperature is set at 50 K. The ROIC signal output is measured as a function of blackbody temperature T_B from 300 to 380 K. By assuming the same spectral response lineshape in Fig. 4.94(a), one can fit the increase of the signal output with T_B using a single fitting parameter, CE $= \eta g$ in Eq. (4.17). The fittings closely match the measured data, and CE is deduced to be 4.0% in the MW and 4.3% in the LW. The uncorrected nonuniformity of CE is 2.6% in both bands. In the fitting, a pixel fill factor of 0.7 is adopted.

The gain of this material has not been measured. To estimate η, we assume $g \propto 1/N$. The gain of the 3 DSL QWIP is therefore 7/3 times larger than that of the 7 DSL QWIP in Fig. 4.92(b) at the same applied field. In both applied fields, g is $\sim 0.0714 \times 7/3 = 0.167$. With this g, η is deduced to be 24.0% for the MW and 25.7% in the LW, indicating relatively high coupling efficiency of the C-QWIP geometry. By using narrow bandpass filters, the FPA was verified to have no observable spectral crosstalk.

Along with FPA signal, the pixel noise was measured at an integration time τ_{int} of 33 ms for the MW and 2 ms for the LW. From this measurement, the NEΔT is deduced in Fig. 4.96. The mean value for MW is 27 mK and that for the LW is 90 mK. The operability with NEΔT less than twice the mean is 98.4% for MW and 95.9% for LW. With a longer τ_{int} of 3 ms, NEΔT was measured to be 77 mK for the LW. Further reduction of LW NEΔT is expected with longer τ_{int}. Figure 4.97 shows an example of quick-look raw

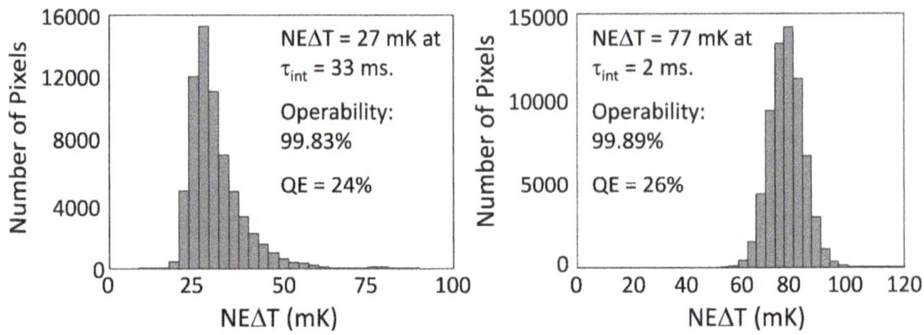

(a) (b)

Figure 4.96. The measured NEΔT (a) for MW and (b) for LW under $f/2.44$ optics. Reproduced with permission from [41].

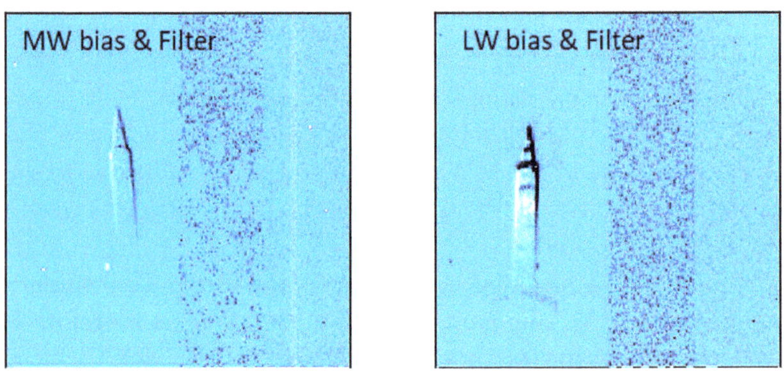

(a) (b)

Figure 4.97. Image of a moving soldering iron taken by a voltage-tunable C-QWIP camera with an on-chip velocity filter at (a) negative bias with MW bandpass filter and (b) positive bias with LW bandpass filter. No images are detected otherwise.

images during camera setup, where an on-chip velocity filter is set to detect only moving targets, such as a moving soldering iron in the image. In these images, the noisy columns are artifacts due to inaccurate timing synchronization of the digital ROIC, which can be eliminated by appropriate timing. The images are acquired with suitable bandpass filters to verify color selectivity. No images are observed when the filters are switched.

4.15 C-QWIP for Polarization Detection

Since linear corrugations only respond to TM mode, whose electric field is perpendicular to the corrugations, C-QWIPs can be used for polarization-sensitive detection [42]. In this sensing mode, the detection of an object is based on polarization contrast against a background instead of intensity contrast. To study this detection scheme, two C-QWIPs with orthogonal orientations shown in the inset of Fig. 4.98 are fabricated in a single chemical etching process. The chemical solution used is $1H_2SO_4 : 8H_2O_2 : 8H_2O$. The C-QWIPs created by this etching solution have regular "V"-shaped sidewalls when the corrugations are parallel to the primary GaAs wafer flat, which is labeled as the y-axis in Fig. 4.98(a), and they have inverted sidewalls when the corrugations are parallel to the x-axis in Fig. 4.98(b). The former detector, Det. A, has corrugation period of $5.5\,\mu$m. The latter Det. B has $7.5\,\mu$m period. Detectors with diagonal orientation tend to have vertical sidewalls and they will not be discussed here. Each detector has linear dimension of $500\,\mu$m and a $50\,\mu$m square on the side for wire bonding.

Figure 4.99(a) shows the experimental responsivity R for linearly polarized light when the incident E field is either perpendicular

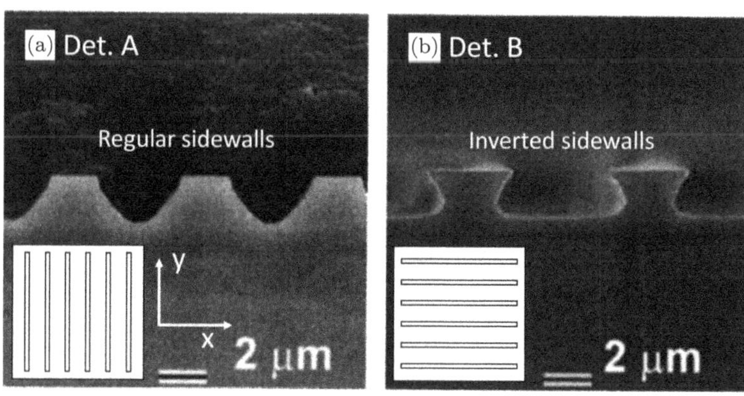

Figure 4.98. The orientations of C-QWIPs A and B and their sidewall profiles. The y-axis is parallel to the primary wafer flat. Reproduced with permission from [42].

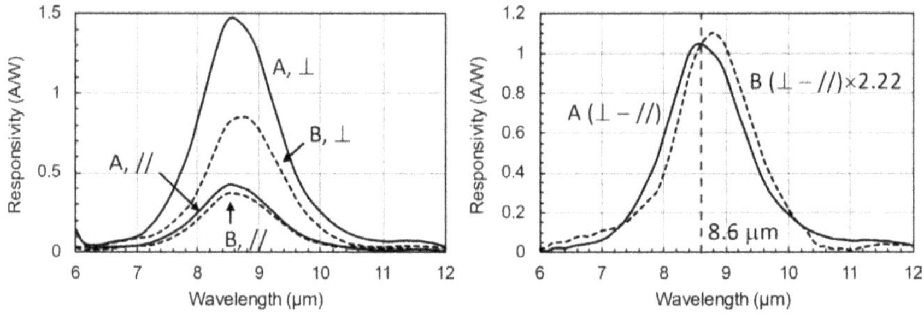

(a) (b)

Figure 4.99. (a) The measured responsivity for linearly polarized light where the incident E field is either $\perp$ or $//$ to the corrugations. (b) The intrinsic TM responsivity after the TE response is subtracted out. R_B is multiplied by 2.22 for comparison.

(TM mode) or parallel (TE mode) to the corrugations. Both detectors show finite TE response when there is no signal expected. This non-zero response is probably due to the stray light reflected from the large substrate sidewalls along the detector chip and is expected to reduce when it is thinned. To yield the intrinsic R due to corrugated coupling, the TE contribution is subtracted out from the TM response, and the difference is shown in Fig. 4.99(b). Without the TE contribution, the TM responsivity, R_A, is 2.22 times larger than R_B at $\lambda = 8.6\,\mu$m, and the two spectra have similar lineshape.

Since the present corrugation length of $\sim$500 μm is much longer than the periods, 2D EM model can be used, in which the tangential H component is used as the input field. However, it is more instructive to use the 3D model to explicitly calculate the polarization dependence. In this case, the input $\boldsymbol{H}$ field is $(\cos\theta, \sin\theta, 0)$ for a polarization angle θ relative to x. The external QE for linearly polarized light from Eq. (4.3) is

$$\eta = \frac{4n}{(1+n)^2}\frac{\alpha n^2}{p^2 Z_0^2}\int_V |E_z|^2 d^3r, \tag{4.23}$$

where n is taken to be 3.0, and Eq. (4.23) is for thick substrate with no AR coating. The modeling geometries for A and B are shown in Fig. 4.100. There are no air gaps at the ends of the corrugation. Therefore, the modeled η is for infinitely long corrugations. The

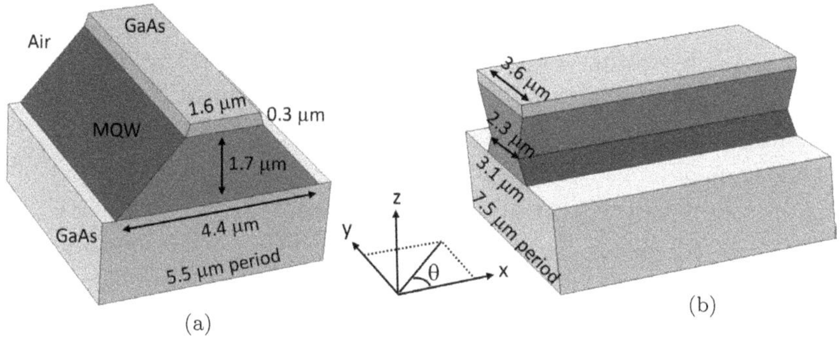

Figure 4.100. The modeled detector geometry for (a) Det. A and (b) Det. B. The inset shows the coordinate system.

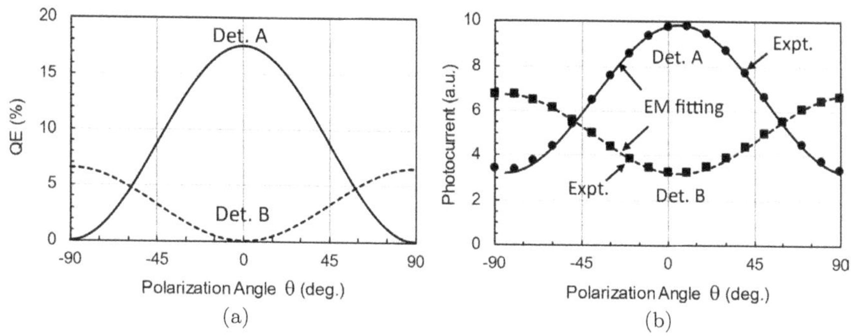

Figure 4.101. (a) The modeled polarized η_A and η_B as a function of the polarization angle θ in the x-y plane. (b) The fitted I_p (curves) using prefactors $s_A = s_B$.

detector is covered by $5\,\mu$m thick air layer, and it is terminated by a perfectly matched layer. The bottom boundary is set to satisfy scattering boundary condition of second order and the incident $\boldsymbol{H}$ field is activated. PBC is imposed on four sides of the unit cell. Therefore, the model represents an infinite periodic array in both x and y.

The polarization dependence of η modeled using $\alpha = 0.22\,\mu\text{m}^{-1}$ at $\lambda = 8.6\,\mu$m is shown in Fig. 4.101(a). Det. A follows the $\cos^2\theta$ law and Det. B follows the $\sin^2\theta$ law as expected, and there is no detector response when E is parallel to the corrugations. For the TM mode, η_A is 17.47%, which is 2.64 times larger than η_B of 6.61%. This η ratio is slightly larger than the measured R ratio of 2.22 at $\lambda = 8.6\,\mu$m in Fig. 4.99(b). The difference can be due to the simplified modeling geometry and the difference in the parameter $g\gamma$ due to the different detector cross-sections that leads to different potential distributions inside the MQW. Part of the smaller η_B is due to the smaller pixel fill factor. Specifically, 61% of MQW material remains in A after etching, and only 37% remains in B. Therefore, the coupling efficiency of A is higher than B by $2.64\times37/61 = 1.6$ times in terms of I_p/I_d ratio.

To characterize the angular dependence, AC photocurrent I_p is measured with a modulated blackbody source through a linear polarizer. The magnitude of I_p is monitored as a function of θ. Figure 4.101(b) shows both I_p's vary as sinusoidal functions as the modeling predicted, and there is a DC offset I_{dc} of about 3.2 arbitrary units caused by stray light as in the R measurement. To relate the modeled η to the experimental I_p, we note that I_p is given by $\eta eg\gamma\Phi + I_{dc}$, where Φ is the in-band optical flux onto a detector. Therefore, $I_A = s_A\eta_A + 3.2$ and $I_B = s_B\eta_B + 3.2$ for certain fitting parameters $s_A = e(g\gamma\Phi)_A$ and $s_B = e(g\gamma\Phi)_B$. To match the measurement in Fig. 4.101(b), both $s_A = s_B = 0.50$. Therefore, the EM modeling is in good agreement with the observed photocurrent ratio between A and B under polarized illumination as expected from the R measurement. In Fig. 4.101(b), the modeled I_p's have been shifted by 5° in the θ axis, which is due to a small misalignment between the polarizer axis and the detector die in the measurement setup.

By taking the experimental I_p ratio of A and B as shown in Fig. 4.102, the intensity dependence is factored out, and the ratio r is a function of θ only, which is verified by experiment [42]. Therefore, by measuring r, θ of a linearly polarized light can be determined. To show one possible application, different materials listed in Table 4.1 are illuminated by unpolarized light from a modulated blackbody and the reflected light is detected by the detector pair. Both illumination and detection are at 45° relative to the normal of the reflecting surface.

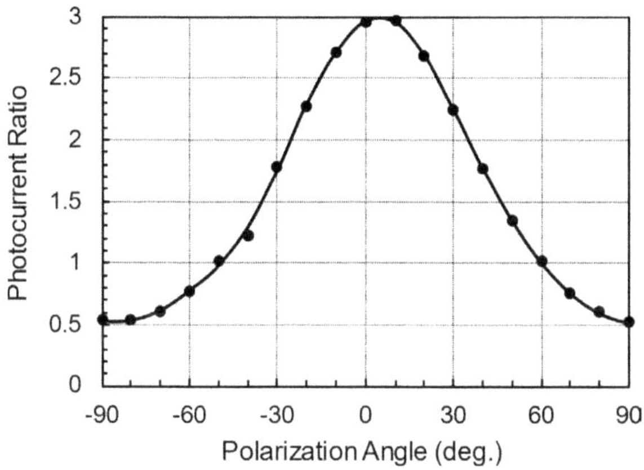

Figure 4.102. The experimental $r = I_A/I_B$ as a function of θ for linearly polarized light. Adapted with permission from [42].

Table 4.1. The current ratio and $|\theta|$ from light reflected off from different materials.

Material	J_A/J_B	θ (deg)
Gold	1.41	48.6
Aluminum	1.36	49.8
Glass	0.844	65.7
Paper	0.694	72.7
Plastics	0.594	79.2

Table 4.1 shows that r is different for different materials. From this difference, one may identify the material as a practical purpose. One can also assign $|\theta| = a\tan(r)$ as shown in Table 4.1. However, since the reflected light also contains an unpolarized component, the assigned $|\theta|$ may not be the true polarization angle of the polarized component. To fully characterize the polarization state of the reflected light, one will need more detectors in the sensing unit.

The detector structures with the same material and corrugation periods have been processed into C-QWIP FPAs for polarization-sensitive detection. The 256×256 FPA has 128 alternate columns of Det. A and Det. B as shown in Fig. 4.103. The pixel pitch is

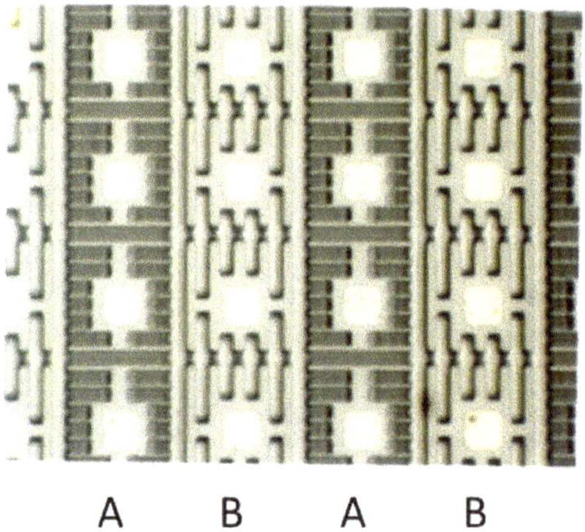

Figure 4.103. The fabricated 256 × 256 polarization-sensitive C-QWIP FPA with regular sidewalls (A) and inverted sidewalls (B).

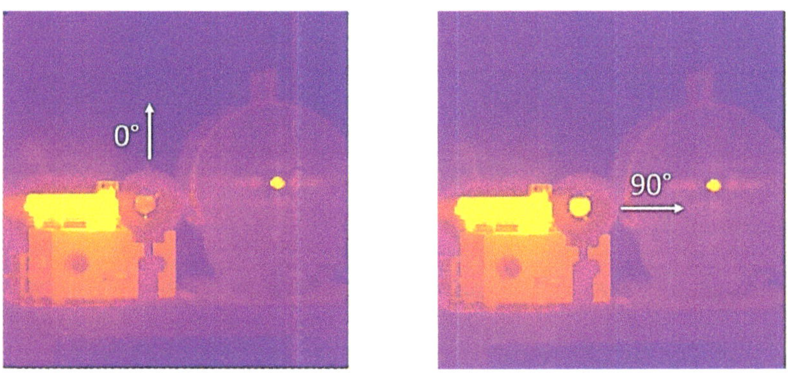

(a) (b)

Figure 4.104. The infrared image of the polarization-sensitive C-QWIP FPA operating at $T = 65\,\mathrm{K}$ when the polarizer axis is either (a) vertical (0°) or (b) horizontal (90°).

$40\,\mu$m and the substrate is unthinned. The near-center island is for indium bump connection. Figure 4.104 shows the infrared images taken by the B columns with a polarizer in front of a hot plate. The image of the polarizer appears to be darker than the rest of the hot plate when the polarizer axis is vertical and it is as bright when it is horizontal, showing its polarization-sensitive properties. Notice the spreading of light intensity from the blackbody source to other pixel areas in these images. This spatial crosstalk is due to the diffraction of light from the inverted C-QWIP geometry having a thick substrate.

References

1. Goossen, K. W. and Lyon, S. A. (1985). Grating enhanced quantum well detector, *Appl. Phys. Lett.*, 47, pp. 1257–1259.
2. Goossen, K. W., Lyon, S. A. and Alavi, K. (1988). Grating enhancement of quantum well detector response, *Appl. Phys. Lett.*, 53, pp. 1027–1029.
3. Hasnain, G., Levine, B. F., Bethea, C. G., Logan, R. A., Walker, J. and Malik, R. J. (1989). GaAs/AlGaAs multiquantum well infrared detector arrays using etched gratings, *Appl. Phys. Lett.*, 54, pp. 2515–2517.
4. Andersson, J. Y., Lundqvist, L. and Paska, Z. F. (1991). Quantum efficiency enhancement of AlGaAs/GaAs quantum well infrared detectors using a waveguide with a grating coupler, *Appl. Phys. Lett.*, 58, pp. 2264–2266.
5. Andersson, J. Y. and Lundqvist, L. (1991). Near-unity quantum efficiency of AlGaAs/GaAs quantum well infrared detectors using a waveguide with a doubly periodic grating coupler, *Appl. Phys. Lett.*, 59, pp. 857–859.
6. Andersson, J. Y. and Lundqvist, L. (1992). Grating-coupled quantum-well infrared detectors: Theory and performance, *J. Appl. Phys.*, 71, pp. 3600–3610.
7. Lundqvist, L., Andersson, J. Y., Paska, Z. F., Borglind, J. and Haga, D. (1993). Efficiency of grating coupled AlGaAs/GaAs quantum well infrared detectors, *Appl. Phys. Lett.*, 63, pp. 3361–3362.
8. Sarusi, G., Levine, B. F., Pearton, S. J., Bandara, K. M. S. and Leibenguth, R. E. (1994). Improved performance of quantum well infrared photodetectors using random scattering optical coupling, *Appl. Phys. Lett.*, 64, pp. 960–962.
9. Rokhinson, L. P., Chen, C. J., Tsui, D. C., Vawter, A. G. and Choi, K. K. (1999). Quantum grid infrared photodetectors, *Appl. Phys. Lett.*, 74, pp. 759–761.
10. Choi, K. K., Dang, G., Little, J. W., Leung, K. M. and Tamir, T. (2004). Quantum grid infrared spectrometer, *Appl. Phys. Lett.*, 84, pp. 4439–4441.

11. Schimert, T. R., Barnes, S. L., Brouns, A. J., Case, F. C., Mitra, P. and Claiborne, L. T. (1996). Enhanced quantum well infrared photodetector with novel multiple quantum well grating structure, *Appl. Phys. Lett.*, 68, pp. 2846–2848.
12. Chen, Y. N., Todorov, Y., Askenazi, B., *et al.* (2014). Antenna-coupled microcavities for enhanced infrared photo-detection, *Appl. Phys. Lett.*, 104, p. 031113.
13. Kalchmair, S., Detz, H. and Cole, G. D. (2011). Photonic crystal slab quantum well infrared photodetector, *Appl. Phys. Lett.*, 98, p. 011105.
14. Choi, K. K., Jhabvala, M. D., Forrai, D. P., Waczynski, A., Sun, J. and Jones, R. (2013). Electromagnetic modeling and design of quantum well infrared photodetectors, *IEEE J. Selected Topics in Quant. Elect.*, 19, pp. 1–10.
15. Choi, K. K., Jhabvala, M. D., Sun, J., Jhabvala, C. A., Waczynski, A. and Olver, K. (2013). Resonator-quantum well infrared photodetectors, *Appl. Phys. Lett.*, 103, p. 201113.
16. Chen C. J., Choi, K. K., Tidrow, M. Z. and Tsui, D. C. (1996). Corrugated quantum well infrared photodetectors for normal incident light coupling, *Appl. Phys. Lett.*, 69, pp. 1446–1448.
17. Chen C. J., Choi, K. K., Chang, W. H. and Tsui, D. C. (1997). Performance of corrugated quantum well infrared photodetectors, *Appl. Phys. Lett.*, 71, pp. 3045–3047.
18. Chen, C. J., Choi, K. K., Rohkinson, L., Chang, W. H. and Tsui, D. C. (1999). Corrugated quantum well infrared photodetector for polarization detection, *Appl. Phys. Lett.*, 74, pp. 862–864.
19. Smith, B. C. (2022). The infrared spectra of polymers V: Epoxies, *Spectroscopy*, 37, pp. 17–19.
20. Choi, K. K., Chen, C. J. and Tsui, D. C. (2000). Corrugated quantum well infrared photodetectors for material characterization, *J. Appl. Phys.*, 88, pp. 1612–1623.
21. Choi, K. K., Leung, K. M., Tamir, T. and Monroy, C. (2004). Light coupling characteristics of corrugated quantum well infrared photodetectors, *IEEE J. Quan. Electr.*, 40, pp. 130–142.
22. Sun, J., Choi, K. K., Jhabvala, M. D. and Jhabvala, C. A. (2012). Advanced substrate thinning process for GaAs-based devices, *J. Micro/Nanolith. MEMS MOEMS*, 11, p. 043003.
23. Choi, K. K., Devitt, J. W., Forrai, D. P., Endres, D., Marquis, J., Bettge, J. and Pinsukanjana, P. (2007). C-QWIP material design and growth, *Proc. SPIE*, 6542, p. 65420S.
24. Choi, K. K., Forrai, D. P., Endres, D. and Sun, J. (2009). Corrugated-quantum well infrared photodetector focal plane arrays, *IEEE J. Quant. Elect.*, 45, p. 1255.

25. Forrai, D. P., Sempsrott, M., Fischer, R., Choi, K. K. and Devitt, J. W. (2007). Characterization of a C-QWIP LWIR camera, *Proc. SPIE*, 6543, p. 654317.
26. Jhabvala, M., Choi, K., Jhabvala, C., Kelly, D., Hess, L., Ewin, A., La, A., Waczynski, A., Sun, J., Adachi, T., Costen, H., Ni, Q., Snodgrass, S. and Foltz, R. (2013). Development of a 2K × 2K GaAs QWIP Focal Plane Array, *GSFC. CP.* 7565.2013.
27. Jhabvala, M., Reuter, D., Choi, K., Sundaram, M., Jhabvala, C., La, A., Waczynski, A. and Bundas, J. (2010). The QWIP focal plane assembly for NASA's Landsat data continuity mission, *Proc. SPIE*, 7660, 76603J.
28. Sundaram, M., Wang, S. C., Taylor, M. F., Reisinger, A., Milne, G. L., Reiff, K. B., Rose, R. E. and Marin, R. R. (2001). Two-color quantum well infrared photodetector focal plane arrays, *Inf. Phys. & Tech.*, 42, pp. 301–308.
29. Schreer, O., Sáenz, M. L., Peppermueller, C. and Schmidt, U. (2006). Dual-band camera system with advanced image processing capability, *Proc. SPIE*, 6542, 65421C pp. 1–7.
30. Chen, C. J., Choi, K. K., Chang, W. H. and Tsui, D. C. (1998). Two-color corrugated quantum well infrared photodetector for remote temperature sensing, *Appl. Phys. Lett.*, 72, pp. 7–9.
31. Simolon, B., Aziz, N., Barskey, S., Hansen, R., Kurth, E., Long, J. and Petronio, S. (2013). Large format two-color CMOS ROIC for SLS detectors, *Inf. Phys. & Tech.*, 59, pp. 78–83.
32. Choi, K. K., Levine, B. F., Bethea, C. G., Walker, J. and Malik, R. J. (1989). Infrared photoelectron tunneling spectroscopy of strongly coupled quantum wells, *Phys. Rev. B*, 39, pp. 8029–8032.
33. Parihar, S. R., Lyon, S. A., Santos, M. and Shayegan, M. (1989). Voltage tunable quantum well infrared detector, *Appl. Phys. Lett.*, 55, pp. 2417–2419.
34. Levine, B. F., Bethea, C. G., Shen, V. O. and Malik, R. J. (1990). Tunable long-wavelength detectors using graded barrier quantum wells grown by electron beam source molecular beam epitaxy, *Appl. Phys. Lett.*, 57, pp. 383–385.
35. Martinet, E., Luc, F., Rosencher, E., Bois, Ph. and Delaitre, S. (1992). Electrical tunability of infrared detectors using compositionally asymmetric GaAs/AlGaAs multiquantum wells, *Appl. Phys. Lett.*, 60, pp. 895–897.
36. Kheng, K., Ramsteiner, M., Schneider, H., Ralston, J. D., Fuchs, F. and Koidl, P. (1992). Two-color GaAs/(AlGa)As quantum well infrared detector with voltage-tunable spectral sensitivity at 3–5 and 8–12 μm, *Appl. Phys. Lett.*, 61, pp. 666–668.

37. Majumder, A., Choi, K. K., Reno, J. L., Rokhinson, L. P. and Tsui, D. C. (2002). Two-color quantum well infrared photodetector with voltage tunable detection peaks, *Appl. Phys. Lett.*, 80, pp. 707–709.
38. Majumder, A., Choi, K. K., Reno, J. L., Rokhinson, L. P. and Tsui D. C. (2002). Electron transfer in voltage tunable two-color infrared photodetectors, *J. Appl. Phys.*, 91, pp. 4623–4625.
39. Chen, C. C., Chen, H. C., Kuan, C. H., Lin, S. D. and Lee, C. P. (2002). Multicolor infrared detection realized with two distinct superlattices separated by a blocking barrier, *Appl. Phys. Lett.*, 80, pp. 2251–2253.
40. Majumdar, A., Choi, K. K., Reno, J. and Tsui, D. C. (2003). Voltage tunable two-color infrared detection using semiconductor superlattices, *Appl. Phys. Lett.*, 83, pp. 5130–5132.
41. Choi, K. K., Jhabvala, M. D. and Peralta, R. J. (2008). Voltage-tunable two-color corrugated-QWIP focal plane arrays, *IEEE Elect. Dev. Lett.*, 29, pp. 1011–1013.
42. Chen, C. J., Choi, K. K., Rokhinson, L., Chang, W. H. and Tsui, D. C. (1999). Corrugated quantum well infrared photodetectors for polarization detection, *Appl. Phys. Lett.*, 74, pp. 862–864.

Chapter 5

Resonator-Quantum Well Infrared Photodetectors

5.1 Introduction

In Chapter 4, we described the use of sidewall reflection to create vertical electric fields for quantum well absorption. Under this light coupling approach, the structure design is intuitive and is applicable to any infrared wavelengths and material bandwidths. The large active volume and the broadband coupling characteristics reduce the doping requirement and thus the dark current. The resulting quantum efficiency (QE) and the detection spectrum do not sensitively depend on the structural parameters and the material refractive indices. It produces extremely low optical crosstalk because of the retro-reflective properties, which enables high-definition imaging. The one-size-fit-all geometry allows mass production of different wavelength focal plane arrays (FPAs) without the knowledge of the material absorption properties as though they possess isotropic absorption. The high-quality infrared images displayed in Chapter 4, which are obtained from nearly the same detector geometry, testify to the advantages and utilities of this coupling approach. However, this approach requires a thick multiple quantum well (MQW) layer to yield substantial QE, and it cannot exceed 50%. Furthermore, the thick layer needs higher voltage to operate and yields small photoconductive gain and thus photocurrent. The material cost is also higher. It will be advantageous to realize a large QE in a thin layer without the use of excessive doping.

Besides optical reflection, optical diffraction is also commonly used for light coupling. Following the early experimental work on grating coupling [1–3], Andersson *et al.* [4–7] designed and optimized the grating structures and demonstrated larger than 90% QE in a $\sim 2\,\mu$m thick material, which represents a breakthrough in quantum well infrared photodetector (QWIP) technology. Although this level of QE is later found to be difficult to obtain in small detector pixels, the benefits of using a thin material layer and small voltages are still significant. Consequently, QWIP FPAs produced today commonly use grating coupling. In this chapter, we will discuss the grating design and other alternative geometries based on diffraction and interference effects. One of the coupling schemes, the resonator-QWIP, will be described in detail.

5.2 Grating-Coupled Quantum Well Infrared Photodetectors

The mathematical tool adopted by Andersson *et al.* for optical design is known as the modal expansion method (MEM) [6]. Similar to the modal transmission-line model described in Section 3.5, MEM is mainly applicable to infinitely extended periodic structures. Under this approach, the exact electromagnetic (EM) modal functions in individual grating cavities are first solved. In Ref. 6, the conductivity for the metal in the grating is assumed to be infinite. With the perfect electric conductor boundary condition applied to the metal/semiconductor interface, the ohmic loss in metal is neglected. The general field in the cavities under normal incidence is expressed as a superposition of modal functions, and it matches the grating diffraction orders outside the grating through appropriate boundary conditions. In numerical calculations, only the grating orders (0, 0), (−1, 0), (1, 0), (0, −1), and (0, 1) are included in the boundary condition matching. The diagonal and higher diffraction orders are ignored. The total field distribution is therefore contributed only from a limited set of wavevector components. Nevertheless, it is a good approximation when the internal wavelength is close to the grating period where those orders dominate. In this section, we will use finite element method (FEM) to analyze the same grating structures and compare the results obtained from both methods. In the FEM analysis, we will use gold as the metal grating material with its (n, κ) shown in Fig. 3.14. The ohmic loss in the metal grating is thus included. Since FEM seeks exact solutions to Maxwell's equations, all diffraction orders and diffraction in all directions are incorporated and it is valid at all wavelengths in principle. In practice, the larger demand for computer memory and computation time limits its accuracy at shorter wavelengths.

Aside from the metal cover difference, Andersson *et al.* used the Drude-Lorentz oscillator model of Eq. (3.2) to describe intersubband transition. In this model, the individual QW is treated as an artificial atom having two energy levels embedded in an isotropic medium. The absolute dielectric constant of the MQW layer ε_a is taken to be [6]

$$\varepsilon_a(\omega) = \varepsilon_\infty + \frac{\omega_p^2 \varepsilon_0 f}{(\omega_0^2 - \omega^2 - i\gamma\omega)}, \tag{5.1}$$

where ε_∞ is the high frequency dielectric constant assumed to be $10.89\varepsilon_0$, $\omega_p^2 = Ne^2/m^*\varepsilon_0$, N is the average doping density in one QW period, f is the oscillator strength of the two-level system, $\hbar\omega_0$ is the energy spacing between the two levels, and γ is the damping coefficient. In this model, f and γ are not calculated but treated as fitting parameters to the experimental spectrum. If we take $N = N_s/L$ where N_s is the areal doping density per QW and L is the QW period, Eq. (5.1) takes the form of

$$\varepsilon(\omega) = \varepsilon_r + i\frac{N_s}{L}\frac{e^2 f}{m^*\varepsilon_0}\frac{\gamma\omega}{(\omega_0^2 - \omega^2)^2 + \gamma^2\omega^2}, \tag{5.2}$$

where $\varepsilon(\omega)$ now denotes the relative complex dielectric constant, and ε_r is the real part of $\varepsilon(\omega)$. With the imaginary part of $\varepsilon(\omega)$ shown in Eq. (5.2), the absorption coefficient α, through Eq. (2.1), is

$$\alpha(\omega) = \frac{\omega}{nc}\frac{N_s}{L}\frac{e^2 f}{m^*\varepsilon_0}\frac{\gamma\omega}{(\omega_0^2 - \omega^2)^2 + \gamma^2\omega^2}. \tag{5.3}$$

With the ω dependence in Eq. (5.3), γ is the full-width-at-half-maximum bandwidth of the α lineshape. At the peak wavelength,

$\omega = \omega_0$, and

$$\alpha_p = \frac{N_s}{L} \frac{e^2 f}{m^* n \varepsilon_0 c} \frac{1}{\gamma}. \tag{5.4}$$

This peak value of α is very similar to that presented in Section 2.6, where intersubband transition is described to be between the ground state and a hypothetical state χ above the barrier. In that case, α is given by Eq. (2.31), which is rewritten here as

$$\alpha(\hbar\omega) = \frac{N_s}{L} \frac{\pi e^2 \hbar f}{2m^* n \varepsilon_0 c} \rho_\chi(\hbar\omega), \tag{5.5}$$

where $\rho_\chi(\omega)$ is the density of state of the level χ embedded in a one-dimensional (1D) QW structure. If one considers level broadening in the Drude-Lorentz model instead of through barrier state coupling in Eq. (2.30), ρ_χ takes the same ω distribution in Eq. (5.3). When ω is near ω_0, the distribution can be simplified to

$$\rho_\chi(\hbar\omega) = \frac{1}{\pi\hbar} \frac{\gamma/2}{(\omega - \omega_0)^2 + (\gamma/2)^2}, \tag{5.6}$$

for the same full-width-at-half-maximum γ. At the absorption peak, $\rho_\chi = 2/(\pi\hbar\gamma)$, and α_p is

$$\alpha_p = \frac{N_s}{L} \frac{e^2 f}{m^* n \varepsilon_0 c} \frac{1}{\gamma}, \tag{5.7}$$

which is the same as Eq. (5.4). Therefore, Eq. (5.1) can be used to estimate the QW absorption coefficient using fitting parameters f and γ if the absorption lineshape is not of interest.

In the detector, there is a composite ohmic metal layer deposited between the top contact layer and the gold grating cover. Since this metal layer is much thinner than the skin depth, and it diffuses into the top contact upon thermal annealing, it can be excluded in EM modeling.

5.2.1 *Laminar Gratings*

The first theoretical grating design to be discussed is a 1D laminar grating presented in Ref. 6. The nominal material structure is shown in Fig. 5.1(a). The grating structures are shown in Fig. 5.2, which

Andersson & Lundqvist (Ref. 6)	Material	Doping (E18 cm^{-3})
Top contact p = 3 μm	GaAs, laminar grating: 0.4 μm without CL, 0.5 μm with CL	1.0
MQW (50 periods) (1.75 μm)	Well: 50Å GaAs Barrier: 300Å $Al_{0.25}Ga_{0.75}As$	1.0
Bot. contact	0.5 μm GaAs	1.0
Cladding layer	3.0 μm AlAs	
Substrate	625 μm GaAs	undoped

(a)

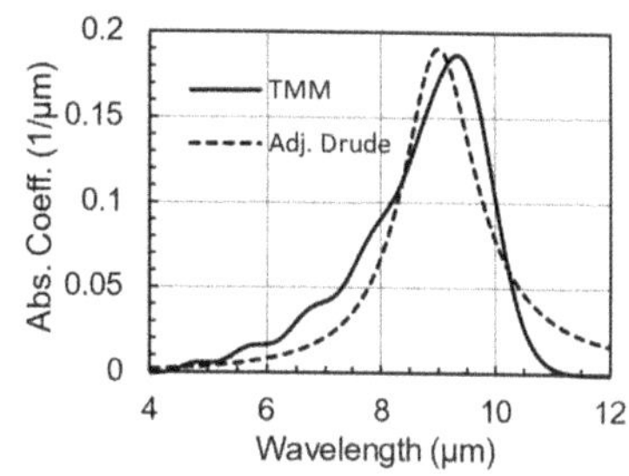

(b)

Figure 5.1. (a) The material structure used in laminar grating design and (b) the corresponding calculated α spectrum using the TMM model and the Drude model with fitted parameters.

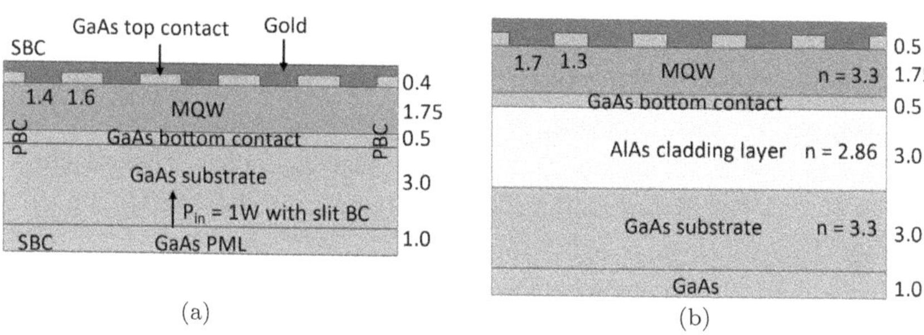

Figure 5.2. The periodic laminar grating structure (a) without an AlAs cladding layer and (b) with a cladding layer. Internal QE is modeled. The dimensions are in microns.

contain periodic grooves filled with gold and the period p is 3.0 μm. In FEM modeling, the refractive indices of MQW and GaAs are taken to be the same as that (= 3.3) in Ref. 6. For the grating detector without the AlAs cladding layer ($n = 2.86$), the groove width w is 1.4 μm and groove height h is 0.4 μm. For the waveguide detector with the cladding layer, $w = 1.7\,\mu$m and $h = 0.5\,\mu$m.

According to the QW material structure, the theoretical α is calculated from the transfer matrix method described in Section 2.4 and shown in Fig. 5.1(b). The calculated peak value of α is

$0.19\,\mu\text{m}^{-1}$ for $N_D = 1.0\times 10^{18}\,\text{cm}^{-3}$. In the FEM modeling, EM wave is launched inside the GaAs substrate. Therefore, the calculated QE is the internal unpolarized QE. The substrate is assumed to be thick so that cavity effects from the substrate are negligible. The external QE for wave launched outside the detector is then the product of the internal QE and the substrate transmission coefficient T_s. With a perfect anti-reflection coating (ARC), the internal QE will be the same as the external QE. To compare with the MEM solutions, the present FEM adopts periodic boundary condition (PBC) at both left and right boundaries so that infinitely extended structures are modeled. For this grating structure, Ref. 6 provides the MEM solutions and they are shown in Fig. 5.3(a). The present FEM solutions are shown in Fig. 5.3(b) for comparison.

The displayed solutions differ not only because of the computation methods but also because of the adopted α spectra. The QEs displayed in Fig. 5.3(a) are modeled using $\alpha(\lambda)$ in the Lorentz-Drude model while those in Fig. 5.3(b) are modeled using $\alpha(\lambda)$ given by the transfer matrix method. By fitting the edge-coupled internal QE in Fig. 5.3(a) to Eq. (3.69) without the T_s factor, the adopted Drude α can be obtained and it is shown in Fig. 5.1(b). This α matches Eq. (5.3) quantitatively with $\lambda = 9.0\,\mu\text{m}$, $f = 0.35$, and $\gamma = 3.7\times 10^{13}\text{s}^{-1}$. Therefore, the adopted α in the MEM modeling has

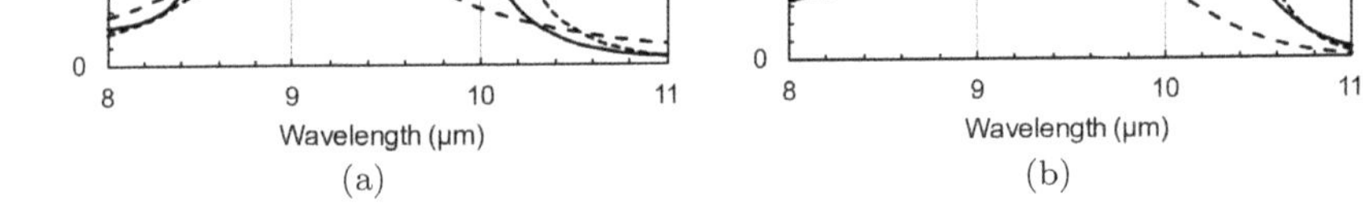

Figure 5.3. The modeled internal η of laminar grating with a cladding layer (solid curves) and without a cladding layer (dashed curves) using (a) MEM and (b) FEM. $\text{IQE}_{\text{EM}} = 0.469\,\mu\text{m}$ for LG and $0.579\,\mu\text{m}$ for LWG. Theoretical internal η_{45} is also shown.

a slightly shorter peak wavelength and a more symmetrical lineshape, although the two magnitudes are very similar.

In Fig. 5.3, the predicted grating QEs share substantial similarities between the two approaches. The predicted peak wavelengths and the peak magnitudes are both very similar for both the linear grating without a cladding layer (labeled as LG) and the linear grating with a cladding layer (labeled as LWG). The longer cutoff in the FEM solution can be attributed to the material difference. Since laminar gratings can only yield maximum QE of 50%, the present internal QE of ~40% from a 1.75 μm thick material shows the high efficiency of grating coupling. The FEM modeling thus supports the optimized laminar grating structure designed by Andersson *et al.* The integrated quantum efficiency IQE defined in Eq. (4.18) are 0.469 μm for the LG structure and 0.579 μm for the LWG structure in Fig. 5.3(b).

These large QEs can be understood from the vertical field E_y distribution in Fig. 5.4 at λ_p. For the LG detector, λ_p is at 10.0 μm in FEM. At this λ, the internal wavelength $\lambda_i = \lambda/n = 10/3.3 = 3.03\,\mu$m, which is nearly the same as the grating period p of 3 μm. From the grating formula $\lambda_i = p \cdot \sin\theta$ for first-order diffraction, the grating diffracts the incident light at 90° at this λ. The diffracted beam thus travels parallel to the surface and attains the longest

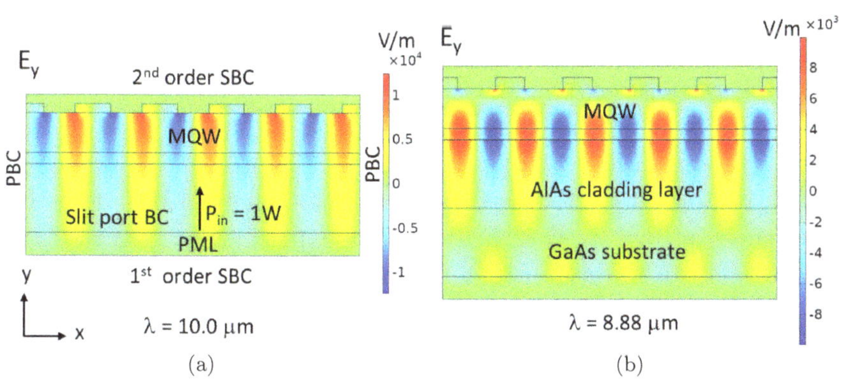

Figure 5.4. The vertical field distributions E_y at the peak wavelengths under 1 W incident power for a laminar grating (a) without a cladding layer and (b) with a cladding layer.

optical pathlength in the absorbing layer. Figure 5.4(a) shows that the parallel propagating light establishes local field maxima at the grating boundary.

For the LWG detector, the maximum QE occurs at a shorter $\lambda = 8.87\,\mu$m in Fig. 5.3(b), corresponding to $\theta = 63.6°$. At this λ, the internal wavevector $k_i = 2\pi n/\lambda$ and its parallel component $k_{//} = 2\pi n(\sin\theta)/\lambda = 2\pi 3.3 \sin(63.6°)/8.87 = (2\pi/3.00)\ \mu\mathrm{m}^{-1}$. Since the internal $\lambda_{//} = 2\pi/k_{//}$ is commensurate with the grating period of $3\,\mu$m, an EM wave can propagate up and down along the MQW layer, which becomes a waveguide. The traveling wave establishes local maxima at the bottom contact/cladding layer interface in Fig. 5.4(b). With a cladding layer, a long optical pathlength can also be achieved by the trapping of light in the MQW layer. Both LG and LWG can thus produce large QE albeit at different wavelengths. Due to the favorable E field direction at $\theta = 90°$, the LG detector achieves nearly the same peak QE as the LWG detector despite α being smaller at the longer λ, and thus it is more efficient. In addition to the diffraction peak, there is also a shoulder in each spectrum in Fig. 5.3(b). It is caused by the large material α that raises the detector absorption in this wavelength region.

Experimental evidence for the modeled QEs is provided in Ref. 4. The experimental LWG detector structure is slightly different from the above design and it is shown in Fig. 5.5(a). The contact thicknesses and the grating structures are different while the grating period $p = 3\,\mu$m and the MQW material design remain the same. From the responsivity lineshape in Ref. 4, the material α can be deduced in conjunction with design parameters, and it is shown in Fig. 5.5(b) and in Fig. 5.1(b).

The responsivity spectrum of a $500\,\mu$m diameter LWG detector is measured in Ref. 4. Together with the experimental gain of 0.48, the internal QE spectrum is deduced and shown in Fig. 5.6(a). It has a maximum value of 30%, which is nearly three times larger than the edge coupling QE. The high efficiency of this coupling scheme is thus experimentally verified. Reference 4, however, does not show the corresponding theoretical MEM solution. Here, theoretical support is obtained from FEM, and the modeled QE is shown in

Andersson & Lundqvist (Ref. 4)	Material	Doping (E18 cm^{-3})
Top contact p = 3 μm	GaAs, tc = 1.3 μm groove w = 1.55 μm groove h = 0.8 μm	1.0
MQW (50 periods) (1.79 μm)	Well: 48Å GaAs Barrier: 310Å $Al_{0.255}Ga_{0.745}As$	1.0
Bot. contact	1 μm GaAs	1.0
Cladding layer	3.0 μm AlAs	
Substrate	625 μm GaAs	undoped

(a)

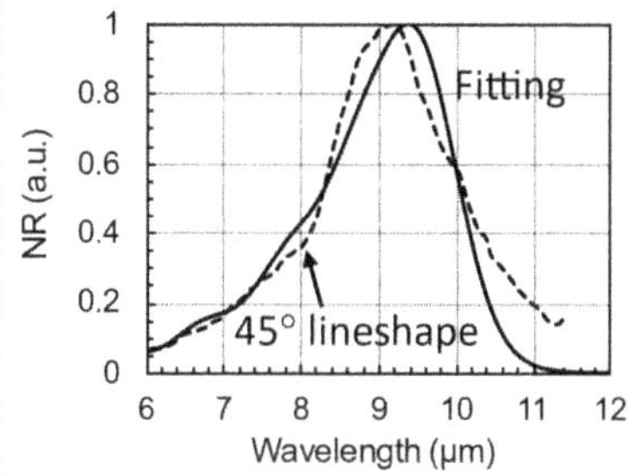

(b)

Figure 5.5. (a) The laminar waveguide detector structure adopted in [4] and (b) the observed and fitted responsivity lineshape.

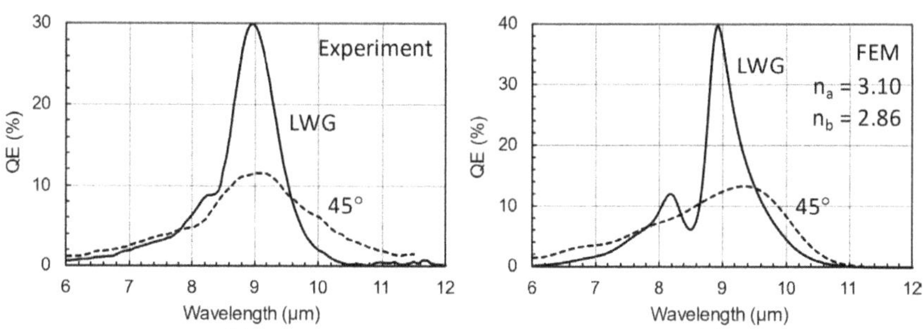

(a) (b)

Figure 5.6. The (a) observed and (b) modeled internal QE of a laminar waveguide detector adopted in [4]. $IQE_{EM} = 0.389\,\mu$m. The value of n_a is the fitted MQW index and n_b is the adopted AlAs index.

Fig. 5.6(b) using the fitted α spectrum. Both the experimental peak values and lineshapes are quite similar. This similarity confirms the designed QE can be experimentally realized. The smaller but broader peak in experiment is attributed to the grating nonuniformity in the fabrication. The experimental IQE is $0.347\,\mu$m, which is close to the modeled $0.389\,\mu$m. Therefore, the spectral broadening does not affect the integrated photocurrent.

In the modeling, the refractive index n_a of the AlGaAs/GaAs MQW layer and the GaAs layers need to adjust from 3.3 to 3.10

in order to fit the observed λ_p. Since the refractive index of AlGaAs barrier is lower than that of GaAs, the lower n_a is expected. Nevertheless, the need for index fitting reveals the fact that optical diffraction and interference are wavelength dependent. Reliable knowledge on the material's optical properties at the targeted wavelengths and operating temperatures are requisite to accurate design of grating detectors and other related detectors. This requirement is a serious challenge to these approaches because there is a lack of reliable data and because the effective refractive index is in fact different for each waveguiding mode, similar to an optical fiber.

A small part of difference in the modeled and measured QE is due to different material α. In order to eliminate its influence, a quantity C known as the light coupling efficiency is adopted. C is the QE ratio between a coupling scheme and 45° edge coupling, *i.e.*,

$$C(\lambda) = \frac{\eta(\lambda)}{\eta_{45}(\lambda)}. \tag{5.8}$$

Since the expression for η in Eq. (4.3) is directly proportional to α in any light coupling scheme, the α dependence is cancelled out in the ratio, and C is a measure of the ability of a coupling scheme in generating vertical E field in the active volume relative to 45° edge coupling. By taking the respective ratios, the theoretical and experimental coupling efficiency $C(\lambda)$ are shown in Fig. 5.7. It shows better agreement between theory and experiment without the material difference. The laminar waveguide detector is thus able to generate up to ~3 times higher vertical intensity than the 45° edge coupling in detectors with 500 μm diameter.

Since FEM can be used to model detectors of arbitrary geometries, it can model detectors with different sizes. To account for all optical scattering and confinement effects, 3D modeling is employed even though the grating grooves are linear. Shown in Fig. 5.8(a) is a LWG detector with linear size $d = 23\,\mu$m in a $D = 25\,\mu$m pitch array. It is made from the same material and grating structures shown in Fig. 5.2(b). The etched groove length is 21 μm, and the entire top pixel area is covered with gold. The gap between mesas is filled with epoxy with $n = 1.5$.

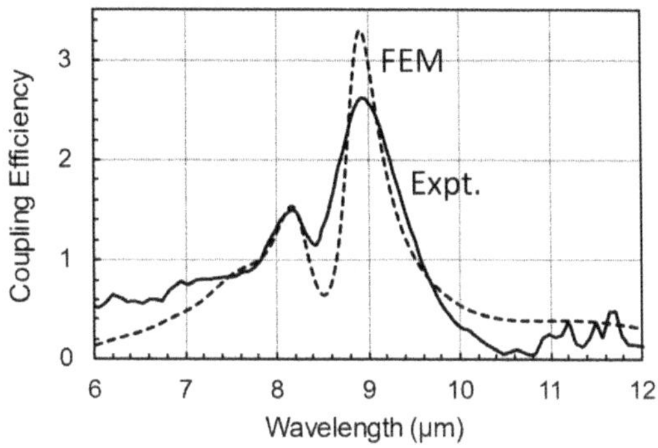

Figure 5.7. The measured and modeled coupling efficiency $C(\lambda)$ of the LWG detector.

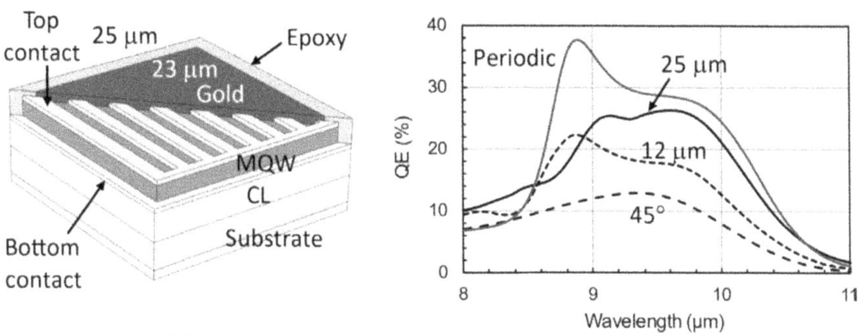

Figure 5.8. (a) The 25 μm pixel pitch LWG detector structure and (b) the modeled unpolarized internal QE for different detector geometries with $\alpha_p = 0.19\,\mu\text{m}^{-1}$.

In the FEM model, an EM wave is launched inside the GaAs substrate backed by a perfectly matched layer domain. This boundary condition eliminates any light reflected back from the bottom boundary of the modeling space and causes additional absorption. Periodic boundary condition is imposed on the four sides of the pixel. The polarization of the incident wave is orientated at $-45°$ relative to the grooves so that both TM and TE responses are modeled at the same time, *i.e.*, the input $\boldsymbol{H}$ is along (1, 1, 0) direction. They both contribute to QE for detectors with mesa sidewalls. The modeled

QE will use the detector area as the optical collecting area. If QE is referenced to the pixel pitch area, the modeled value will be smaller by the pixel fill factor $= (d/D)^2$.

With these model specifications, the modeled QEs of a 23 μm pixel in a 25 μm pitch array and a 10 μm pixel in 12 μm pitch array are shown in Fig. 5.8(b). Even with contribution from the TE mode, the QE of a finite-size pixel is generally smaller than that of an infinite grating. The peak values are 26.4% for 25 μm pitch and 22.4% for the 12 μm pitch, in comparison with 37.9% for infinite gratings. However, these values are still nearly double the 13.1% under edge coupling. The IQEs are 0.58 μm for infinite grating, 0.49 μm for $D = 25\,\mu$m grating, 0.37 μm for $D = 12\,\mu$m grating, and 0.24 μm for edge coupling. Besides the reduced coupling efficiency, the spectral lineshape is also different for different pixel size. This result shows the presence of mesa sidewalls has modified the electric field distribution in the detector.

5.2.2 *Cross Gratings*

Besides laminar grating, cross grating designs are also included in Ref. 6. With and without an AlAs cladding layer, the gratings have the same period of 3 μm. The MQW layer structure shown in Fig. 5.9(a) is the same as that in the laminar grating. For the grating without a cladding layer, labeled as the CG detector, the grating unit cell contains a square box of 1.7 μm wide and 1.2 μm deep. A continuous gold cover fills inside and over the boxes to form a square grating. Similarly, for the grating with a cladding layer, known as the CWG detector, the box width w is 1.8 μm, and box depth h is 1.0 μm. The cross-section of a unit cell of the CWG detector is shown in Fig. 5.9(b), and this grating type is called box grating. In addition to the box grating, in 3D geometry, a complementary grating structure also exists, in which a GaAs contact post replaces the box in the unit cell and the post is covered by a continuous gold layer. This grating type is called post grating, and it is shown in 5.9(c). The gratings discussed in Refs. 4–7 are all box type. Figure 5.10 shows the modeled QE presented in Ref. 6. These QEs have similar lineshapes as the laminar gratings in Fig. 5.3(a) but the peak values

Andersson & Lundqvist (Ref. 6)	Material	Doping (E18 cm^{-3})
Top contact p = 3 μm	GaAs, cross grating: 1.2 μm without CL, 1.0 μm with CL	1.0
MQW (50 periods) (1.75 μm)	Well: 50Å GaAs Barrier: 300Å $Al_{0.25}Ga_{0.75}As$	1.0
Bot. contact	0.5 μm GaAs	1.0
Cladding layer	3.0 μm AlAs	
Substrate	625 μm GaAs	undoped

(a)

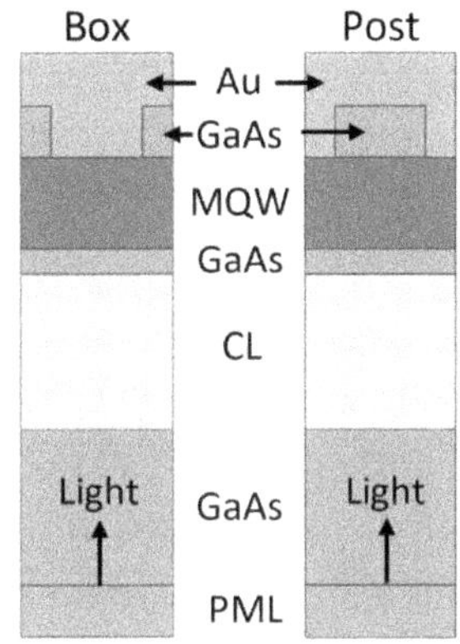

(b) (c)

Figure 5.9. (a) The optimized material structure and (b) box grating structure for cross gratings adopted in Ref. 6. (c) A complementary post grating having the same square dimensions.

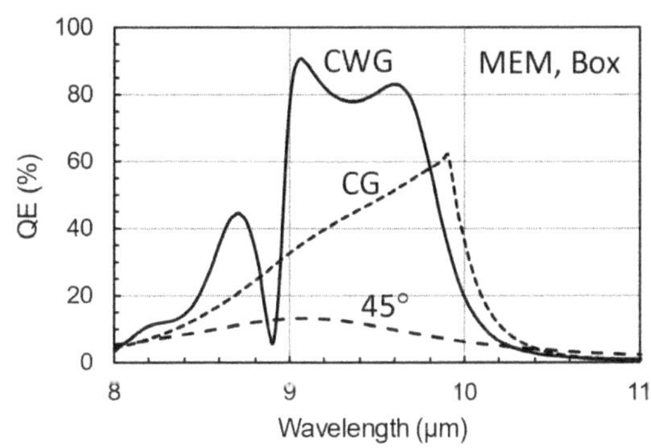

Figure 5.10. The modeled internal QE using MEM in Ref. 6 for box-type cross gratings with (CWG) and without (CG) the AlAs cladding layer.

are nearly twice. The MEM modeling thus predicts the cross grating to be twice as efficient as the laminar grating.

To repeat the modeling with FEM, we note that the present cross grating has square symmetry. Linearly polarized light along x and y directions have the same QE. Therefore, the input H is chosen to be along the (0, 1, 0) direction, *i.e.*, the E field is x-polarized. Periodic boundary condition is set at the four sidewalls of the unit cell. Using FEM, the modeled QEs for box-type gratings are shown in

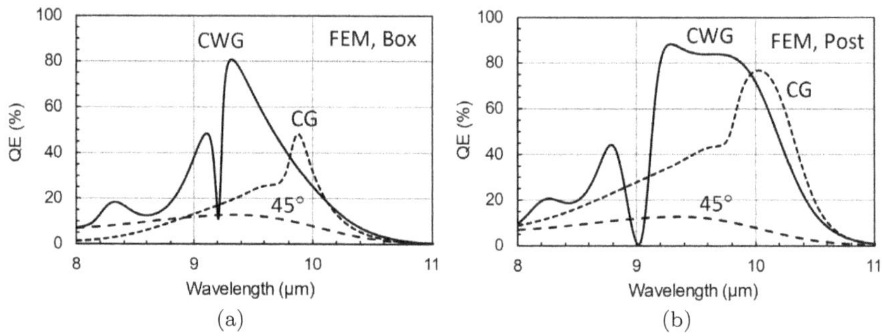

Figure 5.11. The modeled internal QE with (CWG) and without (CG) using FEM for (a) box grating and (b) post grating for the same grating parameters specified in [6].

Fig. 5.11(a). The predicted magnitudes are very similar to the MEM predictions for both CG and CWG detectors. However, their spectral lineshapes are quite different. Interesting, the MEM spectra in Fig. 5.10 are much more resemble to those modeled for the post grating geometries in Fig. 5.11(b), especially when the longer $\alpha(\lambda)$ cutoff in the FEM model is considered. This coincidence is unexplained.

Note that the post grating yields a larger bandwidth than the box grating in this set of grating parameters and it does not need a cladding layer to achieve ~80% QE. Similar to the laminar grating, the light is diffracted at ~90° at $\lambda_p \sim 10\,\mu$m, and it travels parallel to the absorbing layer even without the cladding layer, thus yielding a large QE at this particular λ. On the other hand, the CWG detector traps light in a wider range of λ, and a broader band detector is achieved.

Lundqvist *et al.* [7] fabricated different types of grating-coupled detectors to study their characteristics. Here, we focus only on cross gratings. These gratings are based on different sets of grating and material parameters are shown in Fig. 5.12. The QW doping N_D is $0.37 \times 10^{18}\,\text{cm}^{-3}$ and the Al molar ratio x is 0.29, with which the peak absorption wavelength is around 8.6 μm. The grating period p is 2.8 μm, and the cavity width w is 1.6 μm. The top contact layer thickness t_c is 1.3 μm but the etching depth h is only 0.75 μm, leaving a 0.55 μm contact layer below the cavity.

Lundqvist & Andersson (Ref. 7)	Material	Doping (E18 cm^{-3})
Top contact	GaAs, 1.3 μm	0.7
MQW (50 periods) (2.0 μm)	GaAs Well: 52Å for A; 49Å for B $Al_{0.29}Ga_{0.71}As$ Barrier: 348Å for A; 377Ä for B	0.37
Bot. contact	1.0 μm GaAs	0.7
Cladding layer CL	3.0 μm $Al_{0.78}Ga_{0.22}As$ for A; no CL for B	
Substrate	625 μm GaAs	undoped

(a)

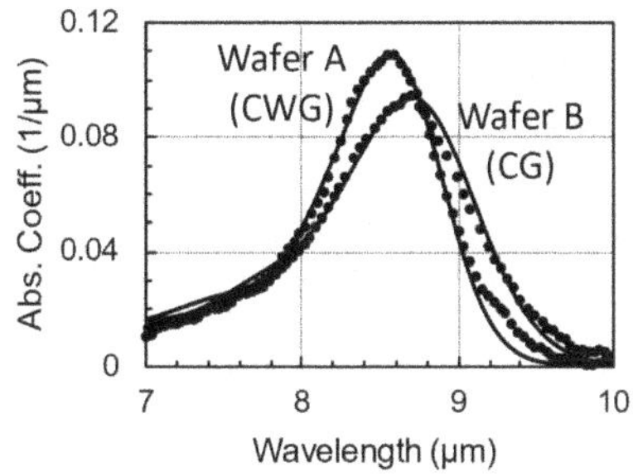

(b)

Figure 5.12. (a) The experimental grating and material parameters for cross gratings and (b) the observed QE lineshapes under edge coupling (circles). The solid curves are theoretical fitting based on the MQW nominal structures.

The detector QE was found to depend on the detector size. When the detector size d decreases from $500\,\mu$m to $25\,\mu$m, the peak QE of the CG detectors decreases from 29% to 13% and that of the CWG detectors decreases from 53% to 26%. Reference 7 presents a summary of the experimental results but without modeling results. The experimental spectra available are from the $500\,\mu$m diameter detectors. These QE spectra are reproduced in Fig. 5.13(a) for the CG detector and Fig. 5.13(b) for the CWG detector. The CG detectors are fabricated on a wafer without a cladding layer (wafer B), while the CWG detectors are fabricated on wafer A with a $3\,\mu$m $Al_{0.78}Ga_{0.22}As$ cladding layer.

From the MQW nominal structure and the provided edge-coupling spectral lineshapes reproduced in Fig. 5.12(b), the α spectrum can be deduced and α_p is $\sim 0.1\,\mu\text{m}^{-1}$ for the specified N_D. The FEM-modeled QEs are plotted with experimental data in Fig. 5.13. Since the refractive indices n_a of MQW and n_b of cladding layer are not precisely known, they are treated as fitting constants to match the observed λ_p's. The QE magnitudes are not sensitive to the indices.

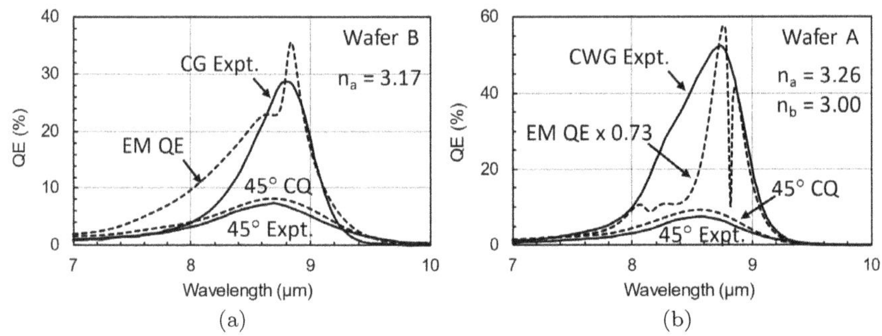

Figure 5.13. The measured and modeled internal QE of a cross grating detector (a) without a cladding layer and (b) with a cladding layer. The modeled QE for the CWG detector needs to be scaled by 0.73 to fit the observed magnitude for the assumed α.

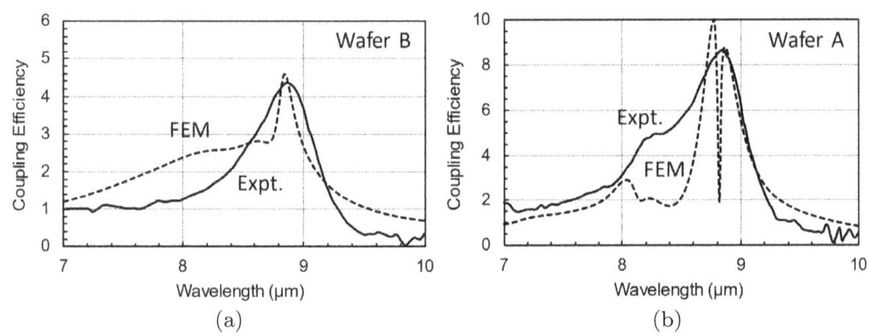

Figure 5.14. The observed and modeled $C(\lambda)$ of (a) a CG detector and (b) a CWG detector.

With the fitted indices, FEM closely predicts the experimental QE = 28.8% of the CG detector in Fig. 5.13(a). For the CWG detector, the peak value η_p is measured to be 52.3%, which is 0.73 times of the modeled 79.3% at 8.76 μm shown in Fig. 5.13(b). Since QE can be affected by α, the coupling efficiency $C(\lambda)$ defined in Eq. (5.8) are plotted in Fig. 5.14 for the two detectors. Without the material influence, FEM is shown to predict the peak values C_p correctly for both detectors.

In this experiment, the CG detector increases the edge-coupled QE by ~4.4 times, while it is 8.5 times for the CWG detector. The CWG C_p is thus more than double the LWG's $C_p \sim 3.0$ in Fig. 5.7. In addition, the modeled ~80% QE of the present CWG detector is the same as that modeled in Fig. 5.11(a) although the present material has ~3 times lower doping. These results show that optical trapping in the waveguide detectors is more beneficial to lower doping materials. When α is small, the light travels a longer distance. If the light does not leave the absorbing layer, it will eventually be absorbed, resulting in the same η.

These examples show that 30–50% internal QE can be obtained from large area grating detectors with a low N_D of 0.37×10^{18} cm^{-3} and a thin MQW layer of 2 μm. With the support of FEM modeling, the effectiveness of the grating designs by Andersson *et al.* is confirmed. FEM predicts a broader spectrum than experiment in Fig. 5.13(a) but a narrower spectrum in Fig. 5.13(b). The former may be related to an energy-dependent photoconductive gain while the latter may be related to material nonuniformity by growing a thick large x cladding layer in the material.

In another experiment [5], Andersson and Lundqvist adopted a higher 1×10^{18} cm^{-3} doping used in the detector design in Fig. 5.10. The material and grating structures are shown in Fig. 5.15(a). The detector mesa has diameter of 500 μm. The MEM-modeled QE and the measured QE from Ref. 5 are shown in Fig. 5.15(b). The agreement in the peak value is excellent. The measured 92% QE represents a major achievement in QWIP technology.

Since Ref. 5 does not include the edge coupling spectrum, the material α is not exactly known. Nevertheless, the MQW structure is nominally the same as that in Fig. 5.5. Therefore, the same α in Fig. 5.5(b) is adopted for FEM modeling. Based on the grating parameters listed in Fig. 5.15, the modeled QE is shown in Fig. 5.16(a). To match the QE peak more precisely, n_a of the MQW is adjusted from 3.30 to 3.28 to blueshift the peak by 0.06 μm. The peak value, however, is modeled at 66.2%, less than the 92% measured or using MEM. The FEM QE thus needs to scale up 1.4 times in order

Andersson & Lundqvist (Ref. 5)	Material	Doping (E18 cm^{-3})
Top contact p = 3 μm	GaAs, tc = 1.3 μm, groove w = 2.1 μm groove h = 0.85 μm	1.0
MQW (50 periods) (1.85 μm)	Well: 50Å GaAs Barrier: 320Å $Al_{0.235}Ga_{0.725}As$	1.0
Bot. contact	0.5 μm GaAs	1.0
Cladding layer	3.0 μm AlAs	
Substrate	625 μm GaAs	undoped

(a)

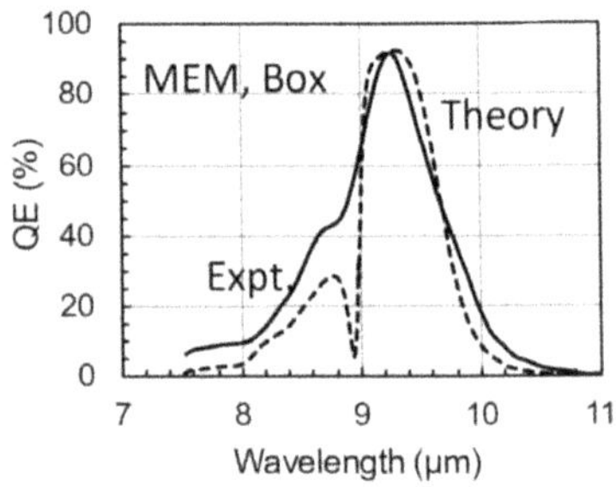

(b)

Figure 5.15. (a) The material and grating parameters for the CWG detector adopted in Ref. 5 and (b) the modeled and measured unpolarized internal QE adapted from Ref. 5.

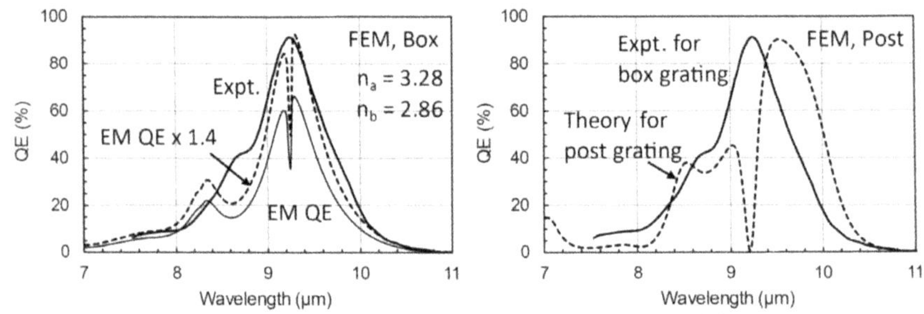

(a) (b)

Figure 5.16. (a) The modeled and fitted internal QE using FEM for the box grating and (b) the modeled QE using FEM for the post grating. n_a and n_b are the refractive indices of the MQW layer and the cladding layer. The adopted α is shown in Fig. 5.1.

to match the experimental value. After this scaling, the modeled lineshape agrees well with the experiment. Unfortunately, due to the lack of the exact α spectrum, the present modeling cannot be directly compared with the MEM result and cannot fully substantiate the measured QE. Nevertheless, FEM does support generally a large QE ~66% and the measured lineshape. Note that if post gratings are used instead, 90% QE can be obtained as shown in Fig. 5.16(b) together with broader bandwidth.

5.2.3 *Cross Grating Characteristics*

From the above discussions, FEM predicts QE of large area gratings satisfactorily. It can then be used to study other properties. An important characteristic is the pixel size dependence in FPA applications. Here, we will focus on CWG detectors of different pixel pitch D and pixel linear size d. For example, a detector with $D = 25\,\mu$m and $d = 23\,\mu$m is shown in 5.17(a). The material and grating parameters are the same as that in Fig. 5.15 except $n_a = 3.30$. PBC is enforced on four sides of the displayed model and scattering boundary condition (SBC) is enforced at the top and bottom surfaces. EM wave is launched using internal slit boundary condition backed by a perfectly matched layer domain. Figure 5.17(b) shows detectors with three different d in a $D = 25\,\mu$m array. The peak QE decreases from 66.2% for an infinite grating in Fig. 5.16(a) to ~40% at this D, and this QE is about 3 times larger than the 13.3% of an edge-coupled detector, *i.e.*, $C_p \sim 3.0$. The IQE under edge coupling is $0.30\,\mu$m, while that under gratings are 0.59, 0.63, and $0.61\,\mu$m for $d = 22$, 23, and $24\,\mu$m, respectively. These gratings thus show a factor of ~2 increase in the edge-coupled IQE. When a gold-covered planar detector with $d = 23$ without grating is modeled, $\eta_p = 4.7\%$ and IQE is $0.11\,\mu$m,

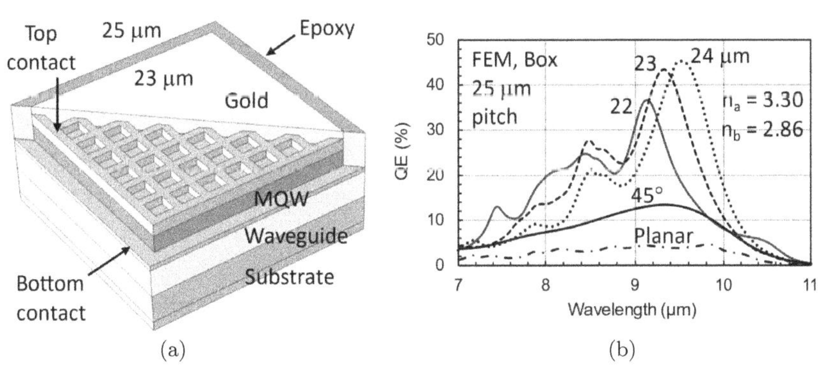

Figure 5.17. (a) A CWG detector structure with $D = 25\,\mu$m and $d = 23\,\mu$m adopted in FEM modeling. The gold cover is partially exposed. (b) The modeled internal QE with different d. The QE of a 45° edge detector and a planar detector with $d = 23\,\mu$m is also shown. $\text{IQE}_{\text{EM}} = 0.59$, 0.63, and $0.61\,\mu$m for $d = 22$, 23, and $24\,\mu$m, respectively.

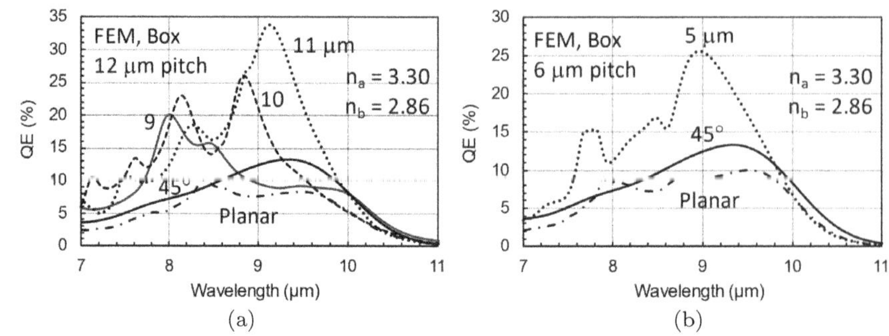

Figure 5.18. The modeled internal QE with different d for (a) a $12\,\mu$m pitch array and (b) a $6\,\mu$m pitch array. The QEs of a 45° edge detector and planar detectors with $d = 9$ and $5\,\mu$m are also shown. IQE $= 0.36$, 0.45, and $0.52\,\mu$m for $d = 9$, 10, and $11\,\mu$m, respectively.

which is due to mesa edge diffraction. C_p of the planar detector is thus 0.35.

When D is reduced to $12\,\mu$m in Fig. 5.18(a), the peak QEs are 20.0%, 26.1%, and 33.8% while the corresponding IQE $= 0.36$, 0.45, and $0.52\,\mu$m for $d = 9$, 10, and $11\,\mu$m, respectively, which are a less-than-double improvement from $\text{IQE}_{45} = 0.30$. Without a grating, a planar detector with $d = 9\,\mu$m exhibits $\eta_p = 10.0\%$, and IQE $= 0.21$, which is nearly half of IQE of the gratings. In small pitch detectors, d plays an important role in determining QE and lineshape.

When $(D, d) = (6, 5)$, $p = 3\,\mu$m, and $w = 2.1\,\mu$m, a square grating only contains one diffraction square in each pixel. QE, however, increases again to 25.5% and IQE $= 0.45\,\mu$m shown in Fig. 5.18(b). For a planar square pixel without the gold cavity, $\eta_p = 10.0\%$ and IQE $= 0.24\,\mu$m. Therefore, the presence of the gold cavity increases planar IQE by a factor of 2.

For imaging arrays with a focusing lens, the effective QE is averaged over different incident angles. Using the same modeling setup in the study of C-QWIPs in Fig. 4.5, one can model CWG QE under different incident angles θ. Figure 5.19 is an example when the internal θ is 5°. It corresponds to an external incident angle of 16.7° for $n = 3.3$, which is the maximum θ for a lens with $f/\# = 1.67$. Figure 5.19 shows that under oblique incidence, the

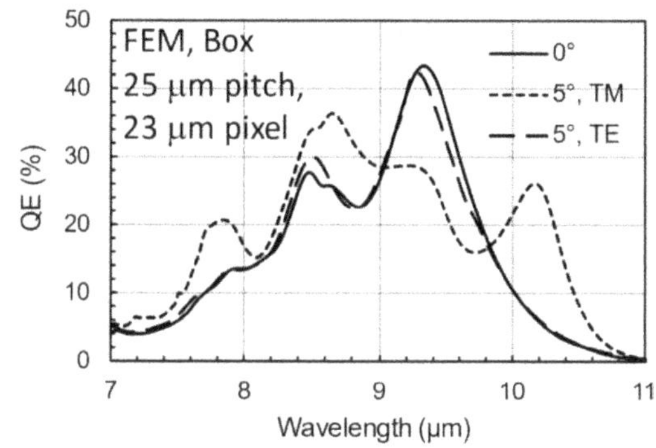

Figure 5.19. The modeled QE at 0° and 5° interior angles.

spectral lineshape of the TM mode is substantially changed while that of the TE mode remains the same within numerical error. For the TM mode, IQE $= 0.74\,\mu$m, similar to $0.63\,\mu$m at $\theta = 0°$. Therefore, the integrated I_p under focused illumination is unaffected by the lens $f/\#$.

Another important characteristic of a light coupling scheme is optical crosstalk (OCT). OCT can be extrinsic or intrinsic. Extrinsic OCT is caused by the presence of the substrate/air interfaces. When the unabsorbed light bounces at an angle inside the substrate of a detector array, it can cause unwanted absorption and raise the detector current level with the same degradation effects of the dark current. Unlike the dark current, the extrinsic OCT current cannot be reduced by reducing the operating temperature. On the other hand, it can be reduced by thinning the substrate so that the bounced light returns to the same detector. In the case of C-QWIPs, when the substrate is thinned to the common contact layer to eliminate the extrinsic OCT, the remaining intrinsic OCT is modeled at the 10^{-4} level, an extremely low value due to the retro-reflective properties of the C-QWIPs. In grating-coupled QWIPs, however, due to the large angle diffraction, OCT can be serious with or without the cladding layer. Even when the substrate is removed, the presence of a $3\,\mu$m cladding layer in CWG detectors can guide the unabsorbed light to other pixels and causes substantial intrinsic OCT. The ability of guiding a wave to other parts of the detector in producing a large

QE in a CWG detector will also give it a large OCT when the other parts are designated to be different detectors.

To study the intrinsic OCT of CWG detectors with $(D, d) = (25, 23)$ and (12, 10), detector arrays of 5×5 pixels are modeled. The center pixel is illuminated with $\boldsymbol{H} = (0, 1, 0)$ A/m, while the rest are not illuminated. The wave is launched at the bottom surface of the GaAs substrate and all external boundaries are set to second-order SBC so that the waves are free to leave the model space. Without substrate reflection in this modeling geometry, extrinsic OCT is excluded, and the modeled OCT is intrinsic and it is due to the presence of a common contact layer and cladding layer in the waveguide detectors. This center pixel illumination scheme models after the focused spot measurement in the usual experiment, in which the incident light is focused onto a single pixel while the responses from the entire array are monitored.

The center cross-sections of the two arrays and the E_z distributions are shown in Fig. 5.20. Figure 5.21 shows the modeled QEs of all

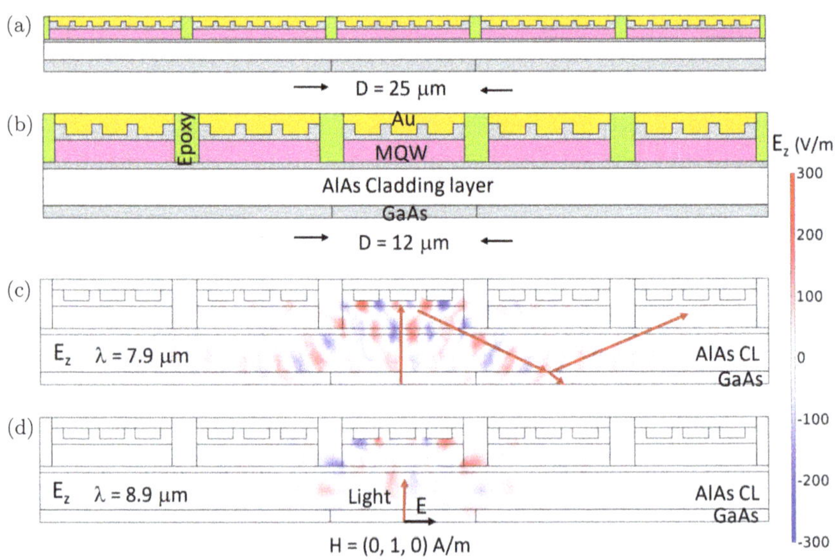

Figure 5.20. The center cross-sections of the 5×5 arrays of (a) $25\,\mu$m pitch and (b) $12\,\mu$m pitch. The E_z field patterns of the $12\,\mu$m array at $\lambda =$ (c) $7.9\,\mu$m and (d) $8.9\,\mu$m.

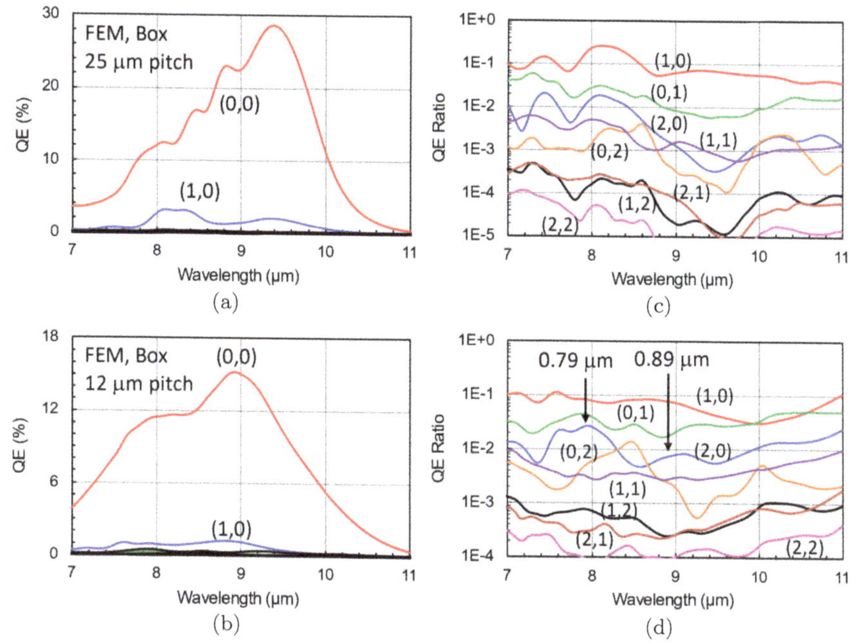

Figure 5.21. The modeled unpolarized internal QE of different pixels in the (a) 25 μm array and (b) 12 μm array. (c) and (d) The corresponding QE ratios in the logarithmic scale.

the pixels and the QE ratios of the neighboring pixels with respect to the center pixel. Figure 5.22 further shows the E_z field distributions in the mid-planes of the MQW layers at the peak wavelengths.

In particular, Figs. 5.20(c) and (d) show that the unabsorbed light at two wavelengths in the center pixel of the 12 μm array is partly refracted into the GaAs substrate and partly reflected into the neighboring pixels at the CL/GaAs interface. The refracted light into the substrate will cause extrinsic OCT when the light reflects at the substrate/air interface. On the other hand, the reflected light from the CL/GaAs interface will cause intrinsic OCT. Since the E_z field can spread to other pixels, each pixel will receive light from its neighbors. The effective detector size will be larger than its physical size d in the detector spatial response function $H(\nu)$ in Eq. (1.14), which will lead to a reduction of the modulation transfer function in spatial frequency ν.

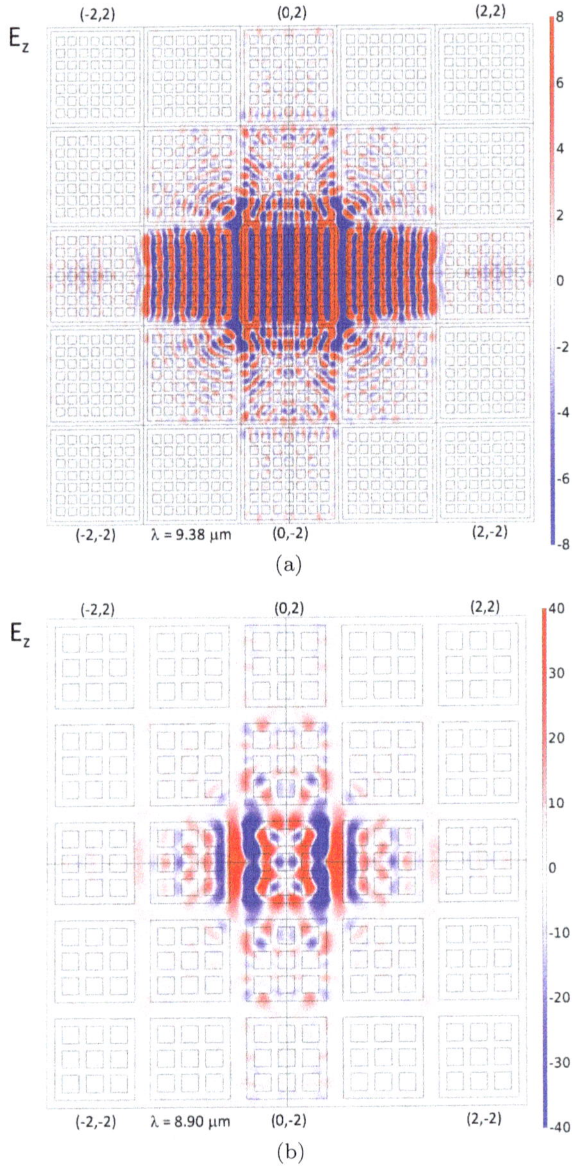

Figure 5.22. The E_z field in the MQW mid-plane at the peak wavelengths of the (a) (25, 23) array and (b) (12, 10) array in the highly expanded E_z-scale.

The QEs of the center pixels in Figs. 5.21(a) and (b) maintain similar lineshapes and peak wavelengths as that in the periodic unit cell (PUC) modeling in Figs. 5.17 and 5.18. In PUC modeling, the input power is assumed to be the same for all unit cells and thus it represents the flood exposure experimental condition. The PUC QEs thus contain extra absorption from light scattered from the surrounding pixels. This explains QE in PUC modeling is generally higher than that in focused spot modeling. Since the wave in each unit cell is in phase, it constitutes coherent OCT. If the phase of the incident light onto each pixel is random such as that from a thermal blackbody, the OCT will be incoherent. In this case, the light intensity in the center pixel is given by the sum of the individually scattered intensities and not the coherent E^2 intensity as in PUC modeling. In the incoherent case, the total QE in one pixel is the sum of all QEs modeled for the entire array. The sum for the 5×5 pixels in the present case is displayed in Fig. 5.23. It is seen that incoherent OCT increases the focused spot QE by 15% in the 25 μm array and 23% in the 12 μm array on average. However, these increases are generally less than that from a coherent light source obtained from PUC modeling.

The amount of OCT in an array is measured by the QE ratio, which is the ratio of QE between a neighboring pixel and the center

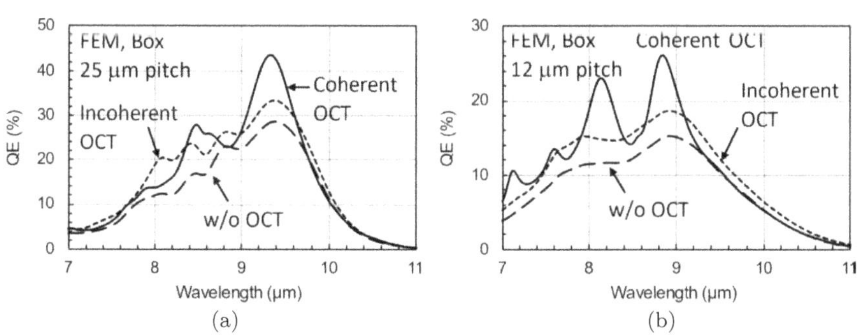

Figure 5.23. The modeled unpolarized internal QE of (a) a (25, 23) CWG detector and (b) a (12, 10) CWG detector under focused spot condition, under flood exposure with an incoherent light source, and under flood exposure with a coherent light source.

pixel under focused spot illumination. These ratios are shown in Figs. 5.21(c) and (d). They are smaller by a factor of ~0.1 when the pixel locates sequentially away from the center pixel. Due to waveguiding effects, this OCT level is far larger than the 10^{-4} level for the C-QWIPs. It is especially higher at certain λ such as that at $\lambda = 7.9\,\mu$m for the (2, 0) pixel in Fig. 5.21(d). The E_z field distribution in Fig. 5.20(c) shows that at this λ, the diffracted beam is readily reflected into the (2, 0) pixel by the cladding layer and raises its absorption.

The QE ratios and E_z distributions are not symmetric with respect to the x and y axes. E_z is much stronger in the (1, 0) pixel than the (0, 1) pixel in Fig. 5.22. It is due to the asymmetric input $\boldsymbol{H}$, which is chosen to be (0, 1, 0) A/m. The incident wave is thus x-polarized. After diffraction in the x-direction, substantial E_z component is created in contrast to almost none in the y-direction, which gives the x-y asymmetry. The finite E_z in the (0, 1) and (0, 2) pixels shows the square grating is capable of converting TE mode in the y-direction into TM mode in that direction, and it is the key to obtain >50% unpolarized QE in cross gratings. It also explains the nearly constant E_z in Fig. 5.22 along the y-axis between cavities for x-polarized incidence. The intensity of the diffraction of light is still symmetric along x- and y-axes, which can be seen by plotting the $|E|^2$ distribution. Only the polarization properties of the diffracted light are different under polarized incidence.

Note that there is also diffracted light in the diagonal pixels. It shows that by seeking exact solutions to Maxwell equations, FEM modeling accounts for all diffraction orders in all directions.

Due to the small difference in the MQW and cladding layer refractive indices, most diffracted light passes through the cladding layer rather than reflected by the layer even at a large angle. Therefore, a more effective means to achieve waveguiding is to omit the cladding layer and remove the substrate completely. An ARC layer can still be retained at the bottom of the GaAs common contact layer as envisioned by Andersson *et al.* The cross-section of a $12\,\mu$m pitch array is shown in Fig. 5.24(a). The larger index contrast

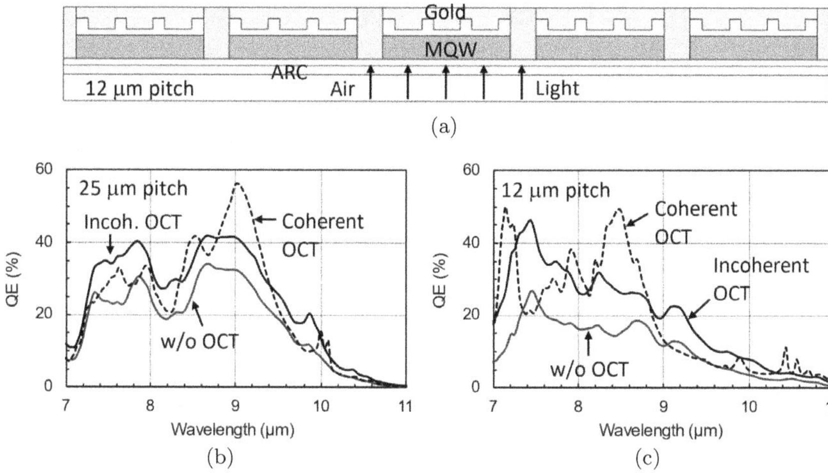

Figure 5.24. (a) The cross-section of a $12\,\mu$m pitch CG detector array without cladding layer and substrate but with an ARC. The modeled external QE of the CG detector with (b) $25\,\mu$m pitch and (c) $12\,\mu$m pitch under focused spot exposure with and without incoherent OCT, compared to the same CG detector under coherent flood exposure. $IQE_{EM} = 0.705\,\mu$m and $0.440\,\mu$m for $25\,\mu$m and $12\,\mu$m, respectively. The adopted α is shown in Fig. 5.1.

between the common contact layer/ARC and ARC/air interfaces reflect light in a larger range of diffraction angles and thus widen the spectral band. The ARC is a simple quarter wavelength design optimized for $9\,\mu$m radiation, which is a dielectric with thickness of $0.682\,\mu$m and $n = 1.817$. The external QE is modeled under both flood exposure with incoherent OCT and focused spot without OCT in Fig. 5.24(b) for the $25\,\mu$m pitch and in Fig. 5.24(c) for the $12\,\mu$m pitch. They are compared with the external QE under PUC modeling where coherent OCT is present. Compare these QE spectra with those in Fig. 5.23: the present QE spectra are broader and taller especially at the shorter wavelengths $\sim 7.5\,\mu$m where the cladding layer is ineffective. Unfortunately, the OCT is also significantly stronger in the present case, which is caused by the same effect that makes the QE larger. The higher index contrasts of the bottom layers make the backscattering of light stronger. At $\sim 7.5\,\mu$m, OCT can double the QE in the $12\,\mu$m FPA. The modeled IQE for the $25\,\mu$m pitch array with OCT is $\mathrm{IQE_{EM}} = 0.705\,\mu$m and that of $12\,\mu$m pitch is $\mathrm{IQE_{EM}} = 0.440\,\mu$m.

5.2.4 *Cross Gratings for Very Long Infrared Wavelengths*

Sarusi *et al.* [8] presents a study on a set of gratings designed for very long wavelength spectral region between 16–17 μm. The material and grating structures are shown in Fig. 5.25. Box gratings are implemented with grating period p from $5.0\,\mu$m to $8.5\,\mu$m. There is no waveguide cladding layer in the material, and the substrate is unthinned and remains to be $650\,\mu$m thick. Detector area is $200 \times 200\,\mu\mathrm{m}^2$, and the detectors are under flood exposure. There is an AlGaAs etch stop layer on top of the MQW layer to allow precise etching of the grating cavities.

From the nominal MQW structure and the experimental edge-coupled responsivity lineshape, the material absorption coefficient can be deduced and fitted, and it is shown in Fig. 5.25(b). The fitted $\alpha_p = 0.086\,\mu\mathrm{m}^{-1}$. The experimental α contains an extra absorption peak near $12\,\mu$m, which is not identified. Nevertheless, the experimental α lineshape is adopted in FEM modeling and the peak α is chosen to be the same at $0.086\,\mu\mathrm{m}^{-1}$. At very long infrared

Sarusi et al.	Material	Doping (E18 cm^{-3})
Top contact p = 5, 5.3, 5.6, 5.9, and 6.2 μm	GaAs, 1.4 μm h = 1.4 μm w = √(p²/2)	0.3
Etch stop	AlGaAs, 800Å	0.3
MQW (50 periods) (3.35 μm)	Well: 70Å GaAs Barrier: 600Å $Al_{0.12}Ga_{0.88}As$	0.3
Bot. contact	GaAs	0.3
Substrate	625 μm GaAs	undoped

(a)

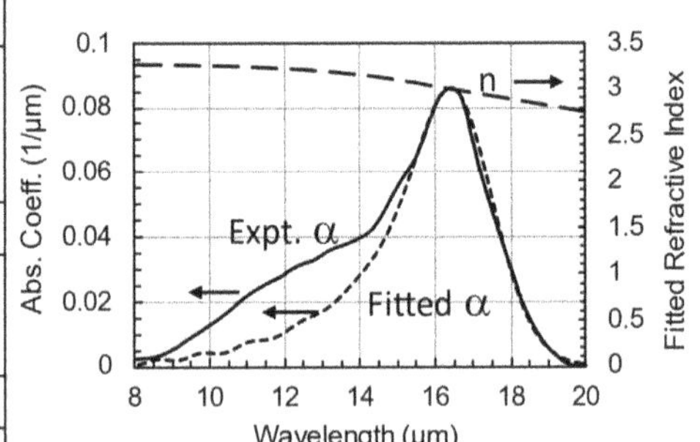

(b)

Figure 5.25. (a) The material and grating structures adopted in Ref. 8 and (b) the experimental and fitted α and n of the MQW structure.

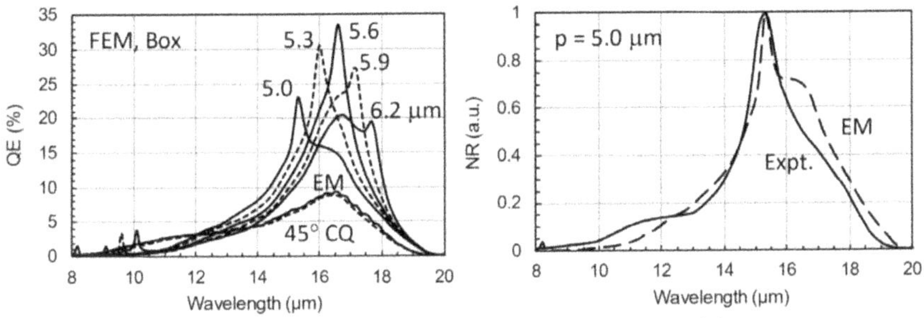

Figure 5.26. (a) The modeled external QE of the CG detectors studied in Ref. 8 and (b) the experimental and modeled responsivity lineshape for the $p = 5.0\,\mu$m grating.

wavelengths, the MQW n is expected to be lower. The fitted n in Ref. 8 is shown in Fig. 5.25(b), which is also adopted. Since there are many grating periods in the detectors, a single periodic grating unit cell similar to that of Fig. 5.9 is modeled. The input H polarization in the port boundary condition is $\boldsymbol{H} = (1, 1, 0)$ so that the input field is diagonally polarized.

Although the experiment includes 10 different periods between 5.0 and $8.5\,\mu$m, there is an abrupt change of the measured detector characteristics when p increases from $6.2\,\mu$m to $6.5\,\mu$m. For example, when p increases from $5.0\,\mu$m to $6.2\,\mu$m in the first set of detectors, there is a smooth shift of peak wavelength from $15.3\,\mu$m to $17.7\,\mu$m. However, when p increases from $6.5\,\mu$m to $8.5\,\mu$m in the second set of detectors, λ_p starts a new trend from $15.4\,\mu$m to $17.1\,\mu$m. In addition, when the detector substrate is thinned from $650\,\mu$m to $150\,\mu$m, the first set QE only increases by ~20% in comparison to the 200% increase in the second set. The large QE increase in the second set might indicate large extrinsic OCT because of different detector layout or packaging. The unexplained $12\,\mu$m absorption in the edge-coupled detector might also be the result of OCT.

QE of all detectors are modeled. However, only the $\eta_{\rm EM}$ on the first set is consistent with experiment, and the modeled spectra are shown in Fig. 5.26(a). Note that the displayed QEs are external QEs including the substrate transmission loss in order to compare with the experiment. The edge-coupled QE is modeled using classical and FEM models. Their results are practically the same and therefore only the classical QE is used as reference.

The modeled QE in Fig. 5.26(a) is seen to be affected by both grating diffraction and material α. At $p = 5.0\,\mu$m, the first-order diffraction angle θ reaches 90° when $\lambda_{\text{int}} = p$ or $\lambda = np = 3.08 \times 5 = 15.4\,\mu$m. The accompanied large E_z thus creates a QE peak around this λ. Besides this peak, a shoulder also appears around 16.7 μm, which coincides with the α maximum. Therefore, a local maximum may appear either $|E_z|^2$ or α is a maximum as intersubband transition depends on the product of these two quantities. When p is larger, the diffraction peak passes through the α maximum, yielding the absolute QE maximum at $p = 5.6\,\mu$m. Afterward, the grating efficiency diminishes as the larger period diffracts the material α wavelengths at a smaller angle.

From the modeled QE, the responsivity can be calculated and normalized to unity at the maximum. The normalized responsivity is then compared to the experimental responsivity lineshape shown in Ref. 8. The result for $p = 5\,\mu$m grating is shown in Fig. 5.26(b). The modeled diffraction peak matches well with the experimental peak but the α shoulder appears to be less prominent in the experiment. The shorter experimental peak near 11 μm is not predicted even when the experimental α is used in the modeling. The E_z generated by the grating is too weak at this wavelength to produce the observed peak. Similar agreement is obtained from other periods in the first set of detectors, and they are shown in Fig. 5.27.

The FEM modeling shows good match in the diffraction peak λ and the lineshape. However, it consistently misses the shorter wavelength peak and overestimates the material α peak. These discrepancies are unexplained. Besides the normalized responsivity lineshape, the modeled magnitude is also consistent with experiment. Figure 5.28 shows the experimental R_{peak} relative to that under edge coupling. It represents the peak value of the coupling efficiency C_p defined in Eq. (5.8). EM modeling correctly describes the trend of C_p with p but overestimates their values by a factor of 1/0.73, which probably is related to the finite size effect of the 200 μm

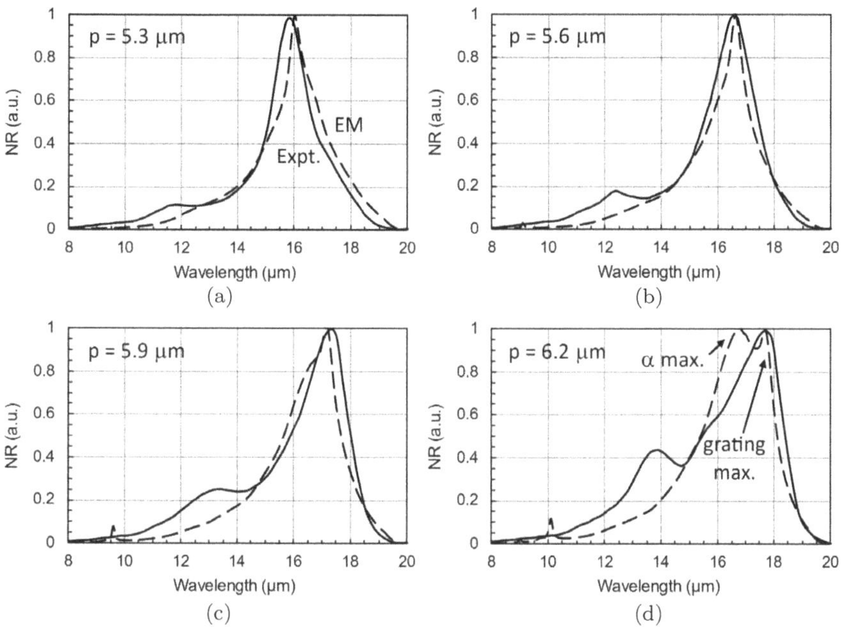

Figure 5.27. The experimental and modeled responsivity lineshape for different p. The double peaks at $p = 6.2\,\mu$m are created by the material α maximum and the grating maximum where $\lambda_{\text{int}} = \lambda/n = 17.7\,\mu\text{m}/2.91 = 6.1\,\mu\text{m} \approx p$ of $6.2\,\mu$m.

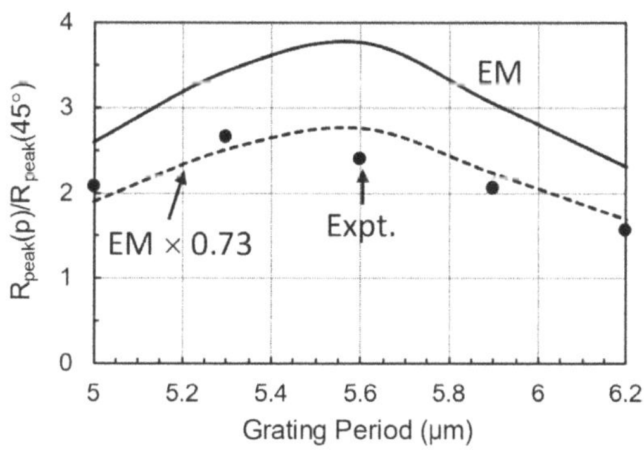

Figure 5.28. The comparison of the peak coupling efficiency C_p between theory and experiment as a function of p. The experimental value is smaller by a factor of 0.73 consistently.

detectors whose QE should be smaller than the modeled infinite grating solution. The finite pixel size effect may also explain the discrepancies between the modeled and observed lineshapes. Based on experiment and modeling on this set of detectors, it can be confirmed that a 200 μm size cross grating detector having a 3.35 μm thick MQW doped at 0.3×10^{18} cm^{-3} can yield an external QE of ~25% on a thick substrate detector without ARC.

When p increases from 6.2 μm to 6.5 μm, the experimental λ_p abruptly jumps from 17.7 μm to 15.4 μm and stays unchanged at $p = 7.0\,\mu$m before it gradually shifts to 17.1 μm at $p = 8.5\,\mu$m. The QE varies with an amplitude similar to that of the first set. When the substrate is thinned from 650 μm to 150 μm, the second set QE becomes 1.5 times larger than the first set. The larger QE is not predicted by the present FEM modeling. Instead, it predicts the QE spectrum continues to decrease with larger grating periods. In Ref. 8, the shorter wavelength peak and the λ_p jump were explained by diagonal diffractions not included in the original MEM formulism. However, this explanation cannot be applied to FEM analysis since its solution is the exact numerical solution to Maxwell's equations. The FEM QE already includes all diffraction orders, and the electromagnetism uniqueness theorem guarantees a single solution to a set of boundary conditions. It will not be feasible in FEM to yield another solution. Therefore, the second set of experimental data remains unexplained in the present analysis. If the larger period grating is indeed capable of coupling light of shorter wavelength into the detector at a 1.5 times higher efficiency, it would be a very welcoming property since gratings of smaller periods are more difficult to fabricate.

5.2.5 *Grating-Coupled Quantum Well Infrared Photodetector Focal Plane Arrays*

Following the promising results from Andersson *et al.*, there were many efforts in producing grating-coupled QWIP FPAs. An in-depth study was led by Kozlowski [9]. The study includes a large number of FPAs and discrete detectors from many wafers. The grating

Table 5.1. The material and grating parameters in the study described in Ref. 9.

Kozlowski	Material	Doping (E18 cm^{-3})
Top contact $p = 3.5\,\mu$m for LG $p = 3.0\,\mu$m for CG	GaAs: $1.0\,\mu$m, $h = 0.7\,\mu$m $w = 1.9\,\mu$m for LG $w = 2.1\,\mu$m for CG	1.0
MQW (50 periods) ($2.70\,\mu$m)	Well: 40 Å GaAs Barrier: 500 Å $Al_{0.25}Ga_{0.75}As$ (35 periods for CWG wafers)	1.0
Bot. contact	$0.9\,\mu$m GaAs	1.0
Waveguide	$3.0\,\mu$m AlAs for CWG wafers	
Substrate	$625\,\mu$m GaAs for unthinned FPA, $30\,\mu$m for thinned FPA	undoped

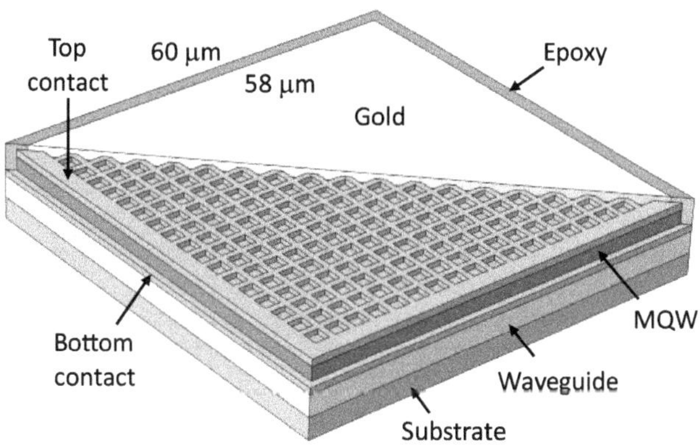

Figure 5.29. The CWG FPA pixel structure used in Ref. 9. The pixel pitch is $60\,\mu$m. The gold cover is partially exposed to show the grating structure underneath.

structures include 1D laminar gratings, 2D cross gratings, and cross gratings with an AlAs cladding layer, also known as a waveguide layer. The pixel pitch of FPAs is $60\,\mu$m, and discrete detectors have a linear size of $200\,\mu$m. No ARCs are employed. Table 5.1 shows the nominal material structure used in all detectors. The FPA pixel structure is shown in Fig. 5.29.

Table 5.2. Discrete detector performance, $200 \times 200\,\mu m^2$ area, focused spot measurement.

Layer Number	1047	1195	1242	1289
Type	1D grating	1D grating	2D grating	2D grating & WG
J (-2 V, 78 K) (A/cm^2)	3.95×10^{-3}	2.15×10^{-3}	4.98×10^{-3}	1.24×10^{-2}
R_p (A/W)	0.040	0.020	0.043	0.376
λ_p (μm)	8.8	8.8	8.8	9.6
g (-2 V)	0.46	0.37	0.35	0.77
ηC_p (i.e., QE)	1.2%	0.77%	1.7%	6.3%
D^*(-2 V, 78 K) (cm$\sqrt{\text{Hz}}$/W)	1.2×10^9	9.0×10^8	1.3×10^9	4.8×10^9

Table 5.3. FPA quantum efficiency under flood illumination, pixel pitch = $60\,\mu$m.

Detector Configuration	Wafer Number	Relative Crosstalk	Typical ηC_p (%)
1D grating with unthinned substrate	1195	Very high	1.0%
2D grating	1242	High	2.3%
2D grating and waveguide	1289	High	15.0%
1D grating with thinned substrate	1195	Low	0.4%
2D grating with thinned substrate	1242	Low	1.2%
2D grating with waveguide & thinned substrate	1289	Low	8.0%

Table 5.2 summarizes the characteristics of discrete detectors, which are fabricated on four different wafers. For these wafer materials, the gain saturates around -2 V top contact bias, after which g decreases slowly. Under focused spot measurement on discrete detectors, the external QE, labeled as ηC_p in Ref. 9 where η is the η_{45} in Eq. (5.8), is about 1% for LG detectors, 1.7% for CG detectors, and 6.3% for CWG detectors. Meanwhile, Table 5.3 summarizes the $60\,\mu$m pixel FPA performance on three wafers under flood exposure condition. When the substrate is thinned to $30\,\mu$m, their QEs are 0.4, 1.2, and 8.0% for the LG, CG, and CWG detectors,

respectively. This level of QE is more than 10 times lower than that measured and modeled in the previous sections and needs explanation.

This large discrepancy is due to a number of factors, including QE definition, detector size, applied voltage, MQW design, and grating design. Firstly, internal QE is usually presented in the above sections assuming that it can be realized with a perfect ARC. It is ~1.4 times larger than the external QE measured in this experiment, which does not use an ARC. Secondly, the discrete detector size of 200 μm in this experiment is not large enough to yield the modeled infinite grating QE. From the previous section, QE at 200 μm can be 0.73 times smaller than the infinite grating QE. Thirdly, QE is measured at -2 V. The applied field E is 40 mV/period for the 50 period MQW. This applied field is too low to yield a tunneling probability (or, more generally, the electron transfer probability) γ of 1, so the measured QE $\eta_d = \eta_{\text{abs}}\gamma$ is only a fraction of the modeled η_{abs}. From the experimental study of QWIP in Section 2.16, E needs to be larger than 60 mV/period to yield a $\eta_d \approx \eta_{\text{abs}}$. Indeed, from the $R(V)$ and $g(V)$ curves shown in Ref. 9, $\eta_d = (h\nu/e)(R/g)$ is increased by a factor of 2.0 when V increases from -2 V to -5 V. Therefore, if $\gamma = 1$ at -5 V, γ at -2 V would be 0.50.

Fourthly, the present MQW structure has a well width of 40 Å and an Al molar ratio of 0.25. The material thus has bound-to-continuum (B-C) energy level structure. With the barrier thickness of 500 Å, the calculated α of the present material is shown in Fig. 5.30(b). It has a large bandwidth of ~4 μm but a small peak value of 0.053 μm^{-1}. This α_p is 3.5 times smaller than the 0.186 μm^{-1} in Fig. 5.1 used in the grating design. It is smaller because the present narrower well contains less doping and thus less ground state electrons, the thicker barrier dilute the average electron density in each QW period, and the broader spectral B-C transitions reduce the oscillator strength at the peak. The lower α_p will reduce η_p by the same factor.

Fifthly, the present grating period is 3.5 μm. It is not optimal for the material absorption spectrum. To evaluate the coupling

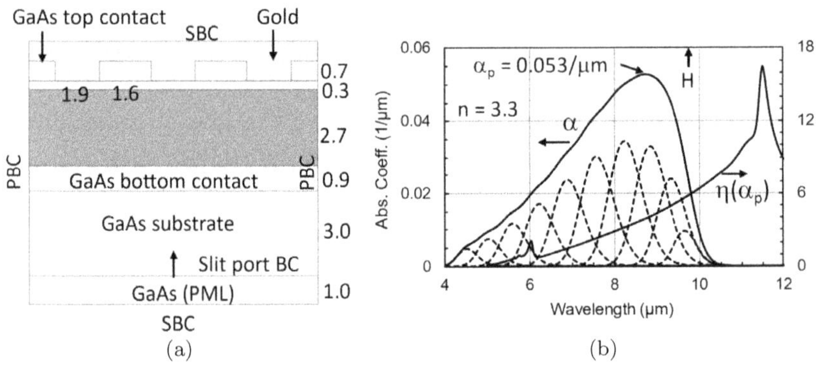

Figure 5.30. (a) The 1D grating structure adopted in Ref. 9. (b) The calculated α spectrum of the adopted MQW structure. H indicates the barrier height and the dashed curves are the α of individual excited states, which are all above H. $\eta(\alpha_p)$ is the modeled external unpolarized QE if α is a constant at α_p. The small modeled peak at $6\,\mu$m is attributed to second-order diffraction.

characteristics of a coupling scheme, one can assume a constant α independent of λ and models the resulting η spectral lineshape. The modeled $\eta(\lambda)$ will be directly proportional to the integrated vertical intensity in the active volume and thus the coupling efficiency C. The first case to be addressed is the $200\,\mu$m discrete detectors with 1D gratings listed in Table 5.2. In Fig. 5.30(b), a constant $\alpha = \alpha_p = 0.053\,\mu\text{m}^{-1}$ is adopted and the external unpolarized η_{abs} is modeled based on the 1D infinite grating structure shown in Fig. 5.30(a) and a substrate transmission of $T_s = 0.714$. The modeled η_{abs} rises gradually from ~0% at $\lambda = 5\,\mu$m to 17% at $11.5\,\mu$m. Therefore, this grating will not detect any light at $5\,\mu$m even though the material α is finite. On the other hand, if the material α is $0.053\,\mu\text{m}^{-1}$ at $11.5\,\mu$m, the grating will produce 17% external η_{abs} even though α is small. With the present α peaked at $8.8\,\mu$m, η_{abs} will have a moderate value at 4.0%. Together with the finite size factor of 0.73 and the γ factor of 0.5, the predicted η_d at -2 V for a $200\,\mu$m 1D grating will be 1.46%, which is not far from the experimental ~1% in Table 5.2. In view of other experimental uncertainties related to material growth and detector fabrication, theory and experiment are consistent.

For the 60 μm FPA LG pixels, 3D modeling is carried out on laminar gratings to evaluate finite size effects and TM and TE couplings. Since experiment shows that this size detector with 30 μm substrate has low levels of OCT, focused spot illumination configuration is adopted for comparison. The detector structure is shown in Fig. 5.31(a). It has a pixel linear size of 58 μm and a cavity length of 56 μm. Other dimensions are listed in Table 5.1. The modeled E_z distribution under TM incident field is shown in Fig. 5.31(b) at λ_p. The field near the pixel edge is seen to be disturbed by the mesa boundary.

With this pixel size, the TE contribution is very small in Fig. 5.32(b), and the contribution is due to diffraction at the mesa boundaries. The total η_{abs} from both TM and TE modes is 1.30%. It is smaller than the infinitely periodic η_{abs} of ~4% shown in Fig. 5.32(a) due to finite size effects. The expected $\eta_d = \eta_{\text{abs}}\gamma$ at -2 V is half of η_{abs}, which gives $\eta_d = 0.65\%$, again not far from the 0.4% in Table 5.3. The modeled spectral lineshapes for both infinite and 60 μm size gratings are close to the 200 μm discrete detector spectrum available in Ref. 9 in considering the size effects on the spectrum. The blueshift of the experimental peak wavelength is attributed to the second-order Stark shift to be discussed in Section 5.4.2.

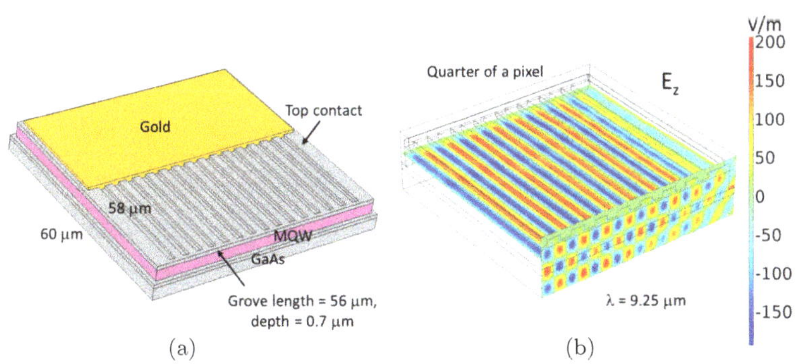

Figure 5.31. (a) The 1D grating pixel geometry in a 60 μm pitch array. The gold cover is partially removed to show the grating underneath. (b) The E_z field distribution at λ_p when the input field $\boldsymbol{H} = (0, 1, 0)$ A/m.

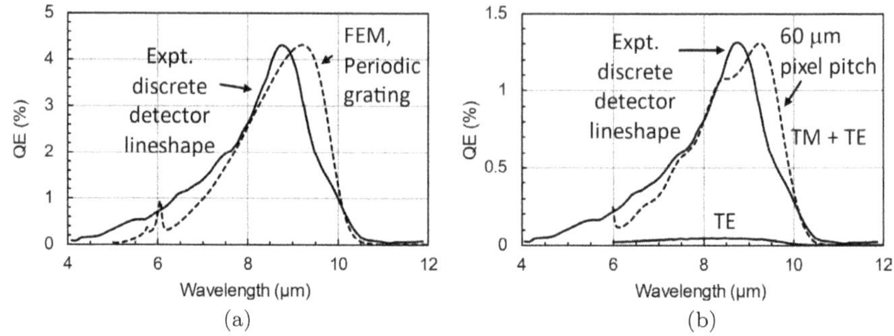

Figure 5.32. The modeled unpolarized external QE spectra of (a) an infinitely periodic 1D grating and (b) a 60 μm pitch 1D grating without waveguide or AR coating. The QE lineshape converted from the R lineshape for the 200 μm detector is also shown.

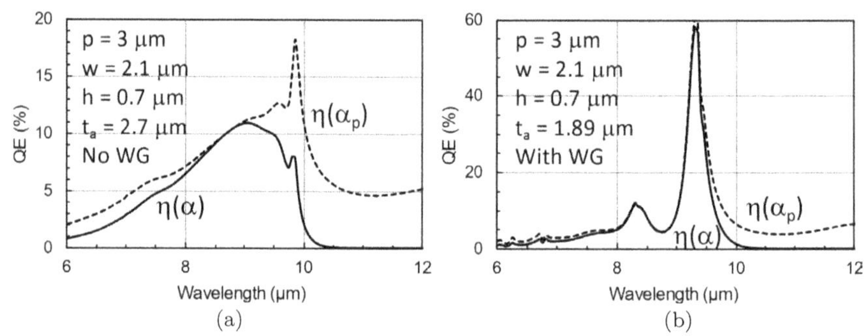

Figure 5.33. The modeled unpolarized external QE spectra of an infinitely periodic cross grating with material α and constant α for (a) without a waveguide and (b) with a waveguide.

The results for the cross gratings listed in Table 5.1 can be similarly analyzed. Figure 5.33(a) shows the modeled external η_{abs} of an infinitely periodic grating without a waveguide. η_{abs} with a constant α_p shows that the present 2D grating with $p = 3.0\,\mu$m is most effective at $\lambda = 9.9\,\mu$m where $\eta_{\text{abs}} = 18.3\%$. With the shorter λ_p at 8.8 μm, the CG detector has a maximum η_{abs} of 10.8% at 9.0 μm. If the same $\gamma = 0.5$ is assumed at -2 V, η_d is expected to be 5.4%, and it is 3.2 times larger than the experimental 1.7% for a 200 μm detector in Table 5.2, and this difference is attributed to finite size effects.

For the infinite grating with a waveguide in Fig. 5.33(b), $\eta(\alpha_p)$ is peaked at 9.3 μm, close to the material peak. With this more optimized CWG detector, η_{abs} is 57.6% at 9.3 μm despite having a thinner MQW layer t_a. The expected η_d at -2 V is then 28.8% with $\gamma = 0.5$. However, it has a very narrow linewidth of ~0.35 μm. The experimental η_d of a 200 μm detector at -2 V is 6.3%, which is smaller than the expectation by 4.6 times. The very narrow CWG spectrum can be susceptible to fabrication nonuniformity that broadens the bandwidth but reduces the peak value. Without information on the spectral lineshape and their voltage dependence from Ref. 9, the discrepancy cannot be quantitatively assessed.

When cross grating FPAs are fabricated with the pixel geometry shown in Fig. 5.29, the measured η_d is 1.2% for thinned CG detectors and 8.0% for CWG detectors at -2 V in Table 5.3. Meanwhile, the modeled external η_{abs} spectra under focused spot illumination for the two FPAs are shown in Fig. 5.34. The peak values are 4.06% and 9.54%, respectively. The CWG detector thus has a much stronger size effect than the CG detector. Assuming the same $\gamma = 0.5$ at -2 V, the

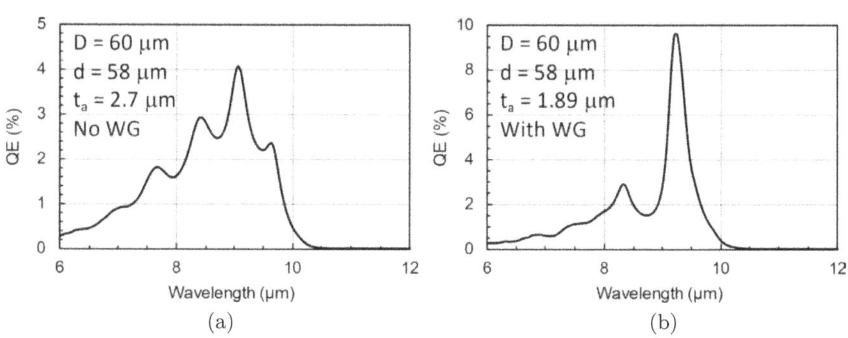

Figure 5.34. The modeled unpolarized external QE spectra of a 60 μm pixel pitch CG detector with the material α (a) without a waveguide and (b) with a waveguide. $\text{IQE}_{\text{EM}} = 0.069\,\mu$m for the CG detector in (a) and 0.077 μm for the CWG detector in (b).

expected η_d will be 2.03% for the CG FPA and 4.77% for the CWG FPA. These values are 1.7 times larger than the measured CG value of 1.2% but 1.7 times smaller than the CWG value of 8.0%. Therefore, in the absence of more accurate information on the voltage dependence of the detectors, EM modeling and FPA measurements agree within a factor of 2 by explicitly modeling the finite-size detectors.

Table 5.4 summarizes the FEM analysis on some of the experimental findings in Ref. 9. This analysis is based on a rather incomplete set of experimental details and therefore can only serve as a qualitive guide. In Ref. 9, only the voltage dependences of the responsivity and gain along with the spectral lineshape of the 200 μm LG detector are provided. Without the knowledge on the α spectrum of the adopted materials and the voltage dependence of R and g, quantitative analysis is infeasible. Nevertheless, the large discrepancy over a factor of 10 between theory and experiment can be substantially reduced to within a factor of 2 in most cases when the relevant factors are considered.

The E_z distributions for the $p = 60\,\mu$m cross-grating detectors are shown in Fig. 5.35 for $\mathrm{H_{inc}} = (0, 1, 0)$ A/m. They show a constant E_z along the y-axis between the cavities. The CGs are thus able to couple linearly polarized light into the detector beyond geometrical optics and double the QE of LGs for unpolarized light.

Table 5.4. The experimental and predicted external η_d based on FEM modeling.

	Experiment	Prediction	Ratio
200 μm discrete detectors, focused spot			
LG	~1%	1.46%	1.5
CG	1.7%	5.4%	3.2
CWG	6.3%	28.8%	4.6
Thinned 60 μm FPA pixels, flood exposure			
LG	0.4%	0.65%	1.6
CG	1.2%	2.03%	1.7
CWG	8.0%	4.77%	0.59

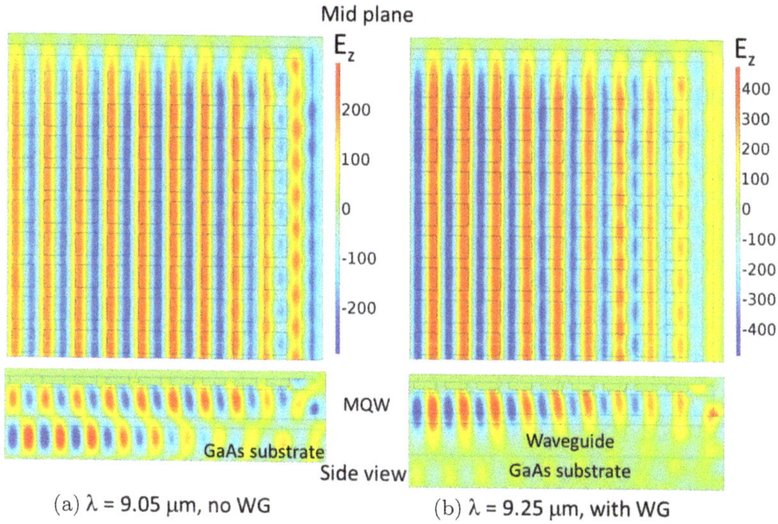

Figure 5.35. The E_z field distributions (a) without a waveguide and (b) with a waveguide in the presence of an input field $\boldsymbol{H} = (0, 1, 0)$ A/m.

In Fig. 5.35(b), the incident light at λ_p is effectively confined by the waveguide with E_z nearly double that without waveguide in Fig. 5.35(a), but its periodic field distribution is more disturbed by the sidewalls deep into the center of the pixel. Therefore, the CWG detector has more pronounced size effects than the CG detector. Based on the E field difference, QE of the CWG detector should be about 4 times larger than that of the LG detector. However, since its MQW layer thickness is thinner, the peak η is thus only 2.4 times larger. The IQE from Eq. (4.18) is 0.077 for the CWG detector, which is only slightly larger than the 0.069 μm for the CG detector because of the much narrower linewidth. The two FPAs should then produce similar photocurrent under a broadband source. These IQE values are about one order of magnitude smaller than the more optimized detectors shown in Fig. 5.24.

In summary, the large discrepancy between designed and observed grating QE is mainly due to the small α of the adopted B-C intersubband transition and the mismatch between grating coupling wavelength and material absorption wavelength. Besides these main factors, QE of a finite-size detector is smaller than the designed QE based on an infinitely periodic structure. The detection QE is also smaller than the modeled absorption QE by a factor γ. This γ is originally used to account for the tunneling probability of a bound photoelectron. However, the experimental QE deduced from $\eta_d = (h\nu/e)(R/g)$ increases with V even for unbound photoelectrons, and its magnitude is always smaller than the modeled η_{abs}. Therefore, a factor γ is also needed for unbound photoelectrons, and it is used to account for the loss of photoelectrons that diffuse in the opposite direction of the applied field and eventually recombine in the upstream QWs. A large field is needed to coerce all the photoelectrons to move in the same direction as the field to realize $\eta_d = \eta_{\text{abs}}$. Therefore, the detection η_d is usually found to be smaller in FPA operation when a small bias is adopted to avoid a large dark current. With all these QE reduction factors, most of the experimental results in Ref. 9 are understood using FEM modeling. With this understanding, one should be able to design and optimize small gratings and determine their realistic η_d under FPA operation without surprises.

With the development of the QWIP technology, the design of material and grating has evolved and continues to improve. However, the improvement is limited due to the drive for small pixels in high-density arrays. As discussed in Section 1.3, one can adopt a small f-number lens to match a small pixel to retain the image resolution. A small pitch can thus be able to reduce the overall FPA size that leads to smaller weight, power, and cost of a camera. In long wavelength infrared (LWIR) detection, pixel pitch as small as 6 μm is recommended [10]. The reduced grating coupling efficiency in smaller pixels negates the improvements on grating and material designs. Because of the diversity of materials and gratings adopted in FPA production, one can expect different QE from different companies.

From two randomly sampled commercial cameras, both with 25 μm pixel pitch, in-house test shows that one exhibits an operating conversion efficiency CE ($\equiv \eta_d g = \eta_{\text{abs}}\gamma g$) $= 1.50\%$ and $g = 0.31$. η_d is then equal to 4.84%. Another exhibits CE $= 0.80\%$ and $g = 0.25$, so that $\eta_d = 3.2\%$. In literature, Thales, Inc. offers Catherine-XP

QWIP camera having 25 μm pitch, $\lambda_p = 8.6\,\mu$m, and AR coating [11]. The responsivity R as a function of V reaches a maximum of 0.42 A/W at −2 V with $g = 0.52$, giving CE = 6.0% and $\eta_d = 11.6\%$. The 15 μm pitch cross grating produced by IRNova, Inc. shows an operating $R = 0.2$ A/W at $\lambda_p = 8.8\,\mu$m [12], yielding CE = 2.8%. In the Thermal Infrared Sensor study [15, 16], a number of 25 μm grating FPAs are produced and characterized. At the highest applied voltage of 1 V, CE = 3.4% for the QmagiQ 020 FPA and 2.2% for the Thales GT037 FPA. Therefore, from this limited sampling of grating FPAs, CE and detection QE of a generally available QWIP FPA are less than 6% and 12%, respectively. This level of QE is less than the expected 40% from Fig. 5.24(b) by a factor of 3 to 4. The lower QE can be a combined result of different grating designs, smaller material α, fewer QW periods in the absorbing layer, and the lack of an ARC.

5.3 Resonator-Quantum Well Infrared Photodetectors

Although grating-coupled FPAs can be operated at lower voltages than C-QWIPs due to the use of thinner MQW materials, the demonstrated CE $= \eta_{\text{abs}}\gamma g$, however, is not much larger than that of the C-QWIPs. In the original infinite grating design, detectors do not have sidewalls. The diffracted wave is free to travel in the absorbing layer. In a finite-size detector, however, the opposite mesa sidewalls introduce optical reflection and interference. Constructive Fabry-Perot interferences between two opposite sidewalls occur only when

$$2d = N\lambda_i \sin(\theta), \tag{5.9}$$

where d is the pixel size, λ_i is the internal wavelength, N is an integer, and θ is the diffraction angle. If $2d/\lambda_i \sin\theta$ is not designed to be an exact integer at the α peak wavelength, destructive interference will occur and QE will be reduced. Since the light traveling at an angle does not only reflect at the sidewalls but also at the detector top and bottom surfaces, a finite-size detector must be viewed as a 3D resonator and not only a 2D Fabry-Perot etalon. The 3D resonator has its own set of resonant cavity modes. The grating can produce strong internal intensity for a large QE only if it is able to excite one or more of these cavity modes upon normal incidence. Otherwise, there is little electric field present in the active volume, and the light will either be reflected back into the air or be absorbed in the metal cover. If the resonant characteristics of the pixel volume are ignored, grating coupling will not be effective. Using numerical EM modeling, it is possible to design an efficient grating-coupled QWIP by adjusting the grating parameters in conjunction with the detector pixel size and thickness. This detector structure can be called as the grating-coupled resonator-QWIP (GR-QWIP), and it is useful in some applications such as polarization-sensitive detection discussed in Section 5.6.

However, even with an optimized design, grating coupling still suffers from several drawbacks. Firstly, the optical interference from the diffracted light is sensitive to the pixel volume. As seen in Fig. 5.17(b), a small change of the pixel area size can shift the resonant wavelength significantly. It can produce different peak wavelengths or increase spectral nonuniformity if the detector size is not consistently produced in the production process. Secondly, given the resonant condition in Eq. (5.9), θ may need to take a less-than-90° value, reducing the achievable QE. Thirdly, the cavity resonance sharpens the grating diffraction peak, leading to very narrowband coupling. Fourthly, the diffracted waves travel at a glancing angle, which easily reflect into other pixels and cause significant OCT. Finally, grating diffraction efficiency generally reduces with grating size. It will be less sensitive in dense arrays.

5.3.1 *R-QWIPs with Localized Surface Plasmonic Coupler*

To overcome some of these drawbacks, a new physical mechanism is introduced to couple light, which is the use of surface plasmon-polaritons (SPPs). The subject of SPP has been discussed in Section 3.3. It is a branch of EM modes that resides at the metal/semiconductor interface. Since the dispersion relation of SPP is the same as that of light in the long wavelength limit, there is no difference in the grating period to produce optical diffraction at 90° or SPP waves at $k_{\text{spp}} = 2\pi/\lambda_i$, where k_{SPP} is the SPP wavevector and λ_i is the internal optical wavelength. On the other hand, when the optical ω approaches the surface plasma resonant frequency $= \omega_p/(1 + \varepsilon_d)^{1/2}$, the grating period p needs to be $p = 2\pi/k_{\text{spp}} \ll \lambda_i$. Therefore, the period of a plasmonic coupler is generally subwavelength. Nevertheless, in infrared wavelengths, the period of a plasmonic coupler is the same as that of the first-order grating. The groove profile is, however, shallower and sinusoidal for better phase matching.

We have seen the creation of SSP can produce a spike in QE in laminar gratings in Fig. 5.3(b) and in cross gratings in Fig. 5.13(a). By producing a surface wave, one can confine the optical energy to the surface without a waveguide layer. However, extended SPP waves are still subject to the Fabry-Perot resonant condition in Eq. (5.9) that leads to QE suppression. The alternative will then be to create localized surface plasmons (LSPs). LSPs are also surface modes but they are not propagating and hence they are not subject to the optical interference caused by pixel boundary reflections.

To create LSPs instead of SPPs, square cavities in a grating are replaced by ring cavities. An example of this resonator-QWIP structure is shown in Fig. 5.36. It has a pixel size $d = 22\,\mu\text{m}$ placed in a $D = 25\,\mu\text{m}$ pitch array and is designed to detect around 8–10 μm light. In the present design, it has 4×4 rings in the pixel. Each ring can be exactly square but the fabricated rings usually have rounded corners. Therefore, the rounded rings are used in the design. In the FPA production, the ring pattern is first formed on the photoresist and etched by selective dry etching through the top contact layer and down to a very thin etch-stop layer on top of the MQW layer. Ohmic contact metals and a thick gold layer are deposited on the top. Using the metals as the etching mask, the MQW layer is etched by non-selective dry etching process to create the mesa structure. The detector array is hybridized to a readout circuit or a fanout circuit, after which the array is backfilled with epoxy for mechanical support. The GaAs substrate is completely removed by mechanical lapping and selective dry etching. No ARC is used unless stated otherwise. The pixel structure in an array is shown in Fig. 5.36.

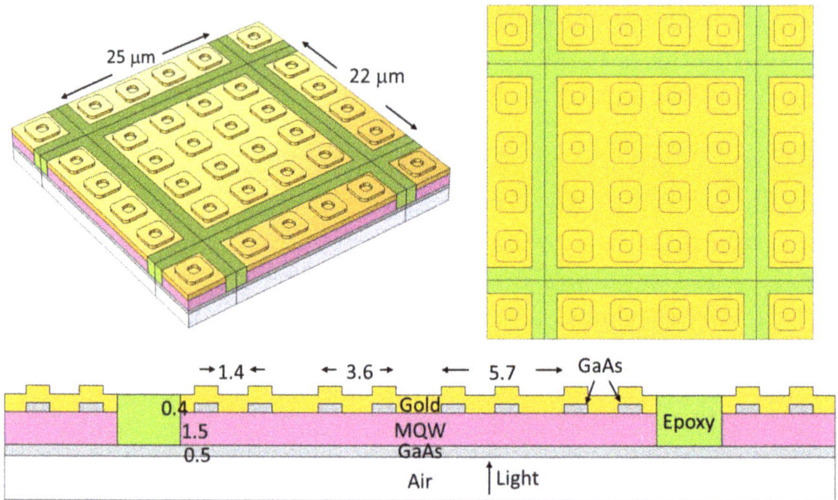

Figure 5.36. 3D perspective, top view, and cross-section of a R-QWIP designed for 8–10 μm detection. The numerical values are detector dimensions in microns.

To design the present R-QWIP structure, the center of the detection band is set around 9 μm. With an assumed $n = 3.239$ for both the MQW and the GaAs materials, the internal optical wavelength λ_i is 2.8 μm. At this wavelength regime, $\lambda_{\mathrm{SPP}} = \lambda_i$. Therefore, a grating used to create SPPs from the incident light will have $p = 2.8\,\mu$m to provide the needed momentum. In the present ring design, if the inner ring dimension s_i is set at $\lambda_i/2$, the outer ring dimension s_o at $3\lambda_i/2$, and the ring spacing p_{ring} at $2\lambda_i$, the cross-section of the ring array would resemble that of a grating. The rings would then support SPP generation with the undesirable pixel size effects. To disrupt the SPP propagation, p_{ring} is adjusted to be larger than $2\lambda_i$. The uneven grating profile will then generate LSPs instead. LSPs excited in nanoparticles are widely used in surface-enhanced Raman spectroscopy for their highly enhanced electric fields. LSPs are used here to enhance the absorption of QWIPs. For initial setting, p_{ring} is set to be $2.5\lambda_i$. With localized field profiles in both horizontal and vertical directions, each ring unit cell can be treated as an individual resonator. The pixel QE will be much less sensitive to the boundary condition set up by the detector volume. To promote the transmission of light incident into the detector, the total detector layer thickness t_{tot} including MQW and bottom contact layers is set at $(2m+1)\lambda_i/4$, where m is an integer. With these thicknesses, the reflected light from the top metal contact is in anti-phase with the reflected light at the substrate, resulting in destructive interference for the reflected light. The detector material will then be its own ARC to enhance substrate transmission without an external coating. It is to reduce OCT and cost.

With this set of initial structural parameters, QE is optimized using FEM by adjusting each parameter individually. To isolate the effects from the pixel size and material α, a unit cell of an infinite ring array is modeled with a constant $\alpha = 0.2\,\mu\mathrm{m}^{-1}$. It turns out that QE is maximized at $\lambda_i = 2.8\,\mu\mathrm{m}$ (*i.e.*, $\lambda = n\lambda_i = 9.07\,\mu\mathrm{m}$) relatively sharply at $s_i = 1.4\,\mu\mathrm{m} = \lambda_i/2$, $s_o = 3.6\,\mu\mathrm{m} = 1.3\lambda_i/2$, and $t_{\mathrm{tot}} = 2.0\,\mu\mathrm{m} = 2.9\lambda_i/4$. On the other hand, external QE only varies in a small range between 80.0 and 86.9% when p_{ring} varies between 5.6 and 6.4 μm (*i.e.*, between $2\lambda_i$ and $2.3\lambda_i$) with a broad maximum at 5.9 μm or $2.1\lambda_i$. The flexibility of the ring spacing indicates that with localized rather than extended state excitations, the excited field is generally compatible with the boundary condition of the detector volume and is a pixel cavity mode. The optimized ring height $h = \lambda_i/7$, which is less than $\lambda_i/4$ for gratings.

To understand the LSP formation in the rings, two complementary grating structures are modeled in Fig. 5.37. One is the box grating having the same inner square as the grating cavity. The unit cell period p_{ring} is 5.7 μm. This p_{ring} instead of 5.9 μm for QE maximum is adopted so that a 4×4 ring array can fit into a pixel size of $22 \times 22\,\mu\mathrm{m}^2$ when the 25 μm pixel pitch detector is considered. Another is the post grating having the same outer square as the grating cavity. A ring array can be considered as a hybrid of these two gratings. Notice that the E_z distributions induced at the same plane by the box and post gratings such as that at 1.2 μm above the substrate are very similar both in shape and in phase. Neither of them is very efficient in light coupling as the maximum E_z under incident $\boldsymbol{H} = (1, 1, 0)$ A/m is in the order of 600 V/m. In

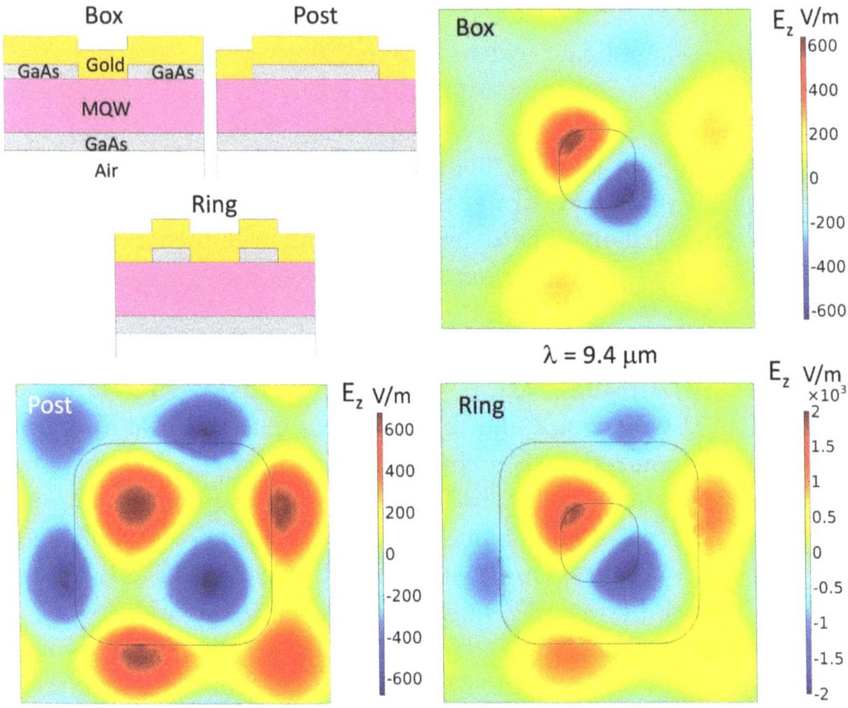

Figure 5.37. The modeled PUC of rings and E_z at $\lambda = 9.4\,\mu$m under incident $\boldsymbol{H} = (1,1,0)$ A/m along a horizonal plane $1.2\,\mu$m above MQW/substrate interface.

contrast, when the two gratings combine to form the ring structure, the maximum E_z increases to $\sim$2000 V/m in Fig. 5.37. These field distribution and magnitude are maintained to be similar in finite-size pixels with $D = 25\,\mu$m and $12\,\mu$m as shown in Fig. 5.38, and they are highly localized in both horizonal and vertical directions. Comparing with the incident intensity $\propto n|E_{\text{inc}}|^2 = 1 \cdot (533.2)^2$, the maximum vertical intensity in the ring resonator $\propto n|E_z|^2 = 3.239(5000)^2$ in Fig. 5.38(a), and it is 285 times larger. Although the average increase in the intensity is much milder in the entire pixel volume, its QE is still larger than either the box or post QE in Fig. 5.39. The maximum external QE is 87.3% at $\lambda = 8.9\,\mu$m with a wide coupling bandwidth of $\sim 2\,\mu$m.

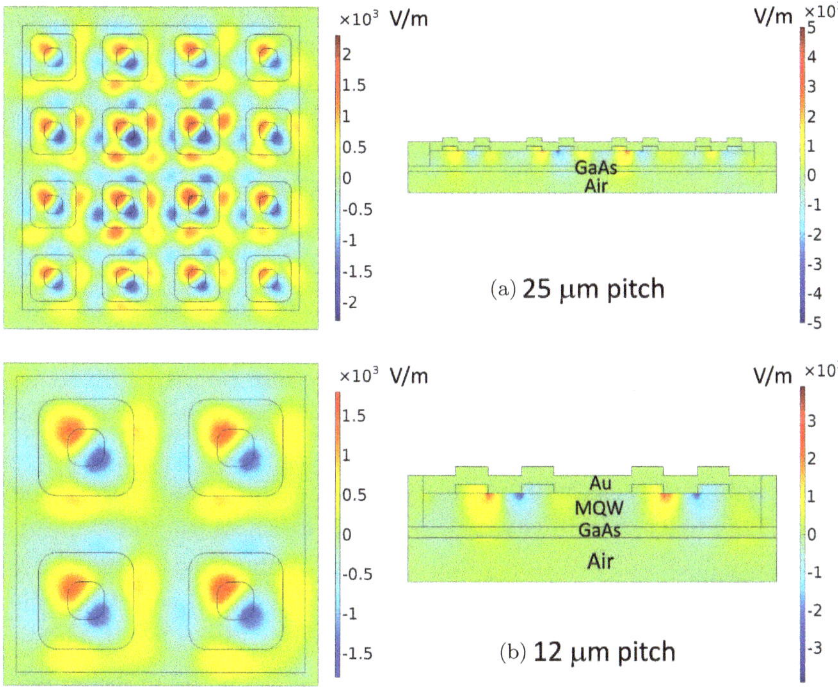

Figure 5.38. The modeled E_z of (a) $25\,\mu$m and (b) $12\,\mu$m pixels with periodic boundary condition at $\lambda = 9.4\,\mu$m along a horizonal plane $1.2\,\mu$m above substrate and the diagonal.

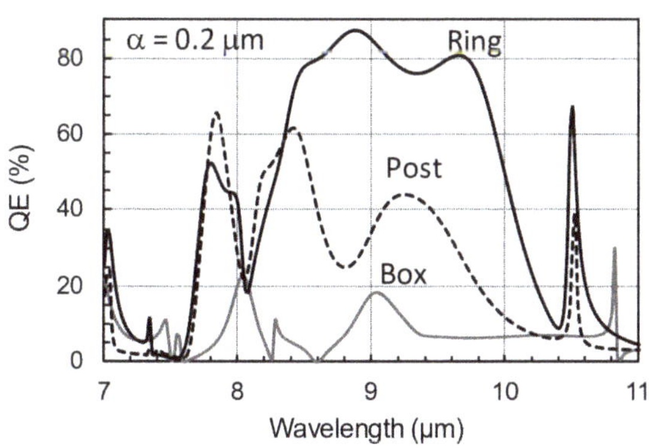

Figure 5.39. The modeled external QE of the box, post, and ring gratings for $\alpha = 0.2/\mu$m with the detector dimensions shown in Fig. 5.36.

The E_z patterns of the box and ring unit cells in Fig. 5.37 are very similar except for their magnitudes. This similarity points to the fact that the field in the ring unit cell is primarily produced by the inner ring. The outer ring merely strengthens the plasma oscillations in the inner ring by providing a favorable boundary condition. Because of the EM shielding of the outer ring, the field enhancement from the inner ring is more stable and insensitive to the environment outside the unit cell. The field enhancement is thus an intrinsic property of the individual ring. The pixel sidewalls affect QE only through the fringing fields and it is much weaker. Each ring can then be viewed as an independent optical resonator with LSP as its cavity mode, similar to a nanoparticle in surface-enhanced Raman spectroscopy.

Accordingly, there will be much weaker size effects for the ring array. This feature is illustrated in Fig. 5.40(a). With the same α, the external QE of a 25 μm pitch detector is modeled to be 81.6% in PUC boundary condition, which is only slightly lower than the 87.3% in the infinite periodic ring modeling in Fig. 5.39. Together with a very similar lineshape, the ring array thus suffers much less QE reduction from finite size effects than gratings. The coupling efficiency C_p at $D = 25\,\mu$m is high at 9.3. Since the LSP field is of short range as seen in Fig. 5.38, OCT should also be much less. Indeed, QE under focused spot modeling on 3×3 pixels shows a

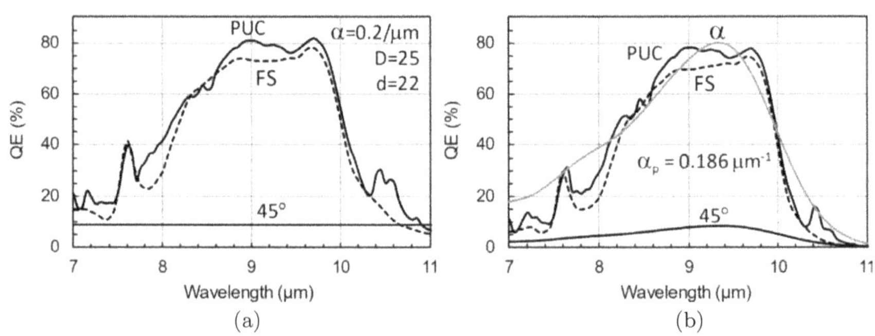

Figure 5.40. QE of a 25 μm R-QWIP with PUC boundary condition appropriate for flood exposure and with SBC appropriate for focused spot exposure for (a) a constant α and (b) a varying α.

maximum at 78.1% in Fig. 5.40(a), nearly the same as the PUC QE, implying little OCT in the 25 μm R-QWIPs. With α in Fig. 5.1, the QE spectrum is shown in Fig. 5.40(b). Its spectral lineshape does not change significantly from that with a constant α, showing that with a large E_z enhancement, QE is less dependent on α.

For the same pixel pitch $D = 25\,\mu$m, QE is modeled with different pixel size d. The modeled spectra are shown in Fig. 5.41(a). Different from grating coupling, all detectors maintain very similar lineshape although their magnitudes differ slightly. A large d yields a smaller QE because E_z fades away from the rings. A larger area thus receives more light but does not produce larger absorption, leading to a smaller QE. This trend also shows the stability of the LSP field against the change of detector area. If QE is multiplied by the pixel fill factor to yield the fraction of light absorbed in the pixel pitch area, nearly the same QE of 60% is obtained for all d in Fig. 5.41(b). With nearly the same spectrum, the integrated photocurrent will be unaffected by detector size fluctuation in array fabrication.

To determine the angular dependence of the R-QWIP QE, the detector is modeled with TM incidence at 16.4° in the air. This angle corresponds to $f/1.7$ and an internal incident angle of 5.0°. The modeled QE spectrum shown in Fig. 5.42 does not change significantly from that under normal incidence. Since TE mode is

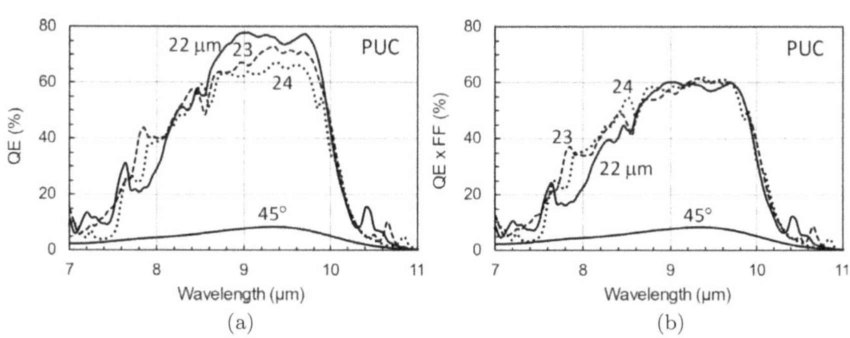

Figure 5.41. The modeled external QE of a 25 μm pitch R-QWIP array for a constant α using (a) d^2 and (b) D^2 as the optical area.

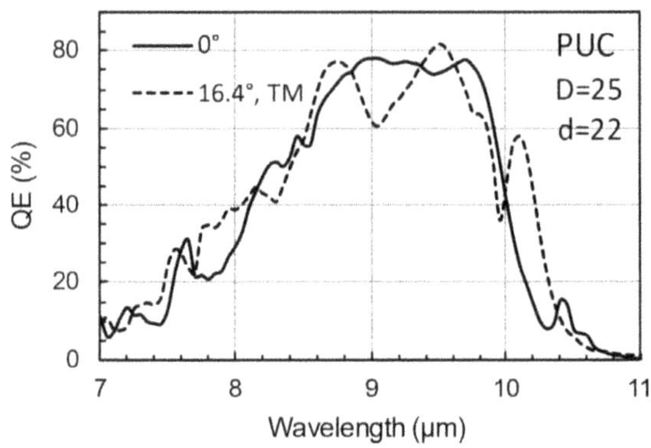

Figure 5.42. The external QE under normal incidence and 16.4° incidence for TM mode.

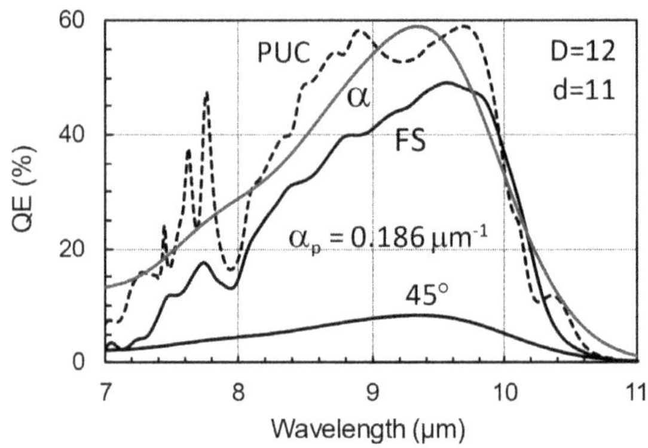

Figure 5.43. The external QE of $D = 12\,\mu$m R-QWIP with PUC boundary condition for flood exposure and with SBC for focused spot exposure using a varying α in Fig. 5.1.

not angular dependent, there will be no loss of optical signal when the lens f-number is as small as 1.7.

For detectors smaller than 25 μm, both C_p and OCT are expected to be worse. At $(D, d) = (12\,\mu\text{m}, 11\,\mu\text{m})$, the modeled QE is 59.0% under PUC and 49.2% under focused spot in Fig. 5.43, indicating a higher level of OCT. QE for $d = 10\,\mu$m and 11 μm is shown in Fig. 5.44, again indicating negligible difference with different d. At $D = 6\,\mu$m, PUC QE is 56.3% and focused spot QE is 34.9% in

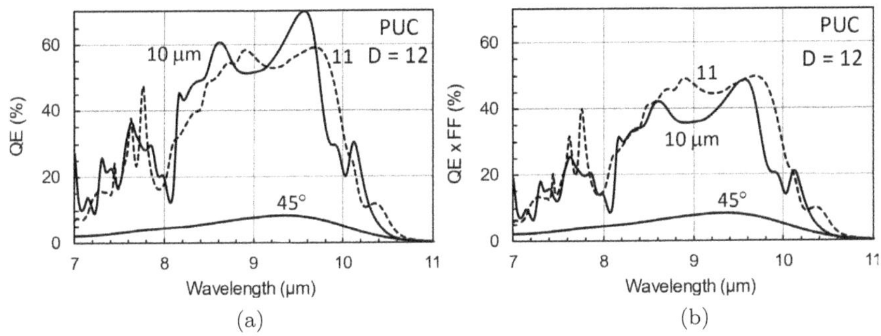

Figure 5.44. The external QE of 12 μm R-QWIP using (a) d^2 and (b) D^2 as optical area.

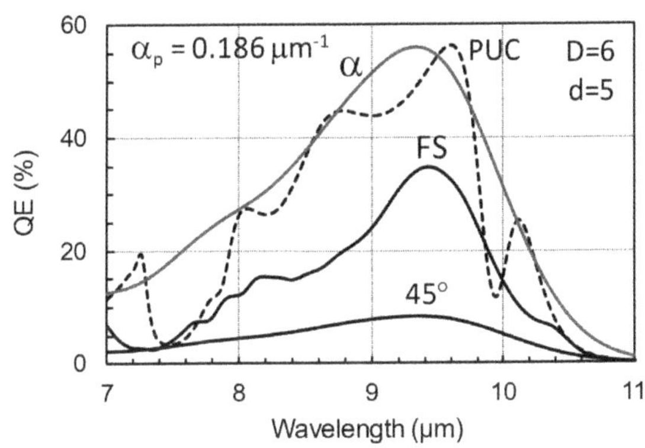

Figure 5.45. The external QE of a $D = 6\,\mu$m R-QWIP with PUC and focused spot for a varying α.

Fig. 5.45. Detectors with $d = 5.0$ and $5.5\,\mu$m still share nearly the same characteristics in Fig. 5.46.

Table 5.5 summarizes the performance of the cross-grating waveguide detectors described in Section 5.3.3 and the R-QWIPs described in this section. All detectors are modeled using the same α shown in Fig. 5.1. There are two minor differences in the detector model. One is the material refractive index n, which is assumed to be 3.300 for the CWG detectors and 3.239 for the R-QWIPs. Another is the MQW layer thickness, which is 1.85 μm for the CWG detectors and 1.50 μm for the R-QWIP. The listed QE are the peak value under normal

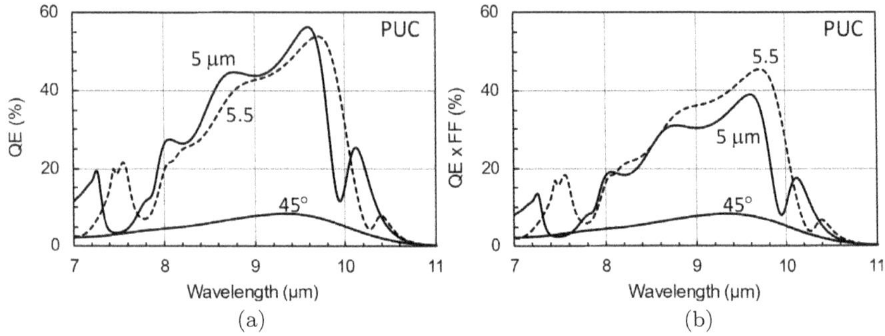

Figure 5.46. The modeled external QE of a 6 μm pitch R-QWIP array using (a) the pixel size as the optical area and (b) the pixel pitch area as the optical area.

Table 5.5. A summary of the external QE and IQE of cross grating waveguide detectors with dimensions (D, d) described in Section 5.2.3 and the R-QWIPs described in this section based on a common α spectrum in Fig. 5.1.

Periodic unit cell modeling for flood exposure condition						
	Grating (25, 23)	R-QWIP (25, 22)	Grating (12, 11)	R-QWIP (12, 11)	Grating (6, 5)	R-QWIP (6, 5)
QE (%)	31.1	78.0	18.7	59.0	18.2	56.3
IQE (μm)	0.451	1.583	0.321	1.270	0.324	0.982
3×3, 5×5, or 7×7 array modeling for focused spot exposure condition						
	Grating (25, 23)	R-QWIP (25, 22)	Grating (12, 11)	R-QWIP (12, 11)	Grating (6, 5)	R-QWIP (6, 5)
QE (%)	20.47	74.7	10.9	49.2	9.16	34.9
IQE (μm)	0.360	1.422	0.240	0.959	0.180	0.560

incidence from the air without ARC. The CWG QE is converted from that in Section 5.2.3 by multiplying the substrate transmission of 0.714.

Table 5.5 shows that R-QWIPs generally have higher QE than grating detectors. They also have lower OCT implied by the similarity between PUC QE and focused spot QE. Even with R-QWIPs, performance decreases with pixel size. The dimensions of the R-QWIP can be scaled to yield similar spectrum in other spectral regimes. However, they have to be finetuned because the material properties of the MQW layer, the GaAs top and bottom contact, and the gold cover are all wavelength-dependent.

While the grating peak QE can be improved by an ARC, it is not always true for R-QWIPs. In the present FEM modeling, the optical absorption of gold is included and this absorption ultimately limits the achievable QE in R-QWIPs. As seen in Fig. 3.15, the gold absorption can be very strong if the condition for SPP creation is met. For C-QWIPs, the detectors are designed to have a thick MgF_2 dielectric layer to reduce the SPP excitation so that plasmonic absorption in gold can be avoided. Unfortunately, for R-QWIPs, the detectors are designed to produce LSPs to increase the surface E field, with which the infrared absorption in gold is also increased along with the MQW layer. With (n, κ) of gold in Fig. 3.14(b), the gold absorption A_{gold} can be computed using Eq. (4.8), where E_z is replaced by the total E and V is the volume of the gold cover. The result is shown in Fig. 5.47(a) for the same $25\,\mu$m pitch R-QWIP considered in Fig. 5.40(a).

The absorption of gold varies between 10% and 20% within the 8–10 μm band, which is far larger than the constant 1.0% in planar gold surfaces from Fresnel's equation under normal incidence.

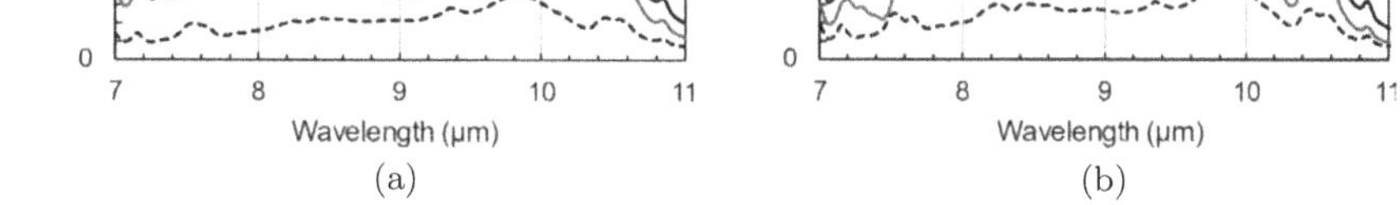

Figure 5.47. The modeled QE of the MQW layer, the absorption in the gold layer, and their total absorption in a $25\,\mu$m pitch R-QWIP under PUC modeling for t_a = (a) $1.5\,\mu$m and (b) $0.5\,\mu$m.

The total absorption in both MQW and gold layers reaches up to 97.9%. Therefore, the maximum QE of R-QWIPs cannot be further improved by an ARC in general. Nevertheless, it can still improve QE in off-peak wavelengths with a proper ARC design. In addition to gold absorption, free carrier absorption in the contact layers discussed in Section 3.2 will also increase with the more intense field in the contacts. The doping in the contact needs to reduce in the very long wavelength regime.

Similar to C-QWIP modeling in Figs. 4.9(a) and 4.16(b), one can study the effects of MQW doping density N_D and thickness t_a on R-QWIPs. The material α is linearly proportional to N_D, and $\eta(\alpha)$ is shown in Fig. 5.48(a) at $\lambda_p = 9.7\,\mu$m. Due to optical confinement and resonance in the detector volume, QE increases rapidly with α but eventually saturates at ~80% due to the gold absorption. Since $\alpha_p = 0.19\,\mu\text{m}^{-1}$ at $N_D = 1 \times 10^{18}\,\text{cm}^{-3}$ for the present MQW material in Fig. 5.1, α and hence QE with different N_D will be known. For example, if N_D is reduced by half, so does α, and QE is 60% from Fig. 5.48(a). When α is very small at the very low N_D, the gold absorption can be larger than that of the MQW. The gold absorption thus places a limit on the maximum LSP field one can achieve in the detector volume.

Likewise, the E field generated by the ring structure in Fig. 5.38 is concentrated in the upper part of the MQW layer. If the lower part of the MQW layer is replaced by the contact material, the

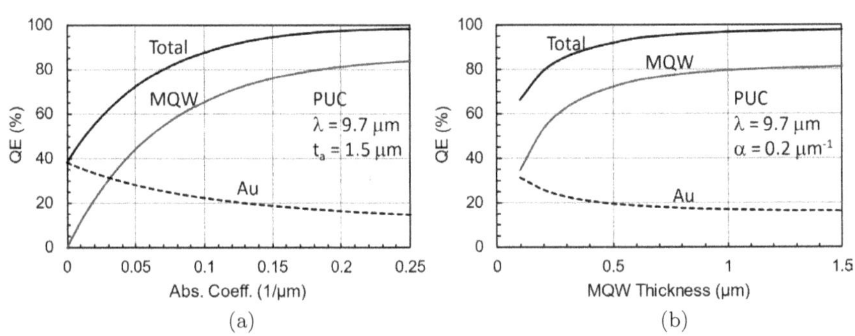

Figure 5.48. The peak QE of a $D = 25\,\mu$m and $d = 22\,\mu$m R-QWIP as functions of (a) α and (b) t_a.

light coupling geometry remains the same, and the impact to QE will be small. Indeed, at λ_p, QE maintains constant at ~80% in Fig. 5.48(b) when t_a reduces from the original 1.5 μm to 1.0 μm. Even when it decreases to 0.5 μm, QE is still 72.3%. The QE spectrum with $t_a = 0.5\,\mu$m is plotted in Fig. 5.47(b), showing very similar spectral lineshape as the thicker sample. By plotting the maximum E_z in the MQW layer, it is found that $E_{\max}$ in the thinner layer is 1.424 times stronger than that with the thicker layer. The larger E_z enhancement in thinner absorbing materials is due to the stronger optical resonance with weaker attenuation in the resonator volume. The stronger E field compensates for the thinner MQW, resulting in similar QE. Assuming the gain g to be linearly proportional to $1/t_a$, g will increase by 3 times while $CE = \eta g$ will be increased by 2.7 times when t_a reduces from 1.5 μm to 0.5 μm. Likewise, a 0.1 μm MQW still produces 35% QE in Fig. 5.48(b) while g increases by 15 times. CE then increases 6.6 times. The CE improvement is seen to be again limited by the gold absorption with thin MQW layers. At $t_a = 0.1\,\mu$m, the gold absorbs as much light as the MQW.

5.3.2 *R-QWIP Biasing Requirement*

Another feature of R-QWIP is the bias polarity preference. By using LSPs for E_z generation and enhancement, the field distribution is asymmetrical along the z-axis. For example, the E_z at the diagonal cross-section of the ring unit cell is shown in Fig. 5.49(a), whose horizonal distribution has been shown in Fig. 5.37. The field is higher at the top surface than the bottom surface of the MQW layer. To know the amount of absorption along the z-axis, the MQW layer with $t_a = 1.5\,\mu$m is divided into 15 layers of equal thickness. The value of E_z is evaluated at the mid-plane of each layer, from which the absorption in each layer is calculated from

$$\eta = \frac{t_a}{15} \times \frac{\alpha n}{2p^2 Z_0^2} \int_p^2 |E_z|^2 d^2 r, \tag{5.10}$$

where $\alpha = 0.2\,\mu\text{m}^{-1}$, $n = 3.239$, $p = 5.7\,\mu$m, $Z_0 = 377\Omega$, and the incident $\boldsymbol{H} = (1, 1, 0)$ A/m. The results for the 15 layers and their total are shown in Fig. 5.49(b). The total QE is the same as that in

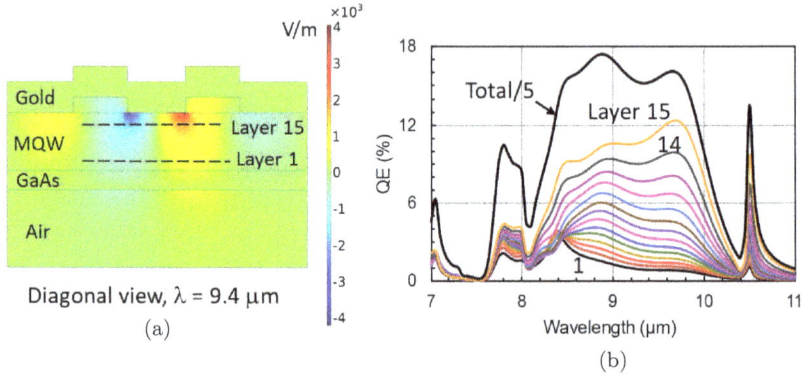

Figure 5.49. (a) The E_z pattern on the diagonal plane of the periodic ring array shown in Fig. 5.37. (b) The calculated QEs of the 15 partitions of the MQW layer and their total. The total QE is scaled down by a factor of 0.2 to fit in the same plot.

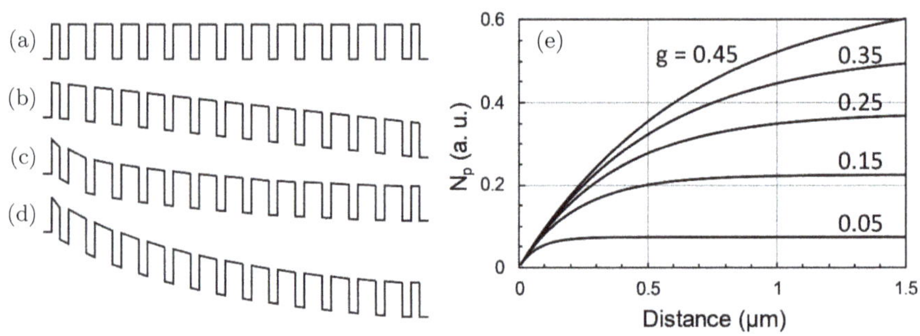

Figure 5.50. A MQW under (a) zero bias, (b) a bias with hypothetical linear potential drop, (c) a small bias with uneven potential drop, and (d) a large bias with uneven potential drop. (e) The number of photoelectrons at distance x from the cathode in arbitrary units.

Fig. 5.39 for the same ring structure, showing the consistency of the modeling. In Fig. 5.49(b), each 0.1 μm layer contributes on average about 6% QE as expected. However, the contribution is largest at the top and it decreases linearly with the distance away from the metal surface. Layer 15 yields 12.3% QE, which is 14.1% of the total QE.

Up to now, we have assumed the potential drop inside the MQW layer to be linear under bias as shown in Fig. 5.50(b). It is usually adequate in describing the first-order effects in carrier transport where the QWs are under uniform thermal excitation, such as that in the dark current fitting in Fig. 2.41. However, under asymmetric optical excitation in R-QWIPs, the potential profile of the MQW affects the responsivity and produces bias asymmetry. The self-consistent potential profile needs rather involved analysis and depends on a large number of parameters such as the satellite valleys offset, intervalley scattering rates, well width, capture probability, operating temperature, optical incident intensity, geometry, etc. A detailed description of the theoretical model is beyond the scope of this text. Interested readers should consult Refs. 15–17 on this subject. Here, we will give a simplified analysis.

In Section 2.15, we discussed that the photoconductive gain g is given by $g = \tau/\tau_{\text{tr}} = \ell/t_a$, where ℓ is the mean free path for electron recombination. After a photoelectron is created, it drifts on average a distance $\ell = gt_a$ before being captured by one of the QWs. There will be an accumulation of photoelectrons N_p above the barriers at distance x away from the cathode emitter.

Assuming that the generation of the photoelectrons per unit length ρ_0 is uniform in the MQW and the discrete nature of QWs is suppressed, $N_p(x)$ is given by

$$N_p(x) = \rho_0 \int_0^x \exp(-x/gt_a)dx = \rho_0 gt_a(1 - e^{-x/gt_a}). \tag{5.11}$$

When $x \gg gt_a$, $N_p(x) = \rho_0 gt_a = \rho_0 \ell$, *i.e.*, when x is much longer than the mean free path, the equilibrium N_p is the accumulated ρ_0 created within ℓ. N_p is then a constant independent of x and it is g times the total number of photoelectrons $\rho_0 t_a$ generated in the detector. The photocurrent $I_p = N_p e v_d$ is thus g times the photocharge generation rate $\rho_0 t_a e$ in the detector volume, where v_d is the electron drift velocity. A constant N_p in the detector can thus occur only when $\ell \ll t_a$ or $g \ll 1$ as seen in Fig. 5.50(e) where Eq. (5.11) is plotted at different values of g. For example, at $g = 0.05$, N_p starts at 0 at $x = 0$ where the cathode is located. It increases and saturates at $x \simeq 0.2\,\mu$m when the generation equals recombination. However, for a large g, N_p continues to increase across the entire layer.

Assuming negligible dark current, the varying $N_p(x)$ under a constant v_d will lead to a location-dependent photocurrent and it will eventually exhaust all the ground state electrons. A varying I_p is thus unsustainable. To yield a constant I_p in the steady state, the local electric E must adjust to yield a location-dependent $v_d(x)$ such that $I_p = N_p(x)ev_d(x) = N_p(x)e\mu E(x)$ is a constant, where μ is the electron mobility. $E(x)$ should then be proportional to $1/N_p(x)$, and its functional form is plotted in Fig. 5.51(a). $E(x)$ is mostly constant in the layer when g is small. The divergence of $E(x)$ at small x is an artifact that $N_p(0) = 0$ in Eq. (5.51) under the continuous medium approximation. Setting $N_p = 0$ at $x = 0$ will need infinite E field to create a finite current. This divergence can be eliminated if the discreteness of the QWs is restored and the origin of x is defined at the center of the first QW where $N_p \neq 0$, and $E(0)$ will be finite. Here, we ignore this divergence for simplicity. At a larger g at higher bias, the E field becomes less uniform. Integrating $E(x)$ gives the potential $V(x)$ in Fig. 5.51(b). It is mostly linear at small g under low bias and curvilinear at large g under high bias. Although this description is on photocurrent, the same is also true for the uniformly generated thermal dark current. The displayed potential profiles are generally consistent with those modeled in Refs. 15–17. Obviously, if the excitation has a different spatial profile, the potential distribution will be readjusted. However, in general, the curvilinear profile will

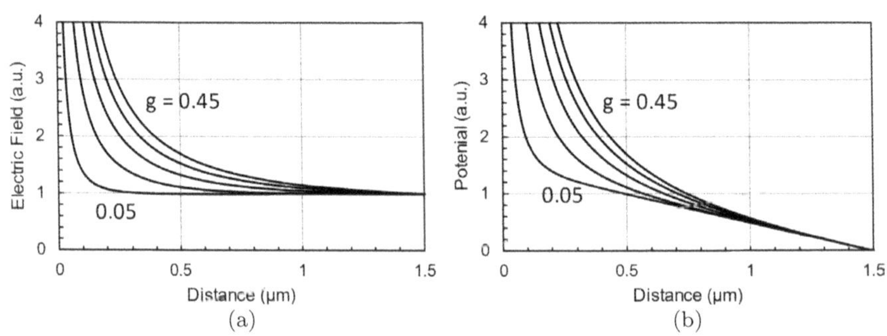

Figure 5.51. The modeled (a) electric field and (b) potential profiles in a MQW layer under uniform excitation condition assuming a constant current flow.

be maintained as long as there is hot-electron accumulation in the detector.

In addition to the conductance component, the capacitance component of the QWIP also affects $V(x)$. Qualitatively, the applied field repels the ground state electrons near the cathode and attracts them at the anode, similar to the dielectric polarization in a capacitor. The change in the ground state electron distribution further increases the field nonuniformity. The presence of space charge that reduces the material α at large bias is seen in Fig. 4.26.

Since the potential drop is larger at the cathode than at the anode, the parameter γ in the $\eta_d = \eta_{\text{abs}}\gamma$ expression is larger at the cathode than at the anode and thus γ acquires a spatial dependence in addition to a voltage dependence. The $\gamma(x)$ dependence causes η_d of the R-QWIP to be larger when the detector is positively biased at the substrate where the QWs near the metal rings have larger potential drop.

In addition to the larger η_d under positive substrate bias, the dark current is also lower at this bias polarity. When the n-type doping is introduced into a QW during the MQW material growth, some of the dopants ride on the growth front and are carried into the next barrier as shown in Fig. 5.52(a). When the electrons from the dopants release into the QW, interfacial electrical dipoles are induced at the top side of the QW as shown in Fig. 5.52(b), resulting in thinner tunneling barrier at the interface. Subsequently, when the QWIP is biased negatively at the substrate as in Fig. 5.52(c), a large thermally assisted tunneling current conducts through this thinner barrier, resulting in a large dark current. In contrast, a thicker barrier is presented to the thermally assisted tunneling process in Fig. 5.52(d) under positive bias, which causes the polarity asymmetry in the dark current. With higher photocurrent and lower dark current under positive substrate bias, the R-QWIPs are preferred to operate under this bias polarity.

Note that although the large potential drop near the cathode helps to increase γ to yield a larger η_d for the R-QWIPs, the dopant migration has lessened this increase. As seen in Fig. 5.52(d), when a photoelectron is promoted to E_2 through an intersubband transition

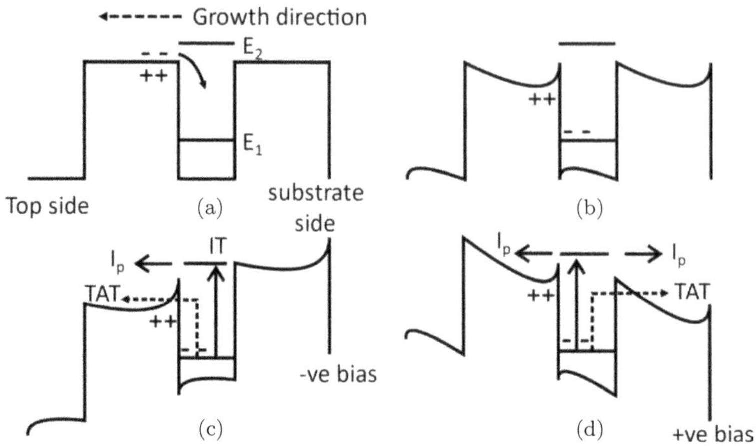

Figure 5.52. The band structure of a QW unit (a) at 0 V bias without dopant migration, (b) at 0 V with dopant migration, (c) under negative bias substrate, and (d) under positive substrate bias.

process, the interfacial dipole field counteracts the applied field in driving the electron to move in the direction of the applied field and thus reduces γ. The reduction is more acute in more heavily doped QWs under a small bias where the dipole field dominates.

5.4 Experimental Studies on R-QWIPs

5.4.1 *R-QWIP Test Detectors*

To test the R-QWIP design shown in Fig. 5.36, two wafer materials, labeled as W3 and W4 in Fig. 5.53(a), are adopted [18]. The W3 wafer has a 1.0 μm thick MQW layer containing 18 QWs. In each QW, only the lower half of the well is doped to $2 \times 10^{18}\,\text{cm}^{-3}$ to achieve the designed average doping in the well, with which dopant migration is minimized. The ring structure is fabricated on the top contact layer as designed. The total thicknesses of the W3 detector including the MQW, the bottom contact, and the etch-stop layer is 1.8 μm, which is slightly thinner than the previously designed 2.0 μm.

The test detectors are fabricated into 40×40 pixel arrays with 25 μm pitch. Each array is hybridized to a fanout circuit that connects all the pixels in parallel. After backfilling the array with

W3 & W4	Material	Doping (E18 cm^{-3})
Top contact	GaAs, 0.4 μm Inner ring: 1.4 μm Outer ring: 3.6 μm Ring spacing: 5.7 μm	1.0
MQW (18 QWs, 1.0 μm for W3, 21 QWs, 1.17 μm for W4)	GaAs Well: (24+24)Å $Al_{0.23}Ga_{0.77}As$ Barrier: 500Å	2.0 at the lower 24Å
Bot. contact	W3: 0.7 μm W4: 0.6 μm	1.0
Etch stop	$Al_{0.3}Ga_{0.7}As$: 0.1 μm	undoped

(a)

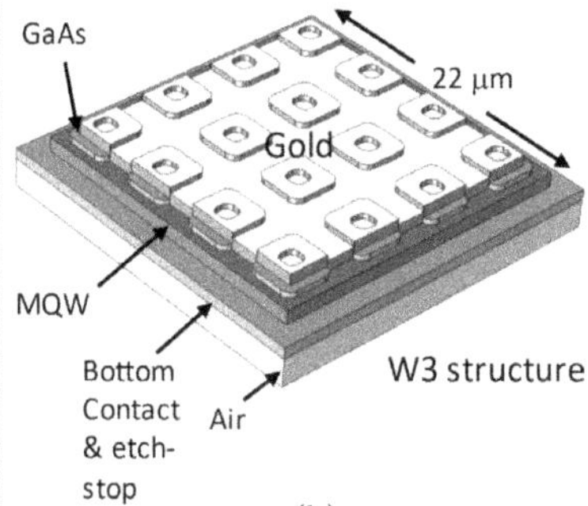

(b)

Figure 5.53. (a) The W3 and W4 wafer material parameters and (b) the fabricated $22 \times 22\,\mu m^2$ detector geometry with a $19 \times 19\,\mu m^2$ metal cover for W3.

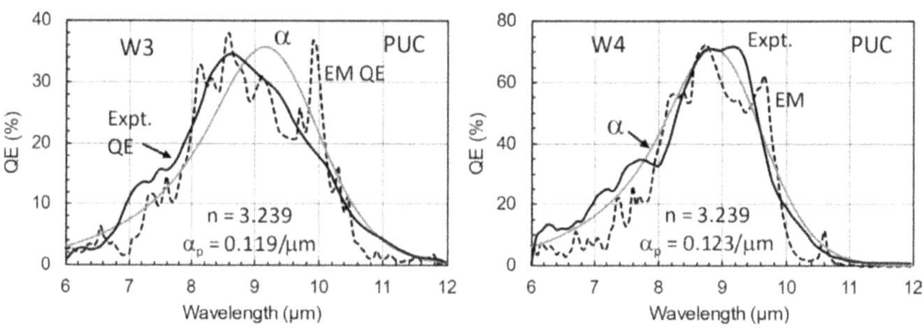

(a) (b)

Figure 5.54. The experimental external QE, the fitted α spectrum, and the FEM-modeled external QE under PUC boundary condition around the pixel for (a) the W3 R-QWIP and (b) the W4 R-QWIP. In the EM model, all the GaAs and AlGaAs materials are assumed to have a wavelength-independent n of 3.239.

epoxy for mechanical support, the GaAs substrate is completely removed by mechanical and selective dry etching, in which the etching process is stopped at the AlGaAs/GaAs interface [19]. The W3 α spectrum fitted to the edge coupling lineshape is shown in Fig. 5.54(a). It has a maximum of $0.12\,\mu m^{-1}$, which is lower than the $0.19\,\mu m^{-1}$ shown in Fig. 5.1. The lower α_p is due to the thicker barriers and the wider spectral width. With λ_p of $9.2\,\mu m$, the material is well matched to the coupling spectral range shown in Fig. 5.47. The wafer is processed into the geometry shown in Fig. 5.53(b). In the process development, the metal contact is evaporated on top of the rings after the rings and mesa are formed. The metal area must then be smaller than the mesa area to avoid shortage to the bottom contact. This incomplete metal coverage on the mesa alters the boundary condition for the R-QWIP. As a result, the QE of this detector geometry is measured to be 34.6% in Fig. 5.54(a), which is smaller than expected. The theoretical QE is then remodeled according to the fabricated geometry, and it is shown in Fig. 5.54(a) as the dashed curve. It has similar magnitude and lineshape as the measured curve. Therefore, both experiment and modeling point to the deteriorating effects of incomplete metal coverage on the rings, which weakens the plasmonic resonances.

To realize the modeled structure in Fig. 5.36, a self-alignment fabrication process is then developed, in which the $22 \times 22\,\mu\text{m}^2$ metal contact consisting of Pd/Ge/Au/Pd/Au metal is first evaporated onto the rings, in which the last thick gold layer is used as a mask for the mesa etch. This self-alignment etching process then produces the same structure as designed. The fabrication process is applied to the W4 wafer, which has a slightly thicker MQW layer of $1.17\,\mu\text{m}$, a thinner bottom contact layer of $0.6\,\mu\text{m}$, and the same AlGaAs etch-stop layer of $0.1\,\mu\text{m}$. The fitted α is similar to that of W3. The measured QE in this case is 71.3% as shown in Fig. 5.54(b). The EM modeling based on the fitted α matches the experimental QE. These examples show that the designed QE can be realized if the detector is fabricated to specification.

5.4.2 *R-QWIPs for 9.5 μm Detection*

The characteristics of R-QWIPs are systematically investigated in an experimental study [20]. In this study, three pixel sizes with $(D, d) =$ (30, 29), (12, 11), and (6, 5) in μm are fabricated on three different wafers, labeled SW1, SW4, and SW7. All wafers have the same 50 Å GaAs/500 Å $Al_{0.218}Ga_{0.782}As$ QWs and the same top layer for the ring structure. The differences are that SW1 contains 10 periods of QWs with $N_D = 0.5 \times 10^{18}\,\text{cm}^{-3}$, SW4 contains 23 periods with

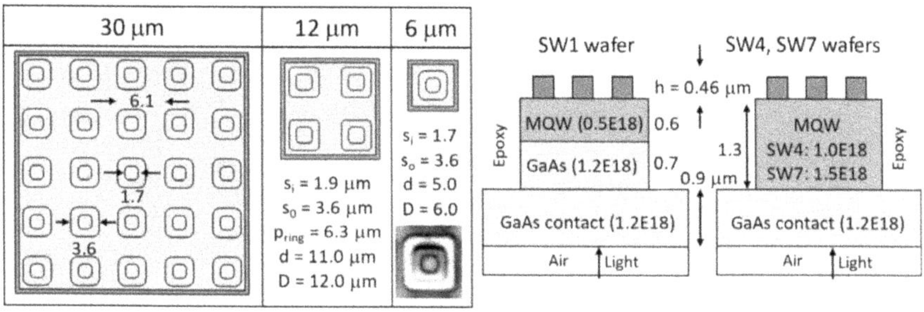

Figure 5.55. The fabricated detector geometry and the schematic material structures.

twice the doping, and SW7 contains 23 periods with three times the doping. To obtain the same resonant detector thickness, SW1 has a thicker GaAs bottom contact layer to compensate for the thinner MQW material. The schematic material and grating structures are shown in Fig. 5.55. These structures are optimized for λ between 9–10 μm. Different number of pixels are connected in parallel so that the overall array sizes are $\sim$1 mm^2 for all pixel pitches.

The SW1 material α obtained from edge coupling is shown in Fig. 5.56(d). It has a maximum of 0.091 μm^{-1} at 9.6 μm. By measuring the responsivity R of the R-QWIPs and the gain g of the wafer material, the detection QE spectrum η_d at each pixel size at different substrate voltage V can be deduced, and they are shown in Fig. 5.56. In this figure, QE increases with V because of the increasing γ in the $\eta_d = \eta_{\text{abs}}\gamma$ expression. At the highest V, QE is 36.4% for the $D = 30\,\mu$m and is about 24% for the smaller D while the corresponding CE for $g = 0.51$ is about 19% and 12%, respectively. These values of QE and CE are rather substantial given their small N_D and thin MQW layer. Because of the broad spectrum of the 6 μm pixel, this small pitch R-QWIP is able to achieve an experimental IQE of 0.468.

The experimental spectra in Fig. 5.56 are substantially blueshifted with increasing V. This trend is in opposite to the usual redshift using edge coupling and C-QWIPs such as that in Fig. 4.23 where γ at the longer λ is enhanced by the bias. While the longer cutoff wavelength due to γ enhancement is still visible in the R-QWIP

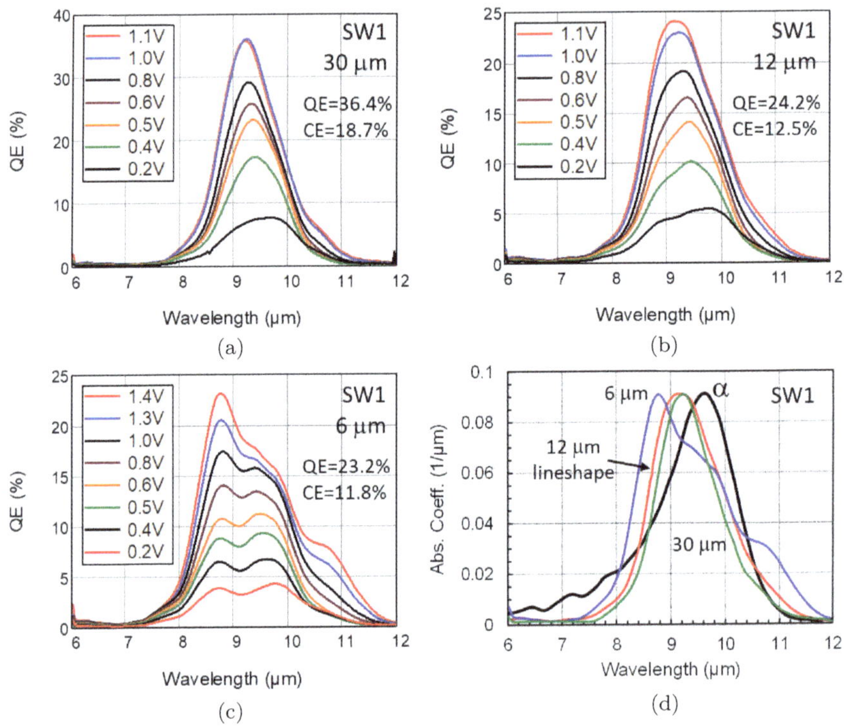

Figure 5.56. The measured external QE at different V for pixel pitch D = (a) 30 μm, (b) 12 μm, and (c) 6 μm. (d) The comparison of material α and the QE lineshapes of different D.

spectrum, the absorption peak nevertheless shifts to the shorter wavelengths at higher bias. Figure 5.56(d) shows the R-QWIP QE lineshapes at the largest V relative to the α spectrum deduced using the flat band energy levels. The shift can be up to 1 μm for the 6 μm pixel. This shift is attributed to the second-order Stark effect that shifts the optical transition spectrum under an intense electric field. Since the present QWs are symmetric, there is no first-order Stark shift. The intersubband transition energy remains the same under moderate electric field. However, the applied field is much more intense near the metal surface under positive bias as shown in Fig. 5.50(d), which creates a strong second-order Stark shift in these QWs. At the same time, the R-QWIP absorption is also concentrated at the metal surface and thus shows a more pronounced Stark shift than other light coupling structures. The blueshift of the α spectrum can be obtained by numerical modeling [20], and it is first observed in QW intersubband transitions by Harwit and Harris [21]. The Stark shifts in R-QWIPs then provide a strong support for the surface wave coupling mechanism and the modeled LSP E field depth profile.

It is quite involved to model quantitatively the Stark shift in R-QWIPs since the electric field changes across the structure and each QW has a different Stark shift. For simplicity, the η_{abs} spectrum will be modeled based on the flat band α shown in Fig. 5.56(d). It turns out that the QE spectral lineshape significantly depends on the material refractive index n. For a given α spectrum, different resonant wavelengths using different n will give different QE spectra that is beyond a simple wavelength shift. Unfortunately, the refractive indices of AlGaAs/GaAs MQWs and GaAs contact layers are not well characterized in the LWIR at low T. In this modeling, we found that the QE spectrum is better fitted with $n = 3.30$ assumed for both materials. With these assumed α and n, the modeled external η_{abs} for the three pixel pitches are compared with the measured η_d at the highest applied V where $\gamma = 1$ in Fig. 5.57. It is found that the modeled peak QEs agree well with the measurement. The modeled QE spectra, however, are redshifted relative to the experimental curves as expected without taking the Stark shift into account. The measured QE is thus supported by EM modeling. The present modeling assumes PUC boundary condition around the pixel pitch boundaries. Therefore, coherent optical crosstalk is present under flood exposure condition. The QE of an infinite periodic ring array is also shown in Fig. 5.57(d) and it can reach 57%. Therefore, there is a significant size effect in this detector and material structures.

The measured and modeled QEs for SW4 are similarly displayed in Fig. 5.58. With $N_D = 1.0 \times 10^{-3}\,\text{cm}^{-3}$ and $t_a = 1.3\,\mu\text{m}$, the QE and CE of the 30 μm SW4 R-QWIP are 64.1% and 30.6% at 3.5 V, respectively, which nearly doubles that of the SW1. The QE spectrum in Fig. 5.58 shows less blueshift than that of the SW1 and agrees with modeling in Figs. 5.59(a) and (b) for $D = 30$ and $12\,\mu$m. However, the modeled QE for the 6 μm detector is significantly larger than

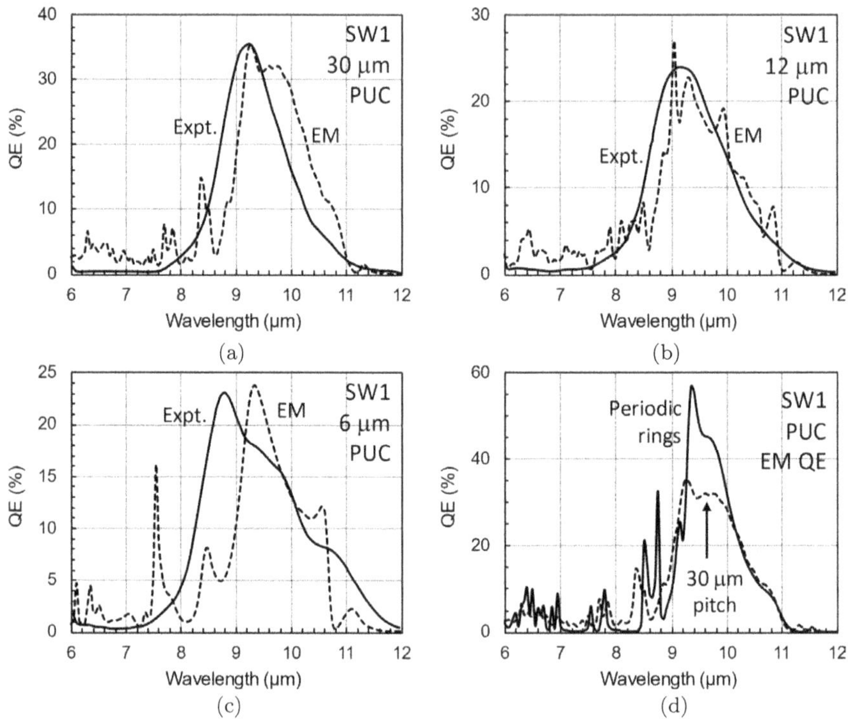

Figure 5.57. The measured η_d under the largest V and the modeled external η_{abs} for $D =$ (a) 30 μm, (b) 12 μm, and (c) 6 μm. $n = 3.3$ is assumed. (d) The modeled QE of an infinite ring array and an infinite 30 μm pitch array.

the observed spectrum in Fig. 5.59(c). The QE spectra in Fig. 5.58 shows a smaller blueshift in the 6 μm spectrum. This indicates that its MQW receives less bias probably due to the presence of a large contact resistence, which leads to the discrepancy. The infinite ring array QE and the 30 μm pitch QE in Fig. 5.59(d) are closer, which indicates less size effects in stronger-absorbing materials.

For the SW7 R-QWIPs, the higher $N_D = 1.5 \times 10^{-3}\,\text{cm}^{-3}$ should yield larger QEs. However, the larger doping also increases the tunneling current I_{tun} at large bias. The exponentially increasing I_{tun} prevents the photocurrent I_p being measured accurately, and the apparent I_p begins to drop beyond 2.7 V. At 2.7 V, the 30 μm SW7 R-QWIP QE and CE are 44.9% and 19.3%, respectively, in Fig. 5.60(a),

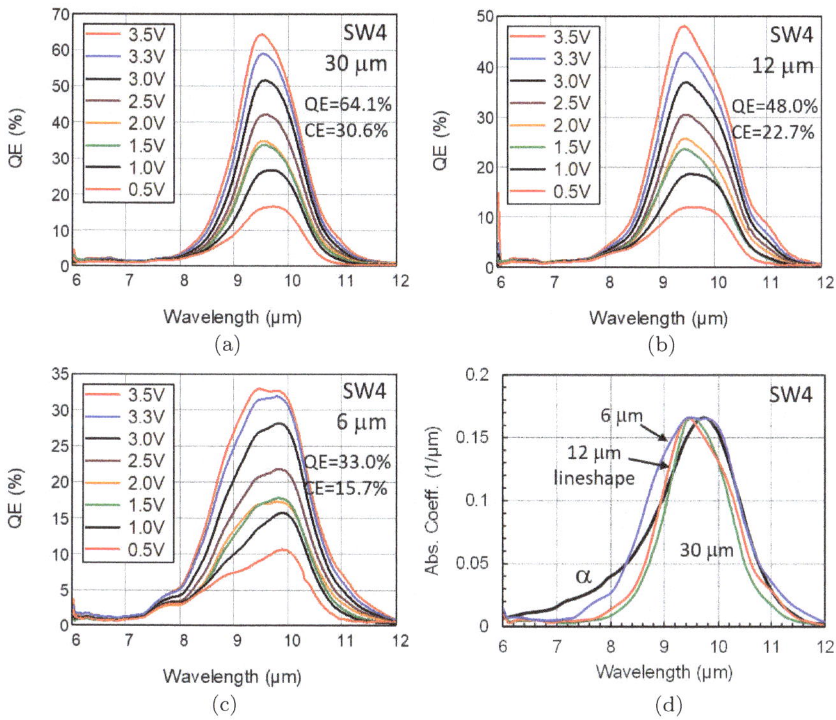

Figure 5.58. The measured η_d of SW4 R-QWIP for $D =$ (a) $30\,\mu$m, (b) $12\,\mu$m, and (c) $6\,\mu$m. (d) The comparison of material α and the QE lineshapes.

which are about the same as those of the SW4 in Fig. 5.58(a) at this same bias. This nearly-the-same QE can be explained by the band bending shown in Fig. 5.52(d). When the doping is higher, more dopants migrate to the adjacent barrier during the material growth and creates larger band bending at the interface. This interface field acts opposite to the applied field under positive bias. The fraction of photocurrent drifting in the direction of the applied field is then reduced, thereby reducing γ. The reduction of η_d from $\eta_{\rm abs}$ is thus larger at the same V. In addition to reducing γ, the resulting band bending inside the QW by the interface dipole also reduces the tilting of the QW potential in an applied field. The Stark shift is thus also reduced concomitantly. The absence of the Stark shift is apparent in the QE spectra of all detectors in Fig. 5.60.

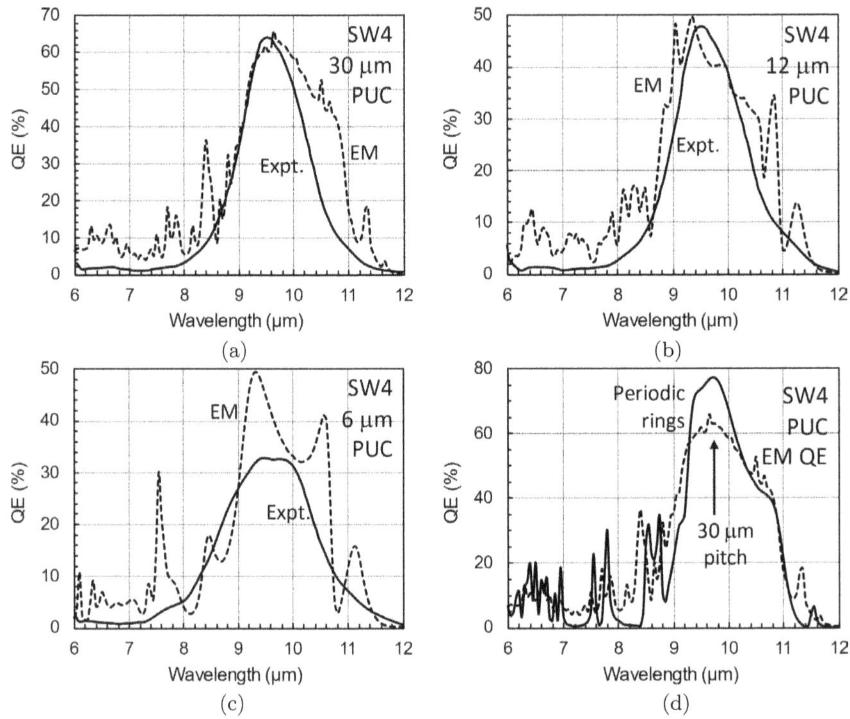

Figure 5.59. The largest measured η_d and the modeled η_{abs} for $D =$ (a) 30 μm, (b) 12 μm, and (c) 6 μm. (d) The modeled QE of ring and 30 μm pitch arrays.

Without the Stark shift, the measured and the modeled QE match well in the spectral lineshape for both $D = 30$ and $12\,\mu$m in Figs. 5.61(a) and (b). However, the modeled η_{abs} magnitude has to be reduced by a factor $\gamma = 0.63$ at 2.7 V to match the measured η_d at this V. For the $D = 6\,\mu$m R-QWIP, the modeled peak value needs to be reduced by a factor of $\gamma = 0.48$ to match the experimental value. The smaller γ may mean a lower pixel connectivity to the external fanout circuit due to a much larger number of indium contacts in the 1 mm $\times$ 1 mm array. In addition, the modeled λ_p of 9.34 μm is displaced by 0.68 μm from the experimental $\lambda_p = 8.65\,\mu$m. This shift is nearly the same as that for SW1 in Fig. 5.57(c). This shift does not have a clear source but could be related to the fact that the fabricated

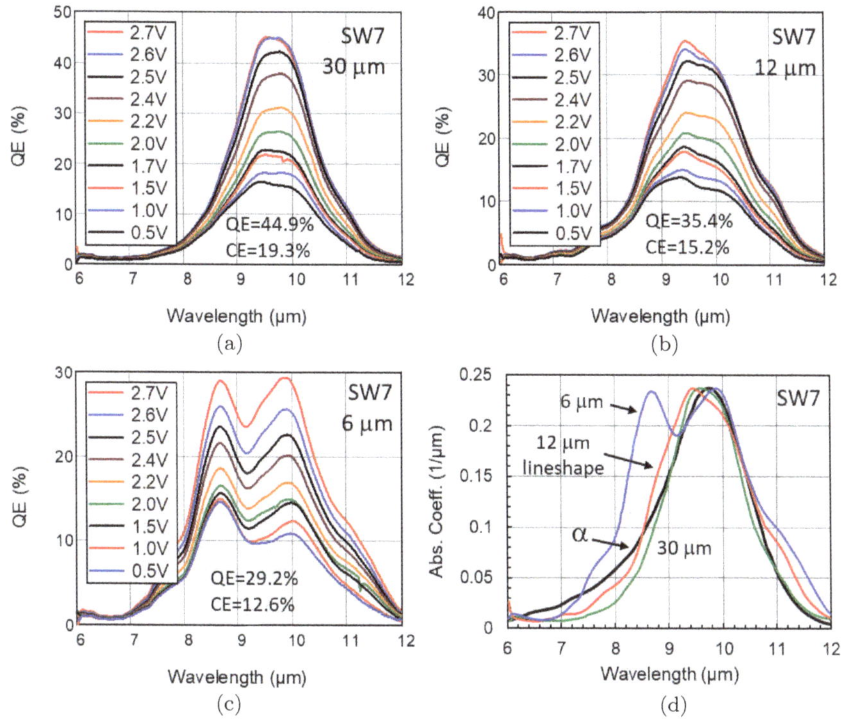

Figure 5.60. The measured η_d of SW7 R-QWIP for D = (a) $30\,\mu$m, (b) $12\,\mu$m, and (c) $6\,\mu$m. (d) The comparison of material α and the QE lineshapes.

pixel size and geometry shown in Fig. 5.55 can be significantly different from the nominal size and geometry in percentage when the pixel size is small.

From this systematic study, it is found that the designed QE can usually be realized and large QE and CE can be extended to small pixels. In fact, it is even possible to design a $3\,\mu$m pitch pixel for $9\,\mu$m narrowband detection. The modeled QE is $\sim$80% although it has not been experimentally verified. However, care must be taken when the doping is either very low or very high, in which cases the MQW deviates from its simple behavior. At low doping, the quadratic Stark shift will change the α peak wavelength under a bias. The design of the R-QWIP structure must account for this material change to yield

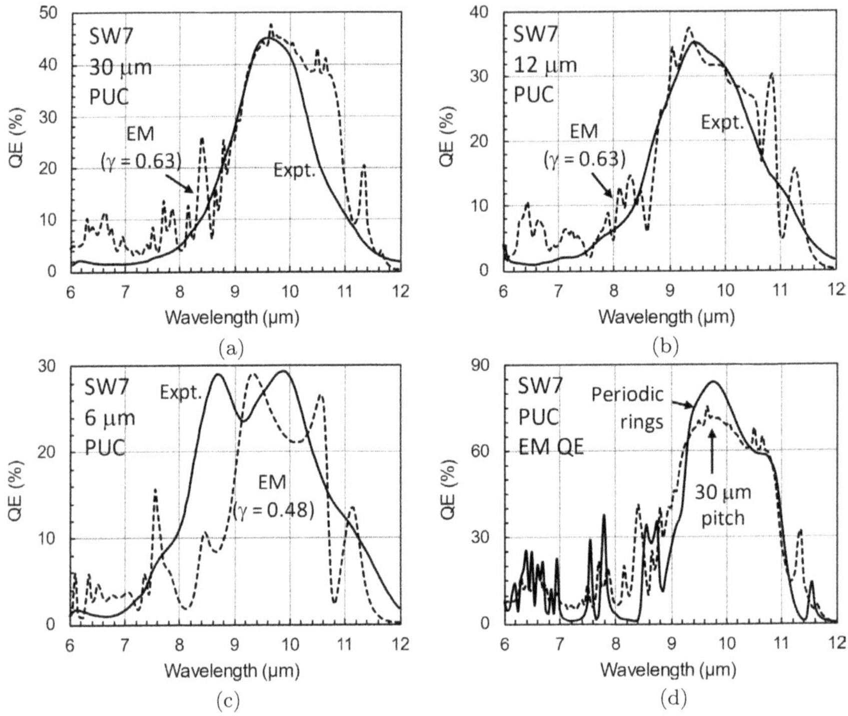

Figure 5.61. The largest measured η_d and the modeled η_{abs} for $D =$ (a) $30\,\mu$m, (b) $12\,\mu$m, and (c) $6\,\mu$m. (d) The modeled QE of ring and $30\,\mu$m pitch arrays.

a large QE at high bias. At large doping, the interfacial dipole can act against the applied field and reduces the photocurrent even if the absorption is large.

To evaluate the potential performance of FPAs made from SW1 and SW4, their window current I_w and dark current I_d are shown in Fig. 5.62 for $D = 30\,\mu$m as an example. I_w is the total DC current measured from the detector when it is exposed to a 300 K blackbody under $f/2$ optics at $T = 10$ K. At this low T, the thermal dark current is totally suppressed and I_w is exclusively contributed by the photocurrent I_p until the tunneling dark current I_{tun} dominates at high voltages.

In addition to the direct measurements, the expected I_p is also computed based on the CE spectrum measured using AC lock-in

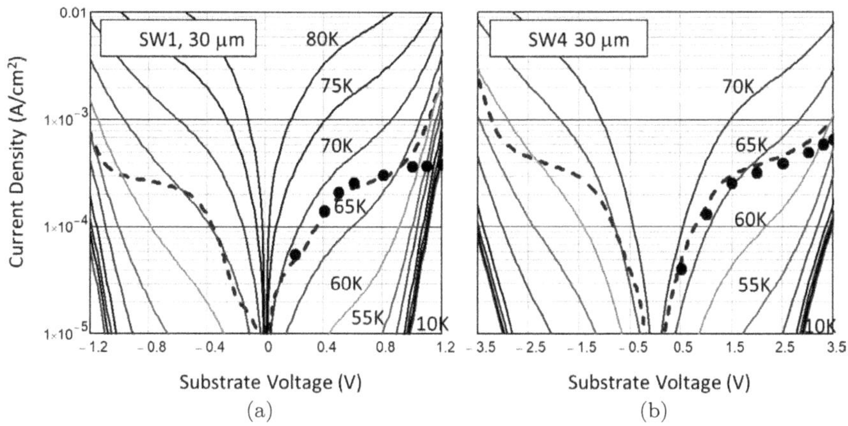

Figure 5.62. The directly measured I_w (dashed curve) and I_d (solid curves) for $30\,\mu$m pitch R-QWIPs for (a) SW1 and (b) SW4 wafers. The circles are I_p from CE measurements. Reproduced with permission from [20].

techniques and the blackbody photon flux density, and they are represented by circles in Fig. 5.62. These calculated values agree with the directly measured I_w in a wide range of V, thus confirming the QE measurements. Similar agreements are also obtained with R-QWIPs with other pitches. At large bias, the calculated I_p is lower than the measured I_w due to the presence of I_{tun}, which cannot be quenched by lowering T. From Fig. 5.62, the highest background limited temperature T_{BLIP} is $\sim$68 K at 0.5 V SW1 with a cutoff wavelength $\lambda_c \sim 10.0\,\mu$m and $\sim$66 K at 1.4 V for SW4 with $\lambda_c \sim 10.5\,\mu$m.

Although modeling PUC of pixels in an array yields accurate peak values of QE, the spectrum often contains high-frequency oscillations that are not observed in measurement such as those in Figs. 5.54 and 5.59. These oscillations are believed to be due to long-range phase coherence and can be weakened in the presence of fabrication nonuniformity across the array and material thickness fluctuation. To eliminate long-range correlations, a truncated geometry (TCG) is modeled in which the pixel unit cell is surrounded only by the first row of rings in the neighboring pixels. An example of the TCG structure for the $12\,\mu$m pitch is shown in the inset of Fig. 5.63(a).

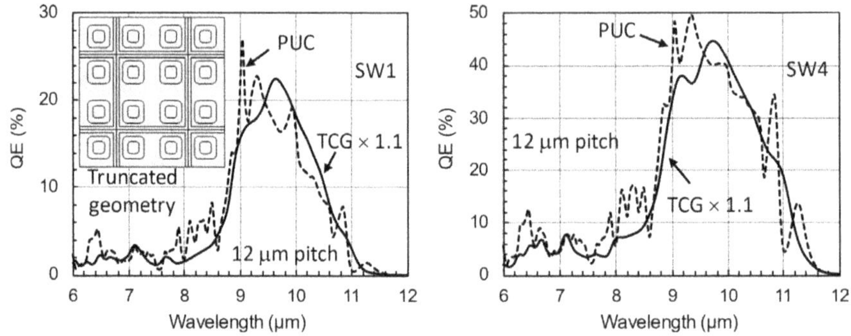

(a) (b)

Figure 5.63. The comparison between modeling a single pixel with PUC and TCG with second-order SBC at their sidewalls. The TCG QE needs to be scaled by 1.1 to match the PUC QE.

Radiation is incident uniformly at the substrate and the diffracted light is free to leave at the truncated boundaries.

The TCG thus contains OCT between the center pixel and the surrounding rings, which is partial to the overall coherent OCT of the entire array. Therefore, the modeled QE of the TCG is expected to be smaller than that under periodic pixel unit cell modeling. In Fig. 5.63, two examples are shown for SW1 and SW4. If the TCG QE is enlarged by a factor of 1.1, it will have a similar peak value and lineshape as that of the PUC modeling but without the high-frequency oscillations. With pixel nonuniformity, the scaled TCG QE is more suitable for QE comparison with experiment.

For other pixel pitches, a different scaling factor is needed. Figure 5.64 shows those with $D = 30\,\mu$m and $6\,\mu$m. For the large $D = 30\,\mu$m, the TCG QE is very close to that of the PUC QE with a difference of only 2%, while the TCG of the $6\,\mu$m pitch makes poorer prediction of the PUC QE by including only the closest rings, and their magnitudes differ by 37%. Therefore, more neighboring rings should be included in the TCG modeling for small pixels.

From the detector photocurrent, dark current, and gain, the noise equivalent temperature difference (NETD) of the array can

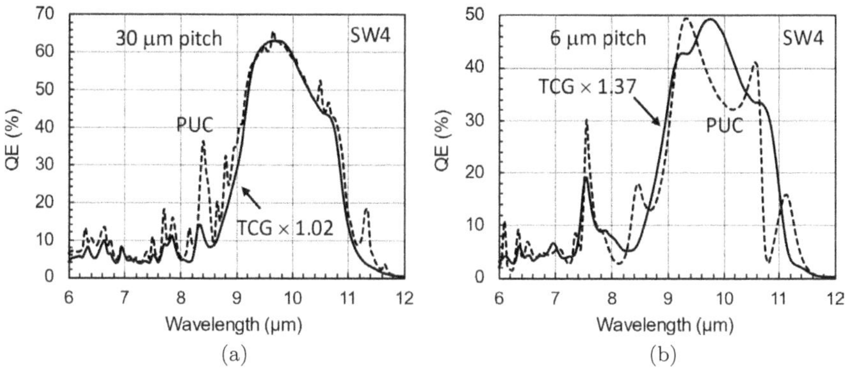

Figure 5.64. The comparison between modeling a single pixel with PUC and TCG when only the closest rings outside the center pixel are included. The TCG QE needs to scale by 1.02 and 1.37 to match the PUC QE with (a) $D = 30\,\mu$m and (b) $D = 6\,\mu$m.

be evaluated using Eq. (1.43). Its temporal component is

$$NETD = \left[\frac{1}{C}\left(\frac{2gN + n_r^2}{N^2}\right)^{1/2}\right]\left(1 + \frac{I_d}{I_p}\right) = NETD_0\left(1 + \frac{I_d}{I_p}\right), \tag{5.12}$$

where C is the thermal contrast, g is the gain, N is the total number of dark and photo electrons collected in an integration time τ_{int}, and n_r is the noise of the readout circuit (ROIC). Figure 5.65 shows the detector NETD for SW1 and SW4. Note that I_p in Eq. (5.12) has been replaced with I_w in evaluating N and I_p/I_d, and hence the NETD value is only meaningful at V less than 1 V for SW1. In Fig. 5.65, the FPAs are operated at $T = 60$ K with F/2 and 300 K background. There is no limitation on charge collection imposed by the ROIC's charge-handling capacity, and n_r is 700 e$^-$ for 30 μm pitch and 500 e$^-$ for 12 and 6 μm pitches.

Since the pixel areas of the 6, 12, and 30 μm pitch arrays are in the ratio of 1: 4.84: 33.6, the photon flux on the detector area is proportionally larger for the larger pixel pitch D, and τ_{int} is proportionally shorter for the same collected N. Figure 5.65 shows SW4 to have a slightly lower NETD than SW1 despite having a

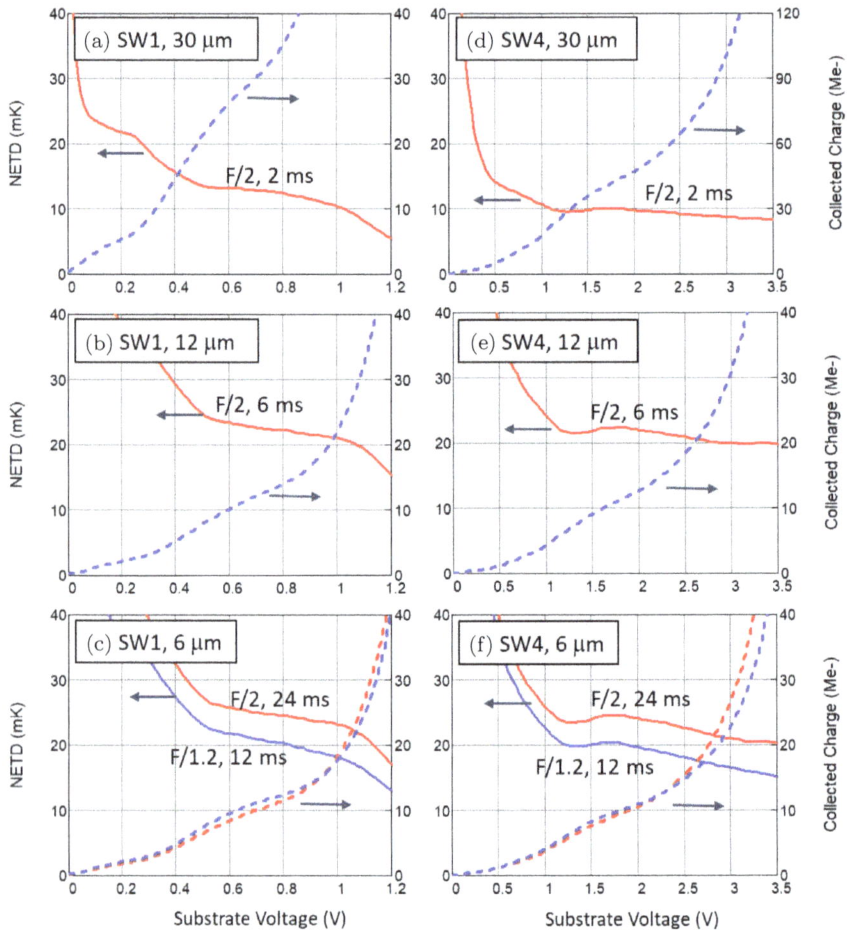

Figure 5.65. The figure shows NETD and collected N for SW1 and SW4 at $T = 60\,\text{K}$. Adapted with permission from [20].

smaller I_p/I_d ratio. In fact, NETD decreases monotonically for all detectors as V increases even though I_p/I_d continues to deteriorate. The small NETD at large V attests to the importance of a large QE in the array sensitivity, which was discussed in Section 4.10. For example, for the $30\,\mu\text{m}$ pitch, NETD of SW4 is $\sim$10 mK in Fig. 5.65(d) compared to the $\sim$12 mK for SW1 in Fig. 5.65(a) even though SW4 has much lower I_p/I_d at 60 K in Fig. 5.62. However,

SW4 requires a larger well capacity as seen in Fig. 5.65. For the similar photon collection efficiency of each pixel measured by $F\lambda/D$, the 6 μm detector under $F/1.2$ yields a slightly lower NETD than the 12 μm detector under $F/2$ for both SW1 and SW4 materials. With similar sensitivity, the 6 μm R-QWIPs will be more advantageous because they can be integrated into a smaller camera unit to reduce size, weight, power, and cost. However, the integration time is twice as long and the pixel OCT is more noticeable as indicated in Fig. 5.45.

However, even with a larger OCT, a smaller pixel is still able to offer a finer resolution than a larger pixel. For a more quantitative analysis, the QE ratios are modeled along the x-axis in a 5×5, 5×5, and 7×7 pixel arrays for $D = 30$, 12, and 6 μm arrays, respectively, as performed in Fig. 5.21. Their values along the center row of each array at $\lambda = 9.6\,\mu$m, which is near the QE peak, are shown in Fig. 5.66(a). Because of OCT, each individual pixel not only senses radiation directly falling on the pixel but also radiation falling on other neighboring pixels. For the 30 μm pitch array, OCT is small at a 2% level from pixel to pixel, but it increases when the pixel pitch is smaller. Because of OCT, the modulation transfer function (MTF) of the detector degrades from the normalized $|H(\nu)|$ given by Eq. (1.14).

One can relate the MTF of an array to the detector QE ratios, which forms the detector spatial response function (SRF). It is like relating the lens MTF to the lens point spread function in Fig. 1.11. By performing discrete Fourier transform on the detector SRF, the MTF of the array can be obtained. Without OCT, the detector SRF is a rectangular function, and the calculated MTF is shown as a dashed curve in Fig. 5.66(b), (c), and (d) for each pixel pitch. It is identical to $|H(\nu)|$ from Eq. (1.14). With the SRF shown in Fig. 5.66(a), the array MTF is reduced. For the 30 μm pitch array in Fig. 5.66(b) calculated using $N = 2^{16}$ and $n =$ sampling distance per 10^{-5} mm increment in Eq. (1.15), the reduction is very small, indicating the 2% OCT is insignificant in degrading the imaging resolution. With higher level of OCT in smaller pixel pitches, the degradation becomes more severe in Fig. 5.66(c) and (d). However, even with a larger relative degradation, the imaging resolution of a

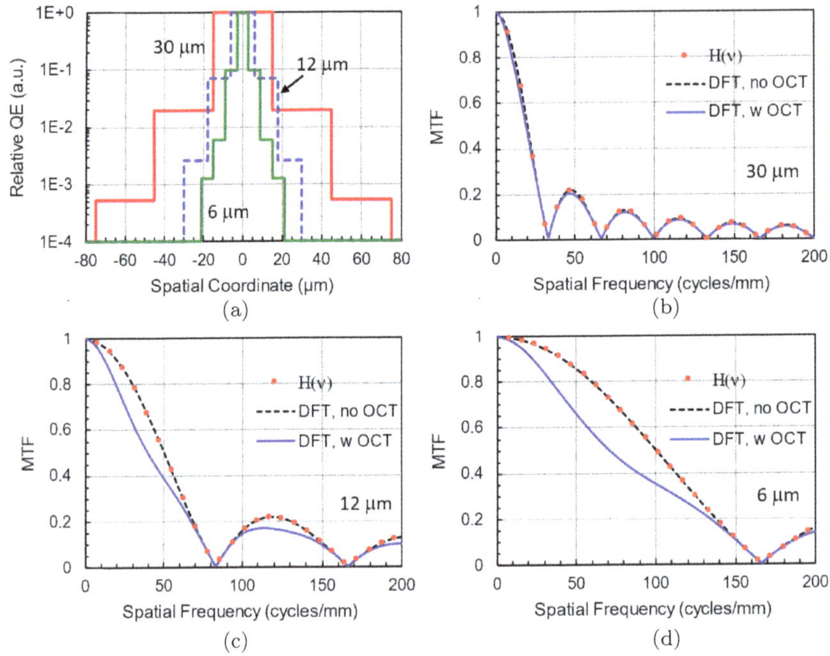

Figure 5.66. (a) The spatial OCT profiles of different pitch arrays and the corresponding MTF for D = (b) 30 μm, (c) 12 μm, and (d) 6 μm.

small pitch array is still superior compared to the larger pitch arrays. For example, at the spatial frequency of 30 cycles/mm, MTF = 0.11, 0.63, and 0.84 for D = 30, 12, and 6 μm, respectively. Therefore, the present R-QWIPs with their exhibited OCT can improve imaging resolution down to 6 μm pitch. In addition, the smaller pitch array also produces less aliasing with a higher Nyquist frequency. Beyond 6 μm pitch, one can also design R-QWIPs with 3 μm pitch with peak QE = 84% at λ = 9 μm and with 2 μm pitch with QE = 20% at 8.4 μm [20], although they have not been demonstrated.

5.4.3 *R-QWIPs for 8 μm Detection*

In certain short-range applications, it is advantageous to detect within a narrowband near 8 μm [22]. The narrowband absorption characteristics of QWIPs are well-suited for these applications. To

NVM5 & NVM9	Material	Doping ($E18\ cm^{-3}$)
Top contact	GaAs, NVM5: 0.40 μm NVM9: 0.35 μm	NVM5:1.2 NVM9:1.5
MQW 21 QWs 1.2 μm	GaAs Well: 48Å $Al_{0.26}Ga_{0.74}As$ Barrier: 500Å	NVM5:1.0 NVM9:1.2
Bot. contact	6000 Å	NVM5:1.2 NVM9:1.5
Etch stop	$Al_{0.4}Ga_{0.6}As$: 1000Å	NVM5:1.2 NVM9:1.5

(a)

$s_i = 1.9$, $s_o = 3.6$, $p = 6.3$, $d = 11$, $D = 12$

(b)

Figure 5.67. (a) The material structures of NVM5 and NVM9 wafer materials and (b) the fabricated detector geometry of $D = 12\,\mu$m detector arrays.

detect near $8\,\mu$m, two similar wafer materials, which are labeled as NVM5 and NVM9, are adopted [23]. The material layer structure and the R-QWIP geometry are shown in Fig. 5.67. The material layers are nearly the same except that NVM5 has a thinner top contact layer for the rings and a lower doping density. The first excited state of these detectors is designed to align with the barrier height to minimize the absorption width. The QE spectral lineshapes of the edge-coupled detectors agree well with the MQW design. The peaks are slightly blueshifted from the flat band spectrum, which are expected from the Stark shift. This agreement shows the materials are grown to specification. The photoconductive gains measured from the edge-coupled detectors increase nearly linearly at low bias and become constant above 1.2 V. The saturated values are ~0.32 for NVM5 and ~0.25 for NVM9. The lower g for NVM9 is likely due to the stronger electron-electron scattering in the higher doping material.

The properties of the R-QWIPs are first characterized in small format $12\,\mu$m pitch arrays with total area of 1 mm × 1 mm. From the measured CE spectra and gains, the η_d spectra are deduced, and they are shown in Fig. 5.68. The maximum η_d of NVM5 occurs at $V = 3.5$ V, where η_d is 24.6% in Fig. 5.68(a). The corresponding CE is 8.1%. At higher bias, η_d decreases. On the other hand, η_d of

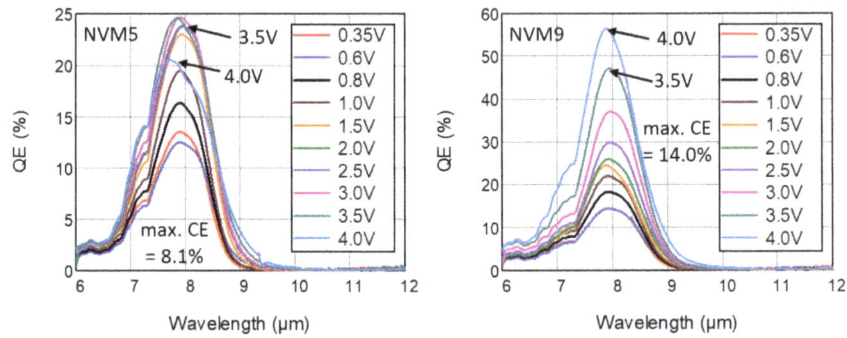

(a) (b)

Figure 5.68. The deduced external detection QE spectrum at different bias from the measured CE and g for the (a) NVM5 R-QWIP and (b) NVM9 R-QWIP with $D = 12\,\mu$m. Adapted with permission from [23].

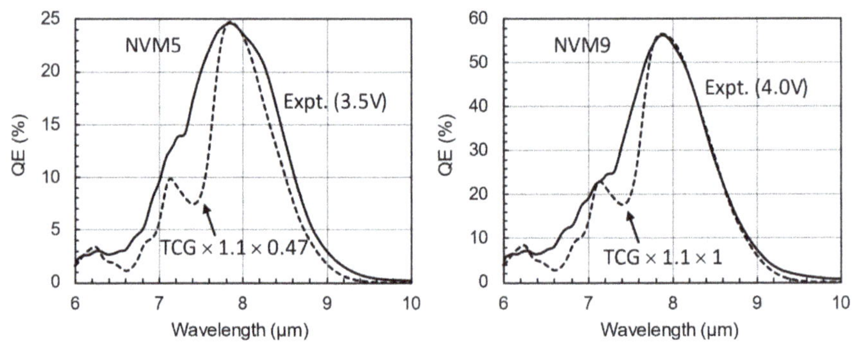

(a) (b)

Figure 5.69. The measured η_d (solid curve) and the fitted $\eta_{\text{abs}}\gamma$ of (a) the NVM5 R-QWIP and (b) the NVM9 R-QWIP at the specified bias. $N = 3.1$ is assumed.

NVM9 increases up to 4.0 V, where η_d is 56.4% in Fig. 5.68(b), and the corresponding CE is 14.0%.

The absorption QE η_{abs} is modeled using the flat band α spectrum, and they are shown in Fig. 5.69. In η_{abs} modeling, the TCG shown in Fig. 5.67 is adopted where the light is incident uniformly onto the bottom surface (*i.e.*, the etch-stop layer surface) and it is free to leave at the model boundaries. The modeled η_{abs} lineshapes of both R-QWIPs agree well with the measured spectra, which will further be improved if the material Stark blueshift is included. Since the TCG η_{abs}, labeled as η_{TCG}, is smaller than the η_{PUC} by a factor of 0.91 at this pixel pitch, the scaled $\eta_{\text{TCG}} \times 1.1$ will be treated as the theoretical η_{abs} under flood illumination. For the NVM5 R-QWIP, one needs to apply a fitting $\gamma = 0.47$ to match the experimental η_d at 3.5 V in Fig. 5.69(a). However, such a fitting parameter is not needed for NVM9 at $V = 4.0$ V in Fig. 5.67(b), indicating that $\eta_d = \eta_{\text{abs}}$ at this large bias.

In Fig. 5.70, the detector window current I_w, dark current I_d, and I_p calculated from the CE spectrum are shown. In addition, large format arrays made from NVM5 are hybridized to ISC1308 ROICs, which are in 1280×1024, 12 μm pitch format. The charge well capacity is 11.5 Me$^-$, and the read noise is 1500 e$^-$. Figure 5.70(a) shows the measured FPA I_p and I_d at 0.35 V and 1.0 V, which are in good agreement with those of the small test arrays. At 0.35 V, the pixel operability is 97.8%, and the nonuniformity (σ/mean) is 12.4% before correction and 0.32% after correction. The operability and nonuniformity are mainly limited by the indium bump size variation on the detector array. At 0.35 V, NETD is measured to be 19.3 mK at half well-fill (*i.e.*, 6 Me), 36 ms integration time, 67 K operating

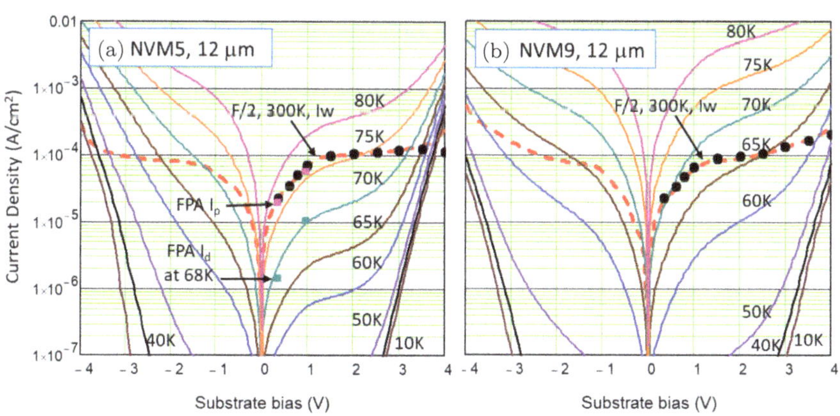

Figure 5.70. The directly measured I_w (dashed curve) and I_d (solid curves) for 12 μm pitch R-QWIPs for (a) NVM5 and (b) NVM9 wafers. The circles are I_p from CE measurements, and the squares are from FPA measurements. Adapted with permission from [23].

temperature, and F/2 optics. Notice that unlike NVM9 detectors, NVM5 detectors have very large I_{tun} above 3.5 V, which prevents η_d to be measured accurately beyond this bias and leads to the lower observed maximum η_{abs}. At $V < 3$ V, the two materials show very similar I_p, implying very similar η_d for both R-QWIP structures.

To estimate the FPA performance at higher voltages under F/1.4 optics, $N = 6 \times 10^6$, and $n_r = 1500$ are adopted in Eq. (5.12) with $C = 0.0209$ in this band. I_d/I_p is obtained from Fig. 5.70. The calculated NETD for NVM5 is 16.3 mK at $V = 0.35$ V, $T = 70$ K, $\tau_{\text{int}} = 20$ ms, and F/1.4. It agrees well with the measured 19.3 mK at 0.35 V, 67 K, 36 ms, and F/2 since both the photon flux and I_p/I_d under F/1.4 are 1.69 times larger than that under F/2, leading to a shorter τ_{int}, while I_p/I_d is nearly the same with different T.

With good agreement at a small 0.35 V, the calculation can be extended to a larger V. Figure 5.71 shows the NETD up to 4 V. With fixed N and n_r, NETD in Eq. (5.12) only depends on $r = I_p/I_d$. Since r decreases with increasing V, NETD increases, which is in opposite direction to a fixed τ_{int} with unlimited N in Fig. 5.65. Although NETD is higher with a fixed N, τ_{int} is shorter as shown in Fig. 5.71 because of the larger I_p. At 2.5 V, τ_{int} is 3.7 ms while NETD is 23 mK for NVM5, showing that sensitive and fast FPAs can be achieved using R-QWIPs. If the read noise n_r can be lowered to 1100 or 500,

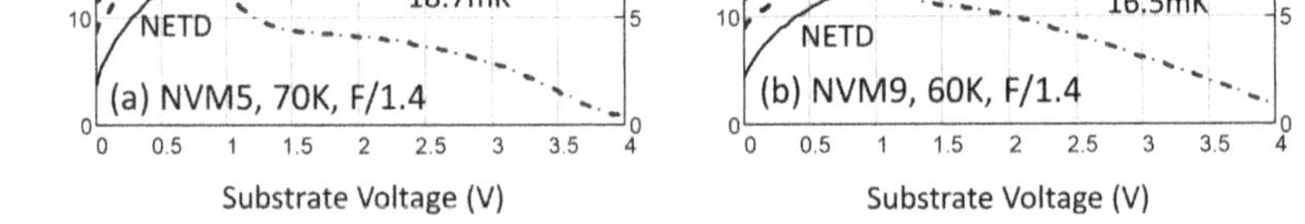

(a) (b)

Figure 5.71. The figure shows NETD and τ_{int} for 6 Me^- collected charges for (a) NVM5 and (b) NVM9 FPAs.

NETD can be further reduced to 21 or 18.7 mK, respectively. For the NVM9 R-QWIP in Fig. 5.69(b), it offers a shorter $\tau_{\rm int}$ at 3.2 ms and a lower NETD at 21 mK for $n_r = 1500$ at 3.0 V. At a higher V of 3.5 V, $\tau_{\rm int}$ decreases to 2.1 ms with NETD at 24.3 mK. This level of sensitivity and operating speed is excellent for small pitch 12 μm arrays. However, it needs to operate at a lower T of 60 K.

5.4.4 *R-QWIPs for 10.6 μm Detection*

From the NETD expression of Eq. (5.12), it can be optimized under two operating conditions. One is ROIC charge limited condition where the collected charge N is limited to half of the well size N_c of the ROIC. Under this condition, the first term in Eq. (5.12) is fixed for given g and N_c, and NETD is only a function of $r \equiv I_p/I_d$. NETD is minimized when r is maximized, which favors a smaller doping density N_D in the QW since I_d is exponentially increased with N_D. On the other hand, if the detection is $\tau_{\rm int}$ limited, the $1/\sqrt{N}$ dependence in the first term favors a larger N_D for a larger I_p. The proper choice of N_D can be investigated in the case of 10.6 μm SF_6 greenhouse gas sensing as an example.

For the present application, the QWIP material is made of 15 periods of 56 Å GaAs/600 Å $Al_{0.2}Ga_{0.8}As$ sandwiched between a 0.5 μm top GaAs contact layer and a 1.4 μm GaAs bottom contact layer. The active layer thickness t_a is thus 1.0 μm and the mesa is etched down to 1.4 μm to provide a suitable resonant volume. The designed inner ring dimension s_i is 2 μm, the outer ring dimension s_o is 4 μm, the ring spacing $p_{\rm ring}$ is 6.8 μm, the pixel pitch D is 30 μm, and the pixel size d is 28 μm. With these design parameters, the modeled $\eta_{\rm PUC}$ under periodic pixel unit cell boundary condition and a constant $\alpha = 0.2\,\mu\text{m}^{-1}$ is shown in Fig. 5.72 (a). It is maximized at 10.6 μm when the refractive index n is 3.00, where $\eta_{\rm PUC} = 73\%$. The $\eta_{\rm TCG}$ modeled under TCG is smaller by a factor of 1/(1.05). The fabricated ring pattern is somewhat different from designed geometry as shown in the inset of Fig. 5.72(b). With the realized pattern, the predicted peak value is about 70%, similar to the design value. Therefore, the R-QWIP QE is not sensitively dependent on the realized geometry.

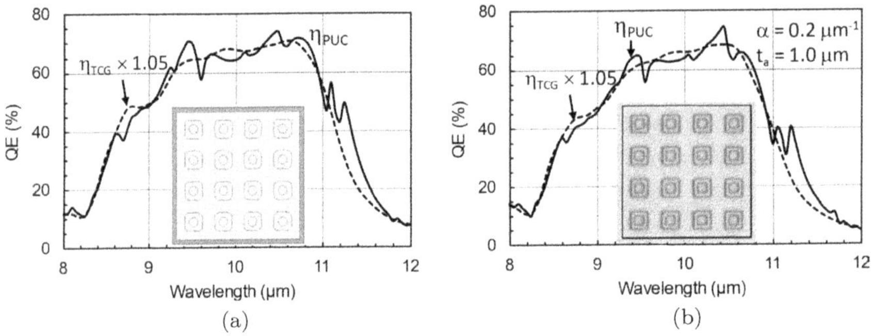

Figure 5.72. The external η_{abs} modeled under PUC and TCG boundary conditions according to the (a) designed ring geometry and (b) realized ring geometry in an experiment. $\alpha = 0.2\,\mu\text{m}^{-1}$ and $n = 3.00$.

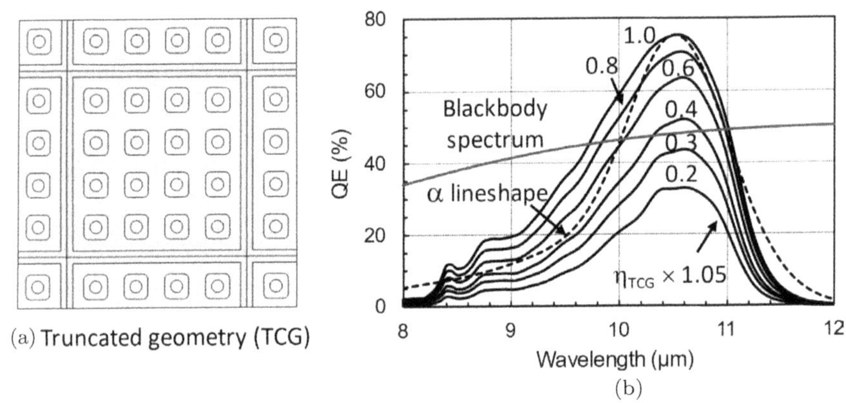

Figure 5.73. (a) The TCG used in modeling $\eta_{\text{abs}} = 1.05\eta_{\text{TCG}}$. (b) The modeled η_{abs} at different N_D. The calculated α lineshape and the 300 K blackbody spectrum in the unit of 5×10^{15} ph/s/cm^2/μm under F/2 are also shown.

With the MQW specification, the α spectrum at different N_D can be calculated and its value at $N_D = N_0 \equiv 1 \times 10^{18}\,\text{cm}^{-3}$ is $0.252\,\mu\text{m}^{-1}$. Its value at a smaller N_D is proportionally smaller. The corresponding η_{abs} of the designed geometry are shown in Fig. 5.73(b) using the TCG geometry in Fig. 5.73(a). Its peak value is 33.0% at $N_D = 0.2N_0$ and rises to 75.9% at N_0. Therefore, QE increases only sublinearly with N_D. To estimate the photocurrent I_p from a 300 K

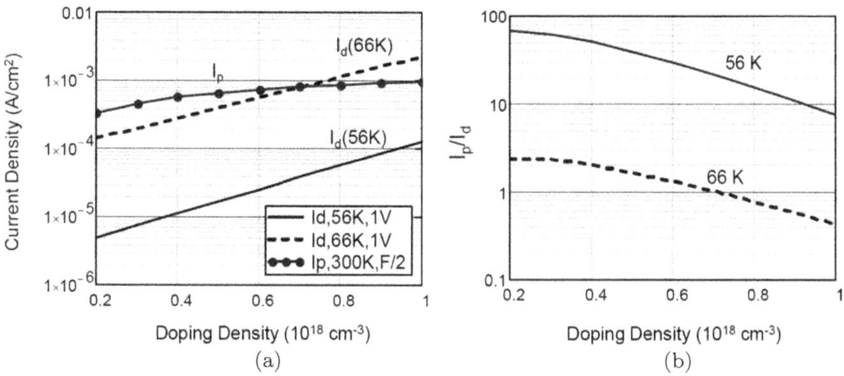

Figure 5.74. (a) The calculated photocurrent and dark current levels and (b) the corresponding ratios of an R-QWIP under 300 K background and F/2 optics at 56 K and 66 K.

blackbody, the photoconductive gain g is assumed to be 0.6 and $\gamma = 1$ at a high V, with which I_p can be calculated from Eq. (4.15). The background photocurrent generated by an R-QWIP under F/2 is shown in Fig. 5.74(a). It rises moderately from 0.3 to $1.0\,\mathrm{mA/cm^2}$ when N_D increases from 0.2 to $1.0N_0$.

To estimate the dark current density I_d, the semi-empirical formula of Eq. (2.51) is utilized, which reads

$$J_d(V,T) = J_0(V)\exp\left(-\frac{hc}{\lambda_c kT}\right)\exp\left(\frac{N_D W}{g_{2D}kT}\right), \tag{5.13}$$

where $J_0(V)$ is the empirical parameter taken to be $2.0 \times 10^4\ \mathrm{Acm^{-2}}$ at the saturation voltage of g. For $W = 56\,\text{Å}$, the estimated I_d at $T = 56\,\mathrm{K}$ and 66 K is shown in Fig. 5.74(a). I_p is always larger than I_d at 56 K for the considered N_d, while it is only true below $0.7N_0$ at 66 K. Figure 5.74(b) shows the corresponding I_p/I_d ratios.

With the known I_p and I_d at different N_D, NETD as a function of N_D can be readily deduced. For N_D selection, it is sufficient to know the NETD ratio relative to a fixed N_D, which is chosen to be at N_0. Figure 5.75 shows that, in the τ_{int} limited case at a low T (e.g., 56 K), a larger N_D always gives a lower NETD. At the higher T (e.g., 66 K), an optimum N_D exists at $0.5N_0$ that yields the best lowest NETD. On the other hand, when the detection is N_c limited,

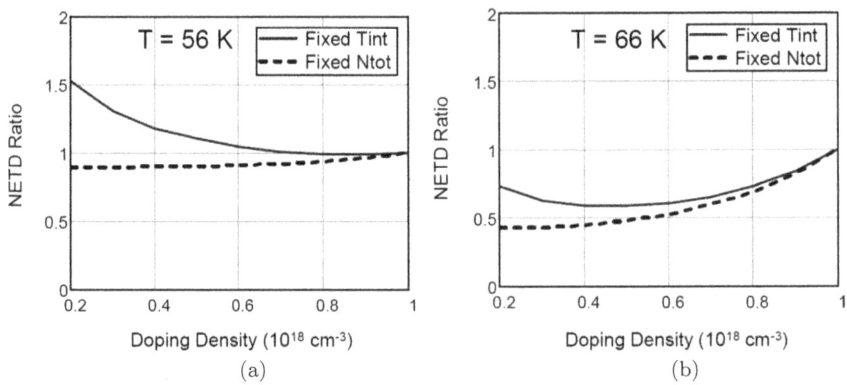

Figure 5.75. The figure shows the variation of $NE\Delta T$ as a function of N_D in cases where τ_{int} is fixed or the total N is fixed at (a) $T = 56\,\text{K}$ and (b) $T = 66\,\text{K}$.

a lower N_D will always give a better result, and it is especially true at the higher T.

To analyze the performance further, a specific ROIC, FLIR 9705 ROIC, is assumed which has $N_c = 18$ Me$^-$ and $D = 30\,\mu$m. If N in Eq. (5.12) is taken to be half-well fill of 9 Me$^-$ and n_r is taken to be 700e$^-$, NETD$_0$ is fixed at 24.2 mK with $C = 0.0154$ for this detection band and $g = 0.6$. Note that this is the lowest NETD achievable with this ROIC, irrespective of N_D. It can only be lowered by reducing g at the expense of a longer τ_{int}. Since I_p for the lowest N_D in Fig. 5.74(a) is 0.33 mA/cm^2, the photocurrent alone will generate 33 Me$^-$ in 2 ms, exceeding the half-well charge. Therefore, the detection tends to be N_c limited, and a smaller N_D will give a better performance. In an experiment [24], $N_D = 0.2N_0$ (labeled as Det. A) and $0.3N_0$ (labeled as Det. B) are adopted. The material α spectra are characterized at $V = 0.6\,\text{V}$ and $T = 77\,\text{K}$. These experimental spectra match closely with the designed lineshape as shown in Fig. 5.76.

The measured photoconductive gain g is shown in Fig. 5.77. As expected, g increases monotonically with bias and saturates at a value of 0.69 for Det. A and 0.57 for Det. B at a bias higher than 1 V. With these gain values and the measured responsivity R of the R-QWIPs, their detection QE η_d can be obtained from $\eta_d = \eta_{\text{abs}}\gamma = h\nu R/eg$.

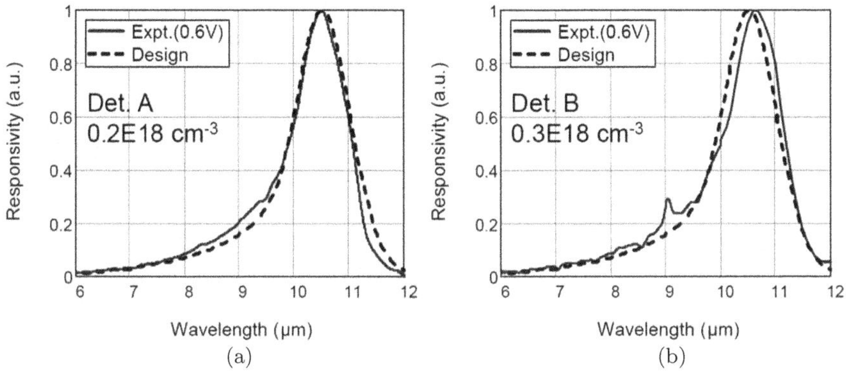

Figure 5.76. The measured responsivity lineshapes (solid curves) of (a) Det. A and (b) Det. B at 0.6 V in planar geometry. Both are very close to the designed α lineshapes (dashed curves).

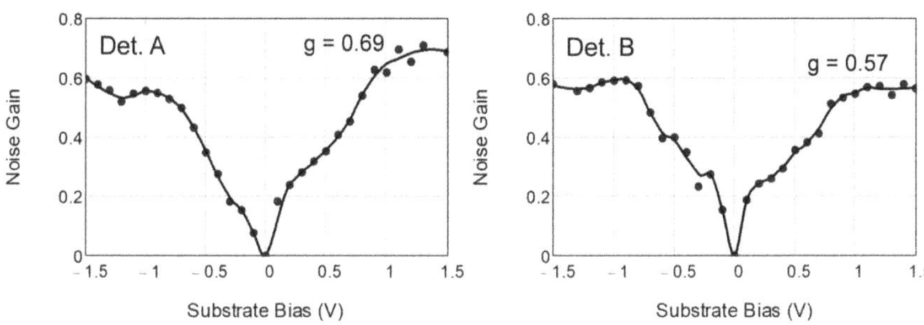

Figure 5.77. The measured photoconductive gain g for (a) Det. A and (b) Det. B.

The measured η_d of the R-QWIPs on fanout circuits are shown in Fig. 5.78. The detectors show the characteristic blueshift of R-QWIPs at increasing V. Det. A has a maximum CE of 20.2% and a maximum η_d of 29.4% at $V = 2.5$ V. This η_d is quite close to the modeled $\eta_{\text{abs}} = 1.05\eta_{\text{TCG}} = 36.2\%$ based on the measured α spectrum at 0.6 V and the processed ring geometry shown in Fig. 5.72(b). To match the measured η_d, γ is 0.813 at 2.5 V. The modeled and the observed spectra are shown in Fig. 5.79(a). The lineshapes differ because of the material Stark shift. For Det. B, which has a higher N_D, it is measured to have a smaller CE of 15.0% and a QE of

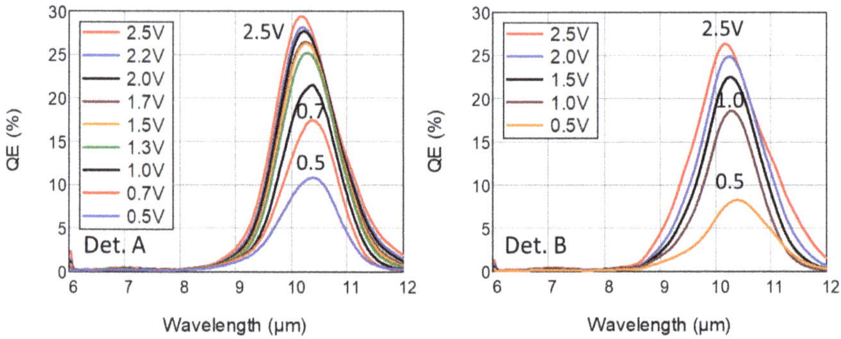

(a) (b)

Figure 5.78. The measured η_d of (a) Det. A and (b) Det. B at different voltages.

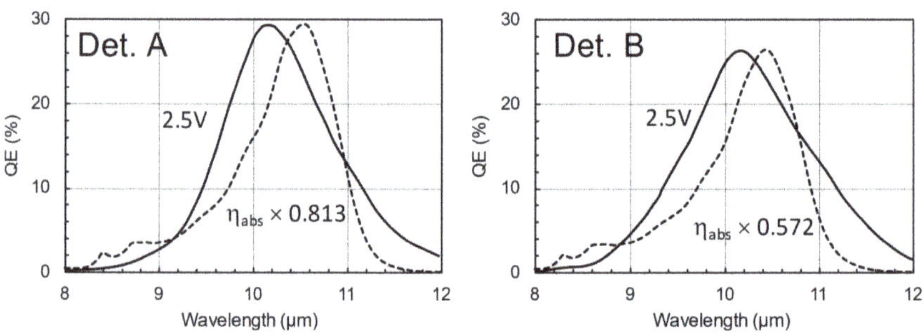

(a) (b)

Figure 5.79. The fitted (dashed curve) and the measured (solid curve) η_d of (a) Det. A and (b) Det. B using a fitting parameter γ.

26.3%. With the larger modeled η_{abs} of 46.2%, the corresponding γ is 0.572.

The current-voltage characteristics of the detectors are shown in Fig. 5.80. The dark current is measured at different T while the photocurrent is measured at $T = 10\,\text{K}$ under F/2 optics and 300 K background. The curves show that the detectors are background limited, defined as $I_p = I_d$, around 65 K. In addition to the directly measured I_p shown as dashed curves, their expected values can also be obtained by numerically integrating the spectral CE with the 300 K blackbody photon flux density, and the results are shown as

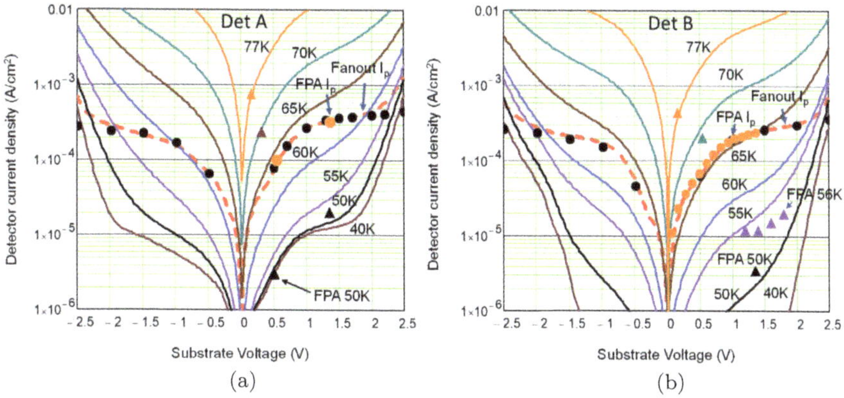

Figure 5.80. The I-V characteristics of Det. A and Det. B measured from fanout circuits and from FPAs.

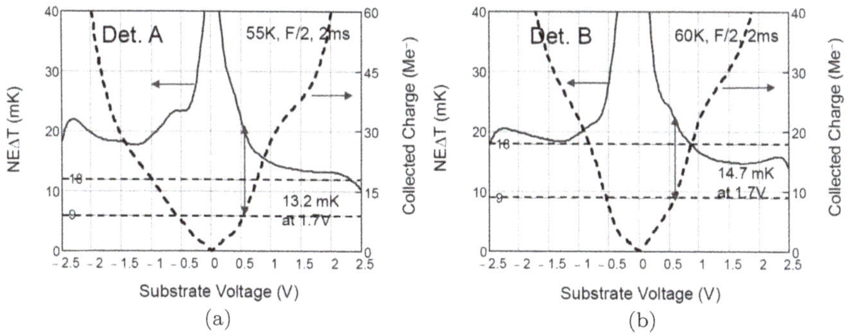

Figure 5.81. The calculated NETD and N of FPAs made from Det. A and Det. B. They are based on the measured I_d/I_p ratio at a fixed τ_{int} of 2 ms under F/2 optics.

black circles in Fig. 5.80. The agreement between the calculated and measured J_p supports the CE measurements. In addition to the fanout detectors, detector pixels from fabricated FPAs are also characterized. The triangles in Fig. 5.60 are I_d and the yellow cycles are I_p. The FPA results are in good agreement with those measured in fanout circuits.

The NETD of the FPAs can be estimated from Eq. (5.12) using the directly measured I_p and I_d. Its value at $\tau_{\text{int}} = 2\,\text{ms}$ and F/2 optics is shown in Fig. 5.81(a) for Det. A at $T = 55\,\text{K}$ and in

Fig. 5.81(b) for Det. B at $T = 60\,\text{K}$. Setting N at 9 Me$^-$, both FPAs can be operated at a small bias of 0.5 V for this τ_{int}, at which g is 0.31 for Det. A and 0.35 for Det. B. Having these smaller g's, both FPAs have smaller NETD of 20 mK than the previous estimated NETD$_0$ of 24.2 mK for $g = 0.6$. Shorter τ_{int} is possible at higher bias, albeit with a higher NETD because of a larger g. Therefore, the present low N_D materials are the proper choices under this operating condition. If the ROIC's N_c is larger, the FPAs will be able to operate at a higher bias such as 1.7 V, at which NETD is about 14 mK for the same $\tau_{\text{int}} = 2\,\text{ms}$. The collected N is about 40 Me$^-$.

5.4.5 *R-QWIPs for Broadband Detection*

To broaden the detection spectrum, the double-well structure described in Section 2.11 can be used. The QWIPs considered in this case have cutoff wavelength at 10.5 μm. The detectors contain either 19 QWs (DW1) or 8 QWs (DW2). Each QW period is made of 50 Å GaAs/60 Å $Al_{0.22}Ga_{0.78}As$/50 Å GaAs/500 Å $Al_{0.22}Ga_{0.78}As$. The total MQW thicknesses are 1.3 μm for DW1 and 0.6 μm for DW2. To yield the same detector geometry, DW2 contains a 0.7 μm GaAs layer underneath the MQW layer. The doping density is 0.5×10^{18} cm^{-3} for both detectors. The pixel pitch D is 25 μm. The band structure of the detector material as well as the R-QWIP geometry design are shown in Fig. 5.82.

The material absorption lineshapes are obtained from edge-coupled detectors, and they are shown in Fig. 5.83. The measured lineshapes are consistent with the material design. The maximum α_{peak} is 0.0717 μm^{-1} at 9.6 μm. Assuming a constant $\alpha = \alpha_{\text{peak}}$, η_{abs} is modeled with different R-QWIP geometries until a more uniform QE spectrum is obtained between 7.5 and 10 μm. Scattering boundary condition is adopted at the pixel unit cell boundaries. It is found that the SBC solution is very close to the PBC solution in this case when the pixel size (20 μm) is substantially smaller than the pixel pitch (25 μm). The optimized detector parameters are shown in Fig. 5.82, and the corresponding QE spectra are shown in Fig. 5.84(a). The detectors exhibit three resonant QE peaks with approximately equal magnitudes within the targeted band. For DW1, whose MQW layer is 1.3 μm thick, the peak QE is about 40%.

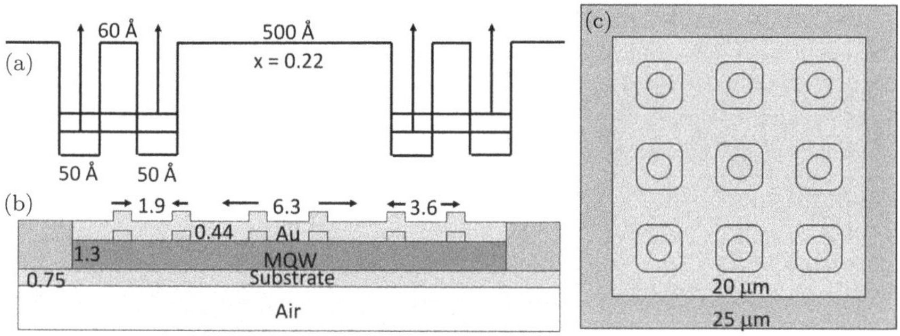

Figure 5.82. (a) The band structure of the MQW having two populated lower energy states employed in the broadband R-QWIP. (b) The center cross-section of the designed R-QWIP geometry. The numbers are in microns. (c) The top view of the detector pixel.

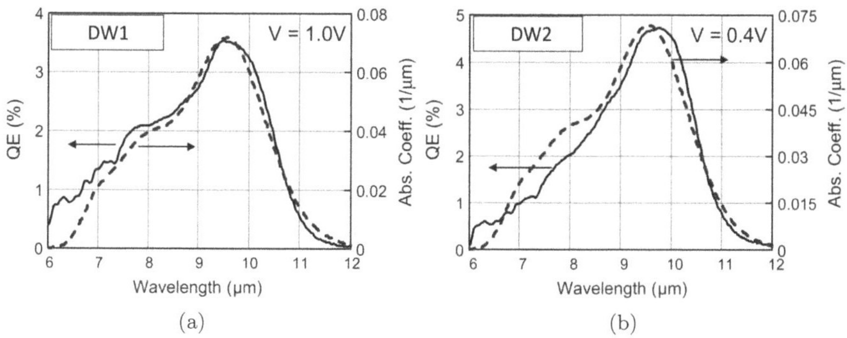

Figure 5.83. The measured η_d (solid curves) and calculated α (dashed curves) of edge-coupled detectors based on double-well materials (a) DW1 with 19 QWs and (b) DW2 with 8 QWs. Adapted with permission from *Appl. Opt.*, 56, pp. B26–B36 (2017).

For DW2 with a 0.6 μm thick MQW layer, QE is about 30%. This coupling bandwidth covers the material absorption spectrum.

Modeling η_{abs} again with the material $\alpha(\lambda)$ yields the predicted QE spectra shown in Fig. 5.84(b), whose peak values are 43.9% for DW1 and 36.5% for DW2. Small format 40×40 pixel arrays are fabricated on fanout circuits for detector characterization. The fabricated R-QWIP geometry is shown in the inset of Fig. 5.85(a). The fabricated pixel size is 21.5 μm with chamfered corners. The ring array is also slightly off-centered. From the measured CE and

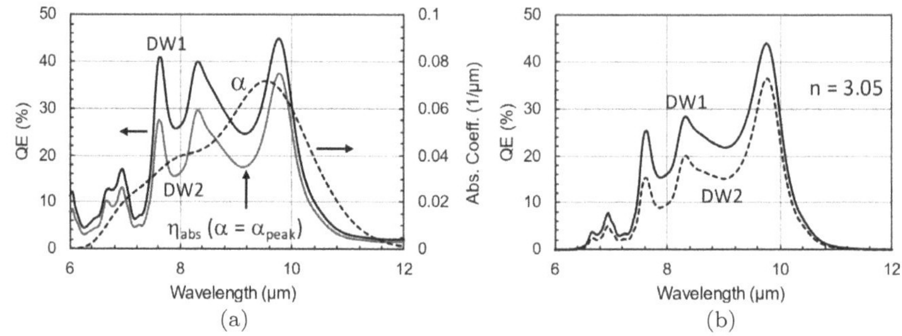

Figure 5.84. (a) The calculated QE of DW1 and DW2 R-QWIPs assuming a constant $\alpha = \alpha_{\text{peak}} = 0.0717\,\mu\text{m}^{-1}$ of the material. (b) The calculated unpolarized external QE assuming the material α spectrum. Adapted with permission from *Appl. Opt.*, 56, pp. B26–B36 (2017).

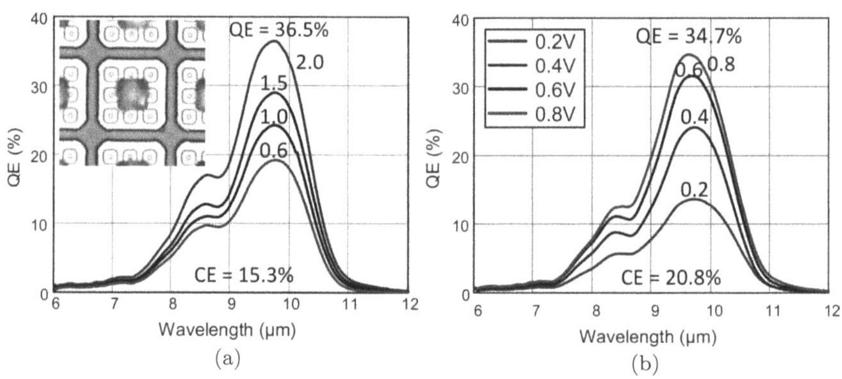

Figure 5.85. The measured η_d of (a) DW1 R-QWIP and (b) DW2 R-QWIP at different bias. Adapted with permission from *Appl. Opt.*, 56, pp. B26–B36 (2017).

gain, each detector QE is deduced and is displayed in Fig. 5.85. For DW1, the peak QE is 36.5% and CE is 15.3% at 2 V. The γ parameter in the $\eta_d = \eta_{\text{abs}}\gamma$ expression is therefore 0.84 at this bias. With this value of γ, the predicted and the measured QE spectra are shown in Fig. 5.86(a). The experimental peaks are broader and less pronounced, which can be attributed to processing nonuniformity and differences between the designed and realized detector geometry.

For DW2, which has fewer QWs, the measured QE is 34.7% at 0.8 V, which is only slightly less than the 36.5% QE for DW1, showing

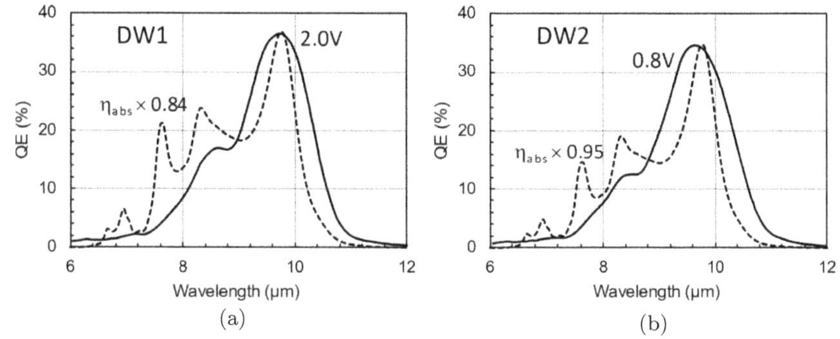

Figure 5.86. The measured η_d (solid curves) of (a) DW1 R-QWIP at 2 V and (b) DW2 R-QWIP at 0.8 V. The modeled $\eta_d = \eta_{\text{abs}}\gamma$ are shown as dashed curves.

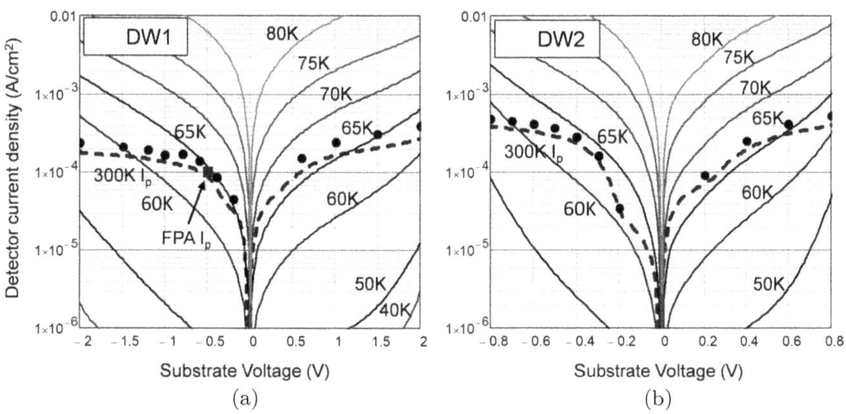

Figure 5.87. The I-V characteristics of (a) DW1 and (b) DW2 in fanout. The solid curves are I_d and the dashed curves are I_p. The circles are calculated from CE measurements. The square is the measured I_p from FPA. Adapted with permission from *Appl. Opt.*, 56, pp. B26–B36 (2017).

the effects of resonance. With a thinner active layer, the gain is larger, which results in a larger CE of 20.8% than that of DW1. With the measured QE, γ is 0.95 at 0.8 V in Fig. 5.86(b).

The current-voltage characteristics of the detectors are shown in Fig. 5.87. Both detectors are BLIP around 66 K at positive bias. The directly measured J_p are slightly lower than those calculated from the spectral CE measurements due to the off-center placement of the fanout detectors relative to the optical window. The field of view was reduced. In addition to the fanout detectors, FPAs with 1024 × 1024 format are fabricated from DW1 wafer material. The ROICs are operated on negative bias of −0.5 V. The photocurrent measured from the FPA agrees with that from the fanout CE measurement as shown in Fig. 5.87(a).

With the good agreement between the fanout and FPA characteristics, FPA performance can be calculated at different bias under F/2 optics and 60 K. The integration time is set at 2 ms, and n_r is 1500. Figure 5.88 shows the calculated NETD and N. For $N = 8.85$ Me$^-$ (half well-fill), NETD is about 22 mK at positive bias for both detectors. It can be further improved to less than 20 mK at higher bias. For the present DW1 FPA camera equipped with a F/2.5 lens and operated at −0.5 V and 55 K, τ_{int} needs to increase to 4.46 ms to achieve 48.6% well-fill (*i.e.*, 8.60 Me$^-$). The measured NETD is 27.2 mK, which is consistent with the estimated 32 mK at $T = 60$ K in Fig. 5.88(a). The FPA operability is 99.5%. Figures 5.89 and 5.90 show infrared images taken under this operating condition.

From the experimental efforts presented in this section, one can witness that QWIP materials are easy to design and grow. As seen in Figs. 5.76 and 5.83, one can design the absorption spectrum of a QWIP material accurately using the standard transfer matrix method. With the advent of EM modeling, the optical coupling

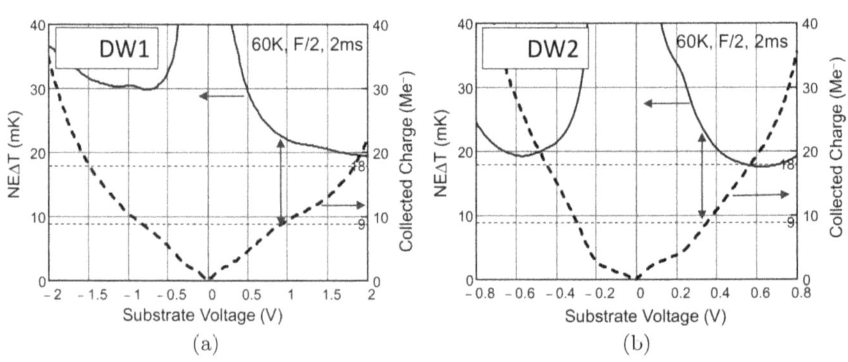

Figure 5.88. The calculated NETD of FPAs and the collected charge of (a) DW1 and (b) DW2 based on the measured I_d/I_p ratio, a fixed τ_{int} of 2 ms, and F/2 optics. Adapted with permission from *Appl. Opt.*, 56, pp. B26–B36 (2017).

Figure 5.89. An infrared image taken by a 1024 × 1024 25 μm pitch DW1 R-QWIP FPA.

Figure 5.90. An infrared image of a scene at 21°C under partly cloudy sky taken by a 1024 × 1024 25 μm pitch DW1 R-QWIP FPA. Adapted with permission from *Appl. Opt.*, 56, pp. B26–B36 (2017).

characteristics of an arbitrary detector geometry can similarly be calculated with high fidelity as seen in Figs. 5.79 and 5.86. Combining these two factors, one can design and engineer the optical properties of QWIP FPAs quantitatively not unlike any other semiconductor technologies. These predictability and manufacturability are great assets of QWIPs in LWIR applications. To obtain a better gauge on the improvement brought by the R-QWIP structures, one can adopt a figure of merit $\chi \equiv \mathrm{CE}/N_D$, where CE is measured at the saturation bias and N_D is in the unit of 10^{18} cm^{-3}. It is a measure of the ability in generating photocurrent for a given electron doping density. Since CE is proportional to I_p and N_D is proportional to I_d, χ will be proportional to the detector I_p/I_d ratio. For detectors A and B in Section 5.4.3 and detector DW1 and DW2 in this section, the experimental χ are 1.01, 0.50, 0.306, and 0.416, respectively. In contrast, the measured χ is about 0.042 ($=0.025/0.6$) in the Thermal Infrared Sensor study using gratings and corrugated-QWIPs [15, 16]. Therefore, the present R-QWIP structures have improved the coupling efficiency by a factor of 10. Note that χ is larger when either N_D or active layer thickness is smaller because these materials promote more photon cycling in the resonator volume.

5.5 R-QWIP for Two-Color Detection

In Section 4.12.2, a voltage-tunable (VT) C-QWIP made of double-superlattice (DSL) unit cells was described. In order to yield a sizable thickness for the C-QWIP geometry, the VT material structure contains 7 DSL unit cells of which the total thickness is 3.2 μm. To avoid spectral crosstalk, highly doped layers are used to connect the DSLs so that only one color can produce photocurrent at each bias polarity. The doped layers, however, reduce photoconductive gain and reduce the spatially averaged absorption coefficient of the material. Multiple DSL units also need a large bias to operate. Since the R-QWIP coupling can reduce the detector thickness substantially, a VT R-QWIP may offer larger QE and CE and a lower operating voltage. To demonstrate a VT-RQWIP, the material structure shown in Fig. 5.91 is adopted.

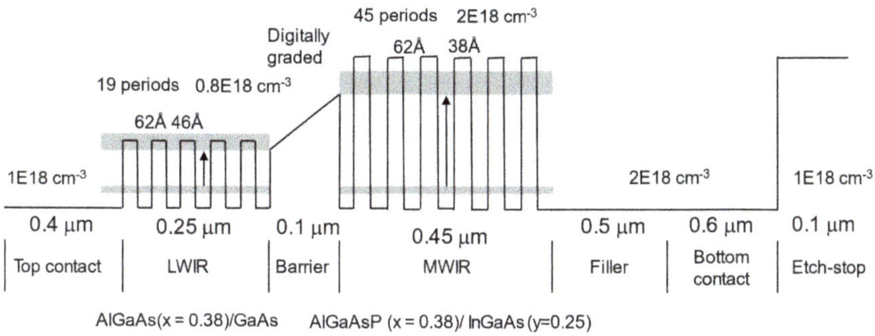

Figure 5.91. The double-superlattice material structure used to produce the voltage-tunable MW/LW 2-color R-QWIPs.

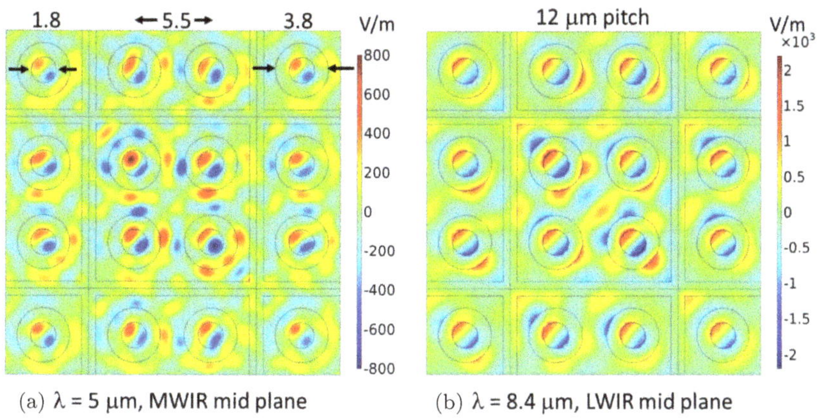

Figure 5.92. The R-QWIP detector structure and the field profiles under 135° polarization angle incidence at (a) the mid-plane of the MWIR layer at $\lambda = 5\,\mu$m and (b) the mid-plane of the LWIR layer at $8.4\,\mu$m.

The detector contains only one DSL unit. Each SL has more QW periods than the previous VT material. The LW SL contains 19 QWs and the MW SL contains 45 periods. The corresponding thickness is $0.25\,\mu$m for the LW and $0.45\,\mu$m for the MW. The material is fabricated into $12\,\mu$m pitch, $11\,\mu$m pixel R-QWIPs. The ring structure and its dimensions are shown in Fig. 5.92. Figure 5.93 shows the y-z cross-section through the center of the rings. The field patterns at the mid-planes of the two layers are also shown at the

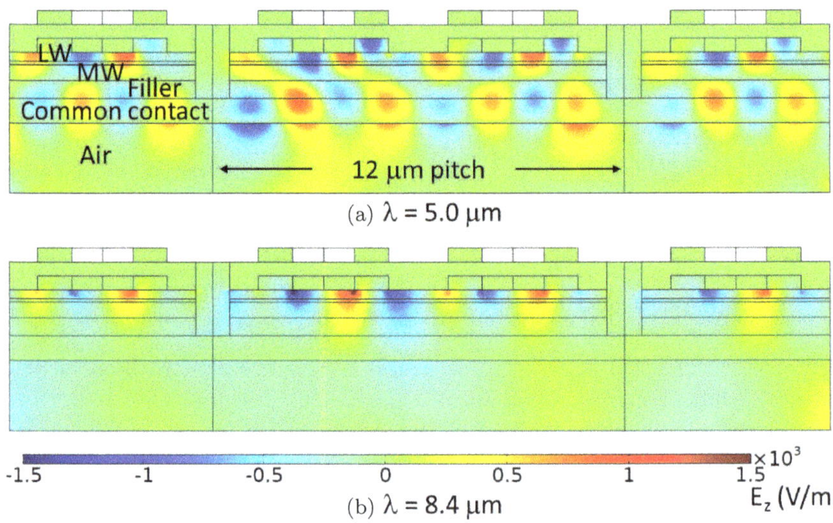

Figure 5.93. The R-QWIP detector material layer and the field profiles at the vertical cross-section across the rings at (a) $\lambda = 5\,\mu$m and (b) $\lambda = 8.4\,\mu$m.

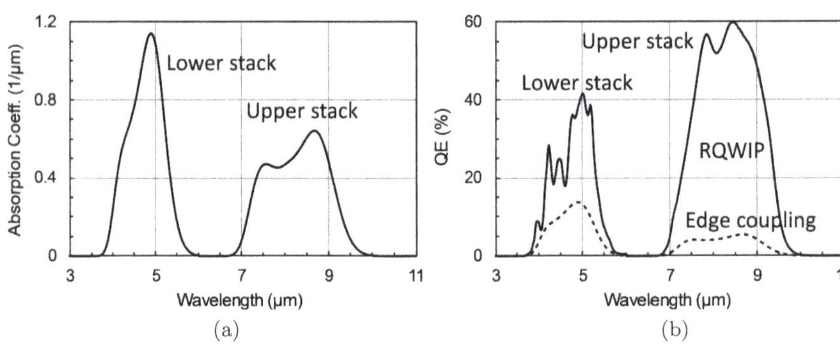

Figure 5.94. (a) The calculated α spectrum according to the superlattice layer structures. (b) The modeled η_{abs} of the R-QWIP assuming $n = 3.05$ and the 45° edge coupling QWIP.

peak absorption wavelength of the respective layers in Fig. 5.93. The incident E field is 533.2 V/m polarized along the diagonal of the pixels. These field patterns show that the rings induce larger E in the LW layer than in the MW layer at the respective peak wavelengths.

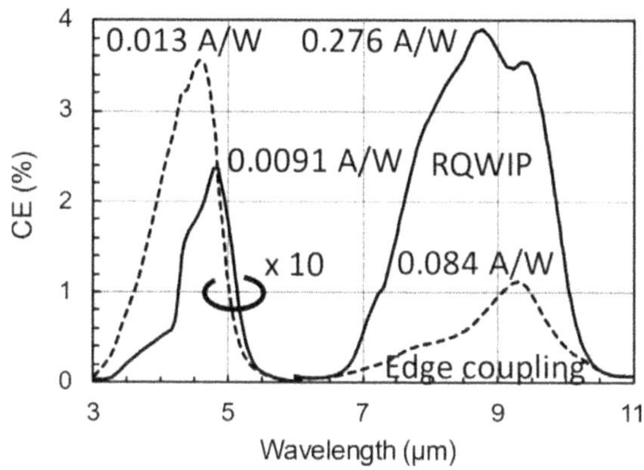

Figure 5.95. The measured CE spectra of the R-QWIP (solid curves) and the edge-coupled detector (dashed curves). The MW CEs are scaled up by a factor of 10 for clarity.

The calculated α spectra of the two layers are shown in Fig. 5.94(a). The peak values are several times larger than the previous values shown in Fig. 4.91(b) because the present α does not take the spatial average. Therefore, the calculated absorption is finite only at the respective active layer locations. Because of the strong E fields produced by the rings, η_{abs} are about 40% for the MW and 60% for the LW despite the very thin layers are being used. These values are several times larger than that under 45° edge coupling, where η_{45} is estimated from the classical expression of Eq. (3.69).

The measured $CE = \eta_{\text{abs}}\gamma g$ of the R-QWIP and the edge-coupled detector are shown in Fig. 5.95. The applied V for the MW is -0.9 V and that for the LW is 0.6 V, whose voltages are both 6.7 times lower than the -6 V and 4 V of the previous C-QWIP detector in Fig. 4.89, respectively. The lower V is consistent with the use of 1 DSL period instead of 7 periods. The corresponding LW R is 0.276 A/W, which is nearly 10 times larger than the C-QWIP LW R of 0.03 A/W. However, the R-QWIP MW R is very small at 0.013 A/W compared to 0.114 A/W of the C-QWIP.

The background photocurrent of the R-QWIP is shown in Fig. 5.96. Consistent with the responsivity measurement, the LW background photocurrent of the R-QWIP is much higher than that of the C-QWIP in Fig. 4.90 while the MW photocurrent is lower.

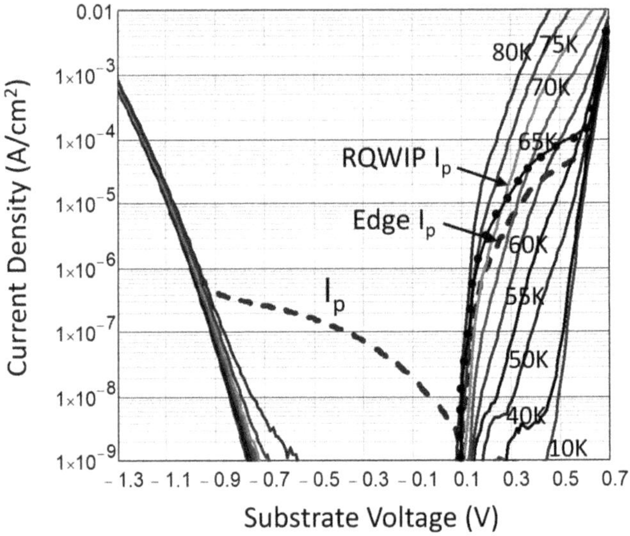

Figure 5.96. The dark current (solid curves) and the 300 K background generated photocurrents (dashed curve and curve with circles) of the detectors. In MW, I_p of the R-QWIP is indistinguishable with the EC I_p in this plot.

The main cause for the small MW response is the many (=45) QW periods in the MW SL compared to the 5 periods in the C-QWIP SL. Most of the generated photoelectrons recombine in the SL before they can reach the thick barrier and be collected in the external circuit. CE is thus low. Another factor is the bias polarity for the MW detection being unfavorable for R-QWIPs. When the substrate is under negative substrate bias in this case, the large potential drop in the MW SL is near the SL/filler layer interface where the photocurrent should be generated. However, the optical E_z field generated by the R-QWIP diminishes rapidly towards the filler layer as seen in Fig. 5.93(a), making the absorption small in that region. The MW response is thus further reduced. Therefore, the MW layer needs to be redesigned.

For example, judging from E_z field distribution in Fig. 5.93(a), high field exists near the filler layer/substrate interface. Therefore, the MW SL should be placed in this location to yield a strong absorption. At the same time, the number of QW periods in the MW SL should be reduced to match the E_z field spatial distribution to improve photoelectron collection. The barrier width between the two SL is adjusted to fulfil the total resonant thickness requirement. With these changes, a sensitive two-color R-QWIP should be achievable.

5.6 R-QWIP for Polarization-Sensitive Detection

In Section 4.15, we discussed using C-QWIPs for polarization-sensitive detection. This detection technique can also be performed by R-QWIPs. To obtain polarization-dependent sensitivity, linear diffraction elements need to be used. Therefore, a GR-QWIP resembles that of a regular grating-coupled QWIP. Usually, a GR-QWIP contains neither a waveguiding layer nor an ARC. Besides using different grating parameters, the pixel volume of the GR-QWIP is also adjusted judiciously so that the extended diffracted waves excite one or more cavity modes precisely and optimally to yield a large QE. The pixel 3D geometry critically determines the achievable QE and bandwidth. In this section, a GR-QWIP is designed to yield polarization-sensitive detection.

Polarization-sensitive detection and imaging are usually used to supplement the usual intensity detection when the thermal contrast is low. In this case, the polarization state of light from different objects may provide a higher contrast for detection. Since smooth man-made objects tend to have a larger degree of polarization, these objects are easier to be identified against a natural background. By using pairs of linear gratings, the degree of linear polarization (DoLP) = $(R_\perp - R_\|)/(R_\perp + R_\|)$ and angle of polarization (AoP) = $(R_\perp/R_\|)$, where R is the pixel response, can be extracted and displayed along the intensity image $(R_\perp + R_\|)$ as demonstrated by Ivanov *et al.* [25].

To fully characterize the polarization state of light, however, the Stokes parameters, S_0, S_1, S_2, and S_4, should be used, where

$$\begin{aligned}
S_0 &= \langle |E_x|^2\rangle + \langle |E_x|^2\rangle = I(0^\circ) + I(90^\circ) = I_0,\\
S_1 &= \langle |E_x|^2\rangle - \langle |E_x|^2\rangle = I(0^\circ) - I(90^\circ) = 2I_1 - I_0,\\
S_3 &= \mathrm{Re}\langle E_x E_y\rangle = I(45^\circ) - I(135^\circ) = 2I_2 - I_0,\\
S_4 &= \mathrm{Im}\langle E_x E_y\rangle = I_{\mathrm{RHC}} - I_{\mathrm{LHC}} = 2I_3 - I_0.
\end{aligned} \tag{5.14}$$

In Eq. (5.14), the brackets represent time-averaged quantities, and S_0 can also be expressed as $S_0 = I(45^\circ) + I(135^\circ) = I_{\mathrm{RHC}} + I_{\mathrm{LHC}}$, where

RHC stands for right-hand circularly polarized light. $I(0°)$ is the horizontally polarized intensity. I_0 is the light intensity measured by a polarization-insensitive detector. I_1 is the light intensity measured by a detector only sensitive to $I(0°)$. Likewise, I_2 is measured by a detector only sensitive to $I(45°)$, and I_3 is measured by a detector sensitive only to RHC. Therefore, if an array contains four subpixels in a sensing unit that measures I_0, I_1, I_2, and I_3 separately, one can extract the Stokes parameters through Eq. (5.14).

Through FEM, four such detectors can be designed and optimized. The material composition and layer thicknesses are listed in Fig. 5.97(a), with which the calculated absorption coefficient is displayed in Fig. 5.97(b). It has a peak wavelength at 9.2 μm. Based on the material thicknesses, the optimized GR-QWIP pixel geometries for 25 μm pixel pitch are shown in Fig. 5.98. The 2D, 90°, 45°, and z gratings are used to measure I_0, I_2, I_3, and I_4, respectively. According to these detector designs, the modeled η_{abs} is shown in Fig. 5.99. For the 2D grating, if PBC is imposed on the 25 μm PUC, the modeled external unpolarized QE can be large at 68.7%. However, this boundary condition is inappropriate for polarization-sensitive arrays since each pixel is surrounded by pixels with different geometries.

	Material	Doping (E18 cm^{-3})
Top contact	0.79 μm GaAs	1.2
Etch-stop	15 Å $In_{0.2}Ga_{0.8}As$	1.2
MQW (27 periods) (1.44 μm)	Well: 48Å GaAs Barrier: 500Å $Al_{0.22}Ga_{0.78}As$	1.0
Etch-stop	100 Å $In_{0.2}Ga_{0.8}As$	1.2
Bot. contact	0.6 μm GaAs	1.2
Etch-stop	0.1 μm $Al_{0.4}Ga_{0.6}As$	1.2
Substrate	625 μm GaAs	undoped

(a)

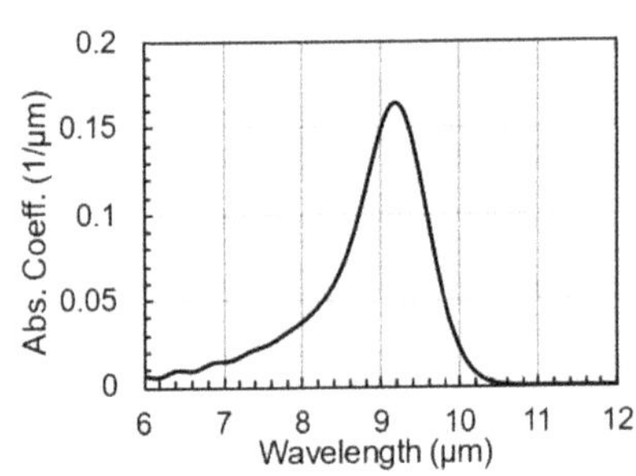

(b)

Figure 5.97. (a) The material layer structure and (b) absorption coefficient of a polarization-sensitive FPA.

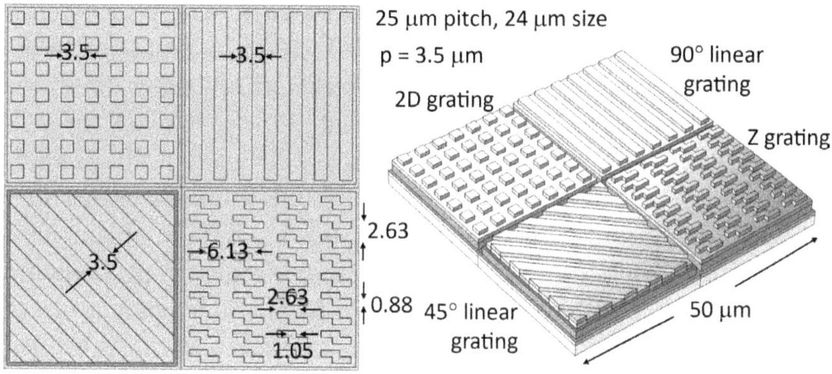

Figure 5.98. A 50 μm pitch polarization-sensitive sensing unit consisting of 2×2 subpixels of 25 μm pitch.

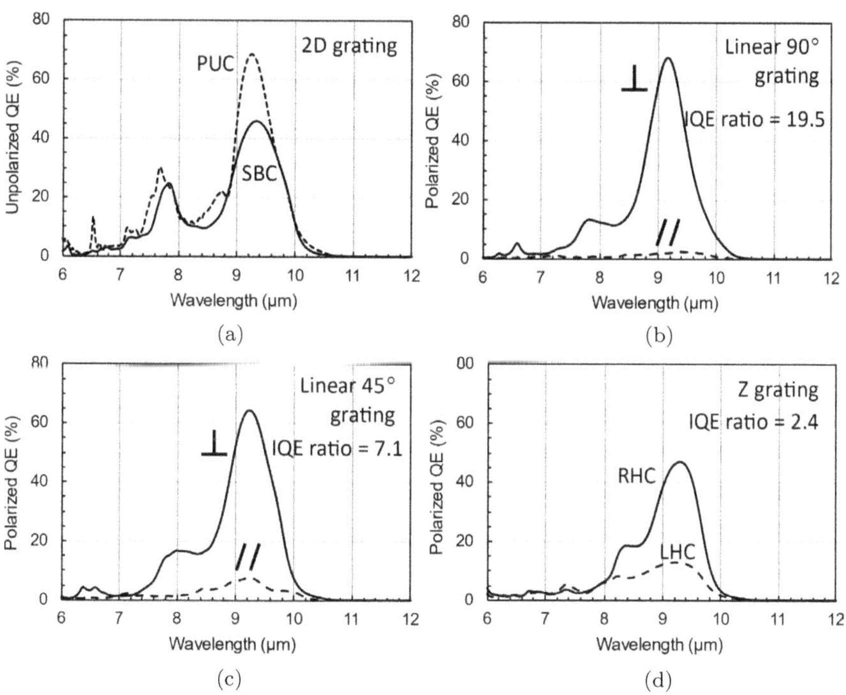

Figure 5.99. The modeled unpolarized and polarized external QE of four different grating structures.

If we assume pixels of different geometries do not cause substantial interference effects, the second-order SBC will be more appropriate, under which the light hitting the four pixel unit cell sidewalls is free to leave and no external light is incident into each pixel through the sidewalls. Under this SBC, the modeled 2D grating QE is lower at 45.7%. This result shows that the PUC QE contains substantial coherent OCT as discussed in Section 5.2.3. OCT is an intrinsic property of an optical geometry that creates sideway propagating waves. Without OCT, the present 2D grating is independent of the incident polarization, and I_0 can be measured precisely. In principle, R-QWIP with a ring pattern can be more effective in measuring I_0. However, it needs different material thicknesses to achieve resonance at λ_p. Therefore, it cannot be incorporated with the linear gratings in the same array.

For the linear 90° grating, the SBC QE under $I(0°) = \langle|E_x|^2\rangle$ is high at 68.3% and low under $I(90°) = \langle|E_y|^2\rangle$ at 2.2%. The IQE (IQE) ratio for the two polarizations is 19.5. Therefore, I_1 can be measured by the 90° linear grating fairly well, with which S_1 can be deduced from Eq. (5.14) quite accurately. The accuracy for S_2 and S_3 decreases as the IQE ratio is 7.1 for the 45° linear grating and 2.4 for the z grating. In modeling the z grating under RHC light, the bottom air boundary is set to second-order SBC with incident $\boldsymbol{H} = (1, +i, 0)$. Since H_y is leading H_x by $\pi/2$ in phase, the light is right-hand polarized. Similarly, the LHC light is represented by $\boldsymbol{H} = (1, -i, 0)$. With the finite $\eta_{\|}$ contribution, the Stokes parameters measured from the present detector design will not be very accurate. Nevertheless, for the purpose of imaging, the acquired information may already be adequate in improving the imaging sensitivity by displaying different combinations of the measured Stokes parameters. This structural design still awaits experimental demonstration.

5.7 R-QWIP for Ultra-Broadband Detection

In Section 5.4.5, a double quantum well structure is adopted, which yields adequate bandwidth for 8–10.5 μm LWIR imaging. In certain spectroscopic applications, it is beneficial to have a wider spectral coverage. For these applications, a triply coupled QW structure can

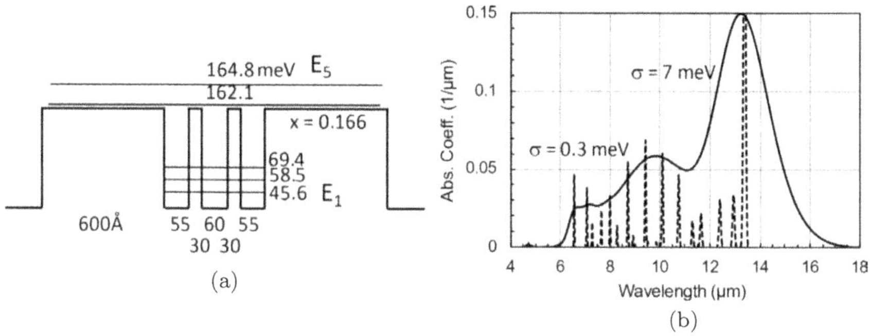

Figure 5.100. (a) The triple QW design for 6–16 μm ultra-broadband detection. (b) The calculated α with two line-broadening parameters.

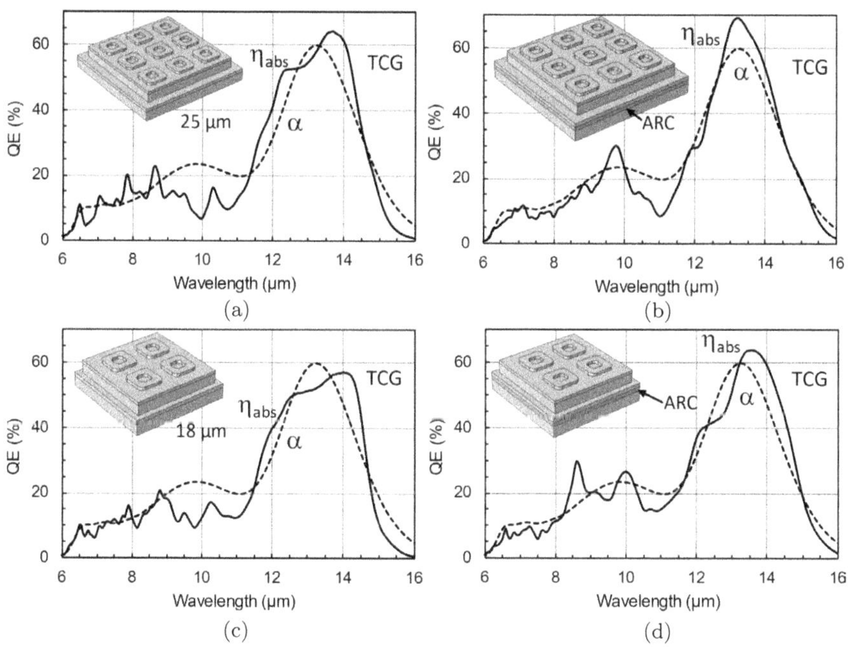

Figure 5.101. The modeled η_{abs} using the TCG for the R-QWIPs (a) with 25 μm pixel pitch, (b) with 25 μm pitch and AR coating, (c) with 18 μm pitch, and (d) with 18 μm pitch and AR coating. The lineshape of α in arbitrary units is also shown.

be utilized. Its structure is shown in Fig. 5.100(a). The calculated α is shown in Fig. 5.100(b) for $N_d = 1 \times 10^{18}$ cm^{-3}. A band offset parameter β_{AlGaAs} of 0.79 in Eq. (2.22) is assumed.

Figure 5.101 shows the external unpolarized QE spectrum of the R-QWIPs with 25 μm and 18 μm pitches. The respective pixel sizes are 21.5 μm and 16 μm. Their structural parameters are: inner ring dimension = 2.64 μm, outer ring dimension = 5.64 μm, ring spacing = 8.4 μm, ring height = 0.66 μm, absorber thickness = 2.16 μm, common contact thickness = 0.84 μm. The n value for QWIP is 3.00.

The detectors can be AR coated to improve QE around 10 μm. For this purpose, the ARC thickness is taken to be 1.44 μm and its n is 1.732. From the similarity between the material α and the detector QE spectra, the R-QWIP structure is shown to provide relatively constant coupling efficiency in a wide range of wavelengths through diffraction and plasmonic interactions with the QW material. An ultra-broadband R-QWIP can therefore be produced for spectroscopic applications. The similar QE spectrum between the two pixel sizes also shows the insignificance of the size effects in this coupling approach.

5.8 Optical Antenna-Coupled QWIPs

Besides gratings and plasmonic resonators described in the previous sections, there are many other optical coupling designs for QWIPs in literature. New concepts are constantly proposed and demonstrated to date. Many of them are worth consideration and show their respective advantages in certain applications. These new designs not only advance the QWIP technology but also enrich the physics of near-field electromagnetics. Using QWIP material as a platform, many novel EM phenomena can be engineered, studied, and demonstrated in the detector spectral response. These engineered optical structures include optical antennas, photonic crystals, and metamaterials. The physical insights gained from each phenomenon can be used to design more capable detectors. Since each approach may involve sophisticated understanding and meticulous structural design, a detailed description of these many approaches will be beyond the scope of this text.

In the following, we mention a set of similar and related structures that can be collectively referred to as optical antennas. In each of these structures, an array of fine structures is etched into the active material and down to the bottom contact layer so that the physical area of the remaining material is smaller than the optical receiving area. If the light impinging on the receiving area establishes resonant antinodes in the remaining material whose amplitude is larger than the incident amplitude, the detector can absorb optical power falling outside its physical area, similar to that of a radio antenna. This optical antenna can then be used to increase the photocurrent to dark current ratio. In addition to a smaller area, some designs can adopt thin absorbing layers without sacrificing their QE. The larger photoconductive gain further raises the photocurrent and the operating temperature. Therefore, the optical antenna is a capable coupling approach.

5.8.1 *Enhanced-QWIPs*

The first example of optical antennas is designed and demonstrated by Schimert and Mitra *et al.* [27–29], and it is called the enhanced-QWIP or E-QWIP. The schematic structure of a 60 μm pixel unit cell is shown in Fig. 5.102. To fabricate an E-QWIP FPA, individual detectors are first isolated by etching trenches around the pixel down to the bottom common contact layer. Planar metal contacts

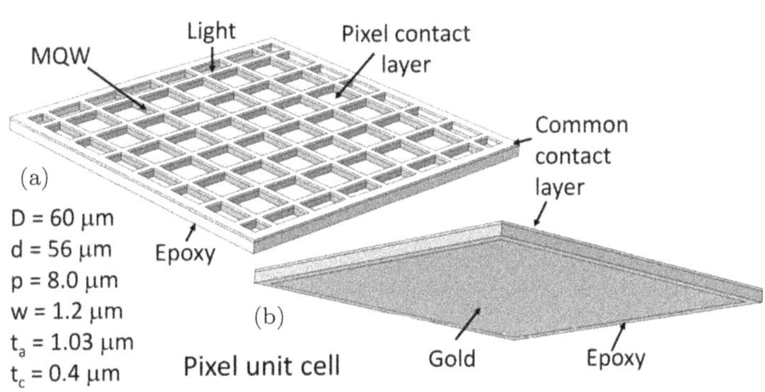

Figure 5.102. The (a) top view and (b) back view of a 60 μm E-QWIP pixel unit cell.

and indium bumps are deposited on the top contact layer. After bump bonding to an individual ROIC and backfilling with epoxy for structural support, the GaAs substrate is removed by selective etching to expose the common contact layer. A 2D cross grid is then etched through the common contact layer into MQW material and down to the pixel top contact layer to form a 3D diffractive resonant optical cavity (DROC).

The DROC is defined by the strip width w, the strip period p, and the cavity height $h = t_c + t_a$, where t_c is the common contact thickness, and t_a is the MQW thickness. Since the DROC is directly facing the light source, there is no substrate reflection loss, and the external and internal QEs are the same. According to Refs. 27–29, there are two resonant modes in the DROC. One is the regular grating diffraction mode, and it is guided by the planar gold surface and travels parallel to the grating. Since this mode travels mostly in air, the period p is close to the free-space wavelength. For $\sim 9\,\mu$m detection, the optimum p is $8\,\mu$m. The second mode is the image mode, which is guided inside the material strips. To know the coupling efficiency of this structure, FEM modeling is performed using constant $\alpha = 0.2\,\mu\text{m}^{-1}$ and $n = 3.15$. The modeled unpolarized η_{abs} is shown in Fig. 5.103. It has a maximum value of 38% at $\sim 9\,\mu$m. If the same α occurs at other wavelengths, QE will be lower for the same

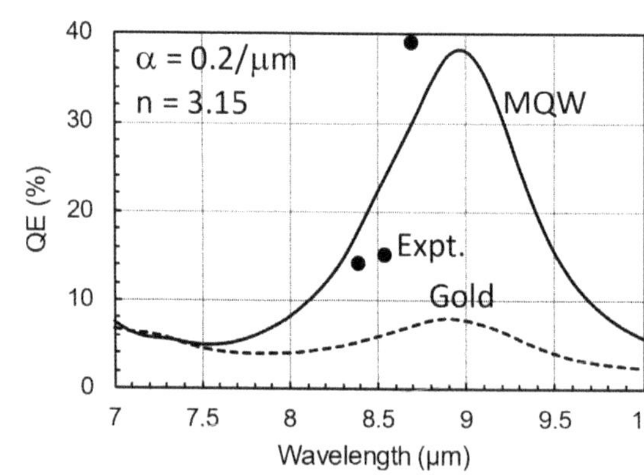

Figure 5.103. The modeled QE of the MQW grid (solid curve) and the metal ground plane (dashed curve) and the measured peak QE values from three different wafer materials.

DROC structure. The maximum QE from three different wafer materials taken from Ref. 27 are plotted in Fig. 5.103. They are consistent with the modeled curve given the unspecified α in the experiment.

Although QE is 38% when it is referenced to the pixel area $A_{\text{pixel}} = (56\,\mu\text{m})^2$, its QE value enlarges by a factor of 3.60 to QE* = 136.8% when it is referenced to the physical area of the grid $A_{\text{grid}} = 8.70 \times 10^{-10}$ cm^2. The larger-than-unity QE* signals the collection of incident photons outside the gird physical area and thus fulfils the signature of an optical antenna. Using QE* instead of QE, one can compare the performance of different antenna designs having the same physical area and thus the same dark current. Therefore, QE* is a useful figure of merit. Note that although a large QE* improves I_p/I_d, the absolute I_p and NETD are nevertheless still governed by QE at a given integration time as analyzed in Section 4.10. Therefore, one should be mindful of the change of QE when QE* is being maximized.

The vertical electric field E_z in the mid-plane of the material strips at $\lambda_p = 8.9\,\mu$m is shown in Fig. 5.104. Infrared absorption

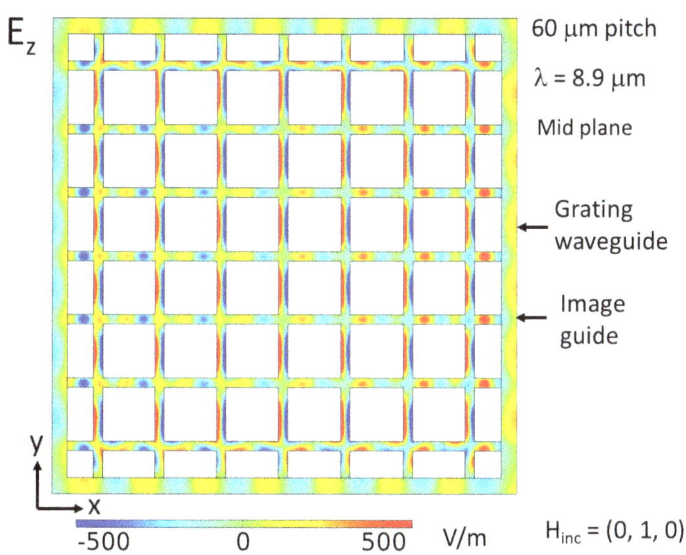

Figure 5.104. The vertical electric field profile under normal incidence under x-polarized light at $8.9\,\mu$m.

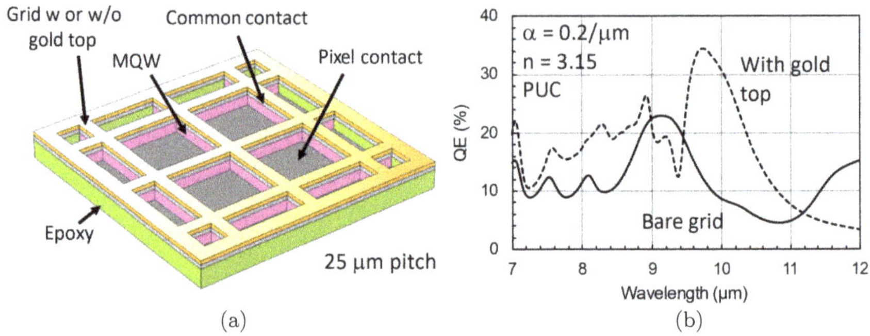

Figure 5.105. The QE of a 25 μm E-QWIP with and without a gold cover on the grid strips.

occurs where E_z is finite in the material strips. From Fig. 5.104, under $\boldsymbol{H} = (0, 1, 0)$ incidence, it is evident that the DROC couples light via two distinct cavity modes. The grating waveguide mode generates absorption in the vertical strips along the y-axis, and the image guide mode generates absorption in the horizontal strips along the x-axis.

Since E-QWIP utilizes extended modes for optical coupling, its QE is affected by the pixel size. In Fig. 5.105, the DROC with 25 μm pitch is modeled and its QE reduces to 22.0% near 9 μm. However, if the grid strips are covered by gold using it as the etching mask, the modeled QE shows a larger value of 35.4% at 9.72 μm. Therefore, E-QWIPs can benefit from having a gold strip cover.

5.8.2 *Quantum Grid Infrared Photodetectors*

A detector structure similar to the E-QWIP is known as the quantum grid infrared photodetector (QGIP) [30,31]. In this detector structure, a laminar grid is etched into the QWIP material using a gold metal strip mask. There is no metal layer underneath the grid. The incoming light is incident from the backside of a thick substrate. The geometry of a detector is shown in Fig. 5.106(a) and Fig. 5.107(a). The top contact thickness is 0.1 μm. The exposed bottom contact thickness is 1.83 μm. The gold cover thickness is 0.2 μm. The MQW thickness is 1.18 μm. The gap between grid lines is 4.65 μm, and strip width varies from 0.97 to 2.77 μm.

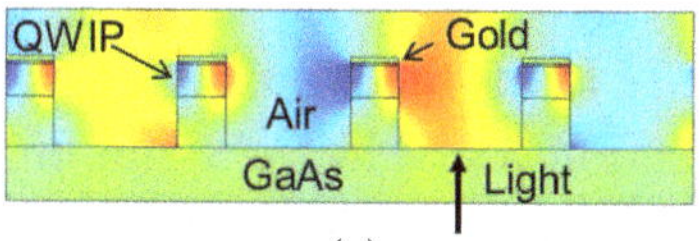

(a)

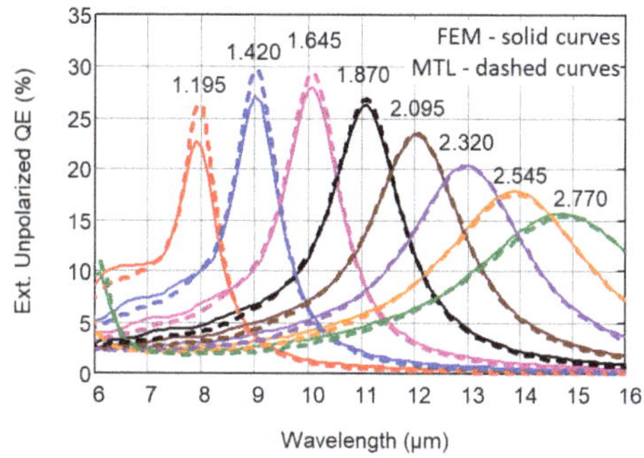

(b)

Figure 5.106. (a) The side view of quantum grid infrared photodetector. (b) The modeled QE at different w in microns.

Instead of using the collective diffraction mechanism, the present light coupling scheme utilizes multipole scattering from the individual metal strips on top of the grid. The maximum coupling efficiency occurs when the width w of the metal strips is

$$w = \frac{N\lambda}{2n}, \tag{5.15}$$

where λ is the incident wavelength, n is the refractive index, and N is an odd integer. $N = 1$ corresponds to the strongest dipole scattering.

The coupling is not strongly sensitive on the spacing between the grid lines. The relationship between w and λ in Eq. (5.15) provides a simple design rule for multicolor detection. Furthermore, the large wavelength separation between different values of N ensures a one-to-one correspondence between w and λ in a wide spectral range. For example, for $\lambda = 15\,\mu$m, $w = 2.40\,\mu$m for $N = 1$, and $n = 3.12$. The first overtone will be at $\lambda = 5\,\mu$m for $N = 3$, and the second overtone will be at $\lambda = 3\,\mu$m. Therefore, with $w = 2.40\,\mu$m, there will be only one detection peak at $\lambda = 15\,\mu$m between 5 and 15 μm. When $w < 2.40\,\mu$m, the overtones will all appear below 5 μm. Similarly, to detect at $\lambda = 5\,\mu$m, w will be 0.80 μm, and its first overtone is at 1.68 μm. Consequently, when w is designed to be between 0.80 and 2.40 μm, the detector detects at a unique wavelength between $\lambda = 5$

and 15 μm. A linear array of these detectors then forms a broadband spectrometer.

With the grid structural parameters listed in Fig. 5.106, the detector η_{abs} is modeled both by the MTL method presented in Section 3.5 and the FEM presented in Section 3.6. The very similar result supports both methods. In this model, a constant relative dielectric constant $\varepsilon_r = 9.722 + i$ is assumed for all wavelengths. At 10 μm, $\text{Im}(\varepsilon_r) = 1$ corresponds to an α of 0.202 μm^{-1}. Using this ε_r, the QGIP shows a relatively constant coupling efficiency across the spectrum. The maximum appears at $w = 1.645\,\mu$m, at which QE = 29% without ARC. When it is referenced to the strip's physical area, QE* is scaled by a factor of (6.285/1.645) to 111%, thus it is sufficiently qualified to be an optical antenna, and QE* will be higher with an ARC. The peak wavelength and w in Fig. 5.106(b) follow a linear relationship with $w = -0.605 + 0.225\lambda$ (in μm), which is slightly different from that given by Eq. (5.15) due to waveguiding effects in the tall grid lines.

In the experiment, a broadband material consists of eight binary superlattices, each separated by a 600 Å thick $Al_{0.19}Ga_{0.81}As$ barrier. Each superlattice consists of eight alternate GaAs QWs of widths 70 and 75 Å, separated by 25 Å $Al_{0.27}Ga_{0.73}As$ barriers. The QWs are doped to 0.4×10^{18} cm^{-3}. Since there is a 0.1 μm doped top contact layer, the active layer reaches down to 1.28 μm below the metal strips, which is approximately equal to the penetration depth of the scattering field as seen in Fig. 5.106(a). To optimize QE, a total grid height of 3.11 μm is needed. Therefore, a thick ~2.5 μm bottom contact layer is grown, and the grid pattern is etched passing the MQW layer to achieve a total grid height requirement. A set of six QGIPs labeled A to F with w = 1.22, 1.43, 1.80, 1.90, 2.06, and 2.50 μm is fabricated, and their dimensions are verified with a scanning electron microscope. Figures 5.107(a) and (b) show the fabricated detector structures and grid lines.

Figure 5.107(c) shows their normalized coupling efficiency defined as $R/R(45°)$ where R is spectral responsivity, with which the effects of the material α at different λ are factored out. To model after the experiment, a finite baseline is added to the theoretical curves

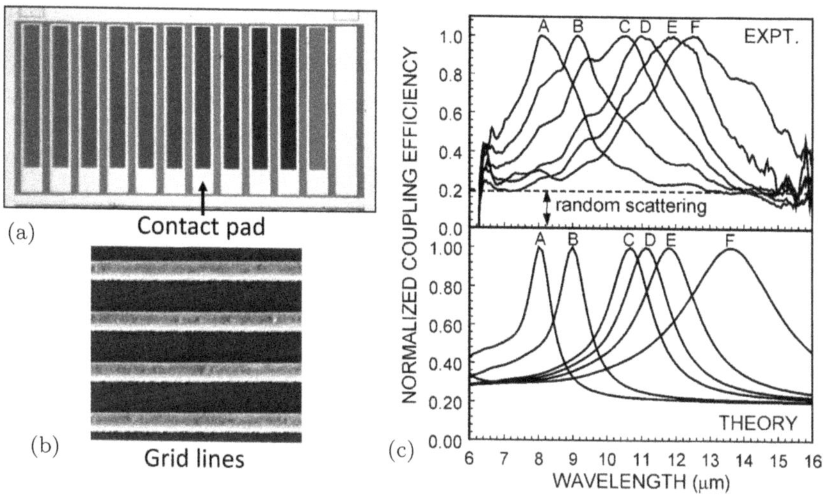

Figure 5.107. (a) The top view of a set of QGIPs with different strip widths. (b) The grid lines under electron microscope. (c) The experimental and modeled spectral lineshapes for different w. Adapted with permission from [31].

to account for stray light scattering in the unthinned substrate. In general, the observed peak wavelength agrees with the prediction, and the device concept is validated. The small wavelength discrepancy of detector F could be due to the relatively poor wavelength selectivity of a wide grid line. The measured spectral widths are wider than the theoretical predictions, which can be attributed to the width fluctuations in the very long (400 μm) strip lines. Better detector design and processing and substrate removal should be able to improve its performance.

5.8.3 *Microstrip Antenna-Coupled QWIPs*

Microstrip antennas (MSA) are widely used in ultra-high radio and microwave frequencies. An MSA usually consists of a square metal patch placed on one side of a dielectric substrate and a continuous metal ground plane placed on the other side. The resonant modes of an MSA are well known [32]. Taking the perfect electric conductor boundary condition along the metal surfaces and the perfect magnetic conductor boundary condition at two sidewalls (magnetic wall approximation) parallel to the incident $\boldsymbol{H}$, the resonant wavelength λ is $\lambda = 2nw$ for the fundamental (1, 0) mode, where n is the refractive index of the dielectric and w is the patch width. To go beyond the magnetic wall approximation, $\lambda = 2n(w + \Delta w)$, where $\Delta w \approx h/2$, and h is the thickness of the dielectric. The longer resonant λ for a given w is due to the presence of fringing fields outside the metal patch. Since E is strictly vertical at the PEC surfaces at both $z = 0$ and $z = h$, the total E field inside a MSA having a thin ($h \ll \lambda/n$) dielectric is primarily pointing at the vertical direction. This E profile is perfectly suitable for QWIP coupling when the thin dielectric is replaced by the MQW material.

The MSA design for QWIPs is first provided by Beck and Mirotznik [33]. In this design, gold square patches of size $w = 1.1\,\mu$m or $1.4\,\mu$m are arranged in square arrays with period p ranging from $2.0\,\mu$m to $6.0\,\mu$m. The MQW thickness t_a is $0.2\,\mu$m and the top and bottom contact thicknesses t_c are $0.1\,\mu$m. The MSA unit cell and a pixel with $25\,\mu$m pitch are shown in Fig. 5.108. In their design, a network of thin transmission lines is used to connect the individual gold patches to a common contact. The optical effects of this network, however, are not modeled. In FPA applications, the planar gold ground plane of the pixel area will be made discrete

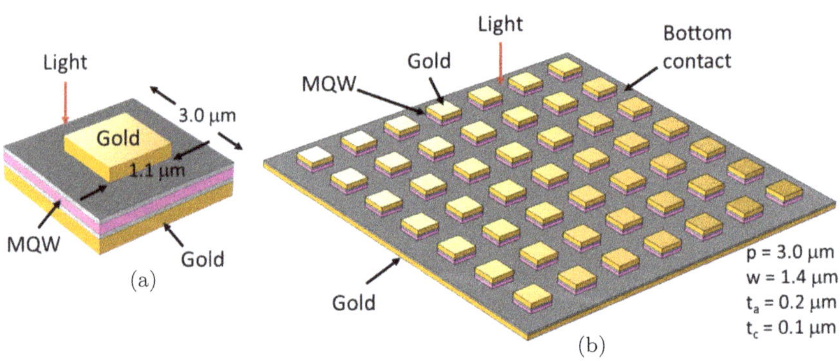

Figure 5.108. (a) The microstrip antenna unit cell with $w = 1.1\,\mu$m and $p = 3.0\,\mu$m. (b) A detector pixel with $25\,\mu$m pitch that contains 7×7 MSA unit cells, which are etched to the bottom contacts. The connection network of the top gold patches is not modeled.

and serves as individual pixel connection to an ROIC, similar to the E-QWIP FPAs. Incident light is falling directly onto the MSA array.

Reference 33 adopts finite difference time domain (FDTD) method to model the QE spectrum, and the peak wavelength is found to be primarily determined by w and varies only in a small extent with p. This result is consistent with the individual antenna coupling, and it does not have substantial collective effects. When t_a increases, the peak is smaller and broader and shifts to longer wavelengths. This shift is explained by the wider spread of the fringing fields at larger t_a that lengthens Δw. For $t_c = 0.1\,\mu$m, a MQW thickness around 0.1 to 0.2 μm is sufficient to yield a wide enough bandwidth to cover the typical QWIP spectral band. The authors also find that there are no appreciable differences in the theoretical spectral QE response between an individual MSA and a collection of MSAs in a 24 μm pitch, 22 μm size pixel. Using a constant optical conductivity σ of 1000 S/m, the maximum QE is modeled to be 30% at 8.9 μm for $w = 1.1\,\mu$m and 34% at 10.9 μm for $w = 1.4\,\mu$m when t_a (= 0.2 μm) and p (= 3 μm) are the same for both detectors.

In this section, MSAs with $w = 1.1$ and 1.4 μm are remodeled with FEM. Given $\alpha = \sigma/(nc\varepsilon_0)$, where σ of 1000 S/m and $n = 3.15$ assumed in Ref. 33, α is $0.12\,\mu\text{m}^{-1}$. For more standardized comparison, α is raised to $\alpha = 0.2\,\mu\text{m}^{-1}$ in FEM modeling. Reference 33 considers only conventional MSA structures, in which the dielectric layer is not etched. Therefore, there is no physical size reduction in these structures. The unetched MSA QE is shown as solid curves in Fig. 5.109(a).

Shown in Fig. 5.109(a), both the $w = 1.1$ and 1.4 μm detectors are peaked at the same wavelengths as those modeled by FDTD. However, the modeled peak values are 40.3% and 46.7%, respectively. They are a factor of 0.83 lower than the FDTD predictions after accounting for the difference in the adopted α. This difference is probably due to the difference in the adopted gold (n, κ). Due to the large induced electrical current in the two gold layers in this detector geometry, the gold properties will affect QE significantly. Based on the gold (n, κ) taken from Ref. 34, the modeled gold absorption is

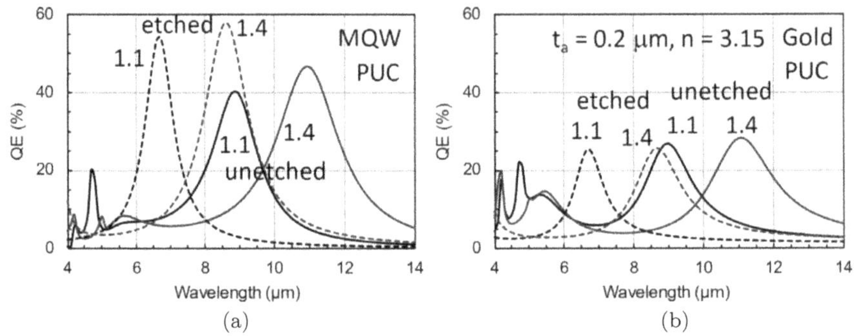

Figure 5.109. The modeled absorption of MSA unit cells in the (a) MQW layer and (b) in the gold layer. The total layer thickness is $(0.1 + 0.2 + 0.1)$ μm. The MQW layer in the MSA can be etched or unetched. The period p is $3\,\mu$m for all detectors. $\alpha = 0.2\,\mu\text{m}^{-1}$.

about 30% at the resonant peaks shown in Fig. 5.109(b), and this presents a limiting factor for the MSA approach.

The modeled peak QE of 40% at $\lambda = 8.9\,\mu$m is quite substantial considering its small absorbing thickness of $0.2\,\mu$m. This large QE is due to magnetic resonance between the two parallel gold layers. When the incident H_y oscillates between these two layers, equal but opposite oscillating currents are induced in the gold layers in the x direction similar to that of a solenoid. To complete the electrical circuit, equal amount of displacement current $\propto \partial E_z/\partial t$ flows along the z-axis at each edge. The generated oscillating E_z then induces infrared absorption in the MQW layer at these locations. Having two coherent currents at both sides of the MQW layer to drive the vertical E field, a larger QE can be obtained in a small volume than most other approaches.

The theoretical designs of Beck and Mirotznik were later experimentally verified by Chen *et al.* [35,36]. In this experiment, the MQW material consists of 8 periods of 65 Å GaAs/200 Å $Al_{0.25}Ga_{0.75}As$ barriers ($t_a = 0.232\,\mu$m). The wells are doped with 2D electron density of 7×10^{11} cm^{-2}, equivalent to 1.1×10^{18} cm^{-3} doping in 65 Å QWs. The top (500 Å) and bottom (1000 Å) contacts are doped at 0.5×10^{18} cm^{-3}. According to this layer structure, the total thickness is $0.382\,\mu$m, which is close to the $0.4\,\mu$m modeled in Fig. 5.109. Large-area detectors ($200\,\mu$m $\times$ $200\,\mu$m) containing many individual MSAs are fabricated with $p = w + 2\,\mu$m and the MSAs are connected by parallel thin ($0.15\,\mu$m) wires. The measured η_d is 33% at $8.9\,\mu$m for $w = 1.1\,\mu$m and $p = 3.1\,\mu$m with a gain of 0.25 at 1 V. The observed peak wavelength is the same as modeled wavelength in Fig. 5.109(a) (where $p = 3.0\,\mu$m). The magnitude is, however, less than the modeled $\eta_{\text{abs}} = 40\%$. The smaller η_d could indicate a photoelectron tunneling probability γ of 0.83 even under a large applied field of 125 mV/period. This less-than-unity γ is consistent with the small gain of 0.25 with 8 periods of QWs and the bound nature of the final state suggested by the authors [35]. The observed 33% η_d for $0.232\,\mu$m thick MQW testifies to the efficient optical coupling provided by this detector geometry.

The unetched MSA probably should not be classified as an optical antenna because its physical active area is the same as the optical receiving area. QE* and QE are the same, and QE* can never be larger than 100%. In Ref. 35, Chen *et al.* takes another important step, in which the MQW active material outside the patch is removed, and it departs from the traditional MSA design. After etching, the group finds that the $w = 1.1\,\mu$m detector is no longer in resonance with the material α. Instead, a larger $w = 1.4\,\mu$m detector ($p = 3.4\,\mu$m) produces a large photocurrent and gives a high T_{BLIP} (at which $I_p = I_d$) than the $w = 1.1\,\mu$m detectors. The T_{BLIP} is 86 K, which is the highest value reported in literature at this peak wavelength.

The λ shift is well predicted by FEM modeling in Fig. 5.109(a). When the external MQW material is removed, the fringing field is reduced. Therefore, the resonant λ_r is shifted to a shorter wavelength. For the $1.1\,\mu$m detector, it has the largest QE when α is peaked at a short λ of $6.7\,\mu$m, which is away from the present α peak. Meanwhile, the $1.4\,\mu$m detector, with its original peak at $10.9\,\mu$m, is blueshifted to $8.6\,\mu$m when the material is etched. If α is peaked at this λ, QE is 57.5%. Even with the present material's α peaked at $8.9\,\mu$m, a large QE of 50% is predicted from the modeled curve. Combined with the smaller dark current by the area ratio of $(3.0/1.4)^2 = 4.6$ and a large gain in the thin material, the detector attains a very large I_p/I_d and

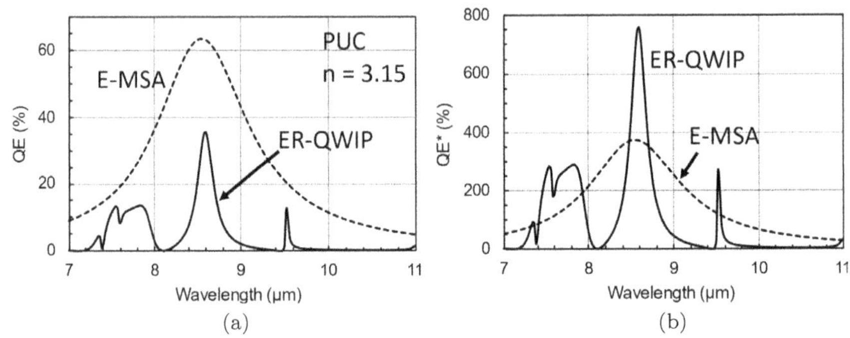

Figure 5.110. The modeled QE of the PUC of E-MSA and ion-implanted ER-QWIP when it is referenced to (a) PUC area and (b) detector-active area.

a very high T_{BLIP}. To know the pixel size effect of the etched MSA structure, the QE of a 25 μm pixel pitch shown in Fig. 5.108(b) is also modeled. The result is nearly the same as that of the single MSA unit cell with a maximum QE of 60.4% at 8.5 μm, consistent with the FDTD result for unetched sample. Therefore, the MSA QE does not depend on the pixel size and it is applicable to small pixels.

The above theoretical analysis is for $p = 3.0\,\mu$m. Although the MSA QE does not depend on p, we nevertheless remodel the experimental $p = 3.4\,\mu$m for quantitative QE. The result is shown in Fig. 5.110(a). It has a slightly larger 63.1% QE at 8.5 μm. At 8.9 μm, QE = 49.3%, nearly the same as the $p = 3.0\,\mu$m detector. With area reduction of $(3.4/1.4)^2 = 5.9$ times, QE* is 291% for the present detector. If α is peaked at 8.5 μm, QE* will be 372% shown in Fig. 5.110(b), indicating a highly efficient optical antenna design. Following the nomenclature of Schimert *et al.* [27], we may call the etched MSA the enhanced microstrip antenna (E-MSA).

5.8.4 *Enhanced Resonator-QWIPs*

The R-QWIP structure can also be made into an optical antenna to improve I_p/I_d and T_{BLIP}. From the small mode volume of the LSP mode at resonance shown in Figs. 5.37 and 5.38, one can reduce the area and the thickness of the MQW layer to achieve smaller dark current and a higher gain without sacrificing QE proportionally.

Since the MQW layer cannot be etched away without modifying the optical resonant characteristics, the active MQW volume can be altered by deactivating the active area by ion implantation. Using this approach, one can choose the size of the active area to achieve the desired balance between QE and I_p/I_d ratio. Ion implantation has been widely used in high-speed III-V industry to create resistive regions for device isolation [37]. Most of these devices and circuits contain active or conducting channels near the surface (<1 μm) and the remainder of the substrate is semi-insulating, where resistivity is near $10^7\,\Omega$cm. For III-V wafer materials, simple masking of ion bombardment with photoresist is sufficient to create excellent device isolation with resistivity higher than the semi-conducting substrate itself. Implant isolation extends into regions under mask edges because of lateral straggle of the implanted ions so that the masked area should be larger than the desired active area. The straggle is typically about half of the projected range. This implant isolation technique has been applied to both II-VI and III-V detector arrays for pixel delineation to avoid surface leakage current at the exposed mesa sidewalls [38–40]. Although QWIPs do not have surface leakage current due to the upward band bending at the sidewalls, a similar technique has been proposed and studied for detector isolation to achieve a planar array surface without mesa sidewalls [41]. In Ref. 41, it has been shown that by using multiple F^+ and H^+ implants with energies up to 200 and 350 KeV, and doses up to 8×10^{12} and 1×10^{15} cm^{-2}, respectively, followed by a 5-minute anneal at 500°C, a 3.2 μm thick QWIP layer can be completely isolated. The detection performance is tested to be unchanged from that with conventional mesa etching. Therefore, implant isolation is viable to reduce the MQW active area without removing the material.

To produce an enhanced R-QWIP (ER-QWIP) array, the first step is to define and etch the ring pattern into the top conducting layer. The active MQW regions in the array is then defined by depositing a photoresist mask for ion implantation with a square size larger than the active area as seen in Fig. 5.111(a). The kinetic energy of the ions is controlled to create an isolation depth larger than the MQW layer of 0.2 μm, chosen to be the same as the E-MSA, while

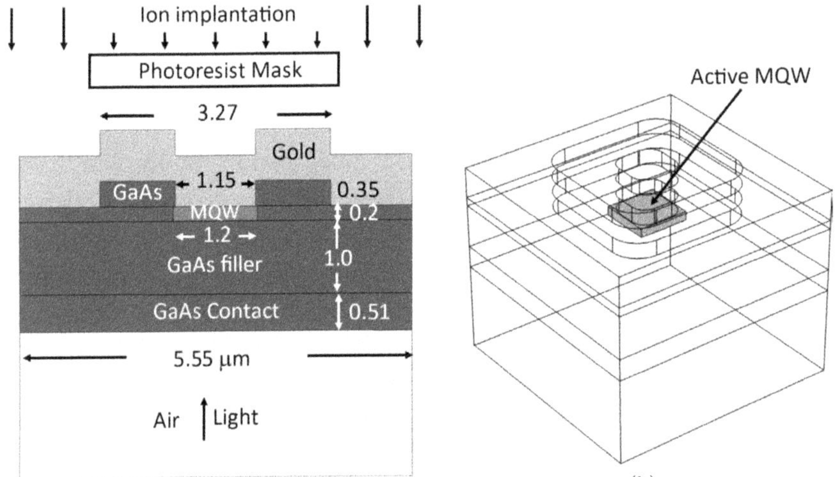

(a) (b)

Figure 5.111. (a) The designed ER-QWIP structure with the suggested size of the ion implantation mask. The filler material is the same as the contact material to yield the proper resonant thickness. (b) The 3D perspective to show the MQW active region.

reducing the size of the active region. In the present example, an active area of $1.2\,\mu\text{m} \times 1.2\,\mu\text{m}$, which is slightly larger than the inner ring, is chosen as an extreme case to yield the highest I_p/I_d ratio. The third step is to perform dry etching to expose the bottom contact layer at the array peripheral. The fourth step is ohmic/reflecting metal deposition on the detector pixels and the ground contact area. No mesa etching is needed with the implant isolation. With these simple steps, a planar ER-QWIP array is produced. According to Ref. 37, a single dose of O^+ implant at 8×10^{12} cm^{-2} and 400 keV is sufficient to isolate $n = 2 \times 10^{18}$ cm^{-3} GaAs to a depth of $\sim$0.6 μm, which is sufficient for the present layer structure.

For the present ER-QWIP structure and the same $\alpha = 0.2\,\mu\text{m}^{-1}$, the modeled QE referenced to the PUC area of $5.55 \times 5.55\,\mu\text{m}^2$ and QE* referenced to the active MQW area of $1.2 \times 1.2\,\mu\text{m}^2$ are shown in Fig. 5.110. The modeled QE is 35.4% at 8.6 μm. It is nearly a factor of 2 smaller than the E-MSA QE of 63.1%, although it is still substantial compared to other coupling schemes. The smaller

QE is due to the large area reduction of 21.4 times versus the 5.9 times in the E-MSA. When the active area is referenced, QE* of the ER-QWIP can be as large as 758%. This large QE* is the result of the structured gold cover in the ring unit cell acting as an efficient antenna in focusing the optical energy into an extremely small mode volume of the LSP. In addition, the substrate/air interface in the R-QWIP also reflects the unabsorbed light back to the MQW layer for additional absorption. The closed boundary of the resonator-QWIP is then more effective in increasing the optical E field than those with open boundaries. With the large gain of ~0.6 expected from a 0.2 μm thick MQW, which contains 3 to 4 QWs, the CE* will be ~454%, and it is over two orders of magnitude larger than the typical CE in determining I_p/I_d ratio. Therefore, the R-QWIP structure can be made into an efficient optical antenna. In the next chapter, we will discuss using similar antenna structures to perform efficient optical transmission.

The large QE* of the optical antennas can be visualized in the E_z field profiles in Fig. 5.112 for the E-MSA and in Fig. 5.113 for the ER-QWIP under the incident $\boldsymbol{H} = (0, 1, 0)$ A/m. For the E-MSA, whose $w = 1.4\,\mu$m and $p = 3\,\mu$m, Fig. 5.112(a) shows that under resonance at $\lambda = 8.6\,\mu$m, H_y is highly concentrated inside the E-MSA and its amplitude at the center of the MQW layer is approximately 18 times larger than the incident H. The fact that $H_y = 0$ at the parallel sidewalls is consistent with the magnetic wall approximation, so does the direction of $\boldsymbol{H}$ at the perpendicular sidewalls in Fig. 5.112(b). Fig. 5.112(c) shows E_z to be strongest near the two parallel sidewalls as expected. Figure 5.112(d) shows the direction of the total E field is nearly vertical inside the cavity. These EM field profiles thus show a highly concentrated optical energy inside the E-MSA cavity and the E field direction is ideal for intersubband transitions.

Figure 5.113 shows that of the ER-QWIP under the same incident condition but from the back of the substrate. The active MQW volume is seen to be slightly smaller than the LSP mode volume at $\lambda = 8.6\,\mu$m to maximize the I_p/I_d ratio.

The maximum E_z of the E-MSA in the active layer is about 250 V/m while that of the ER-QWIP is 2000 V/m for the same

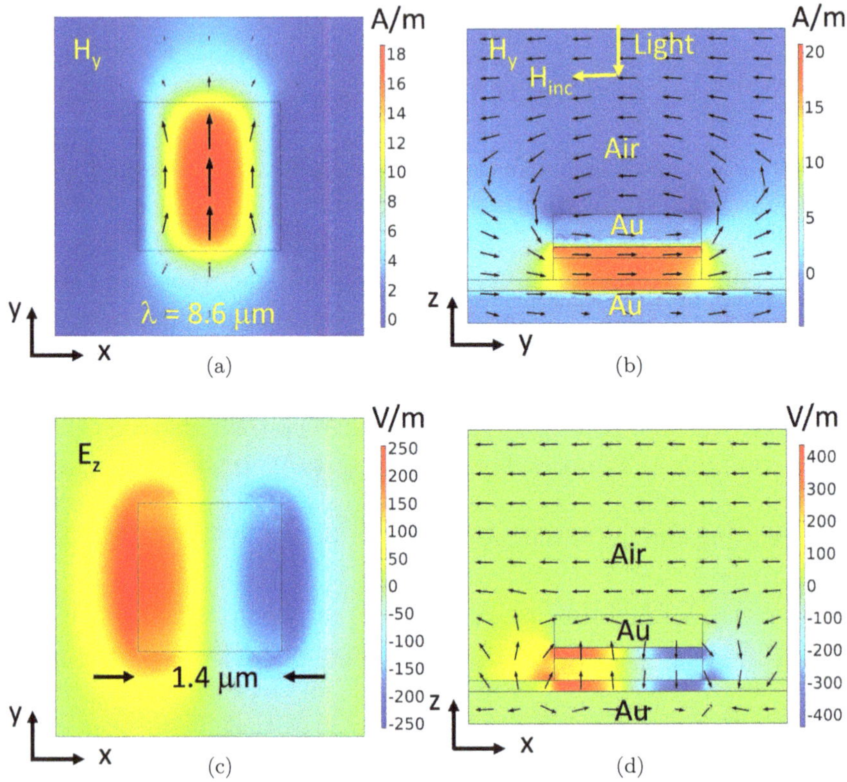

Figure 5.112. (a) H_y at the mid-plane of the MQW layer. (b) H_y in the y-z mid-plane. (c) E_z in the MQW mid-plane. (d) E_z in the x-z mid-plane. Incident $\boldsymbol{H} = (0, 1, 0)$, $w = 1.4\,\mu$m, and $p = 3.0\,\mu$m.

incident E_x of 377 V/m. Since the optical intensity I_o is given by $n\varepsilon_0 cE^2/2$, and $n = 3.15$, I_o is intensified up to 1.4 times in the E-MSA and 88.7 times in the ER-QWIP. The stronger average I_o in the ER-QWIP active volume leads to the larger QE*.

Although a smaller physical volume increases the operating temperature, the resulting sensitivity may suffer. In Eq. (5.12), NETD of an array depends on $(1 + I_d/I_p)$ and NETD_0. Although the first factor is improved by the larger QE* and gain, NETD_0 is proportional to $\sqrt{(g/N_c)}$ in the ROIC charge N_c limited case and $\propto (1/\eta_d \tau_{\text{int}})^{1/2}$ in the τ_{int} limited case. The larger g in a thinner MQW

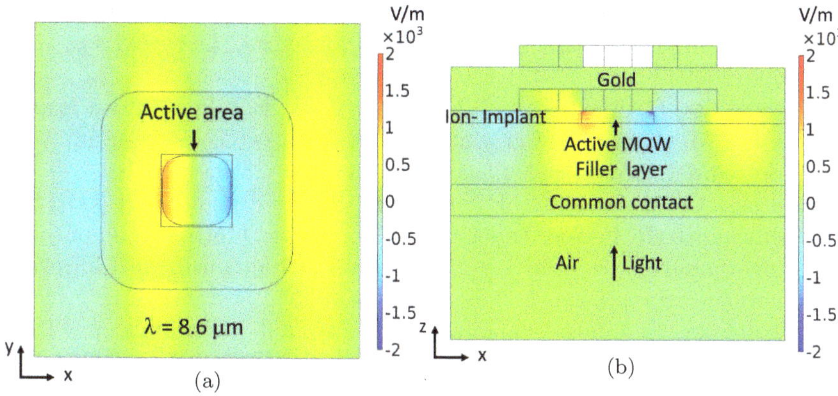

Figure 5.113. E_z (a) in the mid-plane of the MQW layer and (b) in the center cross-section of the x-z plane of an ion-implanted R-QWIP unit cell. Incident $\boldsymbol{H} = (0, 1, 0)$.

layer increases I_p in the first term but also increases NETD_0 in the N_c limited detection. Similarly, any decrease of η_d in this approach will increase NETD_0 in the τ_{int} limited detection. Therefore, the advantages of optical antennas are specific to requirement priorities.

References

1. Goossen, K. W. and Lyon, S. A. (1985). Grating enhanced quantum well detector, *Appl. Phys. Lett.*, 47, 1257–1259.
2. Goossen, K. W., Lyon, S. A. and Alavi, K. (1988). Grating enhancement of quantum well detector response, *Appl. Phys. Lett.*, 53, 1027–1029.
3. Hasnain, G., Levine, B. F., Bethea, C. G., Logan, R. A., Walker, J. and Malik, R. J. (1989). GaAs/AlGaAs multiquantum well infrared detector arrays using etched gratings, *Appl. Phys. Lett.*, 54, 2515–2517.
4. Andersson, J. Y., Lundqvist, L. and Paska, Z. F. (1991). Quantum efficiency enhancement of AlGaAs/GaAs quantum well infrared detectors using a waveguide with a grating coupler, *Appl. Phys. Lett.*, 58, 2264–2266.
5. Andersson, J. Y. and Lundqvist, L. (1991). Near-unity quantum efficiency of AlGaAs/GaAs quantum well infrared detectors using a waveguide with a doubly periodic grating coupler, *Appl. Phys. Lett.*, 59, 857–859.

6. Andersson, J. Y. and Lundqvist, L. (1992). Grating-coupled quantum-well infrared detectors: Theory and performance, *J. Appl. Phys.*, 71, 3600–3610.
7. Lundqvist, L., Andersson, J. Y., Paska, Z. F., Borglind, J. and Haga, D. (1993). Efficiency of grating coupled AlGaAs/GaAs quantum well infrared detectors, *Appl. Phys. Lett.*, 63, 3361–3362.
8. Sarusi, G., Levine, B. F., Pearton, S. J., Bandara, K. M. S., Leibenguth, R. E. and Andersson, J. Y., (1994). Optimization of two-dimensional gratings for very long wavelength quantum well infrared photodetectors, *J. Appl. Phys.*, 76, 4989–4994.
9. Kozlowski, L. (1994). Quantum Well Infrared Photodetector FPA, Final Report for Office of Naval Research, Contract No. N00014-91-C-0163.
10. Holst, G. C. and Driggers, R. G. (2012). Small detectors in infrared system design, *Opt. Eng.*, 51, 096401-1-10.
11. Costard, E. and Bois, Ph. (2007). THALES long wave QWIP thermal imagers, *Inf. Phys. & Tech.*, 50, 260–269.
12. Ivanov, R., Högnadóttir, S., Ramos, D., Evans, D., Visser, D., Rihtnesberg, D., Smuk, A., Becanovic, S., Sehlin, S., Almqvist, S., Smuk, S., Höglund, L. and Costard, E. (2023). QWIP: still the best solution for highend applications in LWIR, *Proc. SPIE 12534, Inf. Techn. and Appl.* XLIX, 125341I.
13. Jhabvala, M., Reuter, D., Choi, K. K., Sundaram, M. *et al.* (2010). The QWIP focal plane assembly for NASA's Landsat Data Continuity Mission, *Proc. SPIE 7660, Inf. Techn. and Appl.* XXXVI, 76603J.
14. Jhabvala, M., Choi, K. K., Waczynski, A., La, A. *et al.* (2011). Performance of the QWIP focal plane arrays for NASA's Landsat Data Continuity Mission, *Proc. SPIE 8012, Inf. Techn. and Appl.* XXXVII, 80120Q.
15. Thibaudeau, L., Bois, P. and Duboz, J. Y. (1996). A self-consistent model for quantum well infrared photodetectors, *J. Appl. Phys.*, 79, 446–454.
16. Cellek, O. O. and Besikci, C. (2004). Detailed investigation of electron transport, capture and gain in $Al_{0.3}Ga_{0.7}As$/GaAs quantum well infrared photodetectors, *Semi. Sci. & Techn.*, 19, 183–190.
17. Memiş, S, Cellek, O. O., Bostanci, U., Tomak, M. and Beşikçi, C. (2006). A detailed ensemble Monte Carlo study of the effect of quantum well width on quantum well infrared photodetector characteristics, *Turkish J. Phys.*, 30, 14.
18. Choi, K. K., Jhabvala, M. D., Sun, J., Jhabvala, C. A., Waczynski, A. and Olver, K. (2013). Resonator-quantum well infrared photodetectors, *Appl. Phys. Lett.*, 103, 201113.

19. Sun, J., Choi, K. K. and Olver, K. (2014). Fabrication of resonator-quantum well infrared photodetector test devices, *J. Micro/Nanolithography, MEMS, and MOEMS*, 13, 013004.
20. Choi, K. K., Sun, J., Olver, K. A. and Fu, R. (2017). Parameter study of resonator-quantum well infrared photodetectors, *IEEE J. Quant. Electron.*, 53, 4000410.
21. Harwit, A. and Harris, J. S. (1987). Observation of Stark shifts in quantum well intersubband transitions, *Appl. Phys. Lett.*, 50, 685–687.
22. Carlon, H. R. (1979). Christiansen effect in IR spectra of soil-derived atmospheric dusts, *Appl. Opt.*, 18, 3610–3614.
23. Choi, K. K., Allen, S. C., Sun, J. G., Endres, D., Olver, K. A. and Fu, R. X. (2018). Small pitch resonator-QWIP detectors and arrays, *Inf. Phys. & Techn.*, 94, 118–125.
24. Choi, K. K., Allen, S. C., Sun, J. G. and DeCuir, E. A. (2017). Resonant detectors and focal plane arrays for infrared detection, *Inf. Phys. & Techn.*, 84, 94–101.
25. Ivanov, R, Smuk, S., Hellström, S., Evans, D., Höglund, L. and Costard, E. (2018). LWIR QWIPs at IRnova for next generation polarimetric imaging, *Inf. Phys. & Techn.*, 95, 177–182.
26. Choi, K. K., Jhabvala, M. D., Forrai, D. P., Waczynski, A., Sun, J. and Jones, R. (2012). Electromagnetic modeling and design of quantum well infrared photodetectors, *IEEE J. Quant. Electr.*, 19, 1–10.
27. Schimert, T. R., Barnes, S. L., Brouns, A. J., Case, F. C., Mitra, P. and Claiborne, L. T. (1996). Enhanced quantum well infrared photodetector with novel multiple quantum well grating structure, *Appl. Phys. Lett.*, 68, 2846–2848.
28. Mitra, P., Brouns, A., Case, F. C. and McCurdy, J. M. (2003). Progress in enhanced quantum well infrared photodetectors, *Proc. SPIE*, 4820.
29. Mitra, P., Case, F. C., McCurdy, J. H., Zaidel, S. A., and Claiborne, L. T. (2003). Multispectral long-wavelength quantum-well infrared photodetectors, *Appl. Phys. Lett.*, 82, 3185–3187.
30. Mao, J., Majumder, A., Choi, K. K., Tsui, D. C., Leung, K. M., Lin, C. H., Tamir, T. and Vawter, G. A. (2002). Light coupling mechanism of quantum grid infrared photodetectors, *App. Phys. Lett.*, 80, 868.
31. Choi, K. K., Dang, G., Little, J. W., Leung, K. M. and Tamir, T. (2004). Quantum grid infrared spectrometer, *Appl. Phys. Lett.*, 84, 4439.
32. Lo, Y. T., Solomon, D. and Richards, W. F. (1979). Theory and experiment on microstrip antennas, *IEEE Trans. Antennas Propagat.*, AP-27, 137–145.

33. Beck, W. A. and Mirotznik, M. S. (2001). Microstrip antenna coupling for quantum-well infrared photodetectors, *Inf. Phys. & Techn.*, 42, 189–198.
34. Olmon, R. L., Slovick, B., Johnson, T. W., Shelton, D., Oh, S.-H., Boreman, G. D. and Raschke, M. B. (2012). Optical dielectric function of gold, *Phys. Rev. B*, 86, p. 235147.
35. Chen, Y. N., Todorov, Y., Askenazi, B., Vasanelli, A., Biasiol, G., Colombelli, R. and Sirtori, C. (2014). Antenna-coupled microcavities for enhanced infrared photo-detection, *Appl. Phys. Lett.*, 104, 031113.
36. Chen, Y. N. (2015). Antenna Coupled Quantum Infrared Detectors, Quantum Physics [quant-ph]. Université Paris Diderot — Paris 7, HAL Id : tel-01390461.
37. Pearton, S. J. (1989). *Ion Implanation for Isolation of III-V Semiconductors*, North-Holland.
38. Musca, C. A., Antoszewski, J., Dell, J. M., Faraone, L. and Terterian, S. (2003). Planar p-on-n HgCdTe heterojunction mid-wavelength infrared photodiodes formed using plasma-induced junction isolation, *J. Electron. Mater.*, 32, 622–626.
39. Chang, S.-H., Fang, Y.-K., Ting, S.-F., Chen, S.-F., Lin, C.-Y. and Wu, C.-Y. (2007). Ultra high performance planar InGaAs PIN photodiodes for high speed optical fiber communication, Sens. *Actuators A Phys.*, 133, 9–12.
40. Dehzangi, A., Wu, D., McClintock, R., Li, J., Jaud, A. and Razeghi, M. (2020). Demonstration of planar type-II superlattice-based photodetectors using silicon ion-implantation, *Photonics*, 7, 68.
41. Hui, S. P. and Pei, S. S. (1994) Planar quantum well photodetector using ion implant isolation, EU. Pat.# 0564163B1.

Chapter 6

Resonator-Pixel Detector and Emitter Technologies

6.1 Introduction

In the previous two chapters, we have applied the finite element method (FEM) to corrugated-QWIPs, resonator-QWIPs, and other coupling structures and obtained quantitative agreements on their spectral absorption quantum efficiencies (QEs) with experiment. From the modeled electromagnetic (EM) field distribution, physical insights into different optical coupling mechanisms and optimization strategies can be obtained. In addition to experimental supports, FEM agrees with other numerical methods such as classical ray tracing, modal transmission-line method, and finite difference time domain method. Therefore, the FEM approach is shown to be reliable and can be readily applied to other detector materials. In this chapter, we will utilize FEM in general and the resonator-pixel (RP) structure in particular to design and improve the performance of other detectors. Since most of the proposed detector structures have not been experimentally verified, the modeled characteristics should not be regarded as factual until they are proven by experiment. FEM provides valid solutions to well-defined material and geometrical structures but its prediction can be inaccurate for many reasons. Firstly, the optical properties of semiconductors and metals depend on growth techniques, preparation methods, and operating conditions. Their properties may not be the same as those published in literature and need to be verified by experiment. Secondly, the adopted materials must be of high optical quality that is free from refractive index inhomogeneity and gain boundary scattering, etc. The propagation of EM waves in all constituent materials must maintain phase coherence in macroscopic scales described by Maxwell's equations. Thirdly, due to the possible presence of intense EM fields in the detector, parasitic absorptions outside the absorbing layer may be significant and can affect the absorber QE. Fourthly, the processed detectors and arrays must have uniform and smooth surfaces/interfaces with the exact dimensions that are being modeled. These requirements can be difficult to all meet in an experiment. Nevertheless, the modeled ideal characteristics are still useful as a guide for experiment and serve as a base for further studies. Material and structural imperfections can be incorporated into the model as needed from experiment. Confidence on the reliability of the model can be gained by modeling conventional detectors and comparing the results to the experimental data. A favorable agreement will lend certain credibility to a new model.

Other than detector applications, the RP technology can also be useful for emitter applications. Due to the internal field enhancement, optical recombination rate in a light-emitting material increases. At the same time, thanks to the antenna reciprocity theorem, a resonator serving as an efficient optical receiver will also be an efficient optical transmitter. The efficient generation and transmission of light in a RP will then increase the output power and directivity of an optical emitting device such as a light-emitting diode or a thermal radiator. In this chapter, we will discuss some of these potential applications.

6.2 GaAs Solar Cells

The first potential application of the RP technology is in GaAs solar cells. By enhancing the cell's absorption QE, the short-circuit current J_{sc} can be increased. In literature, nanostructures have been used to enhance the performance of GaAs solar cells. For example, by using a double-layer anti-reflection coating (ARC; 40 nm TiO_2/90 nm SiO_2) and a planar silver back-reflector on a 100 nm thick GaAs active material, J_{sc} of 16.38 mA/cm^2 was obtained [1]. The solar cell

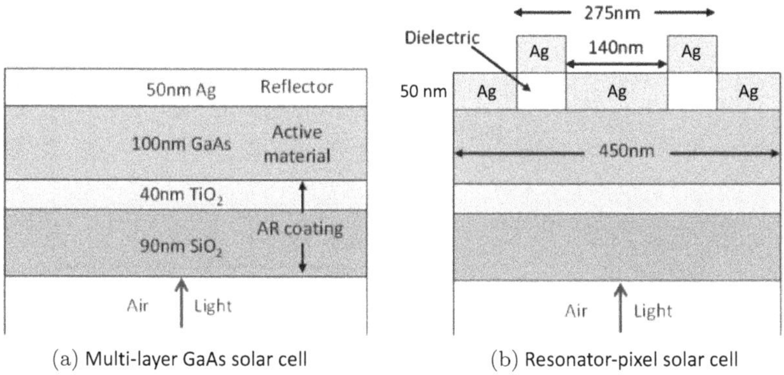

Figure 6.1. The cross-section of (a) a planar GaAs solar cell with a double-layer ARC and Ag back-reflector and (b) the optimized RP solar cell with the same layer structure. Reproduced with permission from *IEEE J. Quant. Electr.*, 57(4), pp. 1–10 (2021).

structure is shown in Fig. 6.1(a). When hexagonally close-packed SiO_2 nanospheres are deposited on top of the ARC, J_{sc} is further enhanced by 11% to 18.14 mA/cm^2. Without the double-layer ARC, the bare cell with a thicker thickness of 135 nm shows a lower J_{sc} of 13.5 mA/cm^2, and it increases to 16.7 mA/cm^2 (a 23% increase) using a quasi-random nanostructure and to 17.0 mA/cm^2 (a 26% increase) using a photonic crystal nanostructure [2].

In the RP approach, the planar Ag reflector in Fig. 6.1(a) is replaced by an array of rings whose unit cell structure is shown in Figs. 6.1(b) and 6.2(a). These rings are made of a wider bandgap dielectric $Al_xGa_{1-x}As$ material on top of GaAs before the Ag reflector is deposited. The complex refractive indices (n, κ) of the materials are shown in Fig. 6.2(b). The AlGaAs layer is assumed to have the same n as GaAs. Both AlGaAs and SiO_2 are assumed to be non-absorbing in the modeled wavelength range.

With the displayed (n, κ), the ring structure is optimized without changing the existing layer thicknesses. The optimized ring dimensions and the array period are shown in Fig. 6.1(b). Many unit cells will be grouped together to form a large solar cell structure. To verify the applicability of EM modeling in this material layer system, the spectral QE and J_{sc} of the planar solar cell in Fig. 6.1(a) are first modeled. The planar QE is modeled in 2D given by the general

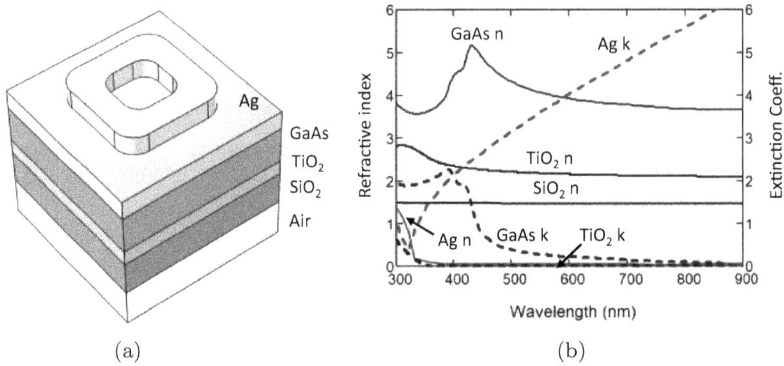

Figure 6.2. (a) A 3D view of the optimized RP unit cell. (b) (n, κ) of individual layers. Reproduced with permission from *IEEE J. Quant. Electr.*, 57(4), pp. 1–10 (2021).

expression for η_{abs}

$$\eta_{\text{abs}}(\lambda) = \frac{\alpha(\lambda)n}{pZ_0^2H^2}\int_{A'} |E(\lambda)|^2 d^2r \quad \text{in 2D,}$$
$$= \frac{\alpha(\lambda)n}{AZ_0^2H^2}\int_{V} |E(\lambda)|^2 d^3r \quad \text{in 3D,} \tag{6.1}$$

where $\alpha(\lambda) = 4\pi\kappa/\lambda$. n is that of GaAs. p is the detector linear size. A' is the cross-sectional area of the GaAs layer. A is the unit cell area. V is the GaAs volume. Z_0 is the vacuum impedance. $\mathbf{H} = (0, 0, 1)$ A/m and $(1, 1, 0)$ A/m are the incident H fields in 2D and 3D, respectively. $|E|$ is the modeled total E magnitude for isotropic absorbing materials.

The modeled QE spectrum is shown in Fig. 6.3(a). With a double-layer ARC, the planar solar cell achieves nearly 100% QE at 370 nm. Therefore, there is very limited room to increase QE further near the peak wavelengths. Under the standard AM1.5 solar flux shown in Fig. 6.3(a), the spectral J_{sc} can be calculated by integrating the photon flux with the modeled QE spectrum, and it is shown in Fig. 6.3(b).

The modeled photocurrent spectrum matches the experimental curve from Ref. 1 without adjusting the solar cell parameters. This agreement confirms the predictable behavior of this material system. The integrated J_{sc} in this wavelength range is 17.44 mA/cm^2, which

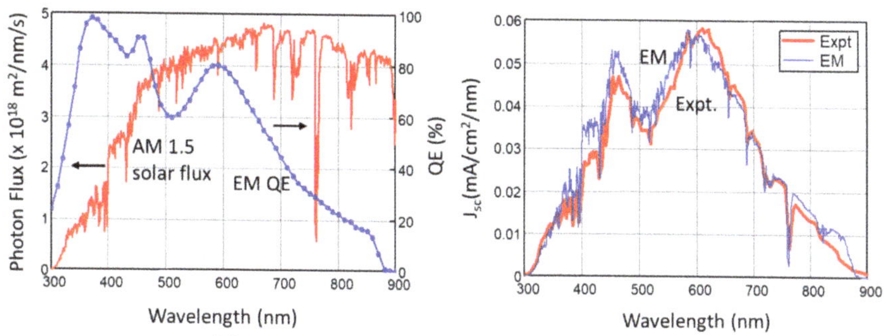

(a) (b)

Figure 6.3. (a) The modeled QE spectrum of the planar solar cell and the AM1.5 photon flux. (b) The calculated and observed spectral short-circuit current density. Reproduced with permission from *IEEE J. Quant. Electr.*, 57(4), pp. 1–10 (2021).

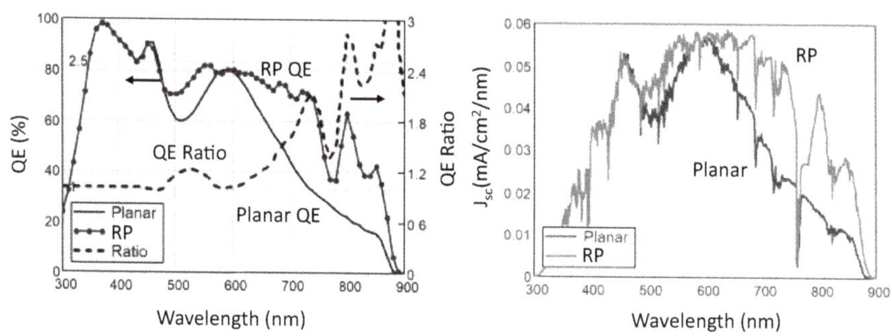

(a) (b)

Figure 6.4. (a) The calculated QE of the *RP*-structure and its ratio in comparison with the planar QE. (b) The corresponding short-circuit photocurrent densities. Reproduced with permission from *IEEE J. Quant. Electr.*, 57(4), pp. 1–10 (2021).

is close to the experimentally observed value of 16.38 mA/cm^2. After confirming the modeling approach, the RP structure is optimized in 3D using Eq. (6.1) and the resulting QE is plotted in Fig. 6.4(a). Below ~450 nm, the RP QE is the same as the planar QE. At these wavelengths, the GaAs κ is large and the incident light is totally absorbed before it reaches the ring structure. Consequently, QE is limited by the double-layer ARC transmission, and the RP structure does not play a role. At the longer λ, the smaller GaAs κ allows more light to reach the back surface and it interacts with the rings, thus QE becomes different. Near the absorption edge, QE is improved by a factor of 2.5. Figure 6.4(b) shows the calculated spectral $J_{\rm sc}$ for both planar and RP structures. It is seen that the RP structure increases $J_{\rm sc}$ substantially beyond 600 nm. Overall, the present design yields a theoretical $J_{\rm sc}$ of 22.0 mA/cm^2, a 26% increase from the planar cell, which compares favorably to the nanostructure approaches.

6.3 Very Long Infrared Wavelength Strained Layer Superlattice Detectors

The second example is III-V GaSb-based strained layer superlattices (SLS). SLS detectors can be operated at higher temperature than InSb detectors in MWIR and thus have wide market acceptance. The longer wavelength SLS detectors, however, are still under development because of their lower detection QE. The lower η_d is originated from the thicker GaSb or InAsSb SLS barriers adopted in smaller bandgap materials. They reduce the electron-hole wavefunction overlap and miniband width. The former reduces the absorption coefficient α while the latter reduces the carrier mobility μ and diffusion length L_d, thus reducing photocarrier collection. The conventional approach of using a thick material to achieve a high η_d is not effective when the thickness increases beyond L_d. Furthermore, a thick absorber yields a large diffusion dark current given by Eq. (2.52), reducing the photocurrent to dark current ratio I_p/I_d. The problem of small I_p/I_d becomes more acute when using VLWIR material to detect SWIR and MWIR since the much smaller photon flux in the shorter wavelength bands from terrestrial objects is compounded by the high dark current of a VLWIR material, leading to a small I_p/I_d. Therefore, SLS detectors need to be improved if they are used for multi-band and hyperspectral imaging. Since the RP uses a thin layer to achieve a large η_d, it can be used to reduce the dark current. Several thin absorbing layers can also be stacked together to form a multi-color detector. In this section, we design a broadband SLS detector and a voltage-switchable, 4-color detector using the RP approach.

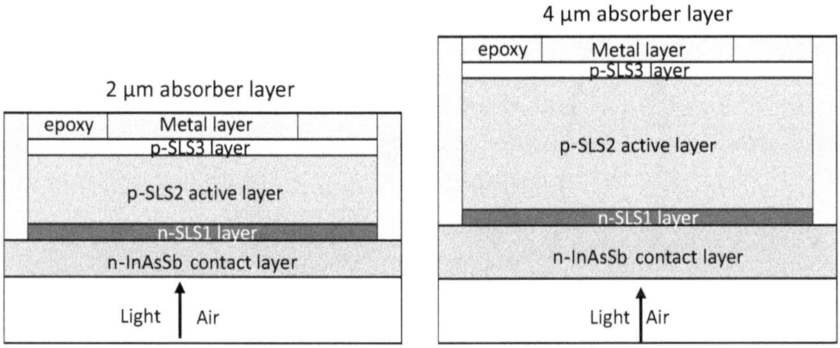

Figure 6.5. Material layer structure of SLS photodiodes with either a $2\,\mu$m absorber or a $4\,\mu$m absorber. The pixel pitch is $25\,\mu$m. Reproduced with permission from *Appl. Opt.*, 56(3), pp. B26–B36 (2017).

6.3.1 *Electromagnetic Modeling of Planar Strained Layer Superlattice Detectors*

As usual, planar detectors are first modeled to validate the modeling reliability. Figure 6.5 shows two SLS photodiodes with different absorber thicknesses t_a. They are of $22\,\mu$m size in a $25\,\mu$m pitch array and consist of a layer of p-InAs/Ga$_{0.75}$In$_{0.25}$Sb SLS (labeled as p-SLS2), having $t_a = 2$ or $4\,\mu$m, and a $0.5\,\mu$m thick wider bandgap n-SLS1 layer to form a photodiode. The photodiode is sandwiched between a $0.5\,\mu$m thick wider bandgap p-type top contact layer (p-SLS3) and a 1 or $1.6\,\mu$m thick n-InAs$_{0.9}$Sb$_{0.1}$ bottom contact layer. On top of the p-contact, there is a $12\,\mu$m wide gold layer for ohmic contact, and the detector is encapsulated by an epoxy material with $n = 1.5$. The detectors are backside illuminated.

Shown in Fig. 6.6 are the theoretical $\alpha(\lambda)$ of the p-SLS2 layer [3] and the deduced $\alpha(\lambda)$ of n-InAs$_{0.9}$Sb$_{0.1}$ contact layer [4]. The latter spectrum is obtained by shifting the experimental n-InAs spectrum by $\lambda = 0.7\,\mu$m in accounting for the smaller bandgap. The wider bandgap n-SLS1 and p-SLS3 are assumed to have the same spectral lineshape as p-SLS2 but absorb in the MWIR band. The SLS n is taken to be a constant at $n = 3.5$, which is commonly adopted. All infrared absorption is assumed to be isotropic. Since the absorption of all layers except the p-SLS2 layer is dissipated as heat and does

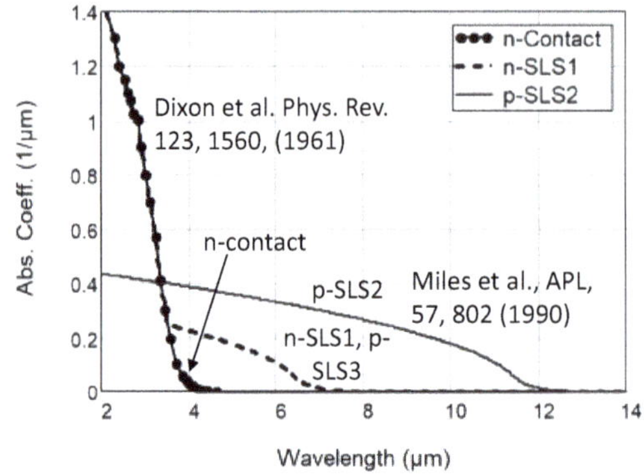

Figure 6.6. Theoretical and experimental α of individual layers obtained from open literature. Reproduced with permission from *Appl. Opt.*, 56(3), pp. B26–B36 (2017).

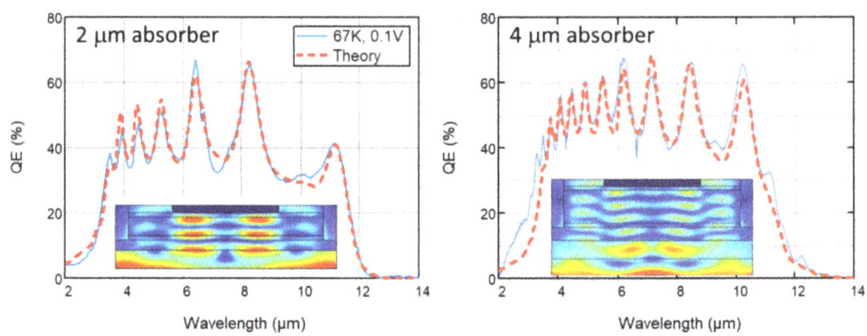

Figure 6.7. Calculated η_{abs} and measured η_d of the p-SLS2 layer. The experimental data is provided by M. Jhabvala of NASA. The insets are plots of $|E|$ fields at $\lambda = 11\,\mu$m. The applied bias is 0.1 V. Reproduced with permission from *Appl. Opt.*, 56(3), pp. B26–B36 (2017).

not contribute to photocurrent due to high doping, only the p-SLS2 layer is active in the detection.

With these material and detector structures, the detector η_{abs} spectra are modeled according to the 2D expression in Eq. (6.1), and they are shown in Fig. 6.7. For the detector with $2\,\mu$m absorber, the modeled η_{abs} matches the experimental η_d both in magnitude and in spectral lineshape without adjusting input parameters. For

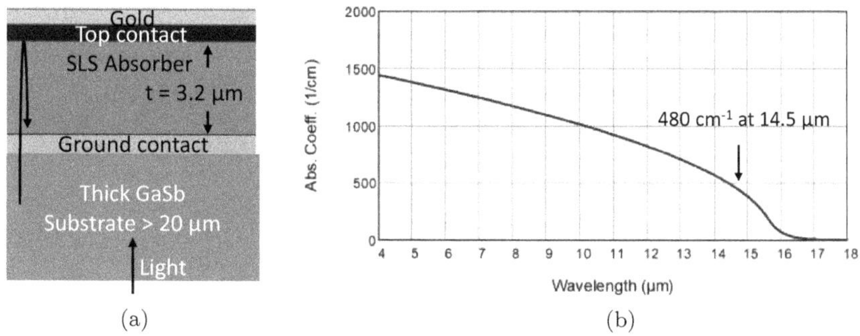

Figure 6.8. (a) The conventional, planar detector structure. (b) The assumed absorption coefficient α spectrum of the 15 μm cutoff InAs/GaSb SLS material.

the 4 μm detector, agreement is reached when α of SLS2 is lowered by 30% from that shown in Fig. 6.6. The apparent α reduction is attributed to the incomplete charge collection in the thicker material. The agreements in these two examples show the predictable detector behavior using EM modeling.

6.3.2 *Conventional Planar Very Long Infrared Wavelength Detectors*

The RP approach is more useful in VLWIR detection because both α and carrier mobility are small. The conventional detector with a planar layer structure is schematically shown in Fig. 6.8(a). It contains a SLS absorber layer, top and ground semiconductor contact layers, a top gold contact, and a GaSb substrate. The ground contact includes a barrier layer in a nBn detector or a *p*-*n* junction layer in a photodiode. The contact layers are assumed to have no infrared absorption in the modeling wavelength range, and all semiconductors have a constant $n = 3.548$. A similar material structure has been studied by Höglund *et al.* [5]. The detector consists of a 3.2 μm thick p^- InAs/GaSb SLS absorbing layer, a n^- junction layer, and a n^+ conduction layer. The latter two are grouped into the ground contact layer in Fig. 6.8(a). In Ref. [5], the absorption coefficients $\alpha(\lambda)$ and η_d, which is also known as the external quantum efficiency EQE(λ), were separately measured in three different materials. Based on the average properties, a very similar α spectrum shown in Fig. 6.8(b) is adopted for the design. It has a cutoff wavelength λ_c at 15.5 μm and magnitudes of 480 cm^{-1} at 14.5 μm and 1000 cm^{-1} at 10 μm.

With the assumed $\alpha(\lambda)$, the classical η_{abs} with 2 passes of light is given by

$$\eta_{\text{abs}}(\lambda, t) = \frac{4n}{(1+n)^2}(1 - e^{2t\alpha(\lambda)}), \tag{6.2}$$

where $t = 3.2\,\mu$m is the absorber thickness. The calculated $\eta_{\text{abs}}(\lambda)$ from Eq. (6.2) is shown as the gray curve in Fig. 6.9(a). It is 32.7% at

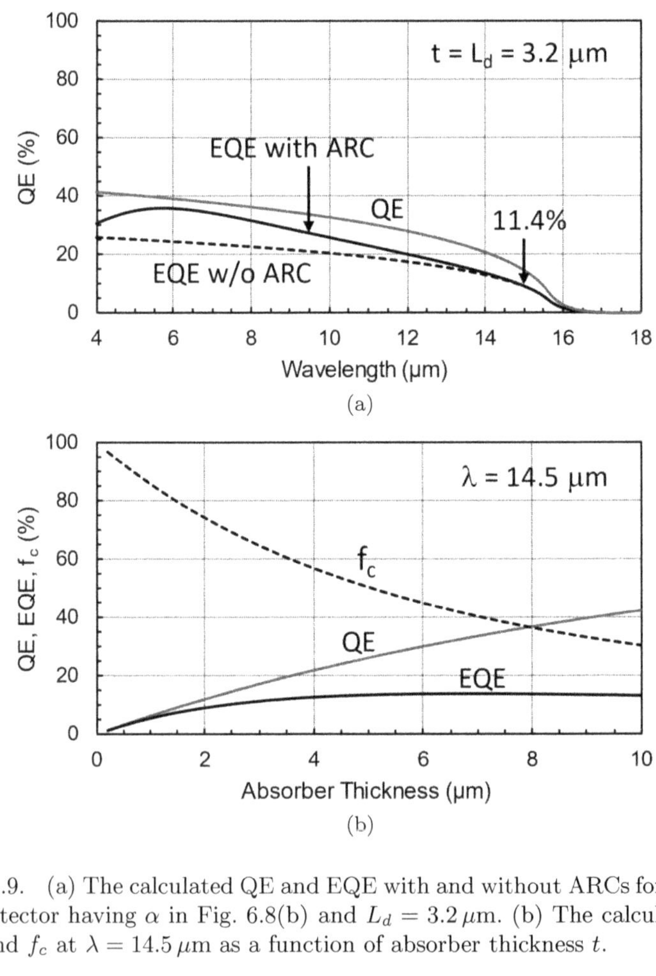

Figure 6.9. (a) The calculated QE and EQE with and without ARCs for a 3.2 μm thick detector having α in Fig. 6.8(b) and $L_d = 3.2\,\mu$m. (b) The calculated QE, EQE, and f_c at $\lambda = 14.5\,\mu$m as a function of absorber thickness t.

10 μm, and it is larger than the measured EQE of 20% in Ref. [5]. This smaller EQE is attributed to the incomplete photocarrier collection with a finite diffusion length L_d. Assuming that the collection efficiency is $\exp(-x/L_d)$ where x is the distance away from the junction, EQE is shown to be

$$\begin{aligned} EQE(\lambda, t) &= \eta_{\text{abs}}(\lambda, t) f_c(L_d, t), \\ &= \eta_{\text{abs}}(\lambda, t) \frac{L_d}{t}(1 - e^{-t/L_d}), \end{aligned} \tag{6.3}$$

where $f_c(L_d, t)$ is the charge collection efficiency assuming uniform carrier generation in the absorbing layer. For simplicity, the same f_c is applied to both planar and resonant detectors. Eq. (6.3) shows that when $t \ll L_d$, EQE $= \eta_{\text{abs}}$, and when $t \gg L_d$, EQE $= \eta_{\text{abs}} L_d/t$. To fit L_d to the experimental EQE, L_d is found to be 3.2 μm, the same as the adopted t. Using this value of L_d, the calculated EQE is shown in Fig. 6.9(a) as the solid curve. It is equal to the experimental value of 20% at 10 μm as expected.

Figure 6.9(b) shows the relationship among QE, EQE, and f_c for $L_d = 3.2\,\mu$m and $\alpha = 480\,\text{cm}^{-1}$. As t increases, EQE increases initially but saturates at 13.6% when $t > 3.2\,\mu$m due to the decreasing f_c. At $t = 3.2\,\mu$m, EQE $= 11.4\%$, not far from the saturated value at large t. Therefore, the optimum thickness for a material with diffusion length L_d is L_d itself, which is adopted in Ref. [3]. An ARC can be used to improve the substrate transmission as one example shown in Fig. 6.9(a) but the effect is rather limited.

Note that with decreasing t, EQE is approaching QE, and thus an effective solution is to increase QE in a thin absorber. Furthermore, when $t < L_d$, the diffusion dark current is directly proportional to t according to

$$I_d = e n_{po} \frac{t}{\tau}, \tag{6.4}$$

where n_{po} is the minority carrier density (in a p-type absorber in this example), and τ is the minority carrier lifetime. Reducing t will also reduce the diffusion dark current proportionally in the small t limit. Therefore, there are advantages in reducing t if a large η_{abs} can be maintained.

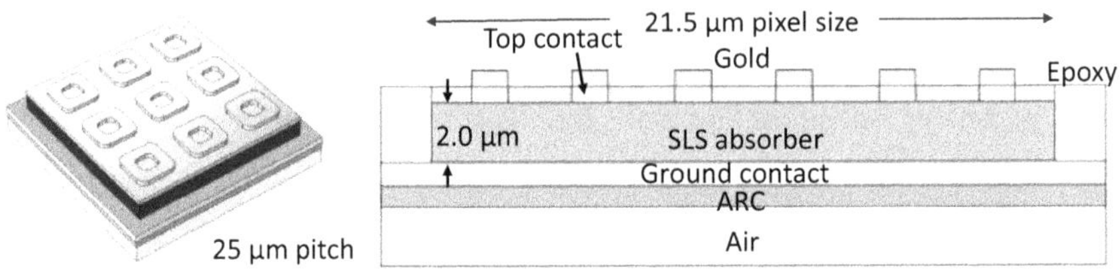

(a) (b)

Figure 6.10. (a) The 3D view of the R-SLS detector. (b) The cross-sectional view.

6.3.3 *Strained Layer Superlattice Resonator-Pixel Detectors for Broadband Detection*

To improve charge collection, the detector should be thinner than $2\,\mu$m in Fig. 6.9(b). Therefore, shown in Fig. 6.10, the optimized RP detector consists of a $2.0\,\mu$m or thinner absorber, a $0.8\,\mu$m ground contact layer, a $0.55\,\mu$m top contact layer etched into a 3×3 ring pattern in a $25\,\mu$m pitch pixel, and a $0.55\,\mu$m gold contact layer (combined with a thin metal wetting layer). The ring height $h = 0.55\,\mu$m, inner size $r_i = 2.2\,\mu$m, outer size $= 4.7\,\mu$m, and ring spacing $s = 7.0\,\mu$m. The ARC, if present, is a $0.74\,\mu$m dielectric material with $n = 1.9$. The detector has an area of $21.5 \times 21.5\,\mu\text{m}^2$.

To yield broadband coverage, the RP structure is designed to be able to redirect light effectively from MWIR to VLWIR. In the design process, the ring structural parameters and the ground contact thickness are adjusted while the EM field distribution and the resulting QE with and without ARC are monitored between $\lambda = 4$ and $18\,\mu$m. The goal is to obtain a large and relatively constant QE across the spectrum. From the distribution of the optical electric field $|E|$ at different λ, the rings are found to interact with light via several physical mechanisms. They redirect light through collective diffraction in MWIR, act as Fresnel zone plates in LWIR, and become plasmonic couplers in VLWIR. Through numerical optimization, the rings exhibit strong optical responses throughout the spectrum and yield broadband coupling.

Figures 6.11(a) and (b) show the QE of the RP-SLS without and with ARC, respectively. The QE of the planar detector without the

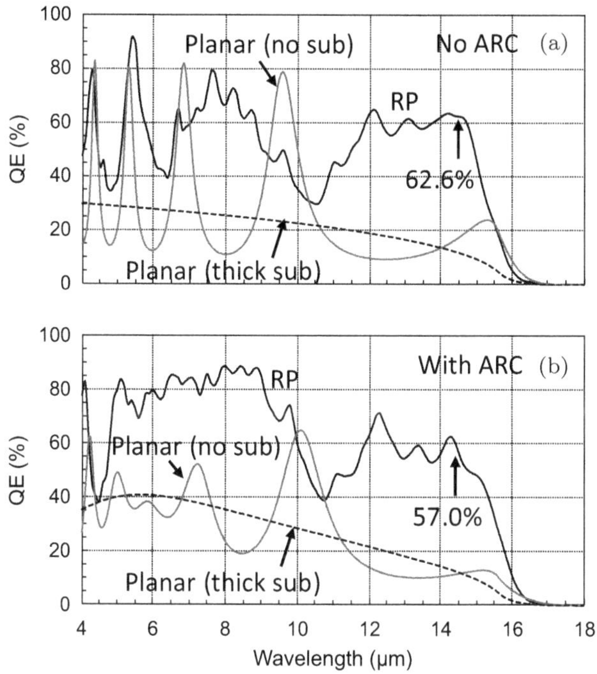

Figure 6.11. (a) The figure shows the absorption QE of a RP-SLS with $t = 2.0\,\mu$m in comparison with a planar detector with thick substrate and with removed substrate. (b) The figure shows those with an external ARC.

rings is also plotted for comparison. The RP-SLS is shown to have a large, relatively constant QE across the spectrum especially with ARC. QE with ARC is about 80% between 5 and 9 μm and about 60% between 12 and 15 μm.

In the material design, the combined thickness t_{tot} of the absorber and the ground contact is 2.8 μm. At this combined thickness, the detector material acts as its own ARC in the VLWIR regime because it is close to $3\lambda/4n$ for λ near 13 μm. The reflections from the material top and bottom interfaces cancel each other to eliminate the substrate reflection. This VLWIR ARC property is only slightly altered as seen in Fig. 6.11(b) when an external ARC designed for $\lambda = 5.6\,\mu$m is added. Using a combination of internal and external ARCs, a broadband RP-SLS with a large QE can be achieved.

Since the incident light is trapped in the detector material, the absorber thickness can be reduced without decreasing proportionally QE and EQE. The reduction of the absorption volume is balanced by the increase in the trapped intensity. Since the dark current decreases linearly with t, I_p/I_d will increase, which will improve the operating temperature. To maintain the same resonant volume, t_{tot} should remain constant. Therefore, with a smaller t, the absorber is partly replaced by the ground contact material at the bottom to maintain the same t_{tot}.

To compare the ARC-coated RP-SLS with the original planar detector reference, their $\text{EQE}(\lambda) = \eta(\lambda) f_c(L_d, t)$ are plotted in Fig. 6.12. Despite having a thinner t, the RP-SLS has equal or larger EQE than the reference. With dark current reduction, I_p/I_d increases by a factor of ~5 for the 2 μm thick R-SLS and ~18 for the 0.25 μm

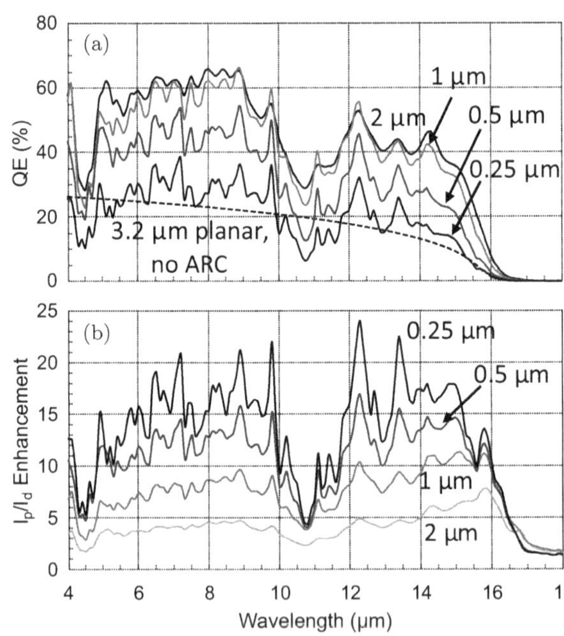

Figure 6.12. (a) The external QE $= \eta(\lambda) f_c(t)$ of the R-SLS with ARC versus the reference detector without ARC. (b) The corresponding I_p/I_d enhancement relative to the reference.

thick R-SLS as shown in Fig. 6.12. A more balanced t is $1.0\,\mu$m, which exhibits a large EQE of ~50% and a large I_p/I_d increase of ~9.

To better understand the light-trapping ability of the resonator, the same RP geometry without ARC is remodeled with α set to 0 in all semiconducting layers. In this case, there is no optical attenuation in the detector except in the gold layer. Under this condition, the photon energy U_s stored in the $2\,\mu$m thick absorber is calculated by integrating the energy density $\varepsilon_r\varepsilon_0|E|^2/2$ over the absorber volume, where $\varepsilon_r = (3.548)^2$. Similarly, the energy density in free space and in a semi-infinite block of material under the same incident condition can be modeled. Since the number of photons is directly proportional to U_s, the large photon number ratios in Fig. 6.13 indicate the high efficiency of the RP structure in trapping photons. At certain resonant wavelengths, the number of photons can be 50 or 100 times more in the resonator than in the planar medium or in free space. The light-trapping ability is mainly limited by the ohmic loss in gold. The minimum ratio is 4 in Fig. 6.13. Therefore, the RP structure can provide broadband QE enhancement.

The nature of the optical interaction of the rings can be deduced from the interference patterns inside the detector. Figure 6.14 shows

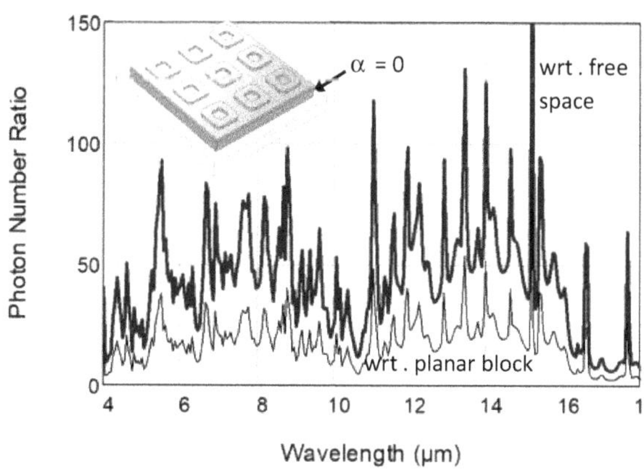

Figure 6.13. The modeled photon number in the RP geometry relative to the planar material block and free space of the same volume. The inset shows the 3D perspective view of the optimized RP modeled in this section.

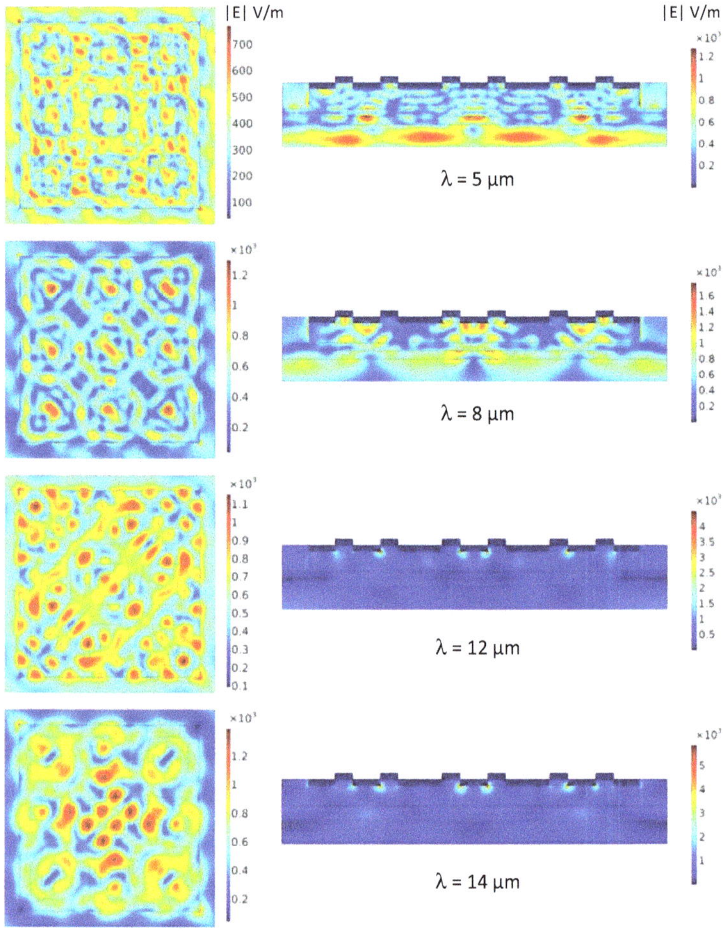

Figure 6.14. The optical $|E|$ field distribution in the mid-plane of the absorber layer and in the diagonal cross-section at different wavelengths under the normal incident with $\mathbf{H} = (1, 1, 0)$ A/m or $|E_{\text{inc}}|$ of 533.2 V/m.

the optical electric field distribution at the mid-plane and diagonal of the detector with $2\,\mu$m absorber and ARC. The incident E is 135° polarized in the x-y plane at different wavelengths. In Fig. 6.14, the field distributions are consistent with small angle grating diffraction at $\lambda = 5\,\mu$m, Fresnel zone plate focusing at $8\,\mu$m, and the excitation of localized surface plasmon polariton at the longer wavelengths.

Through these different interactions, the incident light is redirected and trapped in the detector volume with the maximum $|E| \sim 1200 - 5000\,\text{V/m}$, much larger than the incident $|E_{\text{inc}}|$ of $533.2\,\text{V/m}$. Large QE can thus been obtained.

The QE improvement is similar in smaller detectors. In an $18\,\mu\text{m}$ pitch array, a $16 \times 16\,\mu\text{m}^2$ pixel can accommodate 2×2 rings. With an ARC, the broad detection spectrum can be maintained as shown in Fig. 6.15.

6.3.4 *Narrowband Strained Layer Superlattice Resonator-Pixel Detectors*

In Fig. 6.14, different wavelength regimes exhibit different E field distributions along the vertical z direction. In particular, the distribution is more uniform throughout the detector layer in the

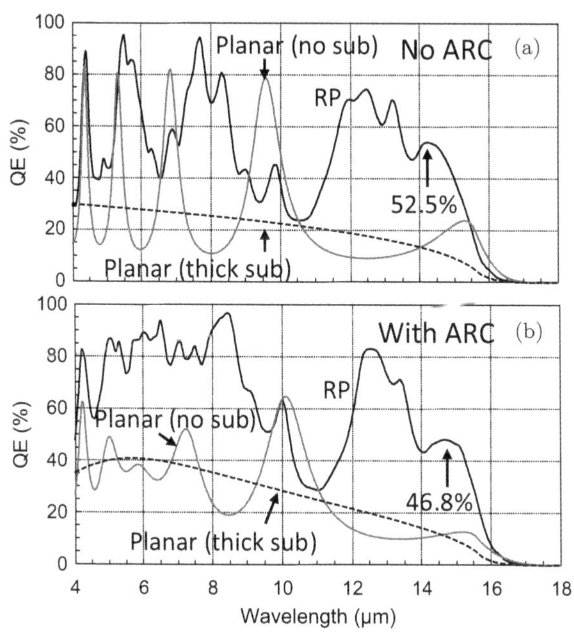

Figure 6.15. (a) The figure shows QE of a RP-SLS with $t = 2.0\,\mu\text{m}$ and size = $16 \times 16\,\mu\text{m}^2$ and a planar detector with or without substrate. (b) The figure shows those with an external ARC.

shorter wavelengths than in the longer wavelengths. These different distributions can be utilized to build a voltage-switchable two-color (VS2C) detector, in which different cutoff materials are placed at different depths. Since the absorbing layer can be very thin, it is feasible to stack four of them together to form two independent VS2C detectors, and their total thickness can be less than a typical single-color SLS detector. Under this approach, half of the array detects two colors in a time-multiplexed fashion while the other half detects the other two colors. In the following, a detector is designed with peak wavelength at 4.0 μm in band 1, 8.0 μm in band 2, 10.5 μm in band 3, and 11.5 μm in band 4. While one VS2C detector (Det. A) detects band 1 and band 3, the other detector (Det. B) detects band 2 and band 4. The adopted materials should have the α spectra shown in Fig. 6.16(a). Band 1 material has the shortest cutoff λ_c at 5.2 μm, and band 2 material has a λ_c at 9.4 μm. Band 3 and band 4 have the material with the same λ_c at 11 μm. The overall material structure is shown in Fig. 6.17, in which Det. B is fabricated on top of Det. A. The diffractive elements on each detector are shown in Fig. 6.18. In the detector fabrication, half of the array area is masked off while the other half is processed. The Det. B material layers are etched away when Det. A is fabricated. The two levels of ground contacts are connected at the external readout circuit.

To yield narrowband detection, the ring structure is replaced by a hexagonal array of circular posts. For Det. A, the hexagonal array is designed to excite traveling surface plasmon polariton (SPP) in the upper stack (band 3) while diffracting light in the lower stack (band 1). The grating period p is first set to be 3.0 μm, which is designed to coincide with the SPP wavelength $\lambda_{pl} = \lambda/(\varepsilon_m\varepsilon_d/\varepsilon_m + \varepsilon_d)^{1/2}$ at $\lambda = 10.5\,\mu$m, where $\varepsilon_m(\lambda)$ and $\varepsilon_d = 12.59$ are the permittivities of gold and the detector, respectively. Because of the short decay length of the plasmonic field along z, the upper stack thickness is 0.1 μm. The barrier (0.2 μm) and substrate (0.5 μm) thicknesses are typical for SLS detectors. The detector parameters are then optimized by adjusting the post period, diameter, height, and the band 1 thickness to produce large QE simultaneously at

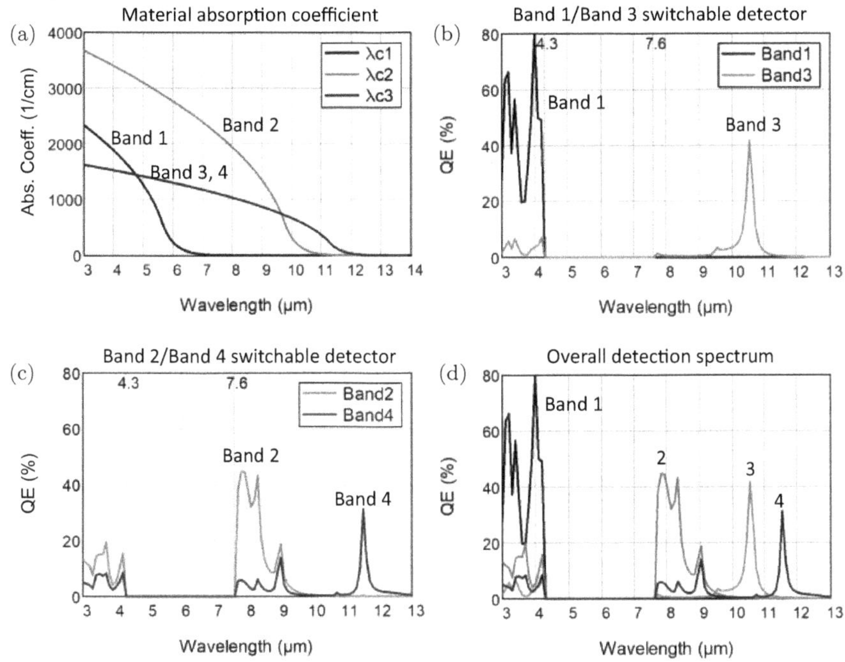

Figure 6.16. (a) The adopted material absorption coefficient spectra. (b) The detection spectrum of the VS2C detector A. (c) The detection spectrum of the VS2C detector B. (d) The overall detection spectrum. An integrated standalone band-stop filter between 4.3 and 7.6 μm has been assumed to block off the absorption peaks in this spectral range.

4.0 μm in band 1 and 10.5 μm in band 3. While p defines the peak wavelengths, the rest of the parameters control their magnitudes. The optimized p is 3.6 μm, shown in Fig. 6.18(a), rather than the initial 3.0 μm. This larger p is necessary to reduce the decay length of E in z to within the band 3 thickness so that spectral crosstalk can be minimized. The modeled QEs in bands 1 and 3 are shown in Fig. 6.16(b), in which distinct detection spectra are achieved. In this spectrum, a standalone band-stop filter is needed to block off the unwanted absorption peaks between 4.3 and 7.6 μm. In addition, the onset of substrate absorption at $\lambda < 3.2\,\mu$m, which is neglected in the modeling, will suppress the shorter wavelength QE in band 1.

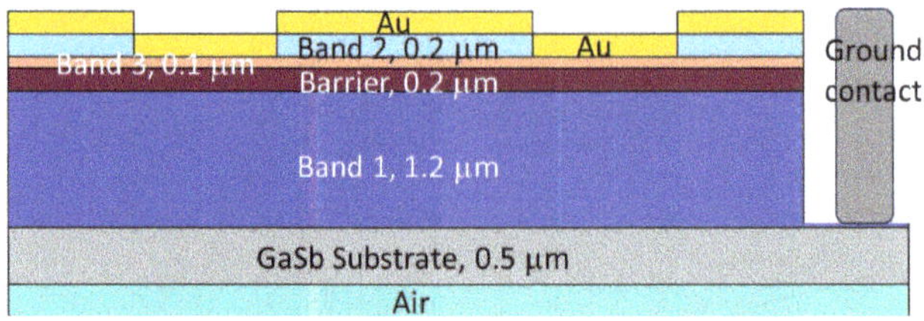

(a) Det. A detecting in band 1 and 3

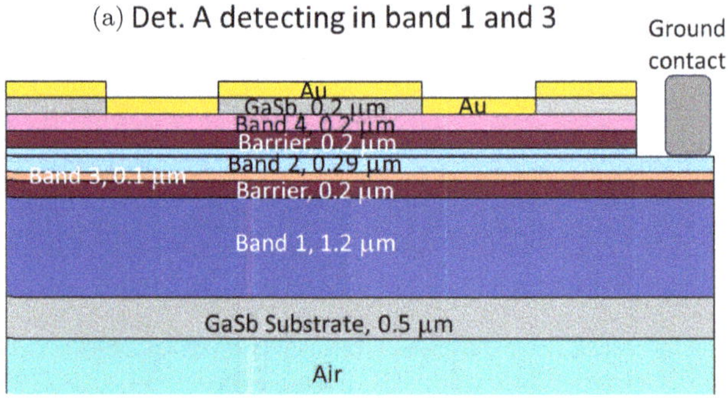

(b) Det. B detecting in band 2 and 4

Figure 6.17. (a) Det. A uses the bottom two SLS layers to detect band 1 and band 3. (b) Det. B uses the top two SLS layers to detect band 2 and band 4.

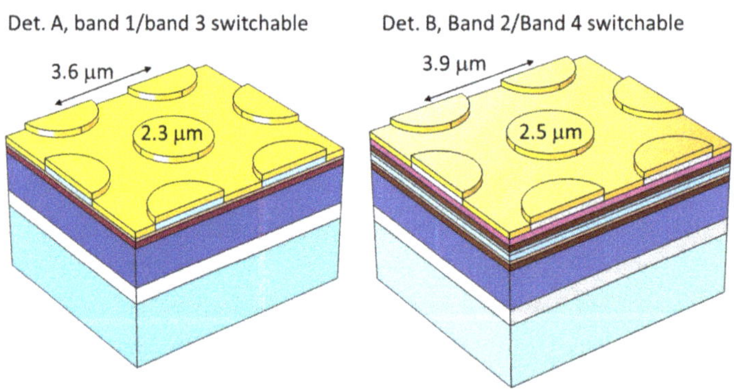

(a) (b)

Figure 6.18. The diffraction elements on top of (a) Det. A and (b) Det. B.

Similarly, Det. B is optimized by adjusting only band 2 thickness and the grating dimensions. The optimized parameters are shown in Fig. 6.17(b) and Fig. 6.18(b). Since the optimized thickness for band 2 is 0.29 μm, which is thicker than the 0.2 μm used for the diffractive elements for Det. A, an extra 0.09 μm layer is etched away when fabricating the A array. The resultant detection spectrum of Det. B is shown in Fig. 6.16(c). Again, the desirable peaks at 8.0 and 11.5 μm are obtained. There is spectral crosstalk in these two bands, which could be reduced by signal processing because of the high degree of spectral correlation when the voltage is switched. In fabricating the B array, the pixel mesas are etched partly into the band 2 layer as noted before so that both band 2 and band 3 layers are used as the common ground contact layer for the B array as shown in Fig. 6.17(b). With this layer arrangement, the combined thickness of all the layers including the substrate is 2.89 μm, which is thinner than the 3.35 μm for the single-color detector shown in Fig. 6.10(b).

6.3.5 *Resonator Pixel-Strained Layer Superlattice Detector Experiment*

In the following, we show a preliminary experiment on an RP-SLS pBn detector. The material structure is shown in Fig. 6.19(a). The absorber is a 0.8 μm unintentionally doped InAs/$InAs_{0.53}Sb_{0.47}$ SLS

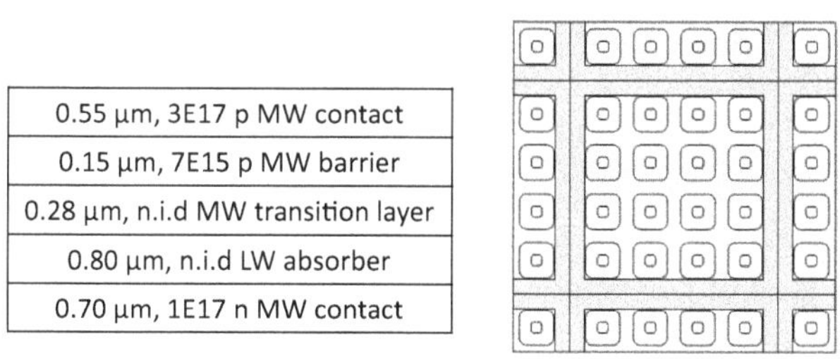

(a) (b) 25 µm pitch

Figure 6.19. (a) The InAs/InAsSb SLS material structure. (b) The resonator-pixel geometry with inner ring dimension of 1.3 μm, outer ring dimension of 4.1 μm, ring spacing of 5.7 μm, and ring height of 0.55 μm. The GaSb substrate is completely removed. No ARC is applied. Light is incident in the substrate side.

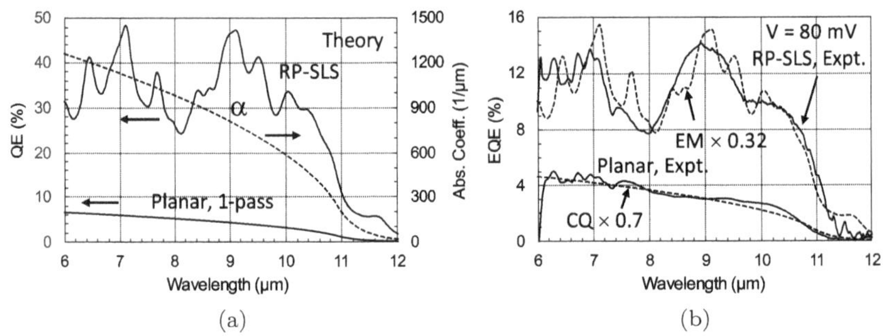

Figure 6.20. (a) The modeled RP-SLS and planar η_{abs} from air without ARC based on the dashed α curve. (b) The experimental and fitted EQE with different charge collection efficiencies.

with λ_c near $11\,\mu$m. The pixel geometry is shown in Fig. 6.19(b) with pixel pitch of $25\,\mu$m. To avoid surface current found in this material, there is no mesa etch except forming the ring structure. The fabricated detector thus differs from the usual design.

The RP QE without sidewalls and substrate is modeled using the truncated pixel geometry shown in Fig. 6.19(b). The adopted α is shown in Fig. 6.20. Along with the resonant detector, planar detectors are also prepared for top illumination. Unlike the RP detectors, the thick substrate of the planar detectors is not removed. Assuming the light is totally absorbed in the thick GaSb substrate and there is no stray light reflected from the substrate sidewalls, there is only one pass of light in the absorber for the planar detector. The resonant and the planar detector absorption QEs are modeled by FEM and ray tracing, respectively. The predicted values are shown in Fig. 6.20(a).

Although the RP detectors are without sidewalls, a large tunneling current is still observed in the fabricated detectors. Therefore, only a small voltage up to 80 mV can be applied to the top contact of the RP detectors. The EQE spectra at different voltages are plotted in Fig. 6.21(a), and the voltage dependence at $\lambda = 9.0\,\mu$m is plotted in Fig. 6.21(b). The largest observed EQE is 14.0%. EQE increases with V because the diffusion length depends on the diffusion coefficient $D = \mu kT/e$ through the Einstein relation and the mobility

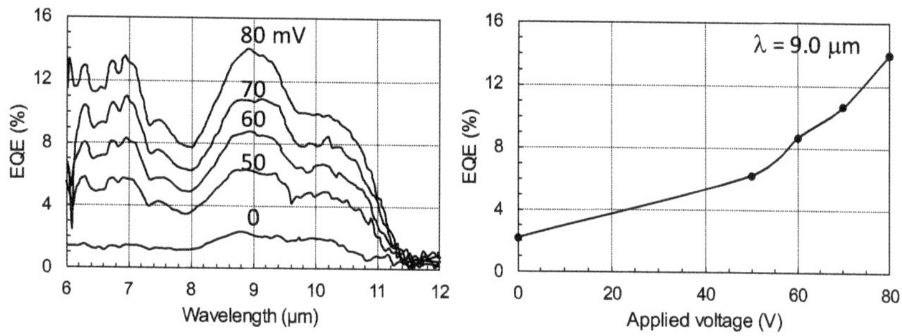

(a) (b)

Figure 6.21. (a) The EQE of the RP-SLS pBn detector measured at different top contact bias. (b) The voltage dependence at $\lambda = 9\,\mu$m.

μ increases with V because of the more conductive superlattice at higher V.

To fit the experimental RP EQE spectrum at 80 mV, the modeled QE is multiplied by the charge collection efficiency $f_c = 0.32$ defined in Eq. (6.3). With this fitting parameter, the modeled spectrum agrees well with the measured spectrum in Fig. 6.20(b). For the planar detector, the measured spectrum agrees with that using the classical expression of Eq. (6.2). The fitted f_c for the planar detector is 0.70 at the same bias. However, with a lower dark current observed in the planar detector, it can be measured up to 300 mV before the detector breaks down. At 300 mV, the EQE spectrum uniformly increases by 2.3 times. The apparent f_c at this V would then be 1.61. The larger-than-unity f_c could indicate additional absorption from stray light in the planar geometry having a thick substrate. The theoretical 1-pass QE thus underestimates the measured QE at 80 mV and needs a larger f_c to fit the observed EQE. The actual f_c would be smaller. If the RP detector follows the same V dependence, its EQE at 300 mV would be 32.2%, which is closer to the 47% η_{abs} prediction with a f_c of 0.69. Although the present RP detector does not demonstrate the predicted QE due to biasing limitation, it is still 4.7 times larger than that of the planar detector at the same bias. Furthermore, the modeled lineshape is very close to the observation and it is different from the planar lineshape, indicating that the detector is operating according to the design. With more research and development, the RP technology should be useful for SLS detectors.

6.4 HgCdTe Resonator-Pixel Two-Color Detector

The RP geometry can also be applied to HgCdTe infrared detectors. The thin layer RP structure is particularly useful to reduce the material thickness of a two-color detector. In a typical back-to-back *n*-*p*-*n* photodiode design [7], the *n*-type MW absorbing layer below the LW layer needs to be thick to absorb the MW radiation in one pass. It is to prevent the MW radiation from reaching the LW layer and causing spectral crosstalk. The LW absorber at the top should also be thick enough to yield high QE near the cutoff where α is small. Therefore, a typical two-color detector can be twice as thick as a single-color detector. This large thickness prevents the reduction of the pixel size for high-density arrays. The RP structure should be able to reduce the pixel thickness and size.

For example, a typical NPN detector may contain a $5\,\mu$m thick MW absorber and another $5\,\mu$m thick LW absorber. Combined with an assumed $0.25\,\mu$m thick *p*-type wide bandgap layer between the two *n*-type absorbers, the total thickness is $10.25\,\mu$m. In Fig. 6.22(a), the height-to-width aspect ratio of a detector pixel with $12\,\mu$m pitch and $11\,\mu$m size is shown if the mesa etching is perfectly vertical. A more isotropic etching will reduce the LW pixel fill factor substantially, and it will be difficult to reduce the pixel size further. On the other hand, with the RP detector geometry, the absorber will be much thinner. One example is shown in Fig. 6.22(b) and (c). It contains a $2.15\,\mu$m thick MW absorber and a $0.1\,\mu$m thick LW absorber. The total layer thickness including the $0.25\,\mu$m wide bandgap layer is $2.5\,\mu$m. This thickness is 4 times thinner than a typical detector. With the thinner material, the detector pixels are easier to produce with high fill factors and at high density.

For $12\,\mu$m pixel pitch, each two-color R-HgCdTe detector contains 4 rings. Using CdTe as the ring material, the inner ring size is $1.3\,\mu$m and outer ring size is $3.3\,\mu$m. The spacing is $5.1\,\mu$m and the height

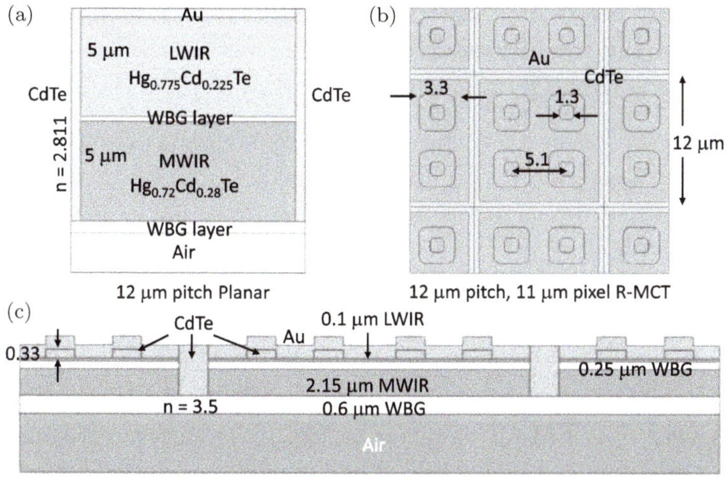

Figure 6.22. (a) The schematic layer structure of a typical n-p-n back-to-back two-color HgCdTe photodiode. (b) The top view of a RP-HgCdTe two-color detector. (c) The RP-HgCdTe cross-section along the centers of the rings.

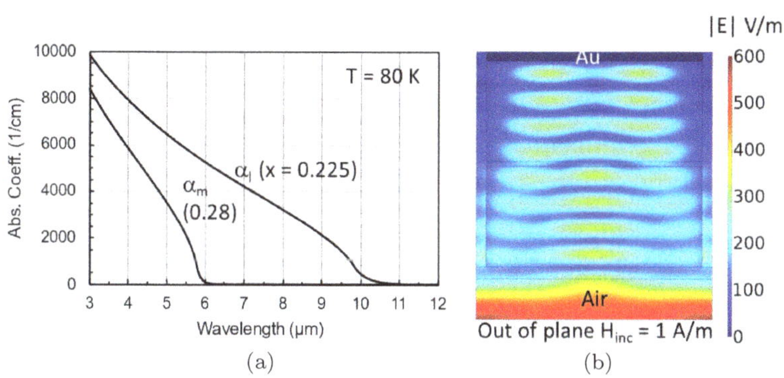

Figure 6.23. (a) The values of α_m and α_l of the two-color detector used in QE modeling. The refractive indices of all semiconductors are assumed to be 3.5. (b) The modeled $|E|$ profile of the planar detector at $\lambda = 9\,\mu$m.

is $0.33\,\mu$m. The bottom wide bandgap common layer is $0.6\,\mu$m thick. ARC is not employed. The R-HgCdTe detector is designed for LW HgCdTe with Cd mole fraction $x = 0.225$ and MW with $x = 0.28$. With these values of x, the absorption coefficients can be calculated and are shown in Fig. 6.23(a). With these α values, the planar detector QE based on the structure shown in Fig. 6.22(a) can be modeled by ray tracing and 2D FEM.

The ray tracing QE is given by

$$\eta_{LW} = \frac{4n}{(1+n)^2} e^{-\alpha_m t_m}(1 - e^{2\alpha_l t_l}), \tag{6.5}$$

and

$$\eta_{MW} = \frac{4n}{(1+n)^2}(1 - e^{-\alpha_m t_m})[1 + e^{-(\alpha_m t_m + 2\alpha_l t_l)}], \tag{6.6}$$

where $n = 3.52$, and t_m and t_l are the MW and LW thicknesses, respectively. The EM QE is given by Eq. (6.1) with $p = 12\,\mu$m. The results are shown in Fig. 6.24(a). The QEs for both colors agree using the two models. The EM QE takes phase coherence into account, thus producing the characteristic Fabry-Perot oscillations. The field profile at $\lambda = 9\,\mu$m is shown in Fig. 6.23(b). Due to diffraction effects in the CdTe trenches, the wavefront is not exactly planar. This allows the EM QE to be equal to the classical QE without trenches (*i.e.*, fill factor is 100%). Due to the thick absorbing layers, the conventional detector achieves large QEs in both colors and they are limited only by the substrate transmission. The spectral crosstalk is also relatively small. This type of QE spectrum is commonly observed [6].

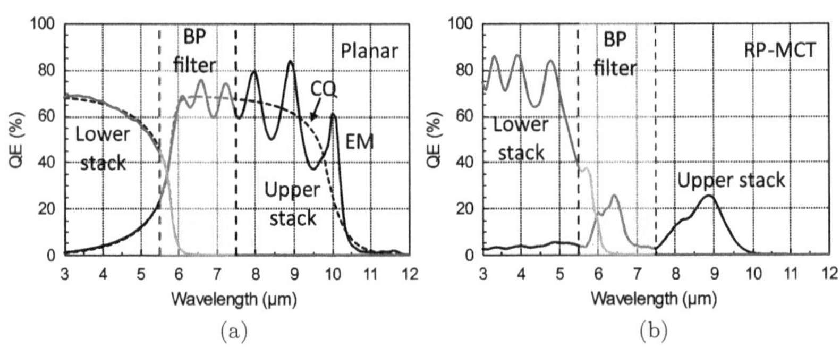

Figure 6.24. (a) The calculated QE using both classical ray tracing techniques (dashed curves) and FEM (solid curves) of $5\,\mu$m thick absorbers. (b) The modeled EM QE of the RP-HgCdTe detector having $0.1\,\mu$m LW and $2.15\,\mu$m MW absorbers.

The present RP-HgCdTe detector is optimized to reduce the spectral crosstalk present in the LW band. Therefore, the MW absorber is designed to be much thicker than the LW absorber. The thicker MW layer absorbs more MW radiation before it passes into the LW layer. The extremely thin LW layer absorbs little light unless the field is strongly enhanced at a particular λ. Obviously, the absorber thicknesses can be adjusted to suit an application QE and crosstalk priorities. Following the similar field patterns in Fig. 6.14, the E field is highly concentrated at the surface when the ring size is subwavelength where localized surface plasmons (LSPs) are exited. The LSPs are used to couple light into the LW absorbing layer.

With the same material α and the displayed detector structure in Fig. 6.22, the modeled QE spectrum of the RP detector is shown in Fig. 6.24(b). Due to diffraction and light-trapping effects, the MW QE of the RP detector is larger than that of the planar detector despite having a thinner layer by a factor of 2.3. Assuming a bandpass filter is employed in both the planar and resonant detectors to block off the radiation between 5.5 and 7.5 μm, the thinner MW RP detector exhibits an integrated QE (IQE) of 1.85 μm between 3 and 5.5 μm while the thicker planar absorber has an IQE of 1.57. Assuming I_p is proportional to IQE and I_d is proportional to absorber thickness, the I_p/I_d ratio of the MW RP detector will be 2.75 times larger.

For the LW band, due to the absorber layer being 50 times thinner, the RP QE is greatly reduced in Fig. 6.24(b). Even with the LSP resonances, QE is only narrowly peaked at two wavelengths, 6.4 and 8.9 μm, and the maximum QE is only 25.3%. With the passband filter, only the 8.9 μm peak is used for detection. On the other hand, the highly confined absorption in the thin layer creates a highly concentrated photocarrier density in comparison to the thermally excited carrier density in the same region, resulting in a large I_p/I_d ratio. The IQE of the RP detector is 0.33 μm while that of the planar detector is 1.57 μm in the 7.5–12 μm band. Despite a lower IQE by a factor of 4.80, the dark current is 50 times lower, resulting in an I_p/I_d ratio 10.4 times higher. The higher LW I_p/I_d will help to improve

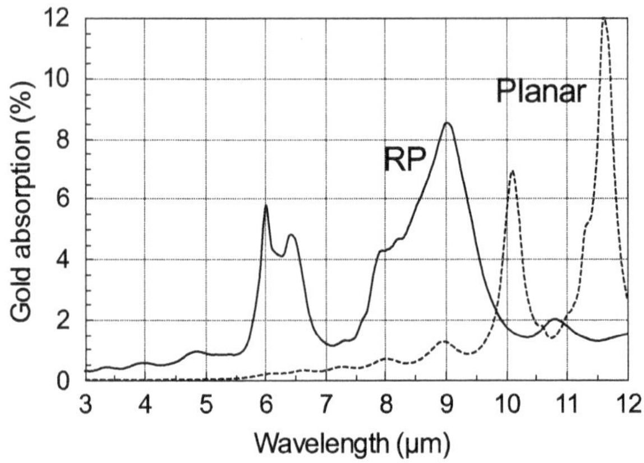

Figure 6.25. The infrared absorption in the gold layer for the planar detector with finite-size gold contacts and for RP detector geometry.

the operating temperature of the detector. With potential to further reduce the pixel pitch to 6 μm or less, the RP technology can be beneficial to HgCdTe two-color detection.

Because of the excitation of LSPs in RP detectors, the infrared absorption in the gold layer is also increased at the resonant wavelengths, similar to that of QWIPs in Fig. 5.47. The gold absorption QE η_{gold} is shown in Fig. 6.25. For the RP detector, the strong LSP response in the gold layer gives the large RP QE in the LWIR wavelengths. At the same time, a large 8.6% of the incident light is also dissipated in the gold layer at 9 μm as shown in Fig. 6.25, which is about a third of the RP QE. For the planar detector, η_{gold} is small, less than 2% below 9.5 μm. However, the 11 μm wide gold contacts in a 12 μm periodic array in Fig. 6.22(a) induce strong plasmonic responses at 10.1 and 11.6 μm and cause large gold absorption in Fig. 6.25 at these two wavelengths. Therefore, the plasmonic resonance of gold also operates in planar structures inconspicuously.

6.5 Resonator-Pixel Emitters

The above examples show that the RP technology is generally useful in improving optical detection. According to Kirchhoff's law or antenna reciprocity theorem, a good absorber is also a good emitter. The excellent absorbing ability of an RP should also make it an efficient optical emitter. In the light-emitting diode (LED) industry, it is well-known that the light generated inside an LED is mostly trapped by total internal reflection [7]. Due to the random direction of emission, only ~2% of light within a small half-cone angle of the critical angle θ_c for total internal reflection can emit into the air. Up to 98% of electrical power is wasted as heat. Since the RP geometry can serve as an excellent optical antenna in receiving light under normal incidence as discussed in Section 5.8.4, the same structure will also be able to reverse the light propagation and achieve efficient normal exitance. In this section, we will discuss using the RP geometry to obtain high-power and high-directivity optical emission from optical and thermal emitters in general.

The EQE of LEDs and interband cascade devices depends on several factors. They are the current injection efficiency (IE), the internal emission quantum efficiency (IQE), and the light extraction efficiency (LEE). IE is mainly an electrical property of the emitter. It arises from the fact that the electrical current in a LED contains several components, which are the bulk electron and hole diffusion currents, the bulk defect-assisted recombination current, and the surface leakage current. Only part of the diffusion currents participate in the carrier recombination process. IE is the portion of the total current that participates in the carrier recombination in the active layer. However, not all the carrier recombinations result in light emission. Some of the carriers recombine through Auger and defect-assisted processes and they are non-radiative. IQE is the fraction of radiative recombination among all the recombination processes, and it is mainly an optical property of the emitter material. IQE varies widely from less than 1% to more than 80% [8], depending on the applied voltage, material band structure, and emission wavelength. LEE, on the other hand, is mainly a geometrical property of the emitter. As mentioned above, LEE is determined by the escape cone angle of light from the material to the air. Its value is relatively fixed at 2% for the typical semiconductor refractive index [7]. The small IQE, particularly in the indirect-gap and mid-infrared materials, and the small LEE in the planar geometry leads to a very small EQE in typical LEDs.

There are many ways in raising IQE and LEE of LEDs. For example, the radiative recombination rate can be increased by increasing the optical electric field in the active medium because of the increased $\mathbf{A}\cdot\boldsymbol{p}$ dipole transition matrix element, where $\mathbf{A}$ is vector potential operator and $\boldsymbol{p}$ is the momentum operator. To increase the optical electric field, one approach is to use SPPs or LSPs created by metal structures [9–16]. For example, Okamoto *et al.* [12] utilized the large field of SPPs generated on the interface between a planar metal layer and an InGaN LED layer and observed 6.8 times larger IQE. Fujiki *et al.* [13] used LSPs in metal nanoparticles to increase the field and achieved 20-fold increase in the molecular fluorescence of organic LEDs. LSP has an advantage over SPP because the resonant wavelength of LSP can be tuned to the material emission wavelength by changing their size, geometry, and spacing. Bakker *et al.* [16] further utilized nanoparticles to increase the IQE of a dye and its LEE by designing the nanoparticles into transmitting nanoantennae. The fluorescence enhancement in this structure can reach 100 times.

In the plasmonic structures, there is coherent cycling of energy between the light-emitting medium and the plasmonic mode such that the light emission rate of the medium into the plasmonic mode is increased, similar to a cavity quantum electrodynamic (cavity QED) system [17]. In this system, the decay rate of an excited atom is enhanced in the presence of a cavity by the Purcell factor in the weak coupling regime. In this analogy, the conduction and valence bands of the emitting medium correspond to the two atomic levels, and the plasmonic mode corresponds to the optical cavity mode. In addition to the increased emission rate, the graininess of the metal film and the geometry of the nanoparticles transduce the evanescent, localized plasmon energy into free propagating optical energy, thereby increasing the EQE.

Although SPP and LSP are able to improve the EQE of LEDs in the visible wavelengths, the improvement is limited by the energy loss in metals [18]. In surface plasmons, part of the optical energy is stored in the EM wave in the LED medium and part of it as the kinetic energy of the moving electrons in the plasma oscillations in metals. Since electrons are subjected to scattering, they lose their kinetic energy to ohmic heating. Therefore, the optical energy emitted by the LED is substantially dissipated as heat in metal, resulting in small IQE enhancement. Furthermore, there is no efficient mechanism to extract the LSP energy into free space. The absolute LEE remains low.

On the other hand, metals such as gold have much smaller ohmic resistance and behave more like a perfect electrical conductor in the infrared wavelengths. This property yields a very long SPP propagating length in the planar surface as seen in Fig. 3.6. Even when strong LSP oscillations are excited at resonances, the ohmic loss is less than 20% for QWIPs in Fig. 5.47 and 8% for HgCdTe detectors in Fig. 6.25. Therefore, the plasmonic approach for IQE enhancement is much more viable in the infrared.

For LEE improvement, we have seen the RP geometry acting as an efficient receiving antenna in collecting normal incident radiation falling onto the pixel pitch area and driving the LSP oscillations in the much smaller inner ring area. The RP geometry thus behaves as an optical antenna in focusing the optical energy into a tight spot. According to antenna reciprocity theorem, the receiving and transmitting properties of an antenna are identical. An antenna has the same directivity in receiving and transmitting radiation power. Therefore, the RP geometry will be able to transmit the optical energy generated in the inner ring efficiently out to the air in the unique normal direction. In this section, we perform EM modeling and design optimization to show that a RP-LED operated under LSP resonant condition will have both large IQE and LEE to achieve a large EQE. Similar to the detector applications, the RP approach is applicable to any infrared-emitting materials with intrinsic isotropic or anisotropic emission properties.

In addition, this technology is not only applicable to active materials but also to passive materials. According to Kirchhoff's law under thermodynamic equilibrium, the emissivity of an object is equal to its absorptivity, which is another terminology for the EQE.

Therefore, the large absorptivity of an RP absorber under resonance will also give it the same large emissivity at that wavelength. This feature can be utilized to produce an efficient, spectrally selective, and spatially directional thermal emitter at an elevated temperature. This emitter structure will be simpler to fabricate than that using 3D photonic crystals on the material emitting surface [19].

Other than increasing the optical throughput and efficiency, the RP emitter uses a very thin active material to reduce reabsorption of light and does not require additional ARC or distributed Bragg reflectors to improve emissivity. It emits at the surface rather than at the edge, and the structured metal surface enables effective thermal management. The fine structures protected behind the emitting surface are also free from air corrosion and environmental contamination. Therefore, the RP geometry has many advantages.

6.5.1 *AlInAs/InAsSb Multiple Quantum Well Emitter for Mid-Infrared Wavelength Application*

The first example is a mid-infrared emitter made of AlInAs/InAsSb multiple quantum wells (MQW) [20]. The present RP emitter is designed to emit at around 4.1 μm. It contains an undoped light-emitting MQW layer sandwiched between a top contact layer and a ground contact layer. The top contact layer is etched into a periodic ring array. A metal layer is then deposited onto the rings to form the structured surface. Optically thin etch-stop layers may be grown at appropriate interfaces to facilitate device processing for ring formation and substrate removal. The contact layers are wider bandgap materials with p-doping ($\sim 2 \times 10^{18}\,\mathrm{cm}^{-3}$) on top and n-doping ($\sim 1 \times 10^{18}\,\mathrm{cm}^{-3}$) at the bottom. The metal contacts are Ti/Au for p-type materials and Ge/Au for n-type materials. Since the top contact layer is essentially removed in etching the ring structure, Schottky contact instead of ohmic contact is formed outside the ring region. The current injection to the light-emitting layer is therefore primarily through the p-type top ring contacts. The pixel sidewalls are encapsulated with a low refractive index material such as epoxy glue. Electrical contacts to the external power supply are made at the

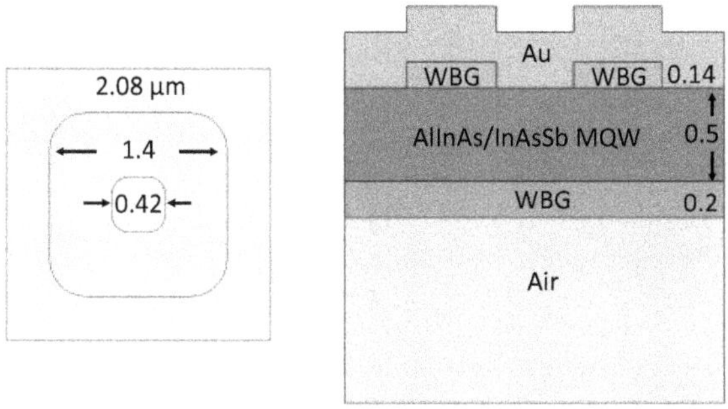

Figure 6.26. The top view and cross-section of a RP emitter unit cell for 4.08 μm emission.

top gold layer and the ground contact layer surrounding the emitter array. Light exits through the ground contact layer into the air. The RP emitter unit cell geometry is shown in Fig. 6.26.

According to the antenna reciprocity theorem, a good transmitting antenna is also a good receiving antenna. Therefore, one can first design a good detector structure and then replace the absorbing material with an emitting material to achieve a good emitter. Assuming $n = 3.735$ for the MQW material and 3.5 for the wide bandgap contact materials, the internal emission wavelength $\lambda_{\text{int}} = \lambda/n = 4.1/3.735 \approx 1.1\,\mu$m. The initial setting of the inner dimension of the ring is then equal to $\lambda_{\text{int}}/2 \approx 0.5\,\mu$m and the outer dimension of the ring is $3\lambda_{\text{int}}/2 \approx 1.5\,\mu$m as in a regular ring design. The ring height is $\lambda_{\text{int}}/4 = 0.27\,\mu$m to eliminate zeroth-order diffraction (*i.e.*, back reflection), and the ring period is $2.5\lambda_{\text{int}} \approx 2.75\,\mu$m. With this small ring period, a typical emitter will contain many unit cells. Periodic boundary condition can be applied to a single unit cell to obtain the valid EM solution. The geometry in Fig. 6.26 is then used for design optimization. If a small-area emitter is employed, the entire structure can be modeled instead.

Based on Eq. (6.1) and with a constant α of 2000 cm^{-1}, η at the designed λ of 4.1 μm is maximized by changing structural parameters. The optimized detector structural parameters are shown in Fig. 6.26, and the modeled η spectrum in Fig. 6.27(b). This RP detector has a QE of 76.3% at 4.08 μm compared to the 12.5% QE for the

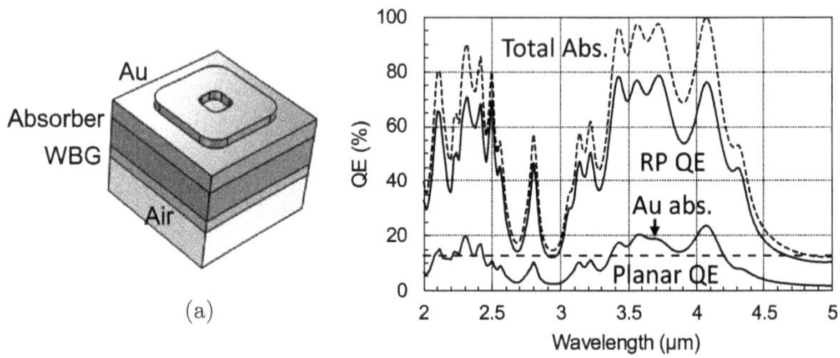

(a) (b)

Figure 6.27. (a) The 3D geometry of the RP detector. (b) The modeled QE with the optimized detector parameters for constant $\alpha = 2000\,\text{cm}^{-1}$. Adapted with permission from *Opt. Express*, 30(2), pp. 780–796 (2022).

planar structure based on ray tracing. The modeling also shows that the gold layer absorbs 23.7% of incident light at the same λ. The maximum total absorption from both layers therefore adds up to 100%. Near 100% absorption is also observed between 3.9–4.2 μm. This large absorption indicates that the incident light transmits almost entirely into the detector at these wavelengths even with an uncoated substrate.

To use the same RP geometry for light emission, we first characterize its properties as a passive resonator, in which the absorber is replaced by a passive dielectric layer (PDL) of the same $n = 3.735$ but with α set to zero. This allows the intrinsic resonant characteristics of the cavity to be studied. Without infrared absorption in the detector volume, the only power attrition is the ohmic loss in the gold layer. Under incident $\mathbf{H} = (1, 1, 0)$ A/m, the energy U_s stored in the PDL is given by

$$U_s = \frac{1}{2}\varepsilon_r\varepsilon_0 \int_{V_r} |\boldsymbol{E}(r)|^2 d^3r, \tag{6.7}$$

where $\varepsilon_r = n^2$, $\boldsymbol{E}(r)$ is the modeled $\boldsymbol{E}$ field under the incident $\mathbf{H}$, and V_r is the PDL volume. Explicit modeling of the energy stored in the gold cover is 4 orders of magnitude lower than that in the PDL, and therefore it can be neglected. The value of U_s indicates the effectiveness of the resonator to hold the incident photons. It can be compared with the energy in the same PDL volume if there is no optical confinement. In the absence of the RP geometry, the transmitted E field into the semi-infinite medium is $E_t = 2E_{\text{inc}}/(1+n)$, where the incident $|\boldsymbol{E}_{\text{inc}}| = Z_0|\mathbf{H}|$, and Z_0 is the vacuum impedance. The ratio of number of photons in the PDL with and without the resonator is then given by

$$\frac{N_r}{N_0} = \frac{\frac{1}{2}\varepsilon_r\varepsilon_0 \int_{V_r} |\boldsymbol{E}(r)|^2 d^3r}{\frac{1}{2}\varepsilon_r\varepsilon_0 E_t^2 V_r} = \frac{\int_{V_r} |\boldsymbol{E}(r)|^2 d^3r}{\frac{4}{(1+n)^2} E_{\text{inc}}^2 V_r}. \tag{6.8}$$

In the steady state, the incident power P_{inc} in air is balanced by the power reflected at the substrate surface P_{refl}, the power transmitted into the resonator but subsequently returned to the air P_{out}, and the power transmitted into the resonator and dissipated in the gold layer P_{abs}. P_{abs} is obtained from

$$P_{\text{abs}} = \int_{V_{Au}} \alpha_{Au} I(r) d^3r = \frac{2\pi\varepsilon_0 c n_{Au}\kappa_{Au}}{\lambda} \int_{V_{Au}} |\boldsymbol{E}(r)|^2 d^3r, \tag{6.9}$$

where (n_{Au}, κ_{Au}) is the complex refractive index of gold shown in Fig. 3.14(b), and V_{Au} is the volume of the gold cover. The total power loss in the resonator P_{loss} is given by $P_{\text{loss}} = P_{\text{out}} + P_{\text{abs}}$. In balance, P_{loss} is equal to the net driving power $= P_{\text{inc}} - P_{\text{refl}}$, and thus it is less than P_{inc}. Since P_{refl} is very small at the resonant wavelengths as indicated in Fig. 6.27(b), $P_{\text{loss}} \approx P_{\text{inc}}$ at resonance. At other wavelengths, $P_{\text{loss}} < P_{\text{inc}}$, depending on the substrate reflection. The quality factor Q of a resonator is generally defined as $Q \equiv \omega U/R$, where ω is the angular frequency, U is the stored energy, and R is the energy loss rate. By counting the energy stored in the PDL and not in the contact layers, Q of the RP is $Q = \omega U_s/P_{\text{loss}}$, which is larger than $\omega U_s/P_{\text{inc}}$ in general. At the resonant wavelengths, $P_{\text{loss}} \approx P_{\text{inc}}$, and $Q \approx \omega U_s/P_{\text{inc}}$. P_{loss} is difficult to obtain unambiguously because P_{out} and P_{refl} travel in the same backward direction and are difficult to separate in the total reflected power. At the wavelengths of interest,

$$Q \equiv \frac{2\pi c}{\lambda}\frac{U_s}{P_{\text{loss}}} \approx \frac{\pi c n^2 \varepsilon_0}{\lambda}\frac{\int_{V_r} |\boldsymbol{E}(r)|^2 d^3r}{P_{\text{inc}}}. \tag{6.10}$$

Since the optical intensity in a resonator with a given Q factor decays as $\exp(-2\kappa t)$ when the driving power is off, where $\kappa = \omega/(2Q)$ is the amplitude decay constant, a large Q will yield a long photon retention time for more light-matter interaction.

Another important parameter is the mode volume V_m. In a resonator, each mode possesses a specific field distribution. Some are more concentrated at certain locations. Others are more diffused as shown in Fig. 6.14. To characterize the physical size of a mode, it is useful to define an effective volume V_m, which is a volume containing all the mode energy if the field in this volume is the maximum field $E_{\max}$. A smaller V_m indicates more intense $E_{\max}$ and more localized light-matter interaction. V_m is modeled using

$$V_m = \frac{\int_{V_r} |\boldsymbol{E}(r)|^2 d^3r}{\max(|\boldsymbol{E}(r)|^2)} = \frac{\int_{V_r} |\boldsymbol{E}(r)|^2 d^3r}{E_{\max}^2} \left(\frac{n}{\lambda}\right)^3 \left(\frac{\lambda}{n}\right)^3 = v\left(\frac{\lambda}{n}\right)^3, \tag{6.11}$$

where v is V_m expressed in the unit of $(\lambda/n)^3$.

The effectiveness of the resonator in enhancing the optical transition is measured by the Purcell factor F_p, which is the increased carrier recombination rate in the mode volume V_m compared to that without the resonator. The Purcell factor is derived from cavity QED. The simplest approach is to study a two-level single electron system as in an atom coupled to a single optical mode in a cavity [17,21]. To better account for the non-uniform $\boldsymbol{E}$ field profile of a cavity mode, the interaction Hamiltonian $H_{\text{int}} = (e/m^*)(\mathbf{A} \cdot \mathbf{P})$ in Eq. (2.3), which is suitable for studying uniform field in a medium, is equivalently reformulated in terms of $\mu \cdot \boldsymbol{E}$, where $\mu = -e\boldsymbol{r}$, and e is the magnitude of an electron charge. The $\boldsymbol{E}$ field operator with spatially varying profile is [21]

$$\boldsymbol{E}(\boldsymbol{r}_E) = i\sqrt{\frac{\hbar\omega}{2\varepsilon_r\varepsilon_0 V_m}} \left(\frac{E(\boldsymbol{r}_E)}{E_{\max}}a - \frac{E^*(\boldsymbol{r}_E)}{E_{\max}}a^\dagger\right)\boldsymbol{u}, \tag{6.12}$$

where $\boldsymbol{r}_E$ is the emitter coordinate, $\hbar\omega$ is the cavity mode energy, a and a^+ are the photon annihilation and creation operators, and $\boldsymbol{u}$ is a unit vector pointing in the direction of $\boldsymbol{E}$. Expressing μ in terms of the electron raising operator $\sigma_+ = |e\rangle\langle g|$ and lowering operator $\sigma_- = |g\rangle\langle e|$, where $|g\rangle$ is the electron ground state and $|e\rangle$ is the excited state, and denoting $\mu_{eg} = \langle e|e\boldsymbol{r}|g\rangle$ and $\mu_{ge} = \langle g|e\boldsymbol{r}|e\rangle = \mu_{eg}^*$,

$$\mu = \boldsymbol{\mu}_{eg}\sigma_+ + \boldsymbol{\mu}_{eg}^*\sigma_-. \tag{6.13}$$

Multiplying Eqs. (6.12) and (6.13) and keeping only the energy-conserving terms,

$$H_{\text{int}} = -i\sqrt{\frac{\hbar\omega}{2\varepsilon_r\varepsilon_0 V_m}}\left[(\boldsymbol{\mu}_{eg}\cdot\boldsymbol{u})\frac{E(\boldsymbol{r}_E)}{E_{\max}}\sigma_+ a - (\boldsymbol{\mu}_{eg}^*\cdot\boldsymbol{u})\frac{E^*(\boldsymbol{r}_E)}{E_{\max}}\sigma_- a^\dagger\right]. \tag{6.14}$$

To simplify Eq. (6.14), the emitter-field coupling parameter $g(\boldsymbol{r}_E)$ is introduced, where

$$\begin{aligned} g(\boldsymbol{r}_E) &= \frac{1}{\hbar}\sqrt{\frac{\hbar\omega}{2\varepsilon_r\varepsilon_0 V_m}}\frac{E(\boldsymbol{r}_E)}{E_{\max}}(\boldsymbol{\mu}_{eg}\cdot\boldsymbol{u}) \\ &= \left(\frac{|\boldsymbol{\mu}_{eg}|}{\hbar}\sqrt{\frac{\hbar\omega}{2\varepsilon_r\varepsilon_0 V_m}}\right)\frac{E(\boldsymbol{r}_E)}{E_{\max}}\frac{(\boldsymbol{\mu}_{eg}\cdot\boldsymbol{u})}{|\boldsymbol{\mu}_{eg}|} \\ &= g_0\psi(\boldsymbol{r}_E)\cos(\xi), \end{aligned} \tag{6.15}$$

where g_0 represents the maximum light-matter coupling strength, $\psi(\boldsymbol{r}_E)$ expresses the position dependence of g, and $\cos(\xi)$ expresses the angular dependence between the directions of the electron dipole oscillation and the cavity electric field. The coupling is the strongest when the mode volume is small to increase g_0, the electronic two-level system is located at the cavity field maximum, and the electron dipole orientation is aligned with the optical field.

In terms of $g(\boldsymbol{r}_E)$, H_{int} is

$$H_{\text{int}} = i\hbar[g^*(\boldsymbol{r}_E)\sigma_- a^\dagger - g(\boldsymbol{r}_E)\sigma_+ a]. \tag{6.16}$$

Combining H_{int} with the Hamiltonians $H_E = (\hbar\nu/2)\sigma_z$ for the emitter and $H_C = \hbar\omega(a^+a + 1/2)$ for the cavity, where σ_z is one of the Pauli matrices, and $\hbar\nu$ is the emitter level spacing, the total Hamiltonian for the coupled emitter-cavity field system $H = H_E + H_C + H_{\text{int}}$, and it is called the Jaynes-Cummings Hamiltonian. Solving H for eigenenergies and eigenfunctions will yield new normal modes of the coupled system and they are known as the dressed states. The eigenstates without H_{int} are known as the bare states.

When H_{int} is turned on, the bare states evolve with time according to the time-dependent Schrödinger equation. If the initial state contains one excited electron and no photons, the excited electron will release one photon over time and then reabsorb it later. The cycle repeats at a frequency of $2|g(\boldsymbol{r}_E)| \equiv 2g$, and the process is known as the Rabi oscillation. When the cavity mode is tuned to the emitter level spacing, *i.e.*, $\nu = \omega$, the first two eigenlevels above the ground level are $|E_-\rangle = (|e, 0\rangle - |g, 1\rangle)/2$ with eigenfrequency $\omega_- = \omega - g$ and $|E_+\rangle = (|e, 0\rangle + |g, 1\rangle)/2$ with eigenfrequency $\omega_+ = \omega + g$.

When the decay of the cavity mode is considered, its angular frequency acquires a complex value $\omega - i\kappa/2$, where $\kappa = \omega/(2Q)$ is the amplitude decay constant. In this case [21],

$$\omega_\pm = \omega \pm \sqrt{g^2 - \frac{\kappa^2}{4}} - i\frac{\kappa}{2}. \tag{6.17}$$

If $g \gg \kappa$, the square root term is real and there are two distinct eigenfrequencies. Therefore, the system retains Rabi splitting along with broadened energy levels due to the finite lifetime of the cavity mode. This system is referred to be in the strongly coupled cavity QED regime.

On the other hand, when $g \ll \kappa/2$, the square root term in Eq. (6.17) is imaginary. The real parts of $\omega_\pm$ are then the same and there is no Rabi splitting. The coupled modes thus still retain their original bare mode identities. Instead, each of the eigenstates has a different imaginary frequency. Taking approximation up to the first-order term in the square root, the two solutions are $\omega_+ = \omega - ig^2/\kappa$ and $\omega_- = \omega - i\kappa$. This system is referred to be in the weakly coupled cavity QED regime, which is also known as the Purcell regime. The ω_- state is identified to be the cavity mode. Because of the presence of the resonant emitter, the energy decay rate of the cavity mode is now 2κ, twice that without the emitter. The photon storage time is thus shorter. At the same time, the ω_+ state is identified to be the emitter excited state. Because of the presence of the optical cavity, its energy decay rate is $2g^2/\kappa$. One can compare this modified rate with that without the cavity. The emission rate in a uniform medium with refractive index n is given by the Einstein A coefficient for spontaneous emission:

$$A = \frac{n\mu_{eg}^2\omega^3}{3\pi\varepsilon_0\hbar c^3}. \tag{6.18}$$

The ratio of the emission rates with and without the cavity in the same medium is the Purcell factor F_p. By substituting g_0 in Eq. (6.15) and $\kappa = \omega/(2Q)$, F_p is

$$F_p = \frac{2g_0^2/\kappa}{A} = \frac{3}{4\pi^2}\left(\frac{\lambda}{n}\right)^3\frac{Q}{V_m} = \frac{3}{4\pi^2}\frac{Q}{v}, \tag{6.19}$$

where v is defined in Eq. (6.11). By increasing Q and reducing v, the emission rate of the emitter can be increased.

Treating the present RP unit cell as an optical cavity, its F_p can be evaluated. By modeling $\boldsymbol{E}(r)$ in the passive resonator, the photon number ratio in Eq. (6.9) and Q factor in Eq. (6.10) are obtained and they are shown in Fig. 6.28. At the designed λ of 4.08 μm, the number ratio has a local maximum of 29.8 while that of Q is 57.1. Similarly, the modeled mode volume v in Eq. (6.11) and the Purcell factor F_p in Eq. (6.19) are shown in Fig. 6.29. The smallest $v = 3.44 \times 10^{-3}$ occurs at 4.08 μm, signaling the presence of very intense field at this λ. This very small v is a signature of the LSP mode. With a large Q and a small v, the Purcell factor F_p is large at 1261.

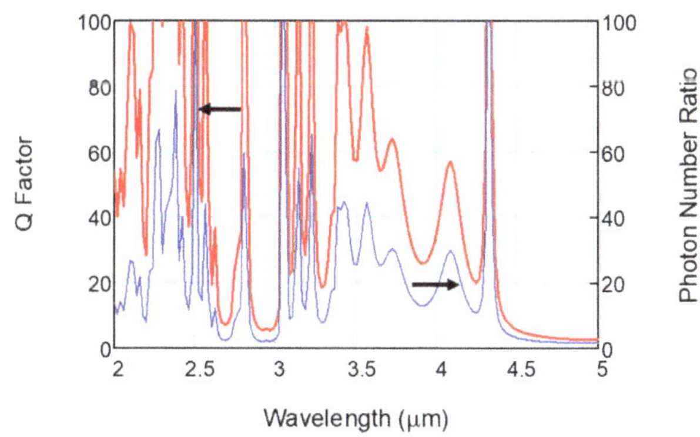

Figure 6.28. The Q factor of the resonator and the number of photons in the dielectric layer of the resonator relative to a semi-infinite block of the same medium. Reproduced with permission from *Opt. Express*, 30(2), pp. 780–796 (2022).

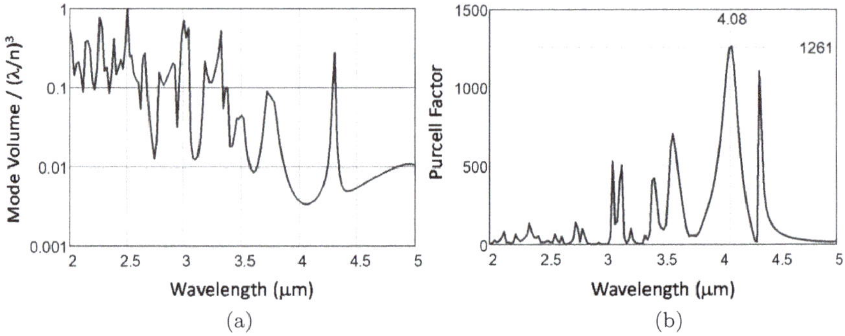

Figure 6.29. (a) The effective mode volume in the unit of $(\lambda/n)^3$. (b) The Purcell enhancement factor for spontaneous emission. Adapted with permission from *Opt. Express*, 30(2), pp. 780–796 (2022).

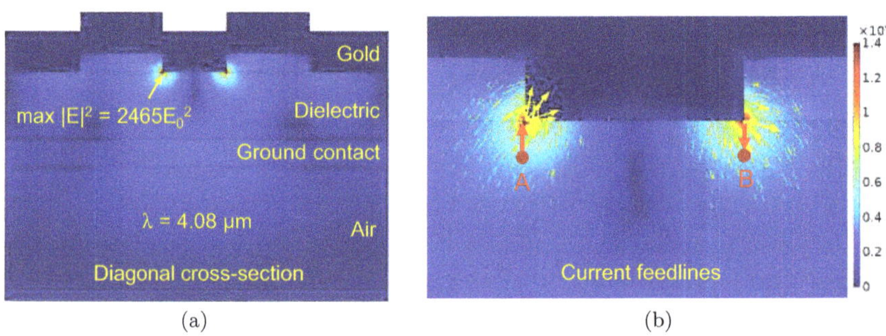

Figure 6.30. The diagonal cross-section of a RP emitter showing the optical $|E|$ when the incident $\mathbf{H} = (1, 1, 0)$ A/m or $|E_{\text{inc}}| = 532.8$ V/m. (b) The E directions near the diagonal corners. A and B are the end points of the current feedlines that are 0.1 μm in length. Adapted with permission from *Opt. Express*, 30(2), pp. 780–796 (2022).

Due to the 4-fold rotational symmetry of the present structure, Q and V_m are independent of the incident polarization angle θ in the x-y plane. For the present choice of incident $\mathbf{H} = (1, 1, 0)$ A/m, $\boldsymbol{E}$ is $(-Z_0, Z_0, 0)$ V/m and θ is 135°. With equal incident E_{0x} and E_{0y} components, the maximum $\boldsymbol{E}$ field is located along the diagonal cross-section of the ring. A plot of the $\boldsymbol{E}$ field in this cross-section in Fig. 6.30(a) shows that $|\boldsymbol{E}|^2$ near the corner of the gold ring can be 2465 times larger than the incident $|\boldsymbol{E}_0|^2$. Since the spontaneous emission rate of an emitter is directly proportional to F_p of the cavity mode, and if the PDL is replaced by a forward-biased LED material, the emission rate near the ring corner will increase by 1261 times as indicated in Fig. 6.29(b). Therefore, the light emission of the RP emitter will be highly concentrated in the corners at $\lambda = 4.08\,\mu\text{m}$ for this polarization. In other locations, the rate increase is much more moderate since $g(\boldsymbol{r}_E)^2$ in Eq. (6.15) is directly proportional to $|\boldsymbol{E}(\boldsymbol{r}_E)|^2$. The average increase of the recombination rate over the entire LED volume is the same as the average increase of photon density in the resonator, which is 29.8 times larger than that of a planar medium. Therefore, with the help of EM modeling, the RP structure is found to increase the carrier recombination rate by releasing its transition energy into the LSP mode in the cavity.

EM modeling can also be used to know the LEE of the RP structure. In QED, spontaneous optical emission is driven by virtual quantum fluctuation (*i.e.*, random occurrence) of free-space vacuum modes. Since the vacuum mode orientations in a homogeneous medium are random, the resulting emissions are isotropic, which causes the LEE in a large refractive index material to be small. On the other hand, in a cavity, the vacuum modes are discrete and they have unique transmission properties that determine the LEE of the structure at each wavelength. To know the transmission of the LSP mode into free space when it is created by optical transition, we need to generate the mode internally in the model.

When a virtual LSP mode appears, its $\boldsymbol{E}$ field drives the electrons in the conduction band to oscillate in a distributed manner according to the field pattern shown in Fig. 6.30(b). The accelerating electrons then release their energy into the LSP mode and fall into the valence band. Accordingly, the electrons primarily move up near one of the ring corners while moving down at the opposite corner. This movement of electrons can be approximated by two current feedlines oscillating in opposite phase as shown in Fig. 6.30(b). These feedlines terminate at point A and point B, which are the locations of the centroids of the opposite charge distribution across the high field region in Fig. 6.30(b). When the current in the feedlines varies at the LSP frequency ω, the equivalent charges at the centroids oscillate

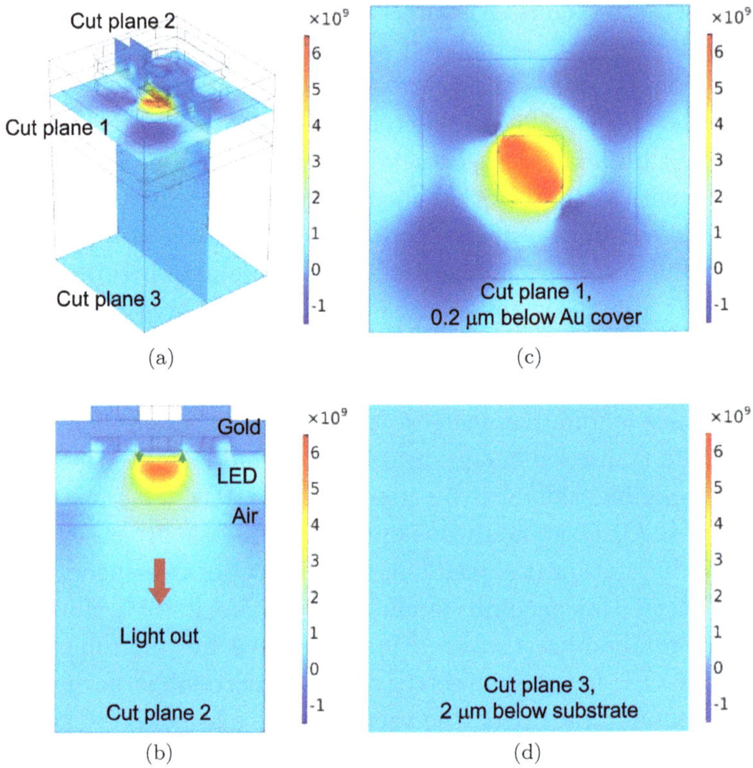

Figure 6.31. (a) The 3D view of the RP emitter unit cell and the power profiles in the unit of Wm^{-2} in different cut planes at $\lambda = 4.08\,\mu$m. The power profiles in (b) cut plane 2, (c) cut plane 1, and (d) cut plane 3. Adapted with permission from *Opt. Express*, 30(2), pp. 780–796 (2022).

harmonically and generate a time-varying dipole field that mimics the oscillating LSP field in the cavity.

Based on this physical picture, a pair of AC current feed-lines $0.1\,\mu$m long is placed in the diagonal positions indicated in Fig. 6.30(b). Each driving current I_0 is arbitrarily set at 10 mA in opposite phase with $\omega = 2\pi c/\lambda$. At $\lambda = 4.08\,\mu$m, the optical power generated inside and outside the unit cell is shown in Fig. 6.31. Figure 6.31(a) defines different cut planes on which the power is shown. Cut plane #1 is the horizontal cross-section $0.2\,\mu$m below the gold cover, cut plane #2 is the vertical cross-section at the center, and cut plane #3 is horizontal cross-section 2 μm below the RP substrate. In Fig. 6.31(b), cut plane #2 shows a strong downward power (in red) radiated from the center of the ring, and a portion of it transmits into the air. In Fig. 6.31(c), cut plane #1 shows there is power propagating upward around the ring (in dark blue) as well as downward at the center of the ring (in red). This power pattern shows the optical energy circulates inside the resonator before it transmits out of the volume. In Fig. 6.31(d), cut plane #3 shows the emitted power is essentially uniform spatially at 2 μm away from the substrate.

In the setup of the EM model in COMSOL, the top and bottom boundaries are set at second-order scattering boundary condition, and the four sidewalls are set at periodic boundary condition. The feedlines are defined at the diagonal edges of a square box in the geometry section, and the AC current is defined as the edge current along each of these two edges. The defined box also helps in increasing the mesh density in this critical region. The total time-averaged driving power P_{in} of the two feedlines is $I_0 V_0$, where $V_0 = \text{Re}(\int E_z dz)$ is the in-phase voltage developed along one of the feedlines and is obtained from EM modeling. The power radiated downward into free space P_{out} can be obtained by integrating $-emw.Poavz$ in COMSOL over the unit cell area and it is shown in Fig. 6.32(a).

The oscillatory P_{in} in Fig. 6.32(a) indicates that the resonator behaves as a radiating antenna, exhibiting varying feed impedance at

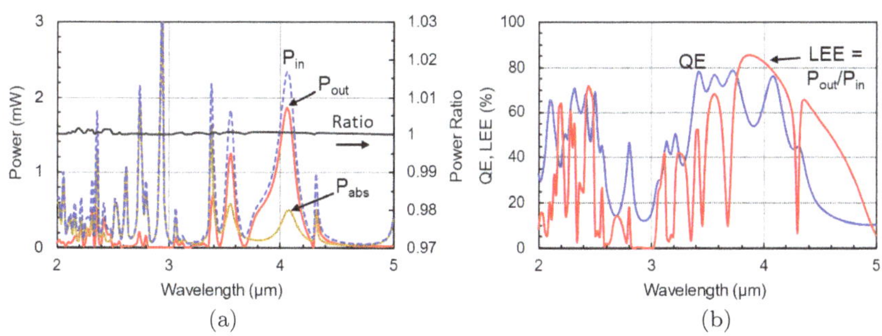

Figure 6.32. (a) The computed electrical driving input power P_{in}, the emitted optical power P_{out} in the air, the absorbed power P_{abs} in the gold layer, and the power ratio $(P_{\text{out}}+P_{\text{abs}})/P_{\text{in}}$. (b) The light extraction efficiency $= P_{\text{out}}/P_{\text{in}}$ of the RP emitter and the absorption QE of the RP detector taken from Fig. 6.27(b). Adapted with permission from *Opt. Express*, 30(2), pp. 780–796 (2022).

different λ. The maximum output is at the LSP resonant wavelength of 4.08 μm. The emitted output power P_{out} can be a substantial fraction of P_{in} at certain λ. Figure 6.32(a) also shows the computed P_{abs} in the gold layer, which is dissipated as heat. The sum of P_{out} and P_{abs} is equal to the driving power P_{in} within the modeling error of 0.1%. We also verify that there is no power transmitting across the four sidewalls, indicating that the radiation is strictly normal to the substrate surface as required by periodic boundary condition applicable to infinite arrays. The LEE of the RP emitter is given by $P_{\text{out}}/P_{\text{in}}$, and it is shown in Fig. 6.32(b). The LEE can be as large as 85.3% at 3.86 μm. It is larger than the 78.0% at 4.08 μm because the plasmonic resonance is weaker at 3.86 μm and the mode has less attenuation in the gold layer. Notice the close similarity between the absorption QE spectrum of the RP detector and the LEE spectrum of the RP emitter in Fig. 6.32(b) as predicted by the antenna reciprocity theorem. A good receiving antenna is indeed a good transmitting antenna, and the efficiencies of both devices are limited by gold absorption in the present case. The two spectral lineshapes are not exactly the same only because the present feedline assignment does not reproduce the cavity mode exactly. Assuming the planar LEE is 2% and the RP LEE is 78%, the LEE improvement is 39 times.

Meanwhile, the RP emitter increases the radiative recombination rate by a factor β of 29.8 times due to the average intensity enhancement. If the original recombination rate is R_{rad} and non-radiative recombination rate is R_{non}, the planar IQE will be $\eta_{PL} = R_{\text{rad}}/R_{\text{tot}}$, where $R_{\text{tot}} = R_{\text{rad}} + R_{\text{non}}$. The RP IQE, η_{RP}, will then be

$$\begin{aligned}\eta_{RP} &= \frac{\beta R_{\text{rad}}}{\beta R_{\text{rad}} + R_{\text{non}}} = \frac{\beta R_{\text{rad}}/R_{\text{tot}}}{\beta R_{\text{rad}}/R_{\text{tot}} + R_{\text{non}}/R_{\text{tot}}},\\ &= \frac{29.8 \times \eta_{PL}}{29.8 \times \eta_{PL} + (1 - \eta_{PL})}.\end{aligned} \tag{6.20}$$

The value of η_{RP} as a function of η_{PL} is plotted in Fig. 6.33(a). Due to the larger RP R_{rad}, η_{RP} increases rapidly with η_{PL}. When η_{PL} is 20%, η_{RP} is 88.1%. Assuming both the planar and RP structure to have the same current IE of 50%, the EQE of the RP emitter will be equal to IE $\times$ IQE $\times$ LEE $= 0.5 \times \eta_{RP} \times 0.78$ so that EQE rises rapidly to 34.2% at $\eta_{PL} = 20\%$.

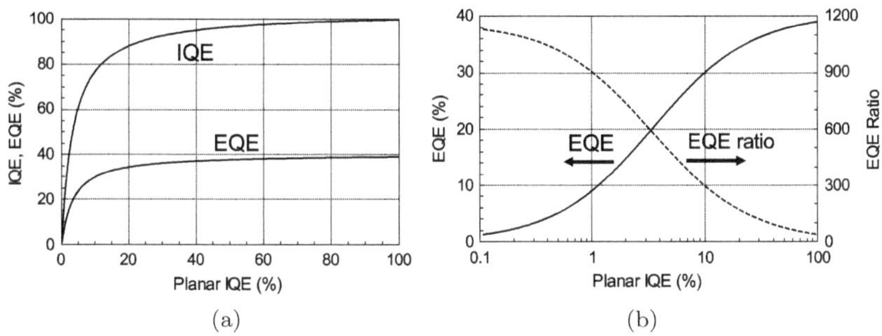

Figure 6.33. (a) The IQE and EQE of a RP emitter as a function of the planar IQE. (b) The RP emitter EQE and its ratio relative to a planar LED at different planar IQE. Adapted with permission from *Opt. Express*, 30(2), pp. 780–796 (2022).

Since the increase of η_{RP} is larger at smaller η_{PL}, we plot the RP EQE and its ratio relative to the planar EQE in logarithmic scale in Fig. 6.33(b). At $\eta_{PL} = 0.1\%$, EQE_{PL} is $1.07 \times 10^{-3}\%$ and EQE_{RP} is 1.21%, which is a difference of 1127 times. On the other hand, when $\eta_{PL} = 100\%$, EQE_{PL} is 1% and EQE_{RP} is 39%, which is 39 times difference. The RP emitter thus improves the planar EQE by more than 39 times. Combined with the directional rather than Lambertian radiation characteristics, the RF emitter will be much more intense at far field, and it will be useful for many mid-infrared applications such as gas sensing and infrared scene generation.

6.5.2 *InGaAs Transceiver for Short Infrared Wavelength Application*

Passive infrared imaging in the SWIR wavelengths under overcast or moonless condition is challenging because of the scarcity of photons (Illuminance/sr $\sim 1 \times 10^{-6}$ lux). The photon flux falling on a $10 \times 10\,\mu m^2$ detector under these conditions is estimated to be ~ 1100 Ph/s/sr between $\lambda = 0.4$ and $1.0\,\mu m$. When the detection wavelength is extended to $1.8\,\mu m$ using $In_{0.53}Ga_{0.47}As$, the available photons increase by an order of magnitude to 3.2×10^4 Ph/s/sr. An InGaAs avalanche photodiode (APD) with current gain can further amplify the photocurrent to exceed the CMOS leakage current for background limited imaging. Low excess noise, high gain, and low operating voltage are some of the desirable attributes for APDs in low light applications.

To improve detection in this environment, it is desirable to reduce the APD dark current I_d to lower noise while maintaining the same high photocurrent I_p. I_d of APD is originated from several current components in the absorbing region. They are the diffusion currents I_{diff} from the neutral regions, the generation-recombination current I_{GR} in the depletion region, the band-to-band and trap-assisted tunneling currents I_{tun} in the depletion region, and the leakage current I_{leak} around the pixel sidewalls [22, 23]. Much efforts have been devoted to reduce I_d in a typical Separate Absorption Grading Charge Multiplication APD. For example, using a double heterostructure, I_{diff} under punch-through biasing condition can be essentially eliminated because of the much lower intrinsic carrier density in the surrounding wider bandgap layers. I_{tun} can be minimized by adjusting the thickness of the multiplication layer and the doping density of the charge layer concomitantly [23]. The peripheral I_{leak} can be reduced or eliminated using various surface treatments, passivation, or adopting a planar detector structure [24]. The remaining dark current is exclusively contributed by I_{GR}. Since I_{GR} is directly proportional to the absorber thickness t_a [22, 23], different attempts have been made to reduce t_a without affecting I_p. Sverdlov and Salvador *et al.* [25, 26] placed a thin absorber with $t_a = 0.1\,\mu\text{m}$ rather than the usual $2\,\mu\text{m}$ [23] in a planar Fabry-Perot resonant cavity and observed greatly improved I_p/I_d ratio, albeit in a limited spectral range. The resonant cavity-enhanced Separate Absorption Grading Charge Multiplication APDs are now commonly employed in optical fiber communications. Verdun *et al.* [27] increases the bandwidth using a guided-mode resonator consisting of linear dielectric grid lines on top of a gold mirror. With a $t_a = 0.09\,\mu\text{m}$, the detector achieves an average QE of 50% over a $0.45\,\mu\text{m}$ bandwidth and collects 50% of the nightglow flux.

In this section, we describe using the RP structure to create LSPs to enhance optical absorption in a similarly thin layer to achieve larger QE and bandwidth. In addition, we also describe using a matching RP LED developed in the last section for active imaging applications. Commercial LEDs and lasers made of InGaAsP alloy emitting at 1.55 μm are widely available for optical communications. However, the output optical power P_{out} and wall-plug power conversion efficiency are rather low at about 8 mW and 4%, respectively [28]. Therefore, their performance can be improved by the RP structure.

The material layer structure of a conventional InGaAs APD is shown in 6.34(a). It consists of a 0.2 μm n^+-InP bottom contact layer, a 0.27 μm undoped InP multiplication layer, an undoped InGaAs absorber of certain thickness, a 0.09 μm p^+-InP top contact layer, and a 0.14 μm Ti/Ag cover layer. The usual Ti/Au contact can also be used but the optical performance will be reduced by 20% from the peak value. The Ti layer in the contact is assumed to be much thinner than its skin depth of 0.07 μm and therefore is excluded in the model. The refractive index of Ag is taken from Ref. [29]. Since the undoped InP layer is under large reverse bias, the photocarriers are amplified in this region through avalanche multiplication. The material properties of InGaAs and InP are shown in Fig. 6.34(b) [30].

Since InP is highly absorptive below λ of 0.9 μm, the InGaAs layer can only be sensitive between 0.9 and 1.8 μm. Based on the values of the absorption coefficient $\alpha(\lambda)$ of the two semiconducting materials, QE of the APD in planar geometry can be obtained from the classical

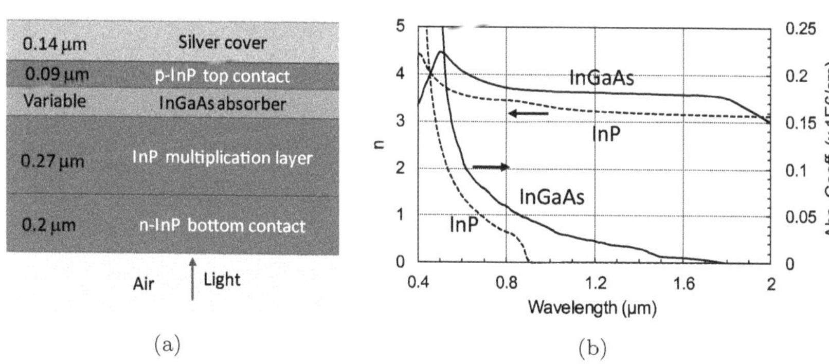

Figure 6.34. (a) The material layer structure of an InGaAs APD. (b) The (n, α) of InGaAs and InP. Adapted with permission from [31].

expression using ray tracing, and it is given by

$$\eta(\lambda) = T_{\text{sub}} \exp[-\alpha_{InP}(\lambda)t_{InP}]\{1 - \exp[-2\alpha(\lambda)t_a]\}, \qquad (6.21)$$

where $T_{\text{sub}} = 4n_{InP}/(1 + n_{InP})^2$, t_{InP} is the total InP thickness below the absorber, $\alpha(\lambda)$ is the InGaAs absorption coefficient, and t_a is the InGaAs absorber thickness. Eq. (6.21) assumes the incident light is reflected at the top metal surface and makes another pass in the absorbing region so that the optical pathlength is $2t_a$.

Since $\alpha(\lambda)$ is small near $1.6\,\mu$m, which is at the peak of the moonless nightglow spectral background, the conventional APD requires a thick t_a of $2\,\mu$m, with which the detector is able to achieve 65.5% QE at this λ and a maximum of 72.5% at a shorter λ in 6.35(b). If t_a decreases to $0.1\,\mu$m, QE will only be 7.9% at $1.6\,\mu$m in 6.35(d). Therefore, one needs to increase QE in thin absorbers if lower dark current and bias are required. The RP-APD is designed to yield strong absorption at the nightglow peak at $1.6\,\mu$m, where the internal wavelength $\lambda_{\text{int}} = \lambda/n = 1.6\,\mu\text{m}/3.59 = 0.446\,\mu$m. The

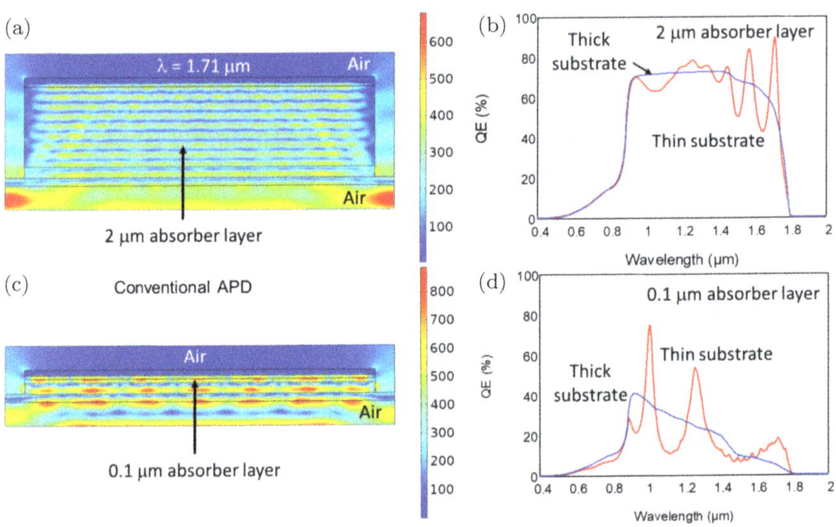

Figure 6.35. The planar APD geometry with (a) $t_a = 2\,\mu$m and (b) the modeled QE using ray tracing for thick substrate detectors and FEM for thin substrate detectors, and with (c) $t_a = 0.1\,\mu$m and (d) its QE. Adapted with permission from [31].

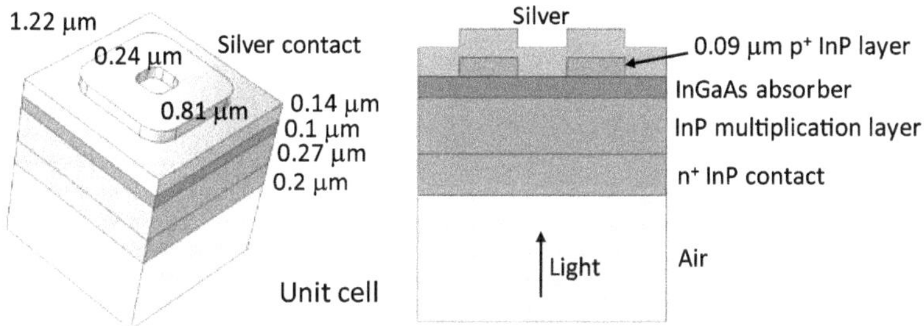

Figure 6.36. The RP-APD geometry with optimized structural dimensions. Adapted with permission from [31].

initial settings for RP detectors are: inner ring dimension $= \lambda_{\text{int}}/2$, outer ring dimension $= 3\lambda_{\text{int}}/2$, ring period $p = 2.5\lambda_{\text{int}}$, and ring thickness $= \lambda_{\text{int}}/4$. The total pixel thickness $t_{\text{tot}} = t_a + t_m + t_s$ is close to $(2m+1)\lambda_{\text{int}}/4$, where t_m is the multiplication layer thickness, t_s is the substrate (common contact) thickness, and m is an integer. Since the designed period p is about 1.1 μm, which is much smaller than the usual pixel size of 10 μm, there are many rings in a detector pixel. EM modeling is therefore performed on a single unit cell with periodic boundary condition as shown in Fig. 6.36.

Before structural optimization, for $m = 2$, if p is set at $2\lambda_{\text{int}} = 0.89\,\mu$m, the cross-section of the rings will have a uniform periodicity, similar to that of a diffraction grating as shown in Fig. 6.37(a). Although this ring geometry can provide effective coupling expected from a grating, its spectral width will be too narrow for nightglow detection as shown in Fig. 6.37(b) and diffractive waves can cause undesirable size effects discussed in Section 5.3. To yield a broader peak, a larger period p of $5\lambda_{\text{int}}/2$ is adopted to disrupt the grating periodicity and create the LSPs. Based on this new RP structure, the unit cell parameters are further adjusted to yield the largest spectral overlap with the nightglow spectrum. The optimized RP-APD structure and its detection spectrum are shown in Fig. 6.37(c) and (d), respectively.

The optimized p turns out to be $2.74\lambda_{\text{int}}$. With the optimized QE spectrum, a 10 μm pitch APD produces 1.785 fA of primary photocurrent under $f/1$ optics in the presence of the nightglow radiance Φ_{ph}. Under the same condition, the 2 μm thick planar detector

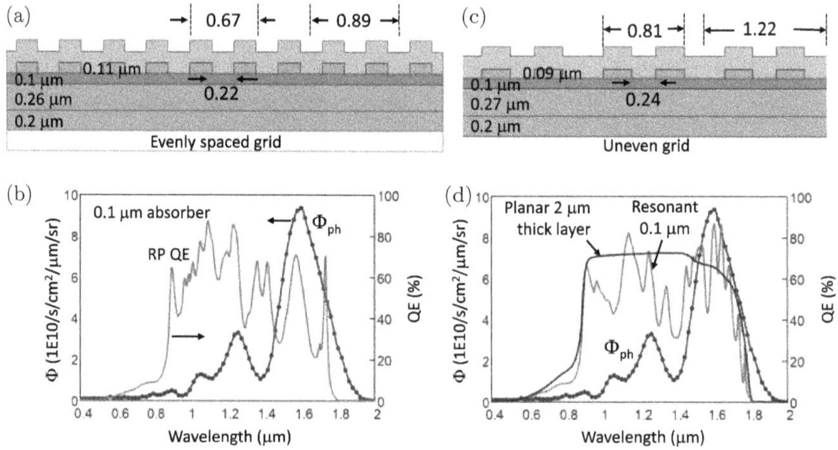

Figure 6.37. (a) The nominal RP-APD structure with $p = 2\lambda_{\text{int}}$ and (b) its modeled QE spectrum. (c) The optimized RP-APD and (d) its QE spectrum. The figures also show the nightglow spectrum Φ_{ph}. Adapted with permission from [31].

produces 1.963 fA. Therefore, the optimized RP-APD is able to produce 0.909 times of the planar I_p with only 1/20 of detector volume. With the proportionally lower dark current I_d, the I_p/I_d ratio will then improve by 18.2 times, enabling lower noise and higher sensitivity detection. With the thinner absorber layer, the applied voltage can also be lower. By increasing t_a from 0.1 μm to 0.2 and 0.3 μm while reducing t_m to maintain the same t_{tot}, I_p can be raised to 1.041 and 1.088 times of the thicker planar I_p, respectively, but the I_p/I_d ratio will not be increased as much due to the increased I_d [31, 32].

Figure 6.38 shows the optical E field distribution at 1.6 μm and under 135° incident polarization. The ring structure is seen to induce a large E field around the ring, which is a signature of LSPs. Since the maximum E field is ~6000/533.2 = 11.3 times higher than the incident field E_{inc} of 533.2 V/m, the corresponding optical intensity $\propto nE^2$ can be ~458 times higher than the incident intensity around the ring corners. The larger overall intensity in the absorbing layer then leads to a larger QE in the RP-APD detector.

To increase the detection sensitivity further, active imaging can be employed, and the RP-LED will be useful for this application.

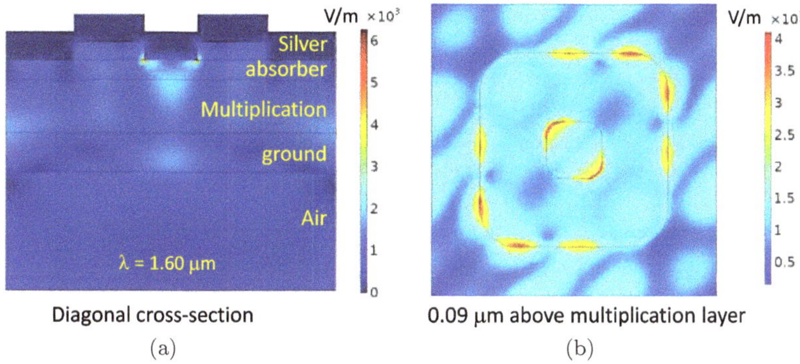

Figure 6.38. The modeled $|E|$ field at $\lambda = 1.6\,\mu$m in the optimized RP-APD. Reproduced with permission from [31].

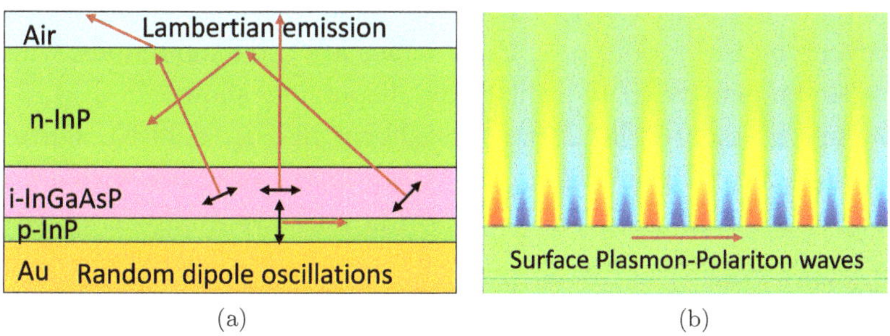

Figure 6.39. (a) The schematic layer structure of a conventional LED. (b) The surface plasmon polariton vertical E_y distribution using homogeneous $n_d = 3.25$.

To understand the enhancement mechanism, we first examine a typical LED structure, which contains a thin *i*-InGaAsP active light-emitting layer but a thick *n*-InP substrate layer as shown schematically in Fig. 6.39(a). Under forward bias, electron-hole pairs recombine in the *i*-InGaAsP layer and emit light. Since there are vast amount of EM modes in the thick device volume, the light emission in the active layer is isotropic. Only half of the emission reaches the top air interface, and only a small part of this half is within the critical angle θ_c for total internal reflection and propagates into the air. As a result, LEE is small at $\sim 1/(4n^2) = 2.4\%$ for the InP substrate with $n = 3.2$ [7]. When the *p*-InP contact layer next to the gold layer is

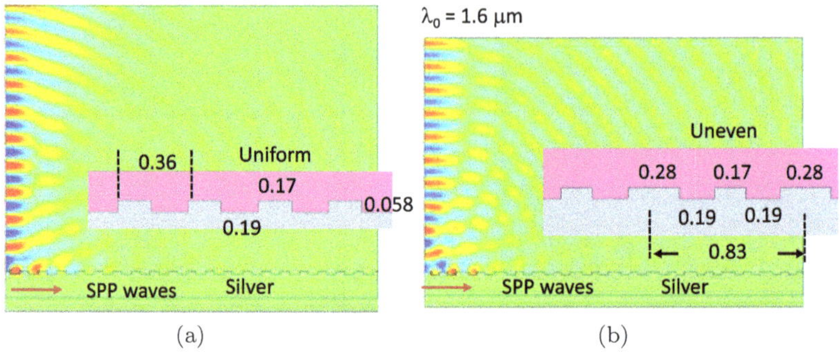

Figure 6.40. The modeled E_x of the scattered wave of (a) a uniform deep grove structure and (b) an uneven groove structure at $\lambda = 1.6\,\mu$m using homogeneous $n_d = 3.25$.

thin, there are also optical transitions near the gold metal surface. However, the EM modes near this surface are the Fano modes, which are also known as SPPs, with the dispersion shown in Fig. 3.4. SPPs are bound to the surface as shown in Fig. 6.39(b). Optical transitions into SPP modes propagate along the interface with a long propagating length shown in Fig. 3.6. They are eventually attenuated by the metal as heat, resulting in no light emission. Therefore, optical transitions near the metal contact do not take part in light emission.

Nevertheless, if the interface is patterned into a periodic grating such as that shown in Fig. 6.40, SPP waves can be scattered by the grating, and the optical transition energy can be converted into propagating EM waves, adding to the light emission. The propagating angle θ of Eq. (3.27) is confirmed by Greffet *et al.* [33] in thermal radiation experiments. The near-normal radiation in a range of λ with $\theta < \theta_c$ can be used to increase the LED output. However, the power increase is quite small, as seen in Fig. 3.9, where the scattering amplitude is small and the SPP wave is only weakly attenuated by the scattering.

However, the scattering of SPP waves can be made much stronger if the surface corrugations are designed to couple with SPP resonantly, in which the corrugations serve as resonant cavities [34–37]. If the SPP wave can be orchestrated to create a strong tangential E field (*i.e.*, E_x) at the top of the corrugations, the array of corrugations can act as a coherently phased antenna array in transmitting the SPP energy into free space as the thermal experiments demonstrated. To illustrate this effect in the LED materials (e.g., InGaAsP/Ag), we model the square corrugations in Fig. 6.40. This example contains either a uniform array of corrugations with $p = 0.36\,\mu$m or an uneven array with $p = 0.83\,\mu$m. The structural parameters shown are taken from the cross-sections of 3D ring arrays discussed later. Using these structures, the SPP wave interacts strongly and resonantly with the corrugations at the designed $1.6\,\mu$m. Most of the generated SPP energy launched from the left is seen to convert into propagating waves at the normal direction within one corrugation period in these arrays. The uneven corrugations yield a more focused beam with greater directivity. The dynamic wave propagation can be observed by performing an animation in COMSOL on emw.Ex in the *Electric field, Surface plot* node. In the *Animation* setting, *Dynamic data extraction, Full harmonic, 36 Frames* are set. When the phase of the wave increases with time, the waves can be seen moving upward from the corrugated surfaces. In the emitted E_x patterns, in addition to a strong upward propagating wave, there is also a weak component radiating in other directions. This off-axis radiation is due to the rapid decay of the advancing SPP wavefront from left to right that leads to uneven interference. It is expected to disappear when the SPP is uniformly excited in every unit cell. The normal emitting wave will add significantly to the LED emission because the emission angle is less than the critical angle. The LEE will further be maximized if the entire volume is thin enough to supports only SPP modes.

The design of a $1.6\,\mu$m RP-LED will be in parallel with the previous RP MQW emitter design in Section 6.5.1, in that an efficient detector is first designed. Since the LED does not need a carrier multiplication layer, this layer is omitted in the design. By adopting a constant absorption coefficient α at 10^4/cm, which is the typical α of InGaAsP alloy in this wavelength range [38], the optimized RP detector is shown in Figs. 6.41(a) and (b), and its QE spectrum is shown in Fig. 6.41(c). The cross-section in Fig. 6.41(b) is the same as that used to model the strong vertical emission in Fig. 6.40(b). The detector has a QE of 89.6% under normal incidence at $1.6\,\mu$m, which

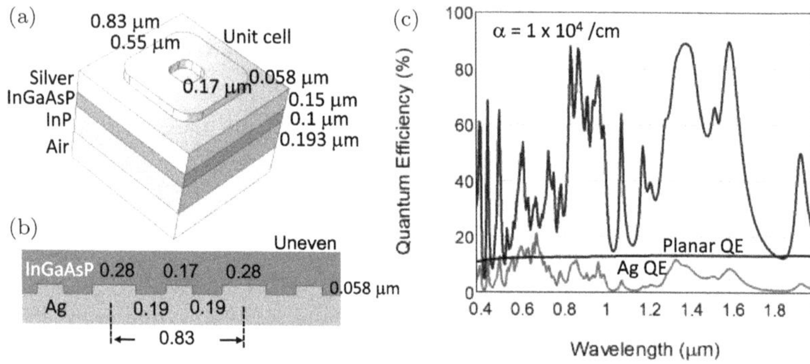

Figure 6.41. (a) The optimized RP-InGaAsP detector design, (b) the ring cross-section, and (c) the modeled QE spectrum. Adapted with permission from [31].

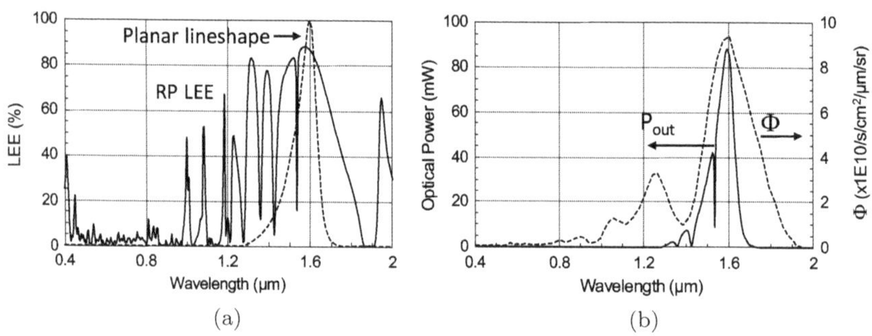

Figure 6.42. The typical emission lineshape of a planar LED and the LEE spectrum of a RP emitter. (b) The comparison of the spectral lineshape between RP-LED and moonless nightglow. Adapted with permission from [31].

is much larger than the 16.2% for the planar structure. The large absorption implies a large emission at this wavelength. In this design, the finite absorption of the InP substrate below 0.9 μm is ignored.

To model the LEE, two AC current feedlines, each having a spatial extent of 0.025 μm, are placed at the diagonal corners of the ring in the resonator. By modeling the output radiation power P_{out} in the air in the z direction under the input electrical power P_{in}, the LEE $\equiv P_{\text{out}}/P_{\text{in}}$ can be obtained. Figure 6.42(a) is the modeled LEE when it is used as an emitter. At 1.6 μm, a LEE of 87.1% is obtained, which is much larger than the 2.4% of a planar LED. By the antenna reciprocity theorem, QE in Fig. 6.4(c) and LEE in Fig. 6.42(a) should have the same spectral lineshape. The two spectra indeed share some similarity. All peaks above $\lambda = 1\,\mu$m in both spectra are found in one-to-one correspondence. At these wavelengths, a good receiving antenna is indeed a good transmitting antenna.

However, their spectral lineshapes are not exactly the same and some of the LEE peaks at shorter λ are absent. These differences are attributed to the limitations of the present feedline assignment that do not produce the cavity modes exactly. At the shorter wavelengths, some of the cavity modes do not have appreciable amplitude at the ring corners so they are not excited by the feedlines, leading to small P_{out}. In this case, the small $P_{\text{out}}/P_{\text{in}}$ is not related to the transmission properties of the modes. Different feedline arrangements will be required for their LEE evaluation.

To estimate the emission spectrum of the RP-LED, the experimental lineshape of a commercial LED [28] is utilized and it is shown in Fig. 6.42(a) as an example. Assuming the LEE of the commercial LED to be constant, the spectral lineshape of the LED will be modified by the RP LEE and will change to that in Fig. 6.42(b). The RP lineshape does not change significantly. The spectral width of the emission remains about $0.1\,\mu$m, and it is narrower than the nightglow spectrum.

Since a bare planar LED has a LEE of $\sim 1/(4n^2) = 2.4\%$ for InP substrate, the RP-LED improves LEE by $87.1/2.4 = 36.3$ times. Combined with the 54.9-times-higher radiative recombination rate because of the higher internal intensity, the RP-LED output power will be larger than the planar LED. If we assume IE $= 50\%$ and $\beta = 54.9$ in Eq. (6.20) for the RP LEE, the RP EQE $=$ IE $\times$ IQE $\times$ LEE is plotted in Fig. 6.43. The RP EQE rises rapidly with the planar IQE. It can reach 40% when $\eta_{PL} = 20\%$, and it has a higher power throughput by more than 36 times. Since the typical commercial LED has a power level of 8 mW and EQE of 4% using special optical packaging [28], the present RP-LED with the same die size should have a power level of $8\,\text{mW} \times 43.6/4 = 87.2\,\text{mW}$, assuming no special packaging, IE $= 50\%$, and planar IQE $> 30\%$.

To illustrate that the present feedlines can generate the intended cavity modes, we show in Fig. 6.44 the field distributions at two

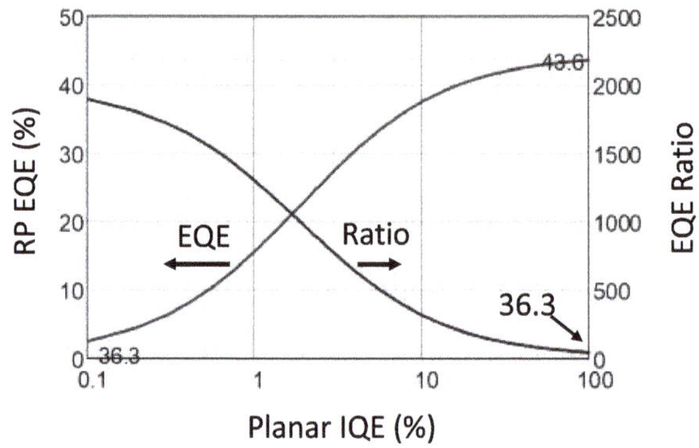

Figure 6.43. The RP emitter EQE and its ratio relative to that of a planar LED at different η_{PL}. Adapted with permission from [31].

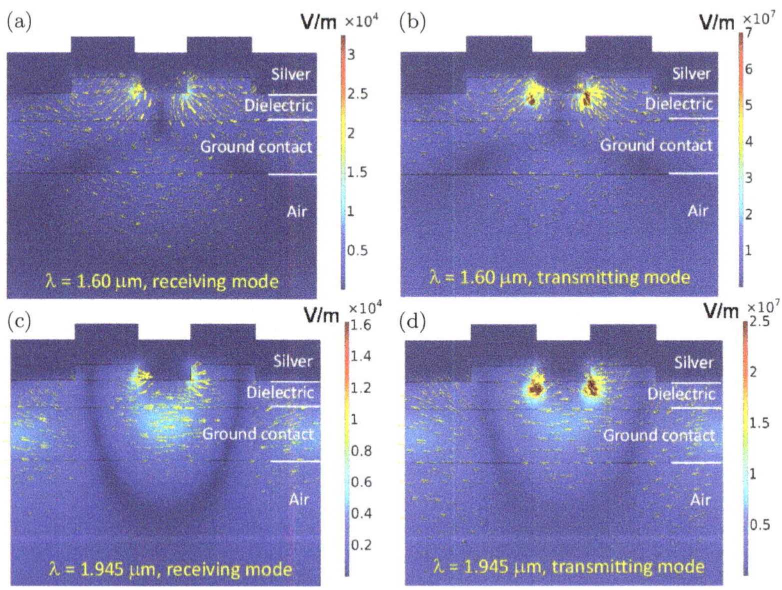

Figure 6.44. (a) and (b) The field patterns in the receiving and transmitting modes at $\lambda = 1.60\,\mu$m, respectively. (c) and (d) Those at $\lambda = 1.945\,\mu$m. Adapted with permission from [31].

LSP-resonant $\lambda = 1.60\,\mu$m and $1.945\,\mu$m. Figure 6.44(a) and (c) are the field patterns when the RP structure is in the receiving mode under external optical excitation. Figure 6.44(b) and (d) are those when the RP structure is in the transmitting mode under internal electrical excitation. The similarity of the field distributions at these two λ testifies to the validity of the present feedline assignment at these two λ.

Although the RP emission is strictly in the normal direction as Fig. 6.31 shows, the present modeling applies periodic boundary condition to the ring unit cell, which is strictly correct only in the infinite arrays. For a finite-size array, beam divergence is expected. We therefore modeled finite-size arrays from 1×1 unit cells to 5×5 unit cells as shown in Fig. 6.45 where the air domain outside the array is surrounded by a spherical perfectly matched layer to absorb all the outgoing radiation. By defining the air as a far-field domain, various far-field quantities at definite angular directions can be computed based on the near-field solution at the air surface within COMSOL. 2D and 3D far-field plots are generated by default. The near-field P_z profiles and the calculated far-field $|E|$ patterns are shown in

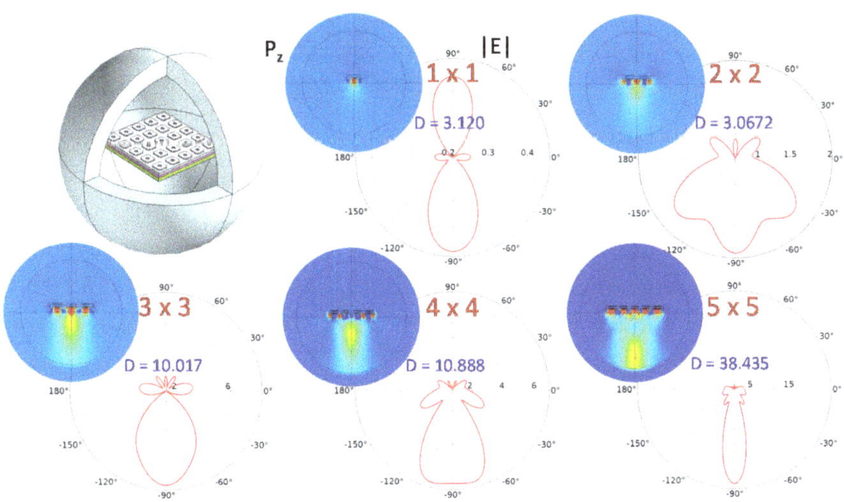

Figure 6.45. The modeling geometry, the near-field P_z profiles and the far-field E profiles of different size arrays. Adapted with permission from [31].

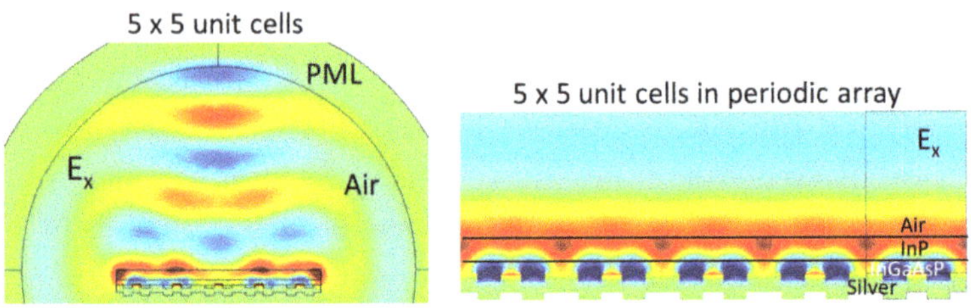

Figure 6.46. The E_x profiles of (a) a 5×5 unit cell array and (b) an infinite periodic array.

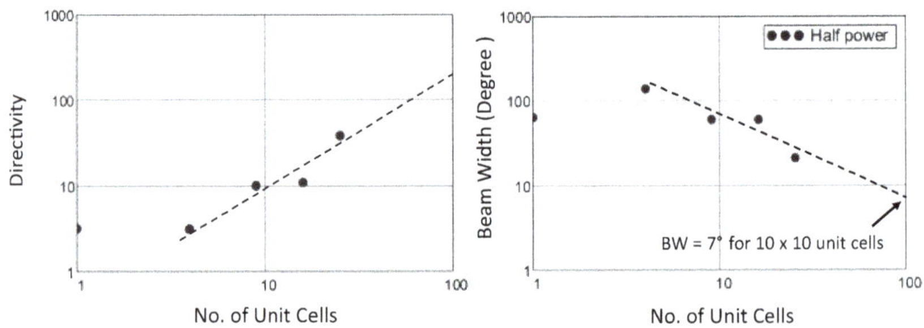

Figure 6.47. (a) The directivity and (b) beam width of the RP-LED with different number of unit cells. Adapted with permission from [31].

Fig. 6.45. The optical beam becomes more directional when the number of unit cells increases. The directivity D of the emitted beam, which is the ratio of the maximum power density over the averaged power density, increases rapidly with increasing array size. The emitted $\boldsymbol{E}$ field also becomes more resemblant to that of the infinite array. Figure 6.46 compares the E_x component of a 5×5 array and an infinite array.

By extrapolating the directivity to 10×10 unit cells, which occupy an area of $8.3\,\mu\text{m} \times 8.3\,\mu\text{m}$, D is predicted to be 200 in Fig. 6.47(a) and the beam width $\Delta\theta$ is predicted to be 7° in Fig. 6.47(b), where $\Delta\theta$ is the full width at half power maximum. When the number of unit cells further increases, the beam width will eventually match that of the infinite array solution. Figure 6.46 shows the waves from both the 5×5 unit cell array and an infinite array transmitting into air without impedance from the semiconductor/air interface at the resonant wavelength of $1.6\,\mu$m. This transmission is similar to the unity substrate transmission at the same λ in the detector mode, which is very different from a planar detector. The less-than-unity LEE is again due to the finite absorption in the metal silver layer. The near-unity normal exitance from resonant cavities is generally observed in thermal excitation experiments, which will be discussed in the next section.

To estimate the far-field intensity, we assume non-uniformity in material growth and device fabrication set at a lower limit on the beam width. The smallest θ is 7°. P_{out} is 87.2 mW from a single RP-LED with $465\,\mu$m in diameter. The spectral bandwidth Γ is $0.1\,\mu$m. $\lambda = 1.6\,\mu$m and photon energy $h\nu$ is 0.775 eV. The RP-LED radiance in the units of Ph/s/μm/m^2 at a distance d will be $P_{out}/\Gamma/h\nu/(2 \cdot d \cdot \tan(\theta/2))^2$. Since the peak nightglow radiance is 3×3^{15} Ph/s/μm/m^2, a single RP-LED will yield the same nightglow radiance at a distance d of 396 m, and it has a spot size of $3.0 \times 3.0\,\text{m}^2$. These emission characteristics will be very useful in applications such as driving aid at night. Integrating the high-sensitivity RP-APD and high-intensity, directional RP-LED will form a powerful SWIR transceiver for light detection and ranging and other 3D applications.

6.5.3 *Wavelength-Selective Thermal Emitter for Thermophotovoltaic Application*

Thermophotovoltaic technology converts thermal radiation of a heated source into electricity directly. It can be used in applications such as heat reclamation and portable energy source in remote sites. The background of this technology can be found in the extensive reviews on emitter materials by Ferrai *et al.* [39] and photovoltaic (PV) materials and ohmic contact materials by Ali Gamel *et al.* [40]. To increase its competitiveness among other energy solutions, the power conversion efficiency (PCE) and production cost must improve. Omair *et al.* [41] achieved a PCE record of 29.1% at an emitter temperature of 1207°C using a highly reflective mirror on the back of the PV cell for efficient photon cycling. To further increase PCE, the spectral emission of the thermal emitter should match better with the optical absorption of the PV cell, with which parasitic power losses can be minimized. These losses include photocarrier thermalization, optical absorption in PV cell contact layers, optical absorption in the sidewalls of the system enclosure, etc. In addition, the PV cell dark current should also be reduced by reducing the layer thickness to increase the open circuit voltage. To lower the production cost, the device should be simple and straightforward to fabricate.

There are numerous approaches in tailoring the emission spectrum of a thermal emitter. Reviewed by Pfiester and Vandervelde [42], the general approaches are based on 1D, 2D, and 3D photonic crystals, metasurface filters, and artificially synthesized materials [43–47]. Fabricating microstructures on a material surface not only modifies the thermal emission spectrum but also changes the emission spatial profile, so both can be used to improve the system efficiency. In general, a microstructure acts as an optical antenna in radiating its thermal energy into space. Although the thermal motion of charge carriers in the structure is random and uncorrelated, the EM modes that can be established and sustained in the microstructure are the resonant cavity modes that satisfy Maxwell's equations. This phenomenon is akin to producing musical tones in flutes by blowing turbulent air. When the accumulated EM fields are radiated from the microstructure into space, quasi-coherent and directional thermal emission is produced. The thermal emission characteristics of a microstructure thus differ from that of a large blackbody. A large blackbody contains a vast number of resonant cavity modes. When these modes are thermally excited, they emit in different energies and in different directions, leading to incoherent, isotropic radiation. In contrast, a microstructure has far fewer resonant modes, and each mode emits in distinct energy and direction, resulting in more coherent and directional thermal radiation.

Accordingly, Schuller *et al.* [48] observed distinct and sharp thermal emission peaks from SiC whisker micro-antennas. These microstructures are dispersed on a heated platinum-coated silicon substrate. The observed peaks are identified to be the transverse magnetic and transverse electric resonant modes of individual micro-antennas. When the microstructures are arranged orderly such as that of a grating, the whole structure becomes an antenna array. Greffet *et al.* [33] demonstrated in linear gratings and Arnold *et al.* [49] demonstrated in cross-slit gratings experimentally that the thermal emission exhibits narrow angular lobes, and the emission angle is wavelength dependent, similar to that of a radiating antenna. In the linear grating, the angular width $\Delta\theta$ is only 1° at $\lambda = 11.36\,\mu$m and $\theta = 60^\circ$, and in the cross-grating, $\Delta\theta$ is around 3.9° at the normal direction ($\theta = 0^\circ$). The emissivity ε is close to 90%. This coherent emission is explained by the coherent surface phonon-polariton waves excited on the SiC surface. The grating channels surface waves into free space with the needed momentum, thus creating the new emission characteristics. The coherence length ℓ_c is approximately equal to the propagating length ℓ_p of the surface wave.

Dahan *et al.* [34, 35] further demonstrated that deep grating grooves can increase ℓ_c by an order of magnitude and observed strong quasi-monochromatic and directional ($\Delta\theta = 1.75^\circ$) thermal emission with $\varepsilon = 0.98$ in the normal direction ($\theta = 0^\circ$). The estimated ℓ_c is 440λ or 5.1 mm, which is comparable to the sample size of 30 mm. The researchers attributed the longer ℓ_c to the increased photon density of states in the coupled resonant cavities formed in the deep grooves. Each coupled resonant cavity in the array also acts as a half-wave antenna in radiating the resonant energy efficiently into space, leading to the large observed ε. Based on this approach, the authors anticipate a quasi-monochromatic thermal source with $\ell_c = 716\lambda$ and an extremely narrow $\Delta\theta$ of 1.4 mrad to be feasible to produce.

The coherence length can be further increased by using surface plasmon-polariton instead of surface phonon-polariton since plasmon-polaritons formed on the metal surface possess much longer propagating lengths. Biener *et al.* [36] fabricated periodic resonant cavities using gold metal on Si substrate and demonstrated a thermal emitter with $\ell_c = 862\lambda$, $\theta = 0^\circ$, and $\Delta\theta = 0.066^\circ$. This measured ℓ_c is shorter than the theoretical value of 2400λ only because ℓ_c in this case is limited by the emitter physical size of 10 mm rather than by ℓ_p. The entire surface in this emitter is in fact spatially coherent in emitting radiation. To increase the emission intensity at higher temperatures, Liu *et al.* [37] adopted a refractory plasmonic material, TiN, and obtained a well-collimated beam at $\lambda = 3\,\mu$m with $\ell_c = 32\lambda$ and $\Delta\theta = 1.8°$. The emission has a very narrow spectral width $\Delta\lambda$ of 70 nm.

In addition to using coherent surface waves, metamaterials can also be used to generate coherent thermal radiation. Perfect-absorber metamaterials are known to produce large and sharp absorption peaks [50]. These materials should then be quasi-monochromatic thermal emitters according to Kirchhoff's law on the equality of absorptivity and emissivity. Indeed, Liu *et al.* [51] fabricated metamaterials using gold metal having cross symbols on Si substrate and verified Kirchhoff's law on the equality of absorptivity and emissivity.

Metamaterials have also been designed to control the emission direction. For example, Bossard and Werner [52] offered two designs. One is based on an EM bandgap metamaterial, which is composed of a periodic metal screen on top of a thin dielectric having a metal backing. The screen permits certain dielectric cavity modes to pass through and steers them to a definite direction upon optical interference. Another metamaterial is based on an array of discrete dielectric resonators. These resonators increase the EM fields at the resonant wavelengths. Since large EM fields yield large absorption and thus large emission, the array acts as a strong phased array antenna in achieving directionality.

The above studies illustrate that the thermal radiation from an object originates from the EM eigenmodes of the object. For a large bulk object, the allowed EM modes are the vast number of standing waves established inside the object's volume. The density of these modes contributes to the characteristic blackbody spectrum. However, thermal radiation is a surface phenomenon, which is sensitive to the surface eigenmodes. These eigenmodes are the surface plasmon-polaritons or surface phonon-polaritons. When they are freed by periodic structures, these surface modes dominate the emission properties. As the emission is now contributed by far fewer modes, it becomes more spatially coherent. When periodic resonant cavities are fabricated on the surface, the cavities strengthen some of the modes that match the resonant condition while rejecting others, further reducing the number of surface modes and leading to a stronger, more spatially coherent, and more directional thermal radiation.

Different from the many studies on tailoring the emission properties of a thermal emitter, there are relatively few studies on using microstructures to tailor and enhance the absorption properties of PV cells. In this section, we will apply the RP approach to achieve wavelength selectivity in both emitters and PV cells, with which the overall power throughput and PCE can be simultaneously improved. The discussion is limited to comparing a thermophotovoltaic (TPV) system with and without the photonic structure, while other common factors such as the emitter and PV cell materials, the device area, the view factor, and the power fill factor, etc., are assumed to be equal. There is no attempt at designing a TPV system with the highest absolute PCE.

The schematic TPV system is shown in Fig. 6.48. It consists of a wavelength-selective and directional emitter, a wavelength-selective PV cell, and a thermal management system for PV cell cooling and fuel pre-heating. Under TPV operation, the thermal energy released from the combustion system heats up the metal block attached to the

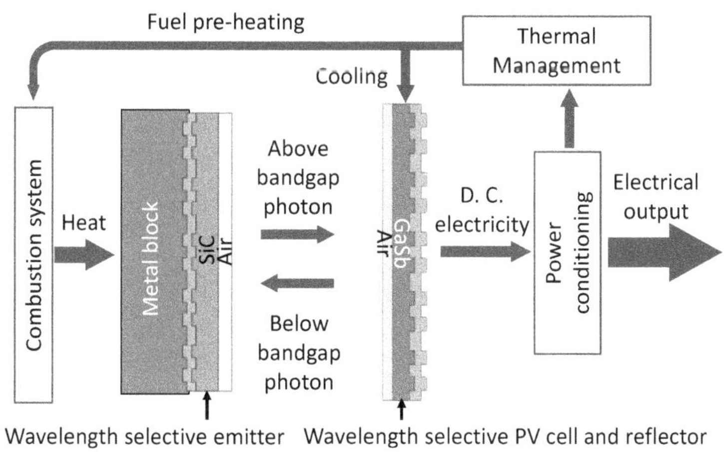

Figure 6.48. The schematic TPV system and the device structures. Reproduced with permission from [53].

SiC emitter and generates a directional beam of thermal energy onto the GaSb PV cell. The PV cell is designed to absorb the emitted spectrum preferentially, thus improving the power throughput and PCE. The absorbed energy is converted into DC power inside the PV cell and then into AC electrical power through external power conditioning. A complete TPV system may also contain a thermal management system that preheats the fuel with recovered heat from the PV cell and unit enclosure.

In addition to the benefits of higher PCE and lower cost, the present approach also offers a more robust system since it relies less on photon recycling, which can be sensitive to operating environment such as enclosure sidewall absorption. Because of the collimated emission of the emitter, the emitter and PV cell can be placed at a larger distance to reduce convectional heating of the PV cell. The present ring pattern, unlike the usual surface tailoring approaches, is located behind the emitting surface. The fine structure can therefore be protected from surface contamination and air corrosion, which makes it more durable.

The RP concept for wavelength-selective thermal emitter is to create desirable resonant cavity modes for thermal emission at specific wavelengths [53, 54]. To understand this approach, we start with a typical optical mirror structure, in which a metallic Ta layer is deposited on top of a SiC dielectric layer. Since the extinction coefficient κ of SiC [55] in Fig. 6.49(a) is very small at ~0.01

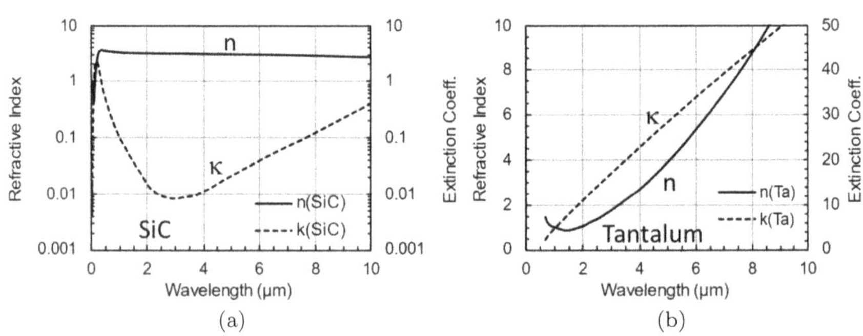

Figure 6.49. The refractive indices of (a) SiC and (b) Ta. Adapted with permission from [53].

near wavelength λ of $2\,\mu$m, the corresponding absorption coefficient $\alpha = 4\pi\kappa/\lambda$ is small at $0.063\,\mu\text{m}^{-1}$. When its thickness t is thin at $0.09\,\mu$m, the optical absorption αt is small at 0.0057. This thin layer thus acts as a transparent dielectric. Meanwhile, Ta is a plasmonic material at this λ since the real part of the dielectric function $\text{Re}(\epsilon) = n^2 - \kappa^2$ is negative in Fig. 6.51(b) [56]. As a result, this material combination forms a highly reflective optical mirror. Since high reflectivity r means low absorptivity $\eta = 1 - r$, by Kirchhoff's law, the structure will also have a low emissivity $\varepsilon = \eta$. To increase ε, η must increase. A simple method is to increase the SiC thickness. However, it will increase η at all wavelengths, resulting in a grey body emitter with a relatively constant $\varepsilon \approx 0.8$. For wavelength-selective emission, a different method is needed.

When SiC thickness $t = 0.09\,\mu$m, the planar structure does not support optical Fabry-Perot resonance for absorption beyond λ of $1.4\,\mu$m, which gives it a small η and thus a small ε at the longer λ. However, by fabricating this layer into the RP structure, LSPs, because of their small mode volumes, can be excited at the Ta/SiC interface, and they can be used for absorption and emission. In particular, the diagonal distance of the inner ring is designed to be half of the internal emission wavelength λ_E, so that the inner ring metal patch acts as a half-wave dipole antenna in generating strong LSP oscillations at λ_E. Since the cutoff wavelength λ_c of GaSb PV cells is about $1.75\,\mu$m, the inner ring dimension s_i would then be $s_i = \lambda_c/(2\sqrt{2}n) \approx 0.20\,\mu$m so that λ_E matches λ_c. Note that the large energy loss in metal when LSPs are created in shorter wavelengths, which is detrimental for visible optoelectronic devices [18], is being utilized here to generate thermal emission in the Ta metal layer. In fact, due to the small extinction κ value of SiC layer near λ_E, the thermal emission from the SiC layer is about 3 times weaker than that from the Ta layer. The structures of emitters A and B shown in Fig. 6.50 are optimized for maximum emission at λ_E in the vertical z direction. They are obtained by maximizing the total normal incident absorption in the SiC and Ta layers at λ_E. The SiC thicknesses t_{SC} are $0.09\,\mu$m and $0.33\,\mu$m for A and B, respectively. They are the two thinnest layers that yield resonant absorption.

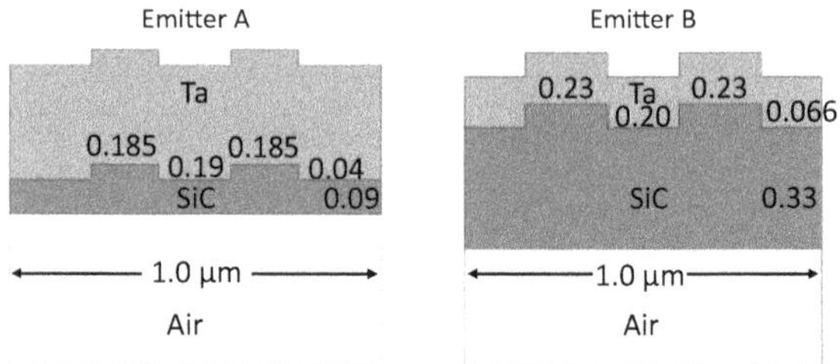

Figure 6.50. The cross-sections of (a) Emitter A with 0.09 μm thick SiC and (b) Emitter B with 0.33 μm thick SiC.

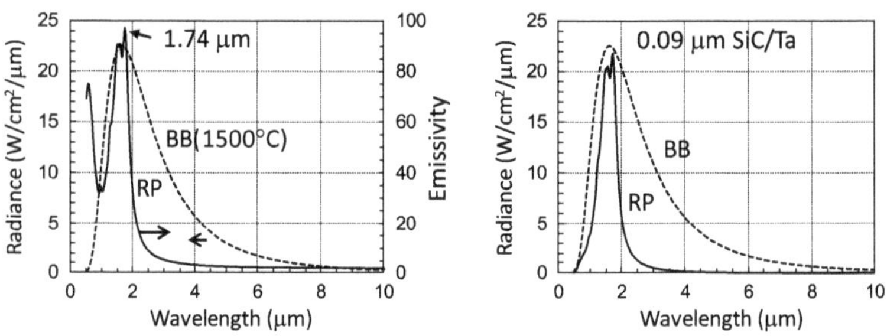

Figure 6.51. (a) The modeled emissivity of Emitter A. (b) Emitter A power radiance compared to that of a blackbody at 1500°C. Adapted with permission from [53].

Figure 6.51(a) shows the RP $\varepsilon(=\eta)$ of A with $t_{SC} = 0.09\,\mu$m. At the resonant $\lambda_r = 1.74\,\mu$m, the emitter ε can be as high as 0.975, close to that of a perfect blackbody. At this wavelength, the SiC/Air interface is transparent to both the incident and transmitting optical radiation. The emissivity peak nearly coincides with the blackbody power radiance P_{BB} maximum at $T = 1500°$C, thus maximizing the power output of the emitter. By multiplying ε with P_{BB}, the power radiance P_{RP} of the emitter can be obtained and is shown in Fig. 6.51(b). Near λ_r, the RP emission power exceeds that of a bulk SiC emitter with a typical wavelength-independent ε of 0.74 while it is much less at other wavelengths. Quantitatively, the RP emitter emits 93.1% of the total thermal power relative to a typical bulk emitter at $\lambda < 1.93\,\mu$m (the absorption edge of the PV cell),

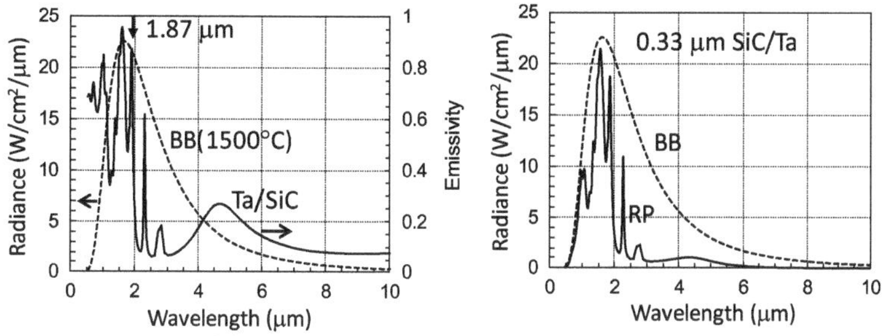

(a) (b)

Figure 6.52. (a) The modeled emissivity of Emitter B. (b) Emitter B power radiance compared to that of a blackbody at 1500°C.

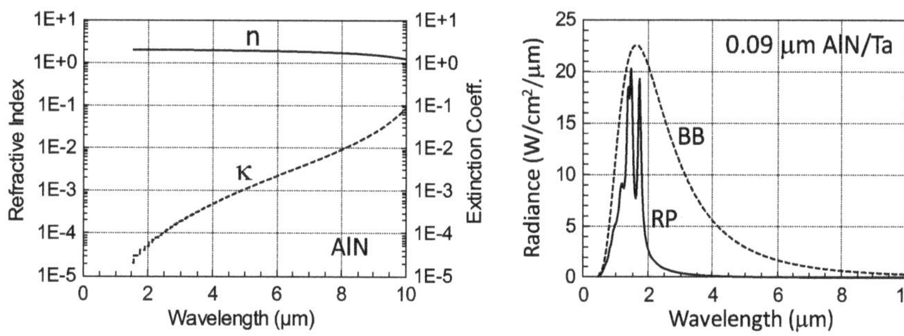

(a) (b)

Figure 6.53. (a) The refractive index of AlN. (b) Emitter C power radiance compared to that of a blackbody at 1500°C.

and only 11.6% at $\lambda > 1.93\,\mu$m. Therefore, the power selectivity is increased by 8.0 times from the bulk emitter.

When the SiC thickness t_{SC} increases to $0.33\,\mu$m, emitter B with a larger resonant volume has more emission peaks as shown in Fig. 6.52(a). While the sharp peaks originate from the cavity modes, the broad peak near $4.5\,\mu$m is one of the Fabry-Perot oscillations from vertical reflections within the SiC layer. This Fabry-Perot peak increases the power emission at $\lambda > 1.93\,\mu$m and thus reduces the power selectivity as seen in Fig. 6.52(b). In addition to these two emitter designs, the SiC material can also be replaced with AlN in emitter A, which forms emitter C. Since AlN has a very small κ around $2\,\mu$m in Fig. 6.53(a), it has negligible absorption or emission

Table 6.1. The characteristics of three considered RP emitters.

	Emitter A	Emitter B	Emitter C
	SiC/Ta (0.09 μm)	SiC/Ta (0.33 μm)	AlN/Ta (0.09 μm)
MP (°C)	2730/2980	2730/2980	2200/2980
$P_{emit}/P_{BB}(<1.93\,\mu m)$	0.674	0.651	0.541
$P_{emit}/P_{BB}(>1.93\,\mu m)$	0.085	0.150	0.052
Power selectivity	7.93	3.34	10.40

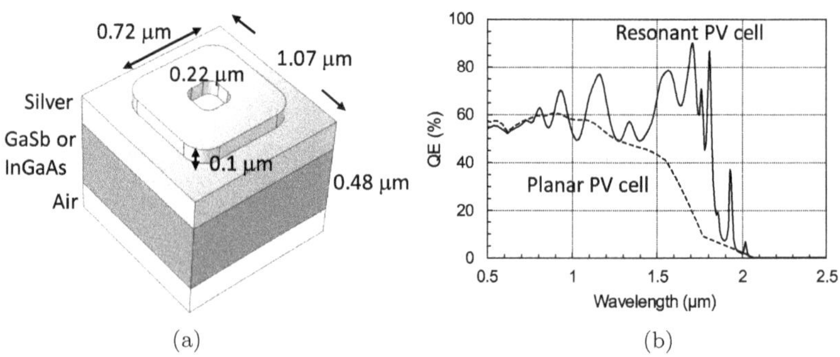

Figure 6.54. (a) The RP PV cell geometry. (b) The modeled RP and planar PV cell QE. Adapted with permission from [53].

at these wavelengths. With emissions only from the Ta cover, the out-of-band emission can be further suppressed and higher power selectivity is achieved in Fig. 6.53(b). Although emitter C yields a higher selectivity, its power output below 1.93 μm is reduced because of the lack of AlN emission.

Table 6.1 shows the salient characteristics of the three emitters referenced to a blackbody emitter with $\varepsilon = 1$. Other than the power performance, SiC also has a higher melting point than AlN, which makes SiC a more durable material for TPV applications. Therefore, we will choose emitter A for further study.

For the PV cell design, a similar RP structure is adopted to enhance the absorption QE near the bandgap of the absorber. The present design adopts GaSb material, and the RP structural parameters are shown in Fig. 6.54(a). Since GaSb and $In_{0.53}Ga_{0.47}As$

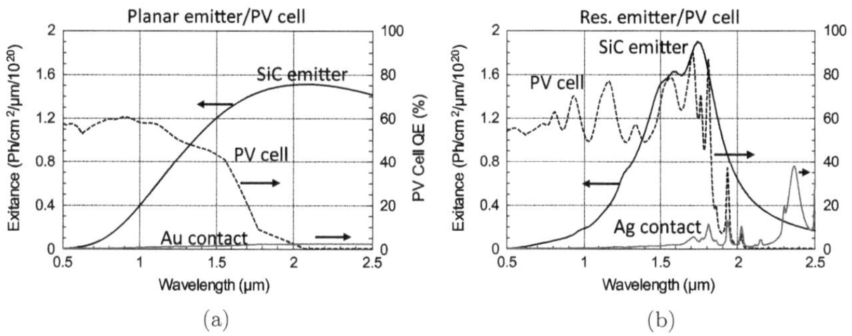

Figure 6.55. The emission and absorption spectra of (a) a planar TPV system and (b) a RP TPV system at $T = 1500°$C. The average ε of the planar emitter is taken as 0.743. Adapted with permission from [53].

share similar refractive indices (3.98 vs. 3.52) and bandgaps (0.72 vs. 0.74 eV), the present design can also be applied to InGaAs material with minor adjustments. The RP structure places a resonant LSP mode near the cutoff wavelength of $\sim$1.8 μm in GaSb, thus raising its QE at these wavelengths by $\sim$8 times as shown in Fig. 6.54(b) for the same absorber thickness of 0.48 μm.

Without using the RP approach, a bulk planar emitter will have a uniform ε of 0.74 and its photon flux spectrum at $T = 1500°$C is shown in Fig. 6.55(a). The exitance maximum is away from the absorption maximum of the PV cell. On the other hand, by using the RP structures on both the emitter and PV cell, the emission and absorption spectra are much better matched as seen in Fig. 6.55(b). The power throughput and PCE can thus be improved.

Specifically, by integrating the emitter exitance spectrum and PV cell η spectrum between 0.5 and 2.5 μm, the short-circuit current density J_{SC} of the RP system is found to be 1.73 times larger than that of the planar system. Here, we assume J_{SC} equals the photocarrier generation rate for both planar and resonant PV cells. In addition to the higher power throughput, the RP system also has less power loss to carrier thermalization because the system emits and absorbs near the GaSb bandgap wavelengths.

The PCE can be determined from the emission and absorption spectra of the system. Figure 6.56 shows the spectral radiance

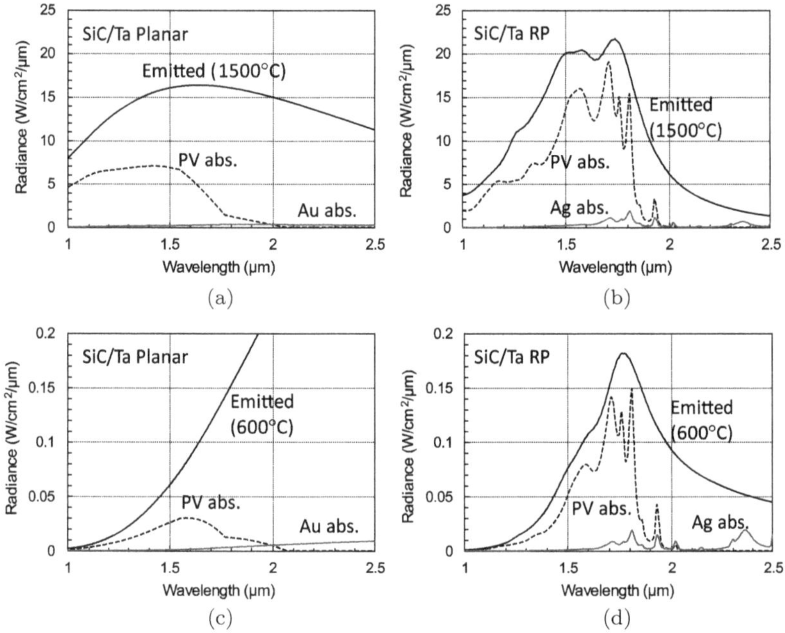

Figure 6.56. The emitted power of the emitter, the absorbed power in the GaSb absorber, and the dissipated power in the metal contact of (a) a planar system at 1500°C, (b) a RP system at 1500°C, (c) a planar system at 600°C, and (d) a RP system at 600°C.

emitted by the emitter, the absorbed spectrum in the PV GaSb active layer, and the absorbed spectrum in the PV metal cover layer at two emitter temperatures of 1500°C and 600°C. The RP system is seen to yield a more targeted thermal emission and a stronger PV absorption in both temperatures than the planar system.

If we assume the open-circuit voltage V_{OC} of the PV cell to be 0.73 V, view factor f to be 1, power fill factor (FF) to be 0.73, which is achievable in GaSb PV cells [57], and the reflected radiance to be completely reabsorbed by the emitter, the relevant powers of the RP TPV system can be calculated by integrating the displayed power spectra. For example, at a typical operating $T = 1500°\text{C}$, the power emitted by the RP emitter, P_{emit}, is 16.83 W/cm^2. The power absorbed by the PV cell, P_{pv}, is 8.40 W/cm^2. The generated electrical power, $P_{el} = J_{SC} \times V_{OC} \times FF = 10.15\,\text{A/cm}^2 \times 0.73\,\text{V} \times 0.73$, is

$5.31\,\mathrm{W/cm^2}$. The dissipated power in the PV cell, $P_{\mathrm{dis}} = P_{PV} - P_{el}$, is $3.09\,\mathrm{W/cm^2}$. The power absorbed in the Ag cover, P_{Ag}, is $0.54\,\mathrm{W/cm^2}$. The reflected power, $P_{\mathrm{refl}} = 7.88\,\mathrm{W/cm^2}$, is the difference between the emitted power and the total absorbed power in PV cell and Ag. The consumed power, $P_{\mathrm{cons}} = P_{\mathrm{emit}} - P_{\mathrm{refl}}$, is $8.94\,\mathrm{W/cm^2}$. The PCE $= P_{el}/P_{\mathrm{cons}} = 0.594$. The fraction of power wasted $= P_{\mathrm{dis}}/P_{\mathrm{cons}} + P_{\mathrm{Ag}}/P_{\mathrm{cons}} = 0.346 + 0.06 = 0.406$. The fraction of integrated power absorbed in the PV cell relative to the consumed power, $\eta' = P_{pv}/P_{\mathrm{cons}}$, is $8.40/8.94 = 0.94$. This value of η' represents the ideal PCE before any power loss other than the Ag absorption is considered.

The modeled PCE of 59.4% represents 0.711 of the maximally allowed $\eta_C = (1 - T_a/T_e) = 83.6\%$ of a Carnot engine, where $T_a = 300\,\mathrm{K}$ is the PV cell temperature and $T_e = 1773.25\,\mathrm{K}$ is the emitter temperature. In practice, there are other factors that reduce the modeled PCE. Besides the view factor and fill factor reduction, the optical absorption in the highly doped PV cell semiconducting contact layers and at the enclosure sidewalls will also reduce PV absorption. Other system factors, such as enclosure thermal insulation, power consumed in cooling the PV cells, power recovered from pre-heating the fuel, fuel combustion efficiency, and DC-to-AC conversion efficiency, etc., will also affect the actual system efficiency. In this text, we restrict our discussion to factors affected by the emitter and PV cell optical designs only.

Interestingly, there are two optical factors that could also affect PCE. One is the absorptivity η of the emitter. To simplify the discussion, we have assumed the reflected light from the PV cell to be completely reabsorbed by the emitter and therefore it does not count toward the consumed power. However, this assumption does not apply to the present RP emitter with its $\eta < 1$ displayed in Fig. 6.51(a). It turns out that the emitter does not need to be a perfect blackbody to achieve photon recycling. Rather, the emitter can partly absorb the reflected light, in which there is no power loss, and partly reflect the light back to the PV cell, in which the light can contribute to the power generation with the same PCE. The cycle will then repeat itself, and PCE does not change. By modeling the emitter and PV cell together in Section 6.5.5, in which the radiance from the emitter to the PV cell settles into a steady state, PCE is confirmed to be the same as the current value. The second factor that may affect PCE is the reduction of J_{SC} due to radiative recombination in the PV cell. Under device operation, the PV cell is forward biased and radiative recombination rate increases. Normally, it will reduce PCE as in solar cells. However, if the resulting luminescent radiation from the PV cell is directed back to the emitter for reabsorption, it will again pose no power loss and thus no PCE reduction. Obviously, this reasoning is true only if all the light travels only between the emitter and PV cell and is not diverted to other locations. In Section 6.5.4, we will show that it is substantially true for the present RP structures.

The RP TPV system can be operated at different T. As T decreases from 1800°C toward 600°C, a larger portion of the long-wavelength photons is being suppressed by the RP emitter. This factor is important in keeping PCE high because the longer λ photons can be absorbed by the metal contact in the PV cell and thus degrade PCE even with 100% photon recycling. In Fig. 6.57(a), the emitted power, consumed power, and the converted electrical power are shown at different emitter T. As T decreases from a high T, the emitted power decreases nearly exponentially. Therefore, it is important to keep PCE constant so that the electrical power

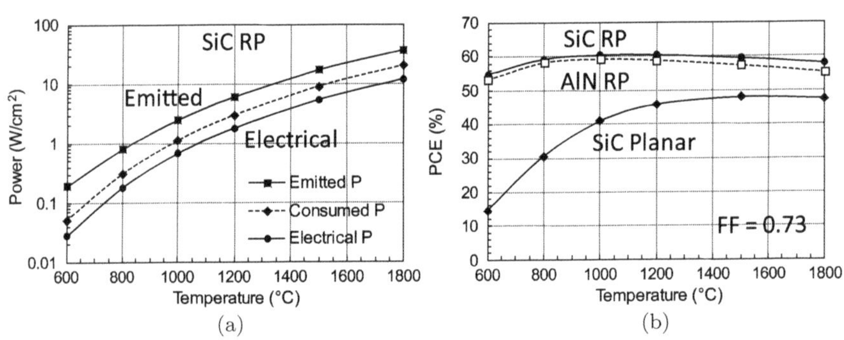

Figure 6.57. (a) The emitted, consumed, and generated electrical power of the SiC RP TPV system at different T. (b) The power conversion efficiency of the SiC RP, AlN RP, and the SiC planar systems at different T. A power fill factor of 0.73 is assumed. Adapted with permission from [53].

converted at the lower T will not be excessively low, and the TPV system is useful for a wider range of operating T.

In Fig. 6.57(b), the RP system is shown to maintain a relatively constant PCE of ~60% between 600°C and 1800°C. In contrast, the planar system can only achieve 47.9% PCE at 1500°C and it decreases to 14.7% at 600°C. This difference can be understood by examining the spectral radiance of the two systems in Figs. 6.56(c) and (d). In addition to the larger ohmic loss in the contact layer in the planar system, the smaller PV absorption power at 600°C also contributes to the lower PCE. The smaller absorption is due to the smaller η near the bandgap, which overlaps less with the redshifted thermal radiation. Due to the large fraction of unabsorbed photons, the planar system is also more reliant on photon recycling.

6.5.4 *Directivity of Resonator-Pixel Emitters*

The spatial power profile of the emitter is also of interest because it is related to the required distance between the emitter and PV cell to maintain a certain PCE. PCE depends on the fraction of emitted photons reaching the PV cell area A_{PV}. For a Lambertian emitter, the photon flux Φ falling on the PV cell is given by $\Phi = A_{em}\pi L f$, where A_{em} is the area of the emitter, L is the radiance in W/cm^2/sr, f is the view factor $= A_{PV}/(\pi s^2)$, and s is the distance between the emitter and PV cell. With a given A_{PV}, PCE decreases as a function of s^2. In Section 6.5.3, we have assumed $f = 1$. However, the Lambertian emission characteristics of a bulk emitter originate from its macroscopic size, in which the thermal motion of carriers can be in any direction. As a result, the electric dipole fluctuations radiate thermal energy in random directions [58, 59], which becomes isotropic upon time averaging. On the other hand, in the microscopic RP emitter, only resonant cavity modes that obey Maxwell's equations can be excited from the random motions. Since these modes in the present RP emitter have fixed orientations, the thermal emission is directional. Furthermore, the spectra of emissivity in Figs. 6.51(a) and 6.52(a) are obtained from the absorptivity under normal incidence. By the antenna reciprocity theorem, the ε spectrum is then also for normal exitance. This argues for the emission from the present periodic ring array to be directed only at the forward direction. The view factor f is thus 1 as adopted. To confirm this argument, explicit modeling of the spatial emission profile is performed.

In Section 3.3, we have modeled SPP modes traveling on a planar surface and a shallow corrugated surface. With a planar metal surface, the SPP waves are bound vertically at the metal interface but free to travel along the surface with a long propagating length. On the other hand, when the surface is modulated at a uniform periodicity, the surface transfers the required momentum to the SSP waves and releases them into space. Using the previous example where a shallow gold grating covered by GaAs was considered, when the grating period p is 0.36 μm, the grating wavevector $k_G = 2\pi/p = 1.75 \times 10^7$ rad/m. Meanwhile, based on the SPP dispersion relation from Eq. (3.22), at the optical frequency $f = 2.4 \times 10^{14}$ Hz, the SPP wavevector k_{SPP} is the same as k_G. According to the grating formula, $k_d \sin\theta = k_{\text{SPP}} \pm Nk_G$, where k_d is the EM wave in GaAs and θ is the diffraction angle, when the SPP wave is scattered by the grating at this frequency, it loses all of its horizontal momentum and travels perpendicular to the interface (*i.e.*, $\theta = 0°$). This scattering condition is illustrated in Fig. 6.58(a) and is modeled using FEM outlined in Section 3.6.1. In the model, a SPP wave is launched at the interface from the left margin of the modeling

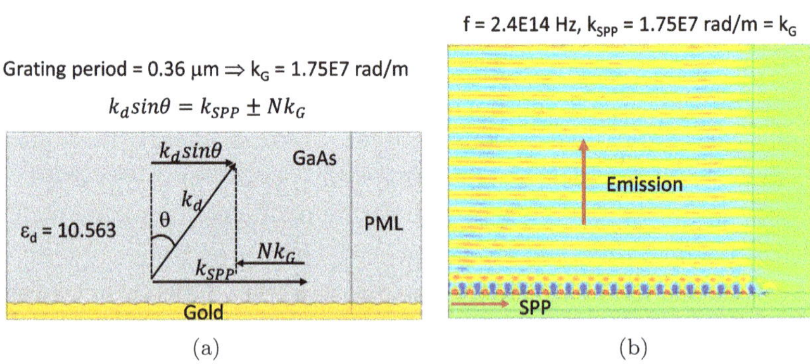

Figure 6.58. (a) A shallow grating made of gold and GaAs. (b) The modeled wavefront of E_x at $f = 2.4 \times 4^{14}$ Hz.

space in Fig. 6.58. Upon scattering, an optical wavefront is seen traveling upward from the interface in Fig. 6.58(b). At the same time, the SPP is attenuated gradually from left to right. When a bulk sample is heated up by a thermal source, both the bulk modes and the surface modes are excited. Since the surface modes are bound, only the bulk modes contribute to thermal radiation, which exhibits the usual Lambertian characteristics. However, when the surface contains periodic corrugations, the surface modes can be freed by the corrugations and add to the thermal spectrum. In this case, the angular emission profile contains sharp peaks at certain angles and frequencies according to the grating formula. These emitting characteristics match the observation in the shallow grating experiments [33, 49] and are reproduced here using FEM modeling.

When the surface is fabricated into periodic coupled resonant cavities with deep grooves, the coupling between surface wave and corrugations is much stronger. When the structures A and B are designed by maximizing their absorption at λ_r, the designs yield the coupled resonant cavity dimensions that interact with the surface wave strongly at the resonant wavelengths. To test their coupling strengths, SPP waves at the respective λ_r are launched from the left in Fig. 6.59 for A and B. The Ta metal surfaces are populated with periodic corrugations having the same cross-sections displayed in Fig. 6.50. In Fig. 6.59, the corrugations are seen to orchestrate the wavefront of the surface wave such that the wave is rapidly channeled into perpendicular propagation within two corrugation periods for A

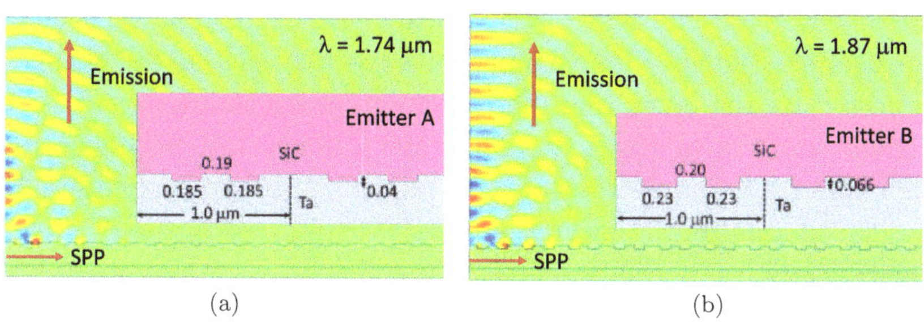

Figure 6.59. The E_x wavefronts of the designed corrugated surfaces of (a) Emitter A and (b) Emitter B using SiC dielectric and Ta metal for TPV applications.

and one corrugation period for B. The much shorter propagating lengths in comparison with that in Fig. 6.58 illustrate the strong coupling between the surface wave and the RP structure. Emitter B exhibits stronger interaction because its grooves are deeper than that of A (*i.e.*, 0.066 μm vs. 0.04 μm). This emitting condition is experimentally verified in coupled resonant cavity thermal experiments [34–37], where strong coherent thermal radiation is emitted perpendicular to the emitting surfaces at unique wavelengths.

Due to its shallower corrugations, emitter A does not emit strictly perpendicular waves when a traveling SPP wave is scattered. To know its directivity when the corrugations are excited uniformly by thermal energy, we use the AC current feedlines as prescribed before. When the RP cavity is excited by normal incident light, the field pattern is very similar to that in Fig. 6.44(a). Therefore, we can use the same AC current feedline arrangement for the present thermal emitter as presented in Fig. 6.30(b). To do so, one can argue that although there are many possible field patterns that can be generated by the random motions of electrons, the resulting fields are generally suppressed by destructive interference in the cavity. Only the cavity mode generated from those motions can be thermally excited in large amplitudes. In creating the mode, the charges in the material must move in an organized pattern governed by the field profile. As before, the distributed charge displacement is approximated by two point charges at the centroids of the opposite charge distributions. The length of the effective dipole will have the same extent as the high-field region in the cavity mode. Judging from the intensity and direction of the field distribution, which is similar to that of Fig. 6.44(a), the effective dipole will have two spatial components connected by the inner ring metal, and they are located at the two diagonal corners of the ring. The cavity mode will be generated when these two dipole components oscillate in opposite phases. In physical thermal radiation, the spatially distributed dipole is driven stochastically by thermal energy to gain the radiation power. In EM modeling, these dipoles are driven by electrical energy. Each feedline has a length of 0.03 μm and carries an arbitrary AC current of 10 mA. The angular frequency ω is $2\pi c/\lambda$.

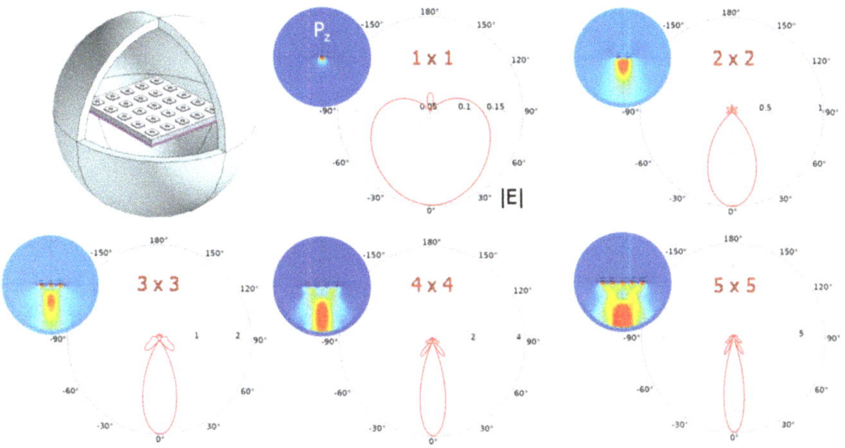

Figure 6.60. The far-field $|E|$ profiles of different-sized ring arrays. The insets show near-field P_z profile. Adapted with permission from [53].

The radiated power spatial profiles of different unit cell array sizes are modeled. Similar to the modeling of a periodic array, the field distributions of these finite arrays under the receiving mode are first modeled to know the current driving configurations. The field distributions and their relative phase are found to be nearly the same as those of the periodic array. Therefore, the same effective dipole arrangement is adopted as in the periodic case. Figure 6.60 shows both their far-field and near-field patterns.

When the number of unit cells increases, the emitted beam is more collimated. The directivity D of the emitted beam increases rapidly with increasing array size in Fig. 6.61. When the array is larger than 10×10 unit cells, which occupy an emitter area of $10 \times 10\,\mu\text{m}^2$, D is extrapolated to be larger than 400. Therefore, the RP thermal emitter is highly directional, which allows large separation of the emitter and PV cell to minimize convectional heating of the PV cell.

6.5.5 *Active Emissivity Modeling*

In modeling the emissivity of A in Fig. 6.51(a) and B in Fig. 6.52(a) in Section 6.5.2, the emitted power of the emitters is not directly modeled. Instead, the value of ε is obtained from absorptivity η

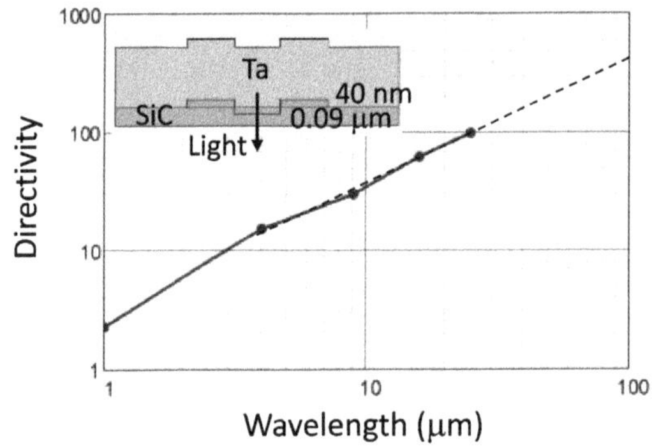

Figure 6.61. The directivity of an RP emitter with different number of unit cells. Adapted with permission from [53].

using Kirchhoff's law. This equality ε and η is widely used in optics. Since the emitted power from an emitter can be modeled using electrical feedlines, we can use the active approach to calculate ε and compare it with η. To proceed, let us recall that a blackbody is a perfect emitter with $\varepsilon = 1$ for all wavelengths. The thermally excited oscillating dipoles in this body are able to emit all their thermal energy into free space. Accordingly, the $\varepsilon(\lambda)$ of an ordinary object is then the fraction of thermal power radiated averaged over all its degenerate cavity modes at λ [58]. This fractional power can be known from the input driving power P_{in} and output radiation power P_{out} of each cavity mode in EM modeling. For emitters A and B, there is only one cavity mode at $\lambda_r = 1.74\,\mu$m and $1.87\,\mu$m, respectively. Therefore, $\varepsilon(\lambda_r)$ of these RP emitters can be obtained by evaluating $P_{\text{out}}/P_{\text{in}}$ of these modes.

For emitter A, because of the rapidly changing E field magnitude and direction in the thin ($0.09\,\mu$m) cavity, it is less accurate in assigning finite-size feedlines for quantitative ε modeling. On the other hand, due to the thicker ($0.33\,\mu$m) cavity in emitter B, its field distribution is more orderly, which makes the effective dipole assignment more accurate. Therefore, emitter B is used for active ε modeling. In Fig. 6.62(a), we first show the modeled absorptivity η (*i.e.*, the EQE) of the structure under normal incidence. The exhibited η is partly contributed by the absorption in the SiC

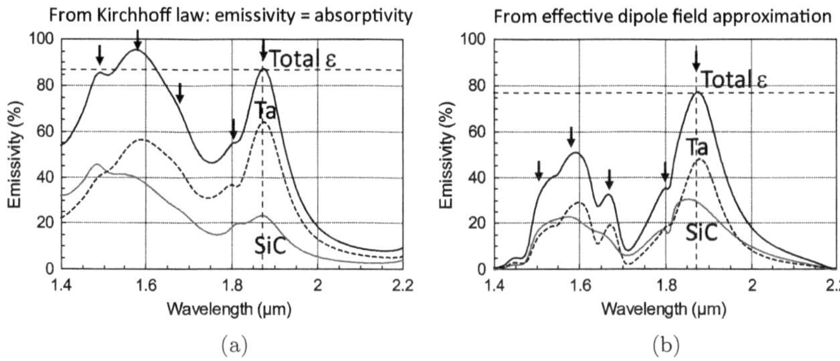

Figure 6.62. The ε calculated (a) from absorptivity modeling and (b) from effective dipole approximation. The black arrows show similar peak wavelengths. Adapted with permission from [53].

dielectric layer and partly contributed by the Ta metal layer. In fact, the larger ohmic loss in Ta gives the layer a higher thermal emission than that in SiC when the structure is heated up. The total η gives the value of ε under passive approach and is considered to be accurate. At $\lambda_r = 1.87\,\mu$m, ε attains a large value of 87.1%.

To model actual emission from B, the modeled field distribution at 1.87 μm under external optical excitation is shown in Figs. 6.63(a) and (c), which exhibit an orderly field pattern. To produce this field distribution, two dipole feedlines are placed in the diagonal positions across the high-field regions. Each current feedline has a length of 0.025 μm and points along the z-axis in opposite directions. Each carries an oscillating current $I_0(\omega)$ of 10 mA. The total time-averaged input driving power $P_{\text{in}}(\omega)$ is $2\times(I_0V_0/2) = I_0V_0$, where $V_0 = \int E_z dz$, and E_z is the resulting in-phase field distribution along one of the feedlines from EM modeling. By computing radiated output power $P_{\text{out}}(\omega)$ under $P_{\text{in}}(\omega)$, the value of $\varepsilon(\omega) = P_{\text{out}}(\omega)/P_{\text{in}}(\omega)$ can be calculated. In addition, with the known $P_{\text{out}}(\omega)$, the radiative resistance $R_{\text{rad}}(\omega)$ of the emitter can also be obtained from $2P_{\text{out}}(\omega)/I_0^2$ along with its ohmic component $R_{\text{ohm}}(\omega) = 2P_{\text{abs}}(\omega)/I_0^2$ from the absorbed power in the emitter.

Since the cavity modes in the resonant emitter are LSP modes, they are common to both SiC and Ta layers. Thermal current

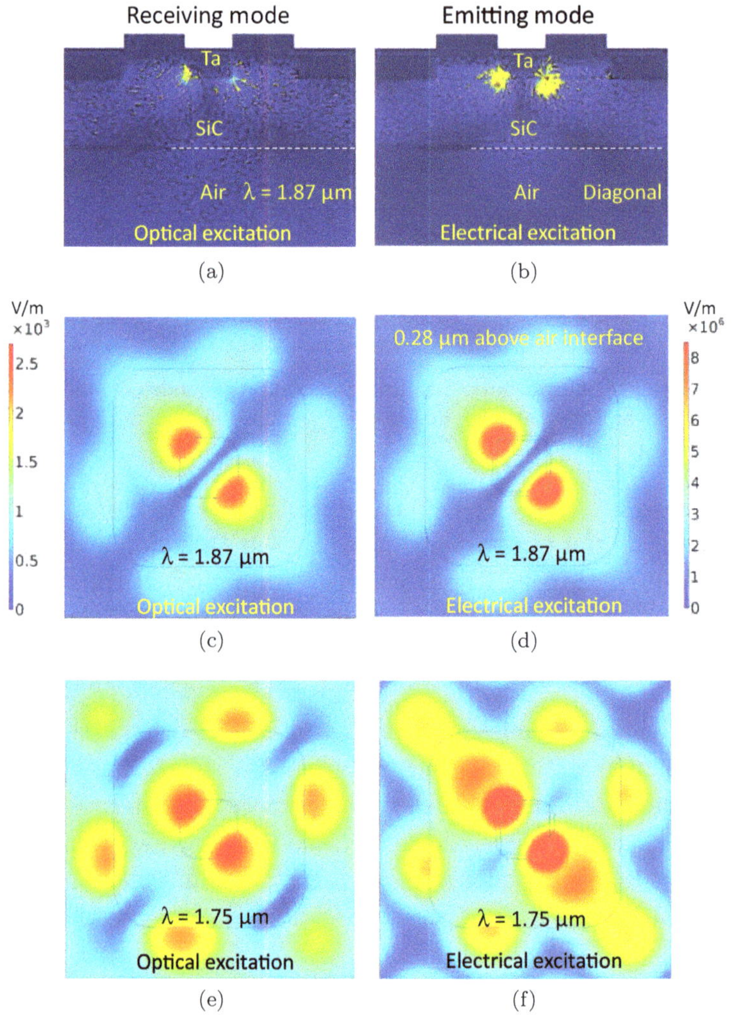

Figure 6.63. The cavity modes (a), (c), and (e) under optical excitations and (b), (d) and (f) under electrical excitations. Adapted with permission from [53].

oscillations in both layers can produce the same cavity modes. To model ε of the SiC layer, the thermal power P_{in} generated in the SiC layer is partly radiated out as P_{out} and partly absorbed by the Ta cover as P_{abs}. The absorption in the SiC layer, however, is not included since energy transfer to the same oscillating dipole will not lead to net power loss. From the ratio of P_{out} and P_{in}, ε of the SiC layer is obtained and shown in Fig. 6.62(b), which is similar to that from η modeling in Fig. 6.62(a). The modeled P_z is generally very similar to that of Fig. 6.31. In particular, the majority of P_z is radiated from the inner ring patch. Since the inner ring diagonal is $0.2\sqrt{2} = 0.283\,\mu$m, it serves as a half-wave dipole antenna at the internal resonant wavelength $\lambda_{\text{int}} = 1.87/3.17 = 0.590\,\mu$m to strongly radiate the thermal energy. Incidentally, the total consumed power P_{cons} obtained by adding P_{out} and P_{abs} is equal to P_{in} within 0.04%, which shows the internal consistency of the modeling.

Similarly, to model the ε contribution of the Ta layer from the SiC surface, we again consider P_{in} due to the oscillating current in the Ta cover to be partly radiated out as P_{out} and partly absorbed by the SiC layer as P_{abs}. The maximum P_{out} occurs at the same λ of $1.87\,\mu$m with a deduced ε of 0.472, quite close to the 0.636 from η modeling. Therefore, the effective dipole approximation also works quite well for this ε component.

Adding ε of both layers gives the emitter ε in Fig. 6.62(b). The effective dipole approximation yields a peak ε of 0.769 at $\lambda_r = 1.87\,\mu$m, which is close to the 0.871 in the absorptivity modeling. It also yields a similar lineshape around λ_r and other peak wavelengths as the arrows indicate. To show that the present effective dipole oscillations approximately produce the intended cavity mode, Figs. 6.63(c) and (d) show the E field profiles under internal electrical excitation. Their similarity testifies to the validity of the present effective dipole assignment at this λ. Note that the larger E field in Fig. 6.63(a) in the air region belongs to the external optical excitation.

To understand the smaller modeled ε particular at the shorter wavelengths, we plot the field distributions at $\lambda = 1.75\,\mu$m in Fig. 6.63(e) under receiving mode and in Fig. 6.63(f) under emitting mode. Although these field patterns bear some resemblance to each other, they are less similar than those at λ_r. With the diverging field patterns away from λ_r, the modeled ε becomes less accurate. For accurate ε modeling at each λ, dedicated feedline assignments are needed. Nevertheless, the present feedline arrangement is still able to reproduce the approximate spectral emissivity near λ_r for this microstructured heterogeneous emitter, which is otherwise difficult to obtain using the active approach.

Finally, we should note that although the ring structure creates LSPs from the thermal energy and transmits its energy vertically away from the surface, it does not necessarily produce a large ε due to optical reflection at the SiC/air interface. The SiC layer needs to be with certain thicknesses so that Fabry-Perot interference in this layer produces transmission maximum at λ_r to overcome the internal reflection. The suitable thickness is automatically obtained when the structure is optimized for maximum absorption under normal incidence. With these thicknesses, the SiC/air interface exerts very little impedance to the transmitted waves across the interface. This situation is illustrated in Fig. 6.64 for both emitters A and B. The parallel E component E_x is seen to be continuous across the SiC/air interface for both emitters, indicating that the internally reflected parallel E to be very small. Figure 6.64 shows that when the periodic unit cells are thermally excited, strong emission can be obtained from both emitters in the normal direction at the resonant wavelengths. As before, the dynamic thermal emissions from the corrugated surfaces can be viewed by performing an animation in COMSOL on emw.Ex in the *Electric field, Surface plot* node. In the *Animation* setting, *Dynamic data extraction, Full harmonic, 36 Frames* are set. When the phase of the wave advances with time, the waves are seen to

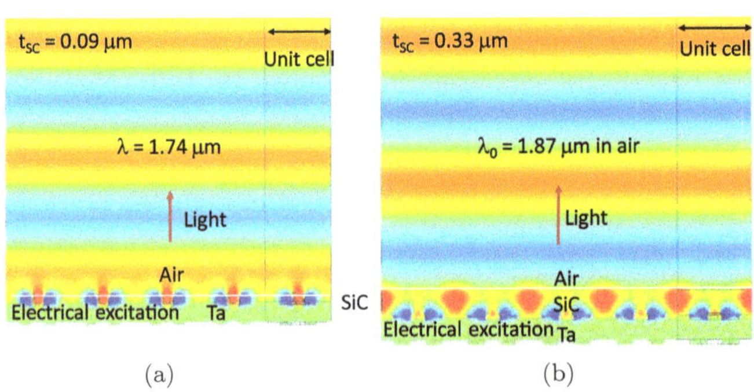

Figure 6.64. The E_x emission profiles of 3D periodic arrays propagating along the vertical z axis for (a) emitter A and (b) emitter B.

propagate upward from the surfaces. The present active modeling of emissivity is thus a demonstration of Kirchhoff's law that a strong absorber is also a strong emitter.

6.5.6 *Near-Field Thermophotovoltaic Effects*

Using a pair of current feedlines to produce an output radiation from emitter B, one can use it to study the emission and absorption characteristics of an integrated TPV system in the same modeling space. The ratio of the absorbed power P_{PV} in the absorber layer to the net radiated power from the emitter, which equals the consumed power, can be obtained. This ratio η' has been defined previously to represent the ideal PCE without energy loss when different powers in a TPV system are discussed.

The purpose of setting up this integrated assembly is to confirm that the emitter does not need to be a perfect blackbody to achieve photon recycling and to study near-field TPV effects. There are studies on the change of power throughput when the gap between the emitter and PV cell is less than the dominant thermal wavelength [60, 61]. For the present emitter, it is less than $2\,\mu$m. When the surfaces of the pair are close to each other, the evanescent field outside the emitter surface such as that in Fig. 6.64 can tunnel into the PV cell and achieve a larger P_{PV}.

To facilitate the modeling, we change the PV cell period from $1.07\,\mu$m in Fig. 6.54(a) to $1.0\,\mu$m so that the emitter and PV cell have the same periodicity. With this small change, we can continue to model only one unit cell with periodic boundary condition. The absorption spectrum of the modified PV cell is shown in Fig. 6.65. It is only slightly different from the optimized PV cell shown in Fig. 5.54(b).

At the same time, the thermal emission is assumed to be from the SiC layer of emitter B shown in Fig. 6.65. These adopted spectra are under the assumption that the two devices are separated far apart so that their power exchange does not maintain phase coherence. Figure 6.66 shows the TPV unit cell arrangements with different separations. The system characteristics are modeled with the gap size from $10\,\mu$m down to $0.01\,\mu$m.

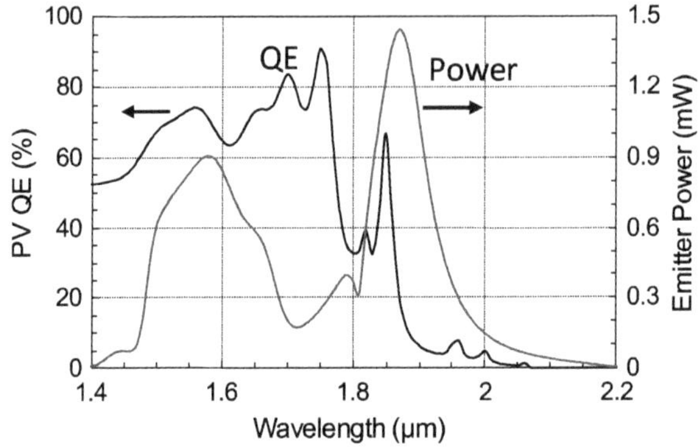

Figure 6.65. The far-field emission power per unit cell and absorption spectra of the PV cell. Reproduced with permission from [53].

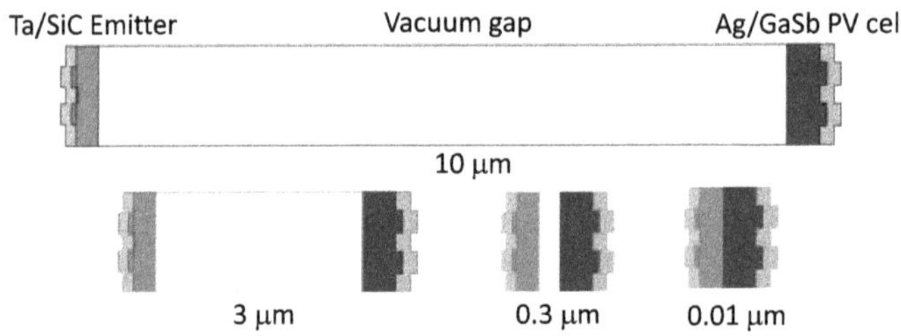

Figure 6.66. The integrated TPV emitter/PV cell with different gap distance from 10 μm down to 0.01 μm. Reproduced with permission from [53].

When the emitter and PV cell are modeled in the same modeling space, the emission spectrum of the emitter and the absorption spectrum of the PV cell are very different from their individual characteristics. Firstly, phase coherence is enforced between emitter and absorber. In addition to optical interference within each device, there are also return waves that add to the interference and produce different sets of resonances. The emission/absorption characteristic becomes a joint property of the integrated system. Secondly, the net emission power from the emitter is the result of multiple power exchanges between the two devices, after which they settle into a steady state. The net emission power thus equals the total consumed power in the PV cell. There is no net return power. Thirdly, if we

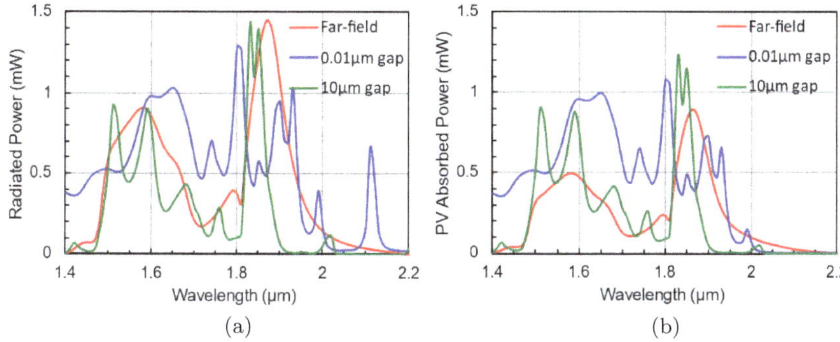

Figure 6.67. (a) The emission spectrum of the emitter in the integrated system. (b) The absorption spectrum of the PV cell in the integrated system. Reproduced with permission from [53].

use the definition of absorption QE, labeled as η', to be the absorbed power over the net emission power (*i.e.*, the consumed power in this case), the total η' from the PV absorber and the Ag metal cover must add up to 100%. The net emission power P_{rad} at the two extreme gap distance of 0.01 μm and 10 μm is plotted in Fig. 6.67(a). It is seen that while the 10 μm gap P_{rad} is closer to the far-field P_{rad}, the 0.01 μm P_{rad} has higher power in more wavelengths.

At the same time, the PV cell also absorbs more power in more wavelengths at 0.01 μm gap shown in Fig. 6.67(b), while the PV cell with 10 μm gap absorbs similar power as in the far-field case. Consequently, both the average emission power and the average PV absorbed power between 1.4 and 1.9 μm band increase as the gap size decreases by less than 2 μm, as seen in Fig. 6.68(b). The power throughput thus increases by a factor of ~2 in near-field TPV. Since the PV cell is responsible for almost all optical absorption above the GaSb bandgap, η'_{PV} is thus large in Fig. 6.68(a) when $\lambda < 1.9\,\mu$m. Conversely, the Ag cover is responsible for all the optical absorption below the bandgap, leading to the large η'_{Ag} at the longer wavelength. Note that although the emitter is able to emit more power into the PV cell through optical tunneling when the gap becomes smaller, the values of η' for both the PV cell and the Ag cover maintain relatively unchanged among different gap values in Fig. 6.68(a). The average

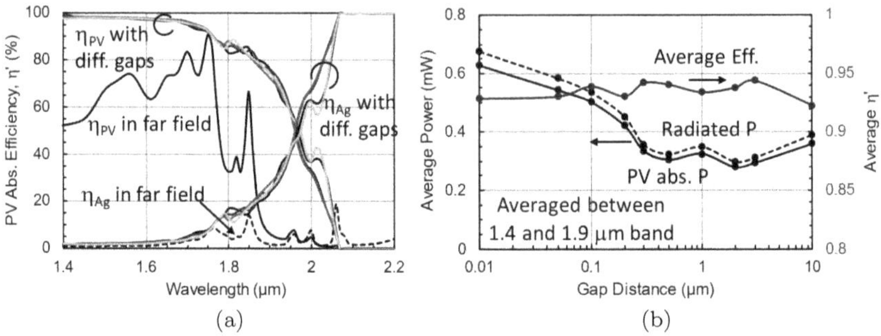

Figure 6.68. (a) The QE η' of the PV cell and Ag cover in the integrated system with gap sizes of 0.01, 0.3, 3, and 10 μm, which show little changes with different gap sizes. (b) The power throughput and the average PV cell QE of the integrated system. Reproduced with permission from [53].

PV cell η' between 1.4 and 1.9 μm is plotted in Fig. 6.68(b) as a function of the gap distance. The average η' maintains a constant value of 0.93–0.95, independent of the gap distance between 0.01 μm and 10 μm. It is remarkably the same as the far-field value mentioned previously when the relevant powers in a TPV system are discussed.

Since η' controls the available power for electrical conversion before the PV cell loss mechanisms set in, the same η' thus indicates the same PCE for both near-field and far-field systems. Therefore, while explicit EM modeling predicts larger power throughput at shorter gaps because of photon tunneling, PCE remains the same as long as there are no changes in the energy loss mechanisms. These predicted trends are consistent with the theoretical and experimental findings in certain near-field TPV systems [62,63]. The modeling also shows the emitter does not need to absorb all the reflected light to achieve the same PCE.

In Fig. 6.67(a), the far-field emission peak at 1.87 μm disappears in the near-field. As the emitter is designed to radiate efficiently at this wavelength, adding another surface close to the emitter changes the resonant condition and reduces its radiation power. Overall, since the RP emitter is designed to have high emissivity in the 1.4 to 1.9 μm band, the increase in the radiation power is much less substantial than other non-resonant planar emitters [60–63].

References

1. Grandidier, J., Callahan, D. M., Munday, J. N. and Atwater, H. A. (2012). Gallium arsenide solar cell absorption enhancement using whispering gallery modes of dielectric nanospheres, *IEEE J. Photovoltaics*, 2, 123.
2. Xiao, J., Fang, H., Su, R., Li, K., Song, J., Krauss, T. F., Li, J. and Martins, E. R. (2018). Paths to light trapping in thin film GaAs solar cells, *Opt. Express*, 26, A341–A351.
3. Miles, R. H., Chow, D. H., Schulman, J. N. and McGill, T. C. (1990). Infrared optical characterization of InAs/$Ga_{1-x}In_xSb$ superlattices, *Appl. Phys. Lett.*, 57, pp. 801–803.
4. Dixon, J. R. and Ellis, J. M. (1961). Optical properties of n-type indium arsenide in the fundamental absorption edge region, *Phys. Rev.*, 123, pp. 1560–1566.
5. Höglund, L., Rodriguez, J. B., Naureen, S., Ivanov, R., Asplund, C., Marcks von Würtemberg, R., Rossignol, R., Christol, P., Rouvié, A., Brocal, J., Saint-Pé, O. and Costard, E. (2018). Very long wavelength type-II InAs/GaSb superlattice infrared detectors, *Proc. SPIE 10624*, 106240I1–7.
6. Rogalski, A. (2005). HgCdTe infrared detector material: history, status and outlook, *Rep. Prog. Phys.*, 68, pp. 2267–2336.
7. Benisty, H., De Neve, H. and Weisbuch, C. (198). Impact of planar microcavity effects on light extraction — part I: basic concepts and analytical trends, *IEEE J. Quant. Electr.*, 34, 1612.
8. Matioli, E. and Weisbuch, C. (2013), Active Region Part A: Internal Quantum Efficiency in LEDs, in Seong, T.-Y., Han, J., Amano, H. and Morkoc, H. (Eds.), *III-Nitride Based Light Emitting Diodes and Applications*, Springer, pp. 121–152.
9. Zhang, J., Zhang, L. and Xu, W. (2012). Surface plasmon polaritons: physics and applications, *J. Phys. D*, 45, pp. 113001–19.
10. Zhong, Y., Malagari, S. D., Hamilton, T. and Wasserman, D. (2015). Review of mid-infrared plasmonic materials, *J. Nanophotonics*, 9, pp. 093791–21.
11. Gu, X., Qiu, T., Zhang, W. and Chu, P. K. (2011). Light-emitting diodes enhanced by localized surface plasmon resonance, *Nanoscale Res Lett.*, 6, 199.
12. Okamoto, K., Niki, I., Shvartser, A., Narukawa, Y., Mukai, T. and Scherer, A. (2004). Surface-plasmon-enhanced light emitters based on InGaN quantum wells, *Nat. Mater.*, 3, pp. 601–605.
13. Fujiki, A., Uemura, T., Zettsu, N., Akai-Kasaya, M., Saito, A. and Kuwahara, Y. (2010). Enhanced fluorescence by surface plasmon

coupling of Au nanoparticles in an organic electroluminescence diode, *Appl. Phys. Lett.*, 96, pp. 043307–3.

14. Huang, K., Gao, N., Wang, C., Chen, X., Li, J., Li, S., Yang, X. and Kang, J. (2014). Top- and bottom-emission-enhanced electroluminescence of deep-UV light-emitting diodes induced by localized surface plasmons, *Sci. Rep.*, 4, 4380.
15. Kim, N., Hong, S., Kang, J., Myoung, N., Yim, S.-Y., Jung, S., Lee, K., Tue, C. W. and Park, S.-J. (2015). Localized surface plasmon-enhanced green quantum dot light-emitting diodes using gold nanoparticles, *RSC Adv.*, 5, pp. 19624–19629.
16. Bakker, R. M., Yuan, H.-K., Liu, Z., Drachev, V. P., Kildishev, A. V. and Shalaev, V. M. (2008). Enhanced localized fluorescence in plasmonic nanoantennae, *Appl. Phys. Lett.*, 92, pp. 043101–3.
17. Steck, D. A. (2016). Quantum and Atom Optics, available online at http://steck.us/teaching (revision 0.11.5, 27 November 2016), ch. 10.
18. Khurgin, J. B. (2015). How to deal with the loss in plasmonics and metamaterials, *Nat. Nanotech.*, 10, pp. 1–6.
19. Lin, S. Y., Fleming, J. G., Chow, E., Bur, J., Choi, K. K. and Goldberg, A. (2000). Enhancement and suppression of thermal emission by a three-dimensional photonic crystal, *Phys. Rev. B*, 62, R2243–2246.
20. Al-Saymari, F. A., Craig, A. P., Lu, Q., Marshall, A. R. J., Carrington, P. J. and Krier, A. (2020). Mid-infrared resonant cavity light emitting diodes operating at 4.5 μm, *Opt. Express*, 28, pp. 23338–23353.
21. Vučković, J. (2017). Quantum Optics and Cavity QED with Quantum Dots in Photonic Crystals, in Fabre, C., Sandoghdar, V., Treps, N. and Cugliandolo, L. F. (Eds.), *Quantum Optics and Nanophotonics*, Oxford University Press, pp. 365–406.
22. Kopytko, M., Sobieski, J., Gawron, W. and Martyniuk, P. (2022). Study of HgCdTe (100) and HgCdTe (111)B heterostructures grown by MOCVD and their potential application to APDs operating in the IR range up to 8 μm, *Sensors*, 22, 924.
23. Zhao, Y., Zhang, D., Qin, L., *et al.* (2011). InGaAs–InP avalanche photodiodes with dark current limited by generation-recombination, *Opt. Express*, 19, 8546–8556.
24. Wang, W., Deng, Z., Li, X., *et al.* (2021). Dark current investigation in individual planar $In_{0.53}Ga_{0.47}As$/InP detector and fine pixel-pitch array with spacing variations, *Results Opt.*, 5, 100181.
25. Sverdlov, B., Botchkarev, A., Teraguchi, N., Salvador, A. and Morkoc, H. (1993). Reduction of dark current in photodiodes by the use of a resonant cavity, *Electron. Lett.*, 29, pp. 1019–1021.

26. Salvador, A., Sverdlov, B., Lehner, T., Botchkarev, A., Huang, F. and Morkoc, H. (1994). Resonant cavity enhanced InP/InGaAs photodiode on Si using epitaxial liftoff, *Appl. Phys. Lett.*, 65, pp. 1880–1882.
27. Verdun, M., Portier, B., Jaworowicz, K., Jaeck, J., Lelarge, F., Guilet, S., Dupuis, C., Haidar, R., Pardo, F. and Pelouard, J.-L. (2016). Guided-mode resonator for thin InGaAs pin short-wave infrared photo-diode, *Appl. Phys. Lett.*, 108, 053501.
28. See, for example, https://www.hamamatsu.com/eu/en/product/light-and-radiation-sources/led/L12509-0155K.html (7 March 2022).
29. Johnson, P. B. and Christy, R. W. (1972). Optical constants of the noble metals, *Phys. Rev. B*, 6, 4370–4379.
30. Adachi, S. (1989). Optical dispersion relations for GaP, GaAs, GaSb, InP, InAs, InSb, $Al_xGa_{1-x}As$, and $In_{1-x}Ga_xAs_yP_{1-y}$, *J. Appl. Phys.*, 66, 6030–6040.
31. Choi, K. K., Oduor, P., Dhar, N. K. and Dutta, A. K. (2022). Resonant InGaAs detectors and emitters for passive and active imaging application, *IEEE J. Quant. Electr.*, 59, 1–13.
32. Choi, K. K., Oduor, P., Dhar, N. K. and Dutta, A. K. (2023). High efficiency metastructure-based emitter and photodetector for imaging, *Proc. SPIE*, 12514, 1251402.
33. Greffet, J. J., Carminati, R., Joulain, K., *et al.* (2022). Coherent emission of light by thermal sources, *Nature*, 416, 61–64.
34. Dahan, N., Niv, A., Biener, G., Gorodetski, Y., Kleiner, V. and Hasman, E. (2007). Enhanced coherency of thermal emission: Beyond the limitation imposed by delocalized surface waves, *Phys. Rev. B*, 76, 045427.
35. Dahan, N., Niv, A., Biener, G., Gorodetski, Y., Kleiner, V. and Hasman, E. (2008). Extraordinary coherent thermal emission from SiC due to coupled resonant cavities, *J. Heat Transfer*, 130, 112401.
36. Biener, G., Dahan, N., Niv, A., Kleiner, V. and Hasmana, E. (2008). Highly coherent thermal emission obtained by plasmonic bandgap structures, *Appl. Phys. Lett.*, 92, 081913.
37. Liu, J., Guler, U., Lagutchev, A., Kildishev, A., Malis, O., Boltasseva, A. and Shalaev, N. M. (2015). Quasi-coherent thermal emitter based on refractory plasmonic materials, *Opt. Mater. Express*, 5, pp. 2721–2728.
38. Seifert, S. and Runge, P. (2016). Revised refractive index and absorption of $In_{1-x}Ga_xAs_yP_{1-y}$ lattice-matched to InP in transparent and absorption IR-region, *Opt. Mater. Express*, 6, pp. 629–639.
39. Ferraria, C., Melinob, F., Pinellic, M., Spinac, P. R. and Venturinic, M. (2014). Overview and status of thermophotovoltaic systems, *Energy Procedia*, 45, pp. 160–169.

40. Mohammed Ali Gamel, M., Lee, H. J., Abdul Rashid, W. E. S. W., Ker, P. J., Yau, L. K., Hannan, M. A. and Jamaludin, M. Z. (2021). Review on thermophotovoltaic cell and its applications in energy conversion: issues and recommendations, *Materials*, 14, 4944.
41. Omair, Z., Scranton, G., Pazos-Outón, L. M., Xiao, T. P., Steiner, M. A., Ganapati, V., Peterson, P. F., Holzrichter, J., Atwater, H. and Yablonovitch, E. (2019). Ultraefficient thermophotovoltaic power conversion by band-edge spectral filtering, *Proc. Natl. Acad. Sci.*, 116, 15356–15361.
42. Pfiester N. A. and Vandervelde, T. E. (2017). Selective emitters for thermophotovoltaic applications, *Phys. Status Solidi A*, 214, 1600410.
43. Chirumamilla, M., Krishnamurthy, G. V., Knopp, K., *et al.* (2019). Metamaterial emitter for thermophotovoltaics stable up to 1400°C, *Sci. Rep.*, 9, 7241.
44. Rinnerbauer, V., Ndao, S., Yeng, Y. X., Chan, W. R. Senkevich, J. J., Joannopoulos, J. D., Soljacic, M. and Celanovic, I. (2012). Recent developments in high-temperature photonic crystals for energy conversion, *Energy Environ. Sci.*, 5, 8815–8823.
45. Jurado, Z., Kou, J., Kamali, S. M., Faraon, A. and Minnich, A. J. (2018). Wavelength-selective thermal extraction for higher efficiency and power density thermophotovoltaics, *J. Appl. Phys.*, 124, 183105.
46. Lin, S. Y., Moreno, J. and Fleming, J. G. (2003). Three-dimensional photonic-crystal emitter for thermal photovoltaic power generation, *Appl. Phys. Lett.*, 832, pp. 380–382.
47. Nakagawa, N., Ohtsubo, H., Waku, Y. and Yugami, H. (2005). Thermal emission properties of $Al_2O_3/Er_3Al_5O_{12}$ eutectic ceramics, *J. Eur. Ceram. Soc.*, 25, pp. 1285–1291.
48. Schuller, J., Taubner, T. and Brongersma, M. (2009). Optical antenna thermal emitters, *Nat. Photonics*, 3, pp. 658–661.
49. Arnold, C., Marquier, F., Garin, M., Pardo, F., Collin, S., Bardou, N., Pelouard, J.-L. and Greffet, J. J. (2012). Coherent thermal infrared emission by two-dimensional silicon carbide gratings, *Phys. Rev. B*, 86, 035316.
50. Landy, N. I., Sajuyigbe, S., Mock, J. J., Smith, D. R. and Padilla, W. J. (2008). Perfect metamaterial absorber, *Phys. Rev. Lett.*, 100, 207402.
51. Liu, X., Tyler, T., Starr, T., Starr, A. F., Jokerst, N. M. and Padilla, W. J. (2011). Taming the blackbody with metamaterials, *Phys. Rev. Lett.*, 107, 045901.
52. Bossard, J. A. and Werner, D. H. (2013). Metamaterials with angle selective emissivity in the near-infrared, *Opt. Express*, 21, 5215.

53. Choi, K. K., Omair, Z., Oduor, P. and Dutta, A. K. (2023). Resonant thermal emitters and PV Cells for thermophotovoltaic power generation, *IEEE J. Photovoltaics*, 13, pp. 716–727.
54. Choi, K. K., Oduor, P. and Dutta, A. (2023). Wavelength-selective thermal meta-emitters for thermophotovoltaic power generation, *Proc. SPIE*, 12513, 1251302.
55. Larruquert, J. I., Pérez-Marín, A. P., García-Cortés, S., Rodríguez-de Marcos, L., Aznárez, J. A. and Méndez, J. A. (2011). Self-consistent optical constants of SiC thin films, *J. Opt. Soc. Am. A*, 28, pp. 2340–2345.
56. Ordal, M. A., Bell, R. J., Alexander, R. W., Newquist, L. A. and Querry, M. R. (1988). Optical properties of Al, Fe, Ti, Ta, W, and Mo at submillimeter wavelengths, *Appl. Opt.*, 27, pp. 1203–1209.
57. Burger, T., Sempere, C., Roy-Layinde, B. and Lenert, A. (2020). Present efficiencies and future opportunities in thermophotovoltaics, *Joule*, 4, pp. 1660–1680.
58. Reiser, A. and Schächter, L. (2013). Geometric effects on blackbody radiation, *Phys. Rev. A*, 87, 033801.
59. Messina, R., Tschikin, M., Biehs, S. A. and Ben-Abdallah, P. (2013). Fluctuation-electrodynamic theory and dynamics of heat transfer in systems of multiple dipoles, *Phys. Rev.*, 88, 104307.
60. Bhatt, G. R., Zhao, B., Roberts, S., *et al.* (2020). Integrated near-field thermo-photovoltaics for heat recycling, *Nat. Commun.*, 11, 2545.
61. Mittapally, R., Lee, B., Zhu, L., *et al.* (2021). Near-field thermophotovoltaics for efficient heat to electricity conversion at high power density, *Nat. Commun.*, 12, 4364.
62. Yang, Y., Chang, J.-Y., Sabbaghi, P. and Wang, L. (2017). Performance analysis of a near-field thermophotovoltaic device with a metallodielectric selective emitter and electrical contacts for the photovoltaic cell, *J. Heat Transfer*, 139, 052701.
63. Lucchesi, C., Cakiroglu, D., Perez, J.-P., Taliercio, T., Tournié, E., *et al.* (2021). Near-field thermophotovoltaic conversion with high electrical power density and efficiency above 14%, *Nano Lett.*, 21, pp. 4524–4529.

Chapter 7

Remote Temperature Sensing

7.1 Introduction

In this chapter, we will describe the use of the optoelectronic properties of a detector array to predict the thermal sensitivity and thermal resolution of an infrared camera and the use of the camera for remote temperature sensing. Remote temperature sensing is an important function for Earth and planetary observations from space. Detecting in the 3–5 μm mid-wavelength (MW) and 8–12 μm long wavelength (LW) atmospheric transmission windows, the Earth observation is relatively free from the solar reflection. The acquired signal primarily originates from the passive radiation emitted from the scene. Therefore, by monitoring the camera output signal, the scene temperature can be readily deduced provided that the scene emissivity [1] and atmospheric transmission [2] are already known. To facilitate camera operation and data analysis, it is advantageous to obtain an analytical correspondence $T(N)$ between the camera digital count output N and scene temperature T with the known source emissivity $\varepsilon(\lambda)$ and atmospheric transmission $\tau_{\text{atm}}(\lambda)$. With a known analytical $T(N)$, different ε and τ_{atm} in flight are easier to be incorporated. It will also allow the camera performance be determined and its limitation be known under different sensing conditions such as different scene temperatures, which is important for mission planning. Furthermore, establishing an analytical $T(N)$ will allow the signal be converted efficiently in frame-by-frame manner to display the temperature profile of a scene for real-time remote temperature imaging. It is more direct than the traditional method in which the conversion is first relating N to the channel radiance R, which is the convolution of the Planck and detector spectral functions [3], and then from R to T through a look-up table [4, 5]. Since the second step is more time-consuming, the traditional method is more suitable for T conversion on the pixel-by-pixel basis.

In this chapter, we will describe an approach to establish the $T(N)$ function based on the optoelectronic properties of the sensor array and the analog-to-digital conversion (ADC) function of the readout electronics. This approach only requires a few experimental data points and is valid over a wide scene temperature range, thus greatly facilitating the camera calibration process. In the following, we will first establish the approach to a camera used on the ground and then extend it to the Compact Thermal Imager onboard the International Space Station and the Thermal Infrared Sensor onboard Landsat-8 satellite. These programs are led by *M.* Jhabvala of NASA and a detailed description of these instruments and programs are published in Ref. [6–15].

7.2 Theoretical Background

Infrared cameras are designed to detect passive radiation emitted from an object. Since this radiation is a function of the object's temperature T and emissivity ε, it is possible to determine the object's T when ε is known or to provide an estimate when ε is within an expected range [1]. To simplify the discussion, we will first present an approach to correlate the digital output N of a camera and the temperature T of a laboratory blackbody source where ε is known to be 1. We will then use this correlation to perform remote temperature sensing assuming unity ε and to provide error estimates when $\varepsilon \neq 1$. To proceed, we start with the photon radiance $L(\varepsilon, T, \lambda)$ emitted by an object. It is given by

$$L(\varepsilon, T, \lambda) = \frac{2c\varepsilon(\lambda)}{\lambda^4} \frac{1}{\exp\left(\frac{hc}{\lambda k(T+273.15)}\right) - 1}, \tag{7.1}$$

where $L(\varepsilon, T, \lambda)$ is in number of photons/s/cm^2/sr/μm, λ is the wavelength of the radiation, and T is in Celsius. After the radiation passes through air with transmission coefficient $\tau_{\text{atm}}(\lambda)$ and the camera lens with a transmission coefficient $\tau_{\text{opt}}(\lambda)$, the radiation is detected by detector pixels in the focal plane array (FPA) and generates photocurrent. The photocurrent together with the dark current is integrated on a capacitor with capacitance C in a readout integrated circuit (ROIC). After a signal integration time τ_{int}, the accumulated charge ye^- on the capacitor increases the capacitor voltage by ye^-/C, where e^- is the electron charge. This voltage is then converted by an ADC convertor linearly into digital counts N. Therefore, by measuring N, one knows y and hence L and then T, if ε is known.

Quantitatively, the total number of electrons accumulated in the capacitor per integration time τ_{int} is

$$y(T) = y_d + A\Omega\tau_{\text{int}} \int_{\lambda_1}^{\lambda_2} \eta(\lambda)\tau_{\text{atm}}(\lambda)\tau_{\text{opt}}(\lambda)L(\varepsilon, T, \lambda)d\lambda, \tag{7.2}$$

where y_d is from the dark current, A is the detector area, Ω is the projected solid angle from the object, λ_1 and λ_2 are the lower and upper spectral limits of detection, and $\eta(\lambda)$ is the detection quantum efficiency (QE). Denoting f-number ($f/\#$) of the optics to be F, $y(T)$ is

$$y(T) = y_d + A\frac{2\pi c\tau_{\text{int}}}{4F^2+1} \int_{\lambda_1}^{\lambda_2} \frac{1}{\lambda^4} \frac{\eta(\lambda)\tau_{\text{atm}}(\lambda)\tau_{\text{opt}}(\lambda)\varepsilon(\lambda)}{\exp\left(\frac{hc}{\lambda k(T+273.15)}\right) - 1} d\lambda. \tag{7.3}$$

With a given object $\varepsilon(\lambda)$ and atmospheric $\tau_{\text{atm}}(\lambda)$, y as a function of T can be calculated from Eq. (7.3) according to camera specifications in y_d, A, τ_{int}, F, $\eta(\lambda)$, and $\tau_{\text{opt}}(\lambda)$. However, since we wish to know T as a function of y, we need to invert Eq. (7.3) to obtain $T(y)$. This operation is not feasible with the wavelength-dependent exponential inside the integral. In general, however, one can fit the computed $y(T)$ values to an algebraic expression in the temperature range of interest. For example, it can be fitted to the form of

$$y(T) = a_1(T + 273.15)^{b_1} + c_1, \tag{7.4}$$

where a_1, b_1, and c_1 are three fitting parameters. Using the fitted Eq. (7.4), the inverse function $T(y)$ is

$$T(y) = \left(\frac{y - c_1}{a_1}\right)^{1/b_1} - 273.15. \tag{7.5}$$

Very often, the detection only covers a narrow spectral band either because of a sharply peaked $\eta(\lambda)$ or because of the inclusion of a narrowband filter. Equation (7.3) in this case can be more advantageously approximated by replacing λ in the exponential function with the center wavelength λ_0 of the band. Equation (7.3) then becomes

$$y(T) = y_d + \frac{a_2}{\exp\left(\frac{hc}{\lambda_0 k(T+273.15)}\right) - 1}, \tag{7.6}$$

where

$$a_2 = A\frac{2\pi c\tau_{\text{int}}}{4F^2 + 1}\int_{\lambda_1}^{\lambda_2} \frac{1}{\lambda^4}\eta(\lambda)\tau_{\text{atm}}(\lambda)\tau_{\text{opt}}(\lambda)\varepsilon(\lambda)d\lambda. \tag{7.7}$$

Using the approximate form of $y(T)$, Eq. (7.6) can be rewritten to

$$T(y) = \frac{hc}{\lambda_0 k}\frac{1}{\ln\left(\frac{a_2}{y-c_2} + 1\right)} - 273.15, \tag{7.8}$$

where $c_2 = y_d$. Although a_2, λ_0, and c_2 are all known quantities from the same information in Eq. (7.3) and therefore no curve fitting is needed, the agreement between Eq. (7.3) and Eq. (7.6) can be improved in the temperature range of interest if we treat a_2 and c_2 as fitting parameters. The wavelength λ_0 can also be treated as a fitting parameter if needed. By expressing Eq. (7.3) in the form of Eq. (7.8) instead of Eq. (7.5), one can more readily keep track of the changes in the viewing conditions and camera settings such as those related to weather condition and integration time. We should note that when c_1 and c_2 in Eqs. (7.5) and (7.8) are treated as fitting parameters, they will have different values in different temperature ranges to yield the best fit. Although they can be roughly regarded as the dark electrons, their values are not exactly the same as y_d in Eq. (7.3).

The photon flux from the scene is sampled by the sensor as the weighted integral in Eq. (7.3) and generates y number of electrons in the capacitor at the end of the integration time. The ADC converts this number y into digital counts N linearly. Their values are then related by

$$y(N) = m_r N + c_r. \tag{7.9}$$

Nominally, if the charge well capacity is Y million electrons (Me$^-$) and the ADC converts this dynamic range into digital counts in K bit resolution, m_r will be $Y/2^K$ Me$^-$/count. On the other hand, the value of c_r depends on a number of factors including a possible A/D DC offset, A/D nonlinearity, and manual background subtraction. The dc offset N_{dos} can be measured by setting the integration time very small, during which the collected y is negligible. If this DC offset is not excluded from the measured N, c_r will acquire a term $-m_r N_{\text{dos}}$. On the other hand, in the background subtraction process, a radiation source with spatially uniform T_0 is placed in front of the camera, and the digital count of each pixel N_b^i is recorded, where i is the pixel index. By subtracting out N_b^i in the subsequent images, all the pixels will register $N = 0$ at $T = T_0$, thereby improving the array spatial uniformity. When N in Eq. (7.9) is interpreted as the count difference from the background, c_r will acquire a term $m_r N_b$. The value of c_r thus depends on the definition of N in Eq. (7.9). Combining Eq. (7.5) or (7.8) with Eq. (7.9), the temperature T of a scene can be known according to N using either

$$T(N) = T(y(N)) = \left(\frac{m_r N + c_r - c_1}{a_1}\right)^{1/b_1} - 273.15, \tag{7.10}$$

or

$$T(N) = \frac{hc}{\lambda_0 k} \frac{1}{\ln\left(\frac{a_2}{m_r N + c_r - c_2} + 1\right)} - 273.15. \tag{7.11}$$

In infrared imaging, it is customary to take a two-point correction, so that all the pixels will give the same counts at two temperature limits. The values of y_d and $\eta(\lambda)$ of all pixels will then be normalized, and a single $T(N)$ function can be applied to all pixels in the array. In this discussion, we will not make the second correction. Instead, we will use the mode counts, *i.e.*, the maximum counts, of the population in front of a uniform source to establish the T and N correspondence as an illustration.

In principle, when all the optoelectronic quantities of the camera in Eq. (7.3) and Eq. (7.9) are accurately known, $T(N)$ can be determined without further experiment. In practice, an overall system calibration is needed when the camera components are assembled together. When fitting Eqs. (7.10) and (7.11) to the experimental data, there are a large number of adjustable parameters, and each may be known to certain margin of error. To facilitate the curve fitting, we can either assume the detector characteristics as given so that the detector parameters (a_1, b_1, c_1) or (a_2, λ_0, c_2) are known and adjust the ADC parameters (m_r, c_r) as free parameters or vice versa. The choice depends on the reliability of individual specifications. The outcome of the $T(N)$ function, however, will not depend on a particular choice. In the following, we will use either of the two choices for camera calibration.

7.3 A 1024 × 1024 Long Wavelength Infrared Strained Layer Superlattice Camera

7.3.1 *Temperature Sensing Calibration*

The first camera makes use of a 1024 × 1024 long wavelength infrared (LWIR) strained layer superlattice FPA. The nominal detector $\eta(\lambda)$ and the lens $\tau_{\text{opt}}(\lambda)$ are shown in Fig. 7.1(a). Their $\eta(\lambda)\tau_{\text{opt}}(\lambda)$ product is shown in Fig. 7.1(b). Using a blackbody source as the object in the laboratory, $\varepsilon(\lambda)$ and $\tau_{\text{atm}}(\lambda)$ are taken to be unity. The detector parameters in Eq. (7.3) are: $A = 16 \times 16\,\mu\text{m}^2$, $\tau_{\text{int}} = 0.76\,\text{ms}$, $y_d = 0.166\,\text{Me}^-$ at 68 K and 0.3 V, $F = 2$, $\lambda_1 = 2\,\mu\text{m}$, and $\lambda_2 = 20\,\mu\text{m}$. The ROIC is FLIR ISC0404 with 18 μm pixel pitch and 10 Me$^-$ charge well. The adopted ARC has 14-bit resolution. The digital counts N thus range from 0 to 16384, and the nominal m_r is 6.104×10^{-4} Me$^-$/count. Since there is no signal processing in the calibration, *i.e.*, no ADC DC offset exclusion and no background subtraction, the nominal $c_r = -m_r N_{\text{dos}}$. This system information fully specifies the camera characteristics. In the

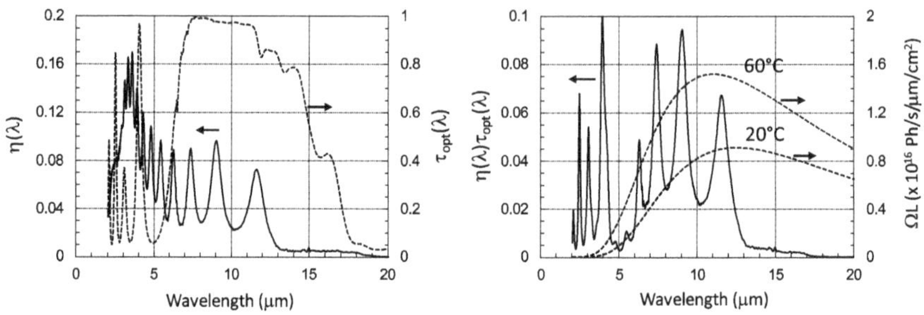

(a) (b)

Figure 7.1. (a) The detector $\eta(\lambda)$ (solid curve) and the lens transmission $\tau_{\text{opt}}(\lambda)$ (dashed curves). (b) The resultant $\eta(\lambda)\tau_{\text{opt}}(\lambda)$ (solid curve) and the blackbody radiation at 20°C and 60°C (dashed curves). Reproduced with permission from [13].

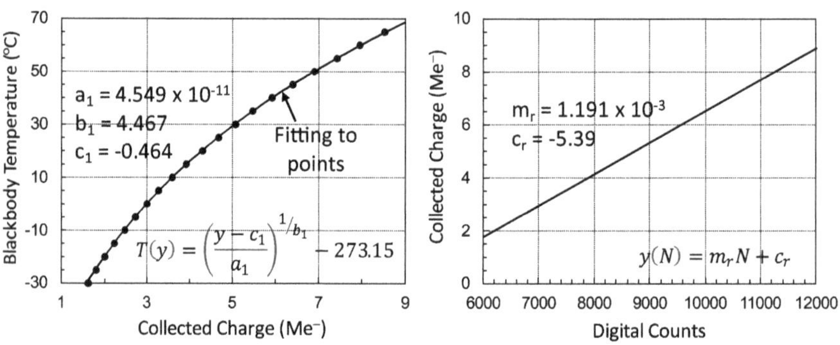

(a) (b)

Figure 7.2. (a) The calculated points (circles) from the exact $y(T)$ of Eq. (7.3) and the inverse fitted function $T(y)$ of Eq. (7.5) (solid curve). (b) The fitted linear ADC function $y(N)$ of Eq. (7.9). Reproduced with permission from [13].

calibration process, however, we will arbitrarily take the detector characteristics and optics as known and adjust the ROIC parameters to fit the experimental data.

With the known detector and optics specifications, $y(T)$ in Eq. (7.3) can be computed. Since the present detector has broad spectral response from 2–18 μm and no band-limiting filter was adopted, narrowband approximation is not applicable. The exact $y(T)$ of Eq. (7.3) is then fitted to Eq. (7.4), and the fitted parameter values are shown in Fig. 7.2(a). After transposing the variables, the computed points from Eq. (7.3) and the fitted inverse function $T(y)$ of Eq. (7.5) are plotted in Fig. 7.2(a). They agree well in the T range between $-30°\text{C}$ and $70°\text{C}$. Hence, from the measured y, the blackbody T can be readily obtained from Eq. (7.5) within this T range.

The ADC conversion of Eq. (7.9) can be determined by choosing two suitable measured data points (N_1, T_1) and (N_2, T_2) as the calibration points. At each T, the corresponding y can be computed from Eq. (7.3), and therefore the data points (N_1, y_1) and (N_2, y_2) on the ADC line will be known. The values of m_r and c_r are then readily deduced. The conversion line using the calibration points at $(N_1, T_1) = (7465.9, 8.41°\text{C})$ and $(N_2, T_2) = (10320, 50°\text{C})$ is plotted in Fig. 7.2(b). The determined m_r is 1.19×10^{-3} and c_r is -5.39. The deduced m_r is about 2 times larger than the nominal value of 6.104×10^{-4}, which may mean a smaller ROIC dynamic range being digitized. A finite and negative c_r indicates the presence of an ADC DC offset and possibly a larger detector dark current represented by c_1 at the camera operating temperature. Nevertheless, the blackbody calibration will be able to correct these uncertainties and establish an accurate correspondence between T and N.

Combining the fitted $T(y)$ and $y(N)$, the function $T(N)$ of Eq. (7.10) can be obtained and it is plotted as the solid curve in Fig. 7.3 along with the experimental mode N at different blackbody T settings. By using the two indicated calibration points, other data points fall onto the fitted function within an error margin $[T(N_{\text{exp}}) - T_{\text{setting}}]$ of $-2.2°\text{C}$ and $+1.4°\text{C}$, and this error margin is within the blackbody setpoint error. The agreement thus shows the validity of the approach, and one can use this calibration function for remote temperature sensing under the approximation $\varepsilon(\lambda) = \tau_{\text{atm}}(\lambda) = 1$.

7.3.2 *Temperature Sensing in Laboratory*

To demonstrate temperature sensing in the laboratory, Fig. 7.4(a) shows an intensity image taken by the camera. The square blackbody source is set at 60°C, and its N is measured to be 11008 counts at the pixel coordinates of (100, 446). Applying the T conversion to

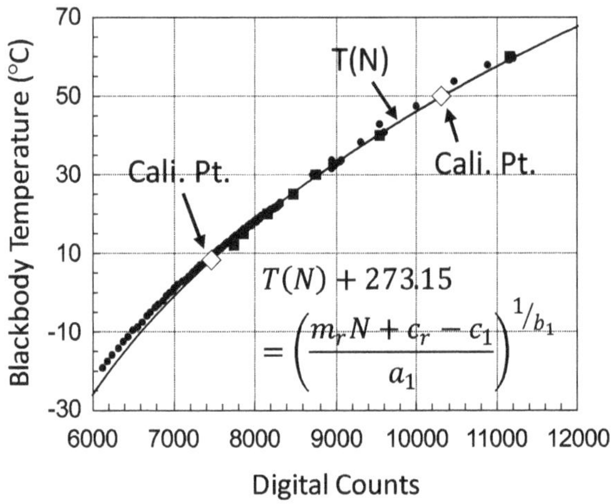

Figure 7.3. The deduced $T(N)$ of Eq. (7.10) (solid curve) and the measured data points (black circles and squares from two calibration experiments). The open diamond symbols represent the two calibration points used to establish the $T(N)$ curve. Reproduced with permission from [13].

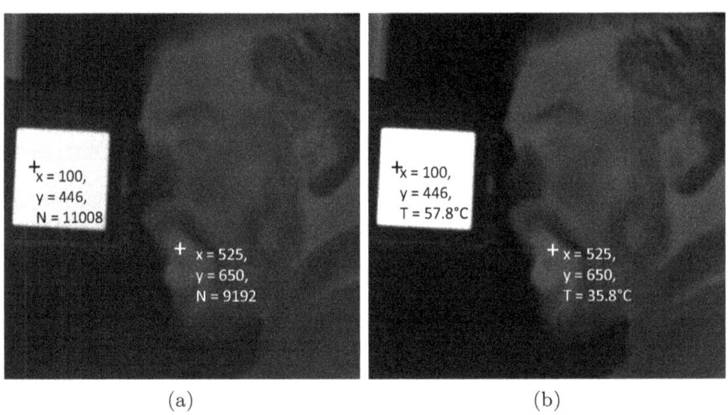

Figure 7.4. (a) The measured digital counts N of an image. (b) The converted temperature T of the same image. Reproduced with permission from [13].

the entire image yields the temperature image in Fig. 7.4(b). The blackbody T is found to be 57.8°C. Although it is lower than the 60°C setting, the discrepancy of −2.2°C is within the data fluctuation in Fig. 7.3. Meanwhile, N of a human body at a different location

(525, 650) is 9192. The converted T is 35.8°C, which is also below the normal range between 36.1°C and 37.2°C of a healthy person. Since the human T is not subject to temperature regulation error, the difference is attributed to the smaller human skin emissivity of 0.98.

If ε of an object is known, its T can still be obtained from the blackbody calibration in Fig. 7.3. When $\varepsilon \neq 1$, Eq. (7.3) is:

$$y(T) = y_d + \varepsilon \left[A \frac{2\pi c \tau_{\text{int}}}{4F^2 + 1} \int_{\lambda_1}^{\lambda_2} \frac{\eta(\lambda)\tau_{\text{opt}}(\lambda)}{\lambda^4 \exp\left(\frac{hc}{k(T+273.15)\lambda}\right) - 1} d\lambda \right]$$
$$= y_d + \varepsilon[a_1(T + 273.15)^{b_1} + c_1 - y_d], \tag{7.12}$$

for the same a_1, b_1, and c_1 as in Eq. (7.5). Using Eq. (7.5) to convert N as though $\varepsilon = 1$, an apparent temperature $T_{ap} = 35.8$°C is obtained for the person's skin. Since both the actual T of Eq. (7.12) and the apparent T_{ap} of Eq. (7.5) arrive at the same y, the two equations can be equated to give

$$T = \left[\frac{(T_{ap} + 273.15)^{b_1}}{\varepsilon} + \frac{(1 - \varepsilon)(c_1 - y_d)}{a_1 \varepsilon} \right]^{1/b_1} - 273.15. \tag{7.13}$$

Substituting $T_{ap} = 35.8$°C and $\varepsilon = 0.98$ into Eq. (7.13), $T = 37.1$°C is obtained, which is a more representative temperature of a healthy person. Therefore, this example shows that a 0.02 difference in ε translates to a 1.3°C difference using this camera.

7.3.3 *Temperature Sensing in the Field*

In conducting thermal imaging in the field, a background subtraction is necessary to remove pixel nonuniformity induced by different operating conditions compared to in the laboratory. Figure 7.5(a) shows the intensity image N_b^i in using a lens cap at $T_0 = 7.14$°C as the background. The mode N is 7397, and the standard deviation σ is 558. At the same time, the intensity image of a scene is shown in Fig. 7.5(b), which has a lower mode value of 6953 and a similar σ of 541. This result indicates the scene has a lower average temperature than the chosen background, and the spatial nonuniformity of the array dominates over the scene's thermal contrast. By subtracting the

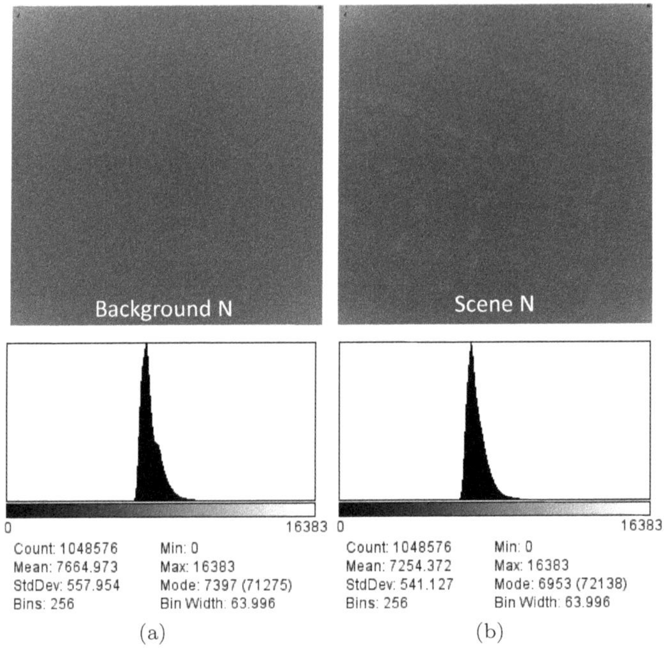

Figure 7.5. (a) The map of digital counts recorded from a constant background of 7.14°C. (b) The map of a scene. The corresponding distribution histograms are also shown. Reproduced with permission from [13].

background N_b^i, Fig. 7.5(c) shows an improved thermal contrast and a smaller σ of 103. The values of N are mostly negative, indicating that the scene is colder than the adopted background.

To convert N after background subtraction into T, a larger $c_r = (7397m_r - 5.39)$, which includes the background counts, is adopted in Eq. (7.10). It amounts to shifting the N axis in Fig. 7.3 such that $N = 0$ at $T = 7.14$°C. Figure 7.6(a) shows the background-subtracted N image, which has values of -239, -296, and -591 at the three marked locations. After applying $T(N)$ of Eq. (7.10), which assumes $\varepsilon(\lambda) = \tau_{\text{atm}}(\lambda) = 1$, the converted T image is shown in Fig. 7.6(b), and the T at the three locations are 2.33°C, 0.89°C, and -5.19°C, respectively. The scene is colder than the lens cap and the temperature of the scene is rather uniform as it's σ is only 2.1°C in Fig. 7.7(b).

(a) N image (b) T image

Figure 7.6. (a) The map of digital count N after background subtraction at 7.14°C. (b) The corresponding T map of the scene. Reproduced with permission from [13].

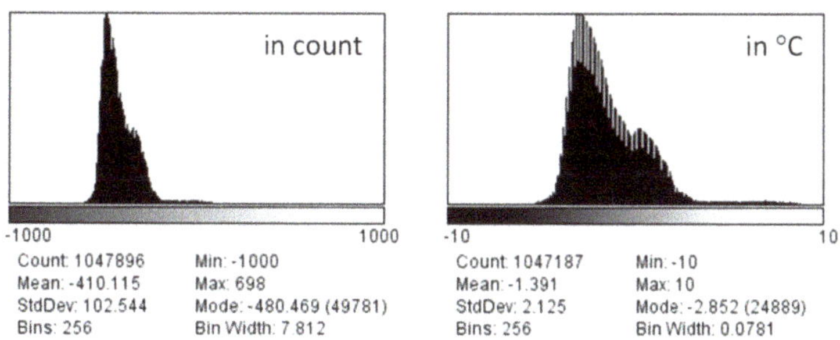

(a) (b)

Figure 7.7. (a) The histogram of ΔN relative to a constant background at 7.14°C. The generally negative ΔN indicates a colder scene than the background. (b) The histogram in T for the same scene. Reproduced with permission from [13].

7.4 Compact Thermal Imager Dual-Band Strained Layer Superlattice Camera

7.4.1 *Mid-Wavelength Infrared Detection in Band 1*

The Compact Thermal Imager (CTI) onboard the International Space Station (ISS) contains a 640×512 strained layer superlattice FPA and its detection spectrum is switchable by reversing the polarity of the bias voltage on the FPA. Under one bias polarity, it detects in MW, which is referred to as band 1. In the opposite bias, the detection spectrum expands to the LW, and it is referred to as band 2. We will describe the detection in band 1 first. In band 1, the detector spectrum with the lens, $\eta(\lambda)\tau_{\text{opt}}(\lambda)$, is shown in Fig. 7.8(a). In pre-flight laboratory blackbody calibration, the relation $\varepsilon(\lambda) = \tau_{\text{atm}}(\lambda) = 1$ is used in the calculation of the collected electrons $y(T)$. On the other hand, onboard the ISS, $\tau_{\text{atm}}(\lambda)$ with an example displayed in Fig. 7.8(a) will need to be taken into account, and the effective detector spectrum will be that shown in Fig. 7.8(b). Since $\tau_{\text{atm}}(\lambda)$ depends on local weather conditions, the converted T based on one transmission spectrum will be subjected to local $\tau_{\text{atm}}(\lambda)$ variations unless it is known and accounted for such as using MODTRAN calculations [2]. The ROIC of this camera is a FLIR ISC0903, which has a pixel pitch of $30\,\mu$m and a well capacity of $18\,\text{Me}^-$. The system A/D resolution is 16-bit. The integration time is 10 ms, and the $f/\#$ of the optics is 3.3.

In laboratory camera calibration, we take an alternative approach in taking adjustable parameters for data fitting, which are the detector dark current and QE magnitude. The ADC, on the other hand, is assumed to perform nominally, *i.e.*, m_r is $18\,\text{Me}^-/2^{16} = 2.75 \times 10^{-4}\,\text{Me}^-/\text{count}$. In the data analysis, a DC offset $N_{\text{dos}} = 21000$ measured at $\tau_{\text{int}} = 1\,\mu$sec is excluded from the N value,

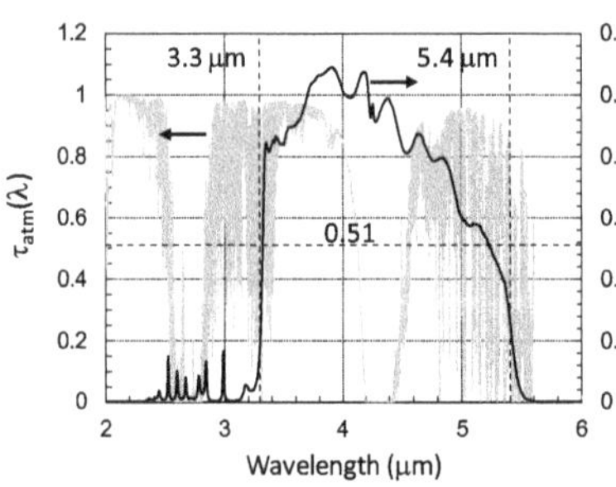

(a)

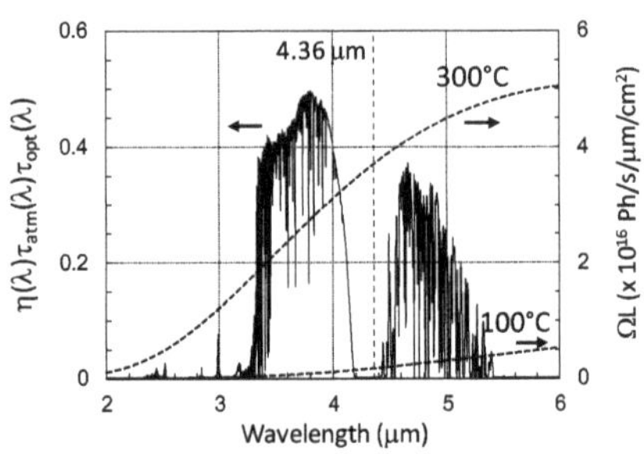

(b)

Figure 7.8. (a) The QE spectrum of the strained layer superlattice sensor with the lens and a typical $\tau_{\text{atm}}(\lambda)$ in MW. (b) The typical $\eta(\lambda)\tau_{\text{opt}}(\lambda)\tau_{\text{atm}}(\lambda)$ (solid curve) at the top of atmosphere and the blackbody radiation at 100 and 300°C (dashed curves). Reproduced with permission from [13].

and since there is no background subtraction, the nominal $c_r = 0$. With the nominal ADC, y_1 and y_2 are known at the two calibration points $N_1 = 20600$ and $N_2 = 3.533 \times 10^6$ (a very large N_2 to be explained in the next paragraph). The two resulting data points (35°C, y_1) and (400°C, y_2) can then be used to fix Eq. (7.3), which has two unknown parameters, y_d and $\eta(\lambda)$ magnitude. After $y(T)$ of Eq. (7.3) is solved, the combined exact $N(T)$ is plotted as the dashed curve in Fig. 7.9 with T being in the y-axis. (We should note that the fitted QE magnitude is 6.75 times lower than that shown in Fig. 7.8(a) but it will not affect the $T(N)$ correspondence.) Although Eq. (7.3) gives an exact $y(T)$ and therefore the $N(T)$ relation, it cannot be inverted to yield the desired $T(N)$. Therefore, we solve the approximate $T(N)$ of Eq. (7.11) directly with the same calibration points $(N_1, T_1) = (20600, 35)$ and $(N_2, T_2) = (3.533 \times 10^6, 400)$. But since Eq. (7.11) has three unknowns (a_2, λ_0, c_2), we will first choose λ_0 to be the center wavelength $\approx (3.3 + 5.4)/2 = 4.36\,\mu$m and solve for a_2 and c_2 and then finetune λ_0 if needed to match the transposed exact $N(T)$ curve. The approximate $T(N)$ with the fitted parameters is also plotted as the grey solid curve in Fig. 7.9(a). It turns out that the two curves agree well between 0°C and 1000°C, λ_0 thus need not

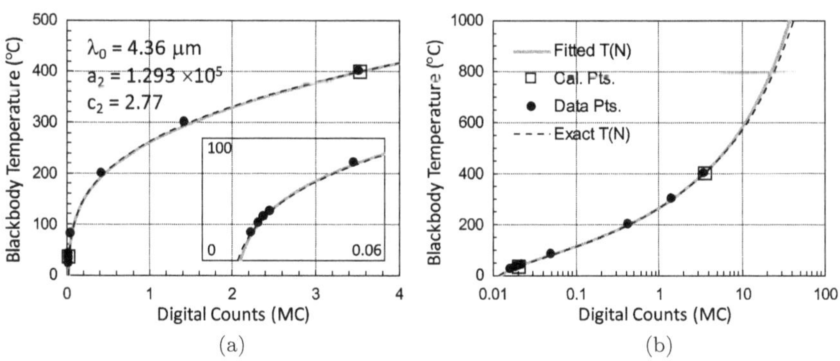

Figure 7.9. The exact $T(N)$ curve from Eqs. (7.3) and (7.9) (dashed curve), the fitted narrowband $T(N)$ curve of Eq. (7.11) (solid curve), and the experimental data points (circles). The inset shows an enlarged section at the low T range. (b) The same plot at wider range up to 100 MC. The squares are the two calibration points to fix the $T(N)$ curves. Reproduced with permission from [13].

be adjusted. This example shows that the narrowband approximation of Eq. (7.11) works well in band 1.

In this measurement, the digital count N can exceed the ADC dynamic range of $2^{16} - N_{\text{dos}} = 44540$ at the elevated T. For these high T's, the integration time is shortened and the measured N is scaled proportionally to the same τ_{int} of 10 ms. Using this procedure, the N value can be recorded up to 400°C in Fig. 7.9(a) with the effective dynamic range of 4 million counts. From Fig. 7.9, the fitted $T(N)$ function agrees with all the experimental data based on just two calibration points. Nonetheless, the camera is able to perform temperature measurement in an extremely wide range of T.

After the CTI camera underwent a laboratory calibration, it was installed on the ISS for Earth observation. Assuming that the camera characteristics remain unchanged in flight, the calibration function $T(N)$ of Eq. (7.11) should still be applicable provided that the parameter a_2 is reduced by the band-averaged $\varepsilon(\lambda)\tau_{\text{atm}}(\lambda)$ and the background subtraction in flight is properly accounted for. By integrating the detection spectra in Figs. 7.8(a) and (b) and taking their ratio, the averaged $\tau_{\text{atm}}(\lambda)$ is 0.51 for the displayed atmospheric transmission. Therefore, a_2 is expected to be reduced by a factor of $\sim$0.51. In flight, the unprocessed digital counts have gone through a "background" subtraction using an image shot at deep space, in which $N_{DS} = 26000$ is recorded. This image shot contains the light counts from deep space which is expected to be nearly 0, the DC offset N_{dos} which is the same at 21000, and the dark counts N_d which are from the detector dark current at the FPA operating temperature and the lens radiance at the lens temperature in flight. From the difference between N_{DS} and N_{dos}, $N_d = 5000$. In essence, with a subtraction of the deep space counts, the DC offset is excluded and the dark counts at a lower operating T is subtracted but the subtraction from the scene radiance is negligible. Accordingly, we expect $c_r = 0$, and $c_2 \approx m_r N_d = 1.37$, which is smaller than the laboratory $c_2 = 2.77$ because of the lower operating T in flight.

Meanwhile, the temperature conversion function in flight, $T_{\text{flt}}(N)$, is obtained by a similar fitting procedure as in the laboratory. We first choose two data points obtained in flight to fix the exact $T(N)$

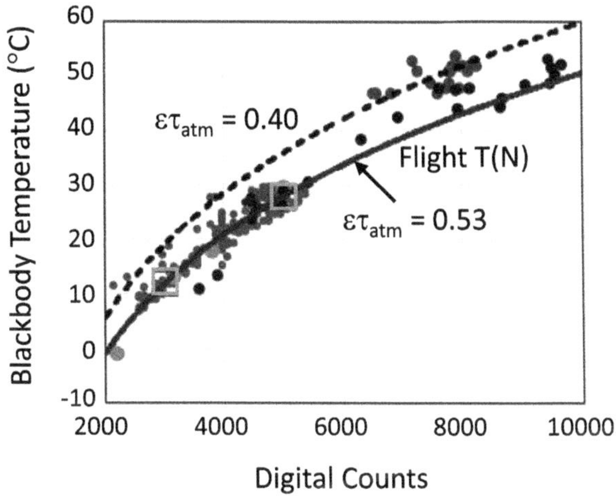

Figure 7.10. The fitted $T(N)$ curve of Eq. (7.11) with $\varepsilon\tau_{\text{atm}} = 0.53$ (solid curve) and the $T(N)$ curve when $\varepsilon\tau_{\text{atm}}$ is reduced to 0.40 (dashed curve). The circles are obtained from the temperature reading derived from other space instruments. The squares are the calibration points. Reproduced with permission from [13].

curve. The approximate $T(N)$ of Eq. (7.11) is then fitted to this exact $T(N)$ with a new set of fitting parameters (a_2', λ_0', c_2'). Figure 7.10 shows several observation (N, T) data sets along with the two chosen calibration points.

The experimental temperature reading is the local temperature obtained from other space instruments observing the same scene near the same time [12, 14]. Specifically, the plotted circles are obtained from the temperature reading derived from sea surface temperature calculation, ECOSTRESS, MODIS, and Lake Tahoe. The resulting $T_{\text{flt}}(N)$ of Eq. (7.11) shown in Fig. 7.10 predicts the trend of the data sets closely. The fitted parameters of $T_{\text{flt}}(N)$ are: $(a_2', \lambda_0', c_2') = (6.8685 \times 10^4, 4.36, 0.1773)$. The flight a_2' turns out to be a factor of 0.53 smaller than the lab a_2, indicating an average product of $\varepsilon\tau_{\text{atm}}$ to be 0.53 for these data sets. It is slightly larger than the 0.51 for the displayed atmospheric transmission spectrum alone. The fitted λ_0' remains the same as λ_0. The flight $c_2' = 0.1773$, which is smaller than the expected 1.37 from the dark counts. This discrepancy is due to the fact that c_2' obtained from curve fitting need not be the same as the physical dark counts. The fluctuations of the data may be attributed to different local τ_{atm} at the time of CTI measurement. Taking this assumption, a smaller $\varepsilon\tau_{atm} = 0.40$ will be able to explain the upper bound of the data points.

To visualize the difference between $T_{lab}(N)$ and $T_{flt}(N)$, the two functions and the experimental data are plotted in Fig. 7.11(a). Explicitly, with $m_r = 2.75 \times 10^{-4}$ Me$^-$/count,

$$T_{lab}(N) = \frac{hc}{\lambda_0 k} \frac{1}{\ln\left(\frac{1.293\times10^5}{m_r N - 2.77} + 1\right)} - 273.15, \tag{7.14}$$

and

$$T_{flt}(N) = \frac{hc}{\lambda_0 k} \frac{1}{\ln\left(\frac{6.869\times10^4}{m_r N - 0.177} + 1\right)} - 273.15 \tag{7.15}$$

$T_{lab}(N)$ can be brought to match $T_{flt}(N)$ by taking the following transformation

$$T'(N) = \frac{hc}{\lambda_0 k} \frac{1}{\ln\left(\frac{0.53\times1.293\times10^5}{m_r (N+9440) - 2.77} + 1\right)} - 273.15. \tag{7.16}$$

The quantity 9440 can be regarded as the effective background subtraction in flight relative to the ground. The converted function

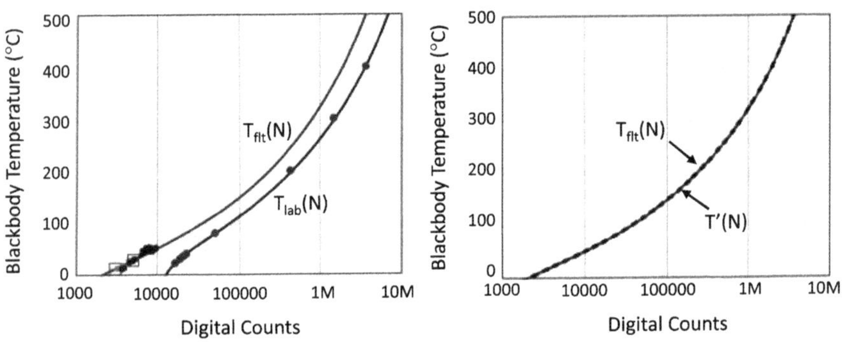

(a) (b)

Figure 7.11. (a) The lab $T(N)$ and the flight $T(N)$ functions based on the (N, T) data shown as symbols. (b) The converted $T'(N)$ (dashed curve) to match flight $T(N)$ (solid curve) by knowing the atmospheric transmission coefficient and the relative background subtraction. Reproduced with permission from [13].

$T'(N)$ is plotted with the flight $T(N)$ in Fig. 7.11(b). They are identical as expected. Consequently, one will be able to project the flight $T(N)$ based on laboratory calibration by accounting for atmospheric attenuation and different background subtraction.

After obtaining $T_{\text{flt}}(N)$, we convert the intensity image taken from the ISS. Figure 7.12(a) shows one of the scenes having mode N of 3784. By applying $T_{\text{flt}}(N)$ conversion, the temperature image of Fig. 7.12(b) can be produced. The mode T is 19.4°C. With this T map, the temperature at any location can be known within $\varepsilon\tau_{\text{atm}} = 0.53$ assumption. For example, when N is measured to be 4133, the corresponding T will be 22.02°C.

Figure 7.13 shows a scene containing fire with N histograms. The land's counts are predominantly 6400 while those of the hottest spot are 41913. According to $T_{\text{flt}}(N)$, the corresponding land's T is 35.9°C, which may be expected in the general area around the fire. On the other hand, the fire's T itself is only 105.8°C, much lower than the expected T of order 1000°C. Since the measured N is within the

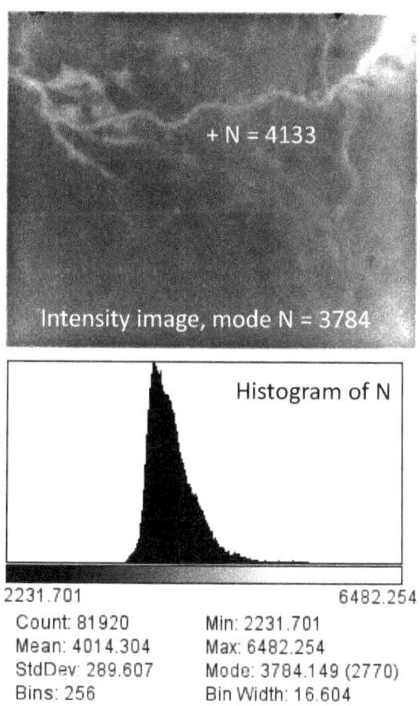

(a)

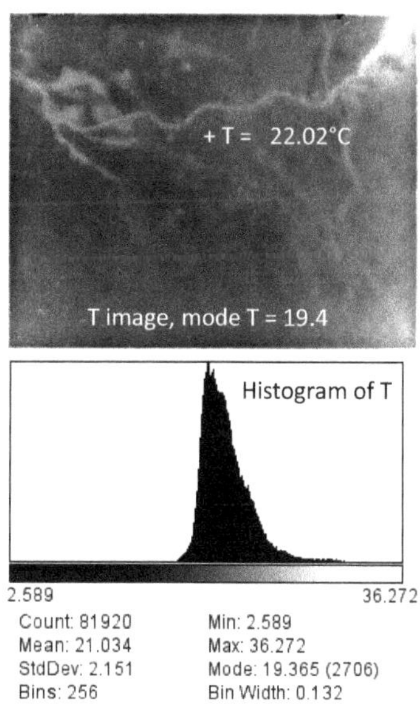

(b)

Figure 7.12. The (a) N and (b) converted T images taken by the CTI camera in band 1. Reproduced with permission from [13].

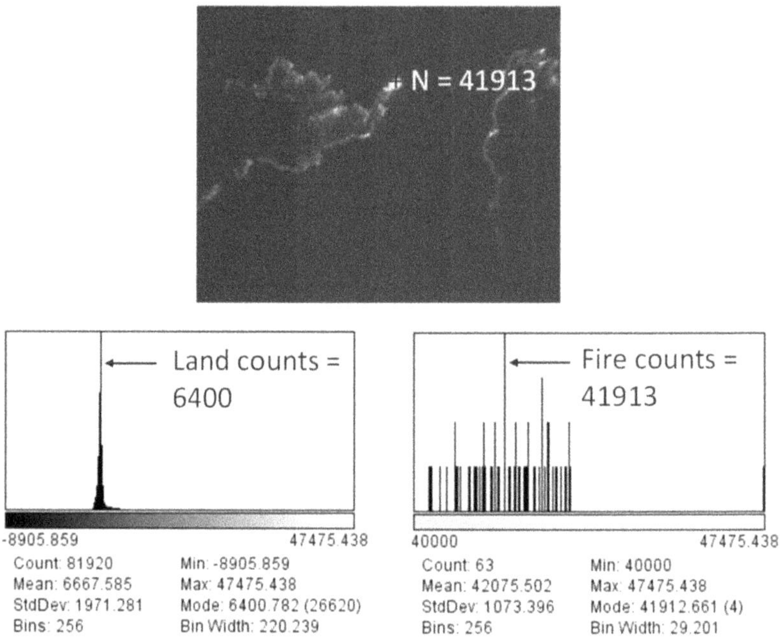

Figure 7.13. The intensity image of a fire scene taken by the CTI camera in band 1 and the histograms of N showing the distribution peak and tail. Reproduced with permission from [13].

dynamic range of 44540, the present T should be reliable and it can be interpreted as the average T of the area of $100 \times 100\,\mathrm{m}^2$ spanned by a single pixel [10].

7.4.2 *Long Wavelength Infrared Detection in Band 2*

The CTI camera has a longer cutoff wavelength in band 2 under an opposite applied voltage. Figure 7.14(a) shows the supplier's QE spectrum [16] after integration with the lens and a notch filter centered around $6.8\,\mu$m. Similar to band 1, the QE needs to be reduced by a factor of 5.2 to yield the observed digital counts at the calibration points if the ADC specification is adopted. The band 2 $T(N)$ has gone through similar blackbody calibration in the laboratory as in band 1, but it will not be repeated here. Instead, we will establish the T and N correspondence directly from the flight data.

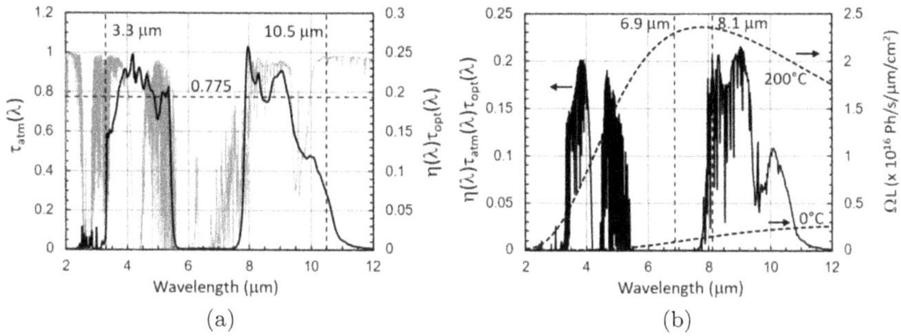

Figure 7.14. (a) The QE spectrum in band 2 of the detector after integrating with optical notch filter and lens. (b) The effective detector QE spectrum with the assumed atmospheric transmission and the blackbody spectrum at 0°C and 200°C. Reproduced with permission from [13].

With a notch filter, the detector QE exhibits a double-peaked spectrum, and it spans from 3.3 μm to 10.5 μm. The narrowband approximation is thus less applicable. Nevertheless, we will continue to use this approximation, albeit within a smaller temperature range. We will also follow the same fitting approach as in band 1, in which the detector parameters are adjustable, and the ADC parameters are fixed at $m_r = 2.75 \times 10^{-4}$ Me$^-$/count and $c_r = 0$ following the deep space image subtraction of $N_{DS} = 19000$. Based on the measurement on very bright objects, such as large wildfires and solar glint from ocean surfaces, the maximum unprocessed N was found to be about $N_{\max} = 38000$ imposed by the ADC. Therefore, the actual dynamic range in band 2 is about $N_{\max} - N_{DS} = 19000$.

By choosing two calibration points indicated in Fig. 7.15, one can fix y_d and η magnitude in Eq. (7.3) using the spectral lineshape in Fig. 7.14(b), $\tau_{\text{int}} = 1.8$ ms, and $\varepsilon(\lambda) = 1$, thereby knowing the exact T and N correspondence from Eq. (7.3) and Eq. (7.9). This exact $T(N)$ curve is plotted in Fig. 7.15(a) as the dashed curve. The exact $T(N)$ can be fitted to an approximate expression of Eq. (7.11) using a_2 and c_2 as adjustable parameters and taking λ_0 to be the center wavelength of 6.9 μm. The fitted $T(N)$ curve passes through the two calibration points with (a_2, c_2) shown in Fig. 7.15(a). However, this fitted curve does not match well with the

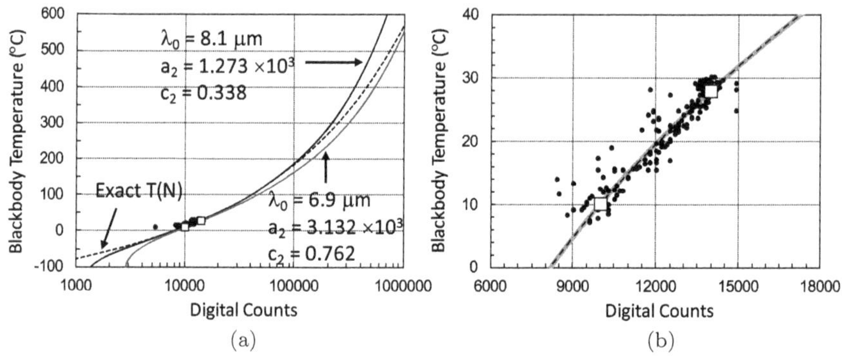

Figure 7.15. (a) The fitted $T(N)$ of Eq. (7.11) (solid curves) with two λ_0, the exact $T(N)$ (dashed curve) and the two calibration points (open squares). (b) The fitted $T(N)$ with $\lambda_0 = 8.1\,\mu$m (solid curve), the exact $T(N)$ (dashed curve), and the temperature reading derived from sea surface temperature calculation (circles). Reproduced with permission from [13].

exact $T(N)$ outside the calibration range with $T < 500°$C. This is due to the fact that band 2 produces a larger LW photocurrent around terrestrial temperatures as suggested by the blackbody spectrum and QE in Fig. 7.14(b). Therefore, λ_0 needs to move to the longer λ at the lower T range. By increasing λ_0 to $8.1\,\mu$m, a closer agreement is obtained between $T = -50°$C and 300°C. Nevertheless, the two fitted curves are essentially indistinguishable in the calibration range between 0°C and 40°C in Fig. 7.15(b). The data points from the sea surface temperature calculation [12] follow this trend closely in this T range. From this example, the narrowband approximation is seen to be also applicable to band 2 even though its spectral bandwidth is as wide as $7\,\mu$m. The wider band 2 spectrum does reduce the useful range from ~1000°C in Fig. 7.9(b) to ~300°C in Fig. 7.15(a).

In Figure 7.14(a), the band 2 spectral QE matches well with the atmospheric transmission $\tau_{\text{atm}}(\lambda)$ so that the spectrally averaged τ_{atm} is 0.775 for this detector. In Fig. 7.16, we show the $T(N)$ curves for different $\varepsilon\tau_{\text{atm}}$ from 0.6 to 0.9. The data from all the data sets [12,14] fall within this τ_{atm} range.

Figure 7.17 shows a slightly displaced fire scene shown in Fig. 7.13 for band 1. The mode land count is 15423, while the fire N is 19639.

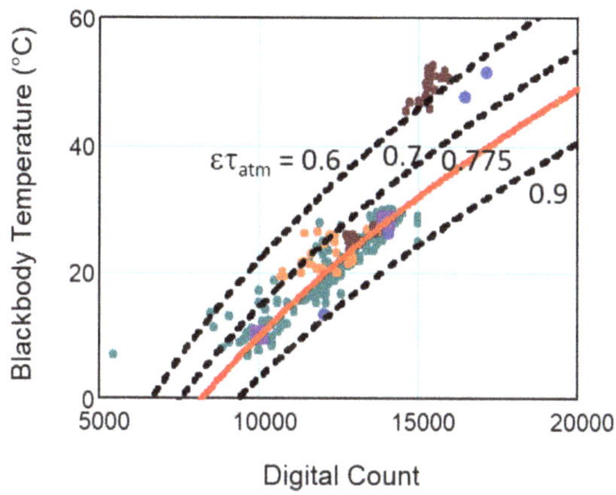

Figure 7.16. The fitted $T(N)$ (solid curve) based on the available data sets. The two purple squares are the selected calibration points. The dashed curves are the $T(N)$ under different average $\varepsilon\tau_{\text{atm}}$. The circles are obtained from the temperature reading derived from sea surface temperature calculation (green), ECOSTRESS (blue), MODIS day (brown), and MODIS night (orange). Reproduced with permission from [13].

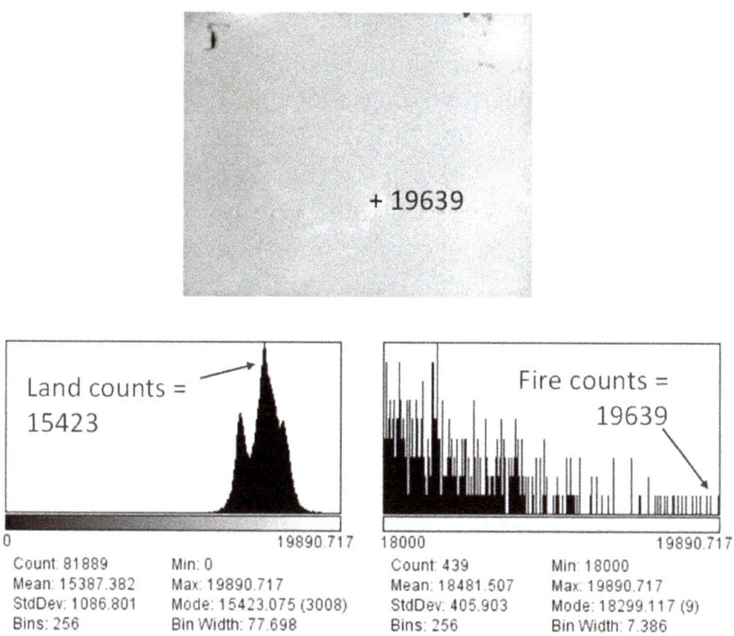

Figure 7.17. The intensity image and histograms taken by the CTI camera in band 2. Reproduced with permission from [13].

According to the fitted $T(N)$, the land's mode T is 33.5°C, which is close to the mode T of 35.9°C measured in band 1. On the other hand, the fire T is only 47.8°C in band 2, which differs significantly from $T = 105.8$°C measured in band 1. This discrepancy is attributed to the very limited dynamic range at ∼19000 in band 2, with which the fire signal saturates the ADC.

7.5 Camera Sensitivity

Using the detector's spectral line shape and ADC resolution, one is able to establish an analytical $T(N)$ function for remote temperature sensing based on just a few calibration points. This $T(N)$ function can be safely extended beyond the calibration range, if the detector characteristics remain the same. The correspondence can be readily corrected for different weather conditions, object emissivity, and signal processing and allows the temperature sensing be conducted in real time. At the same time, it can be used to ascertain the detector and electronics specifications. In this section, we use the deduced analytical functions to identify several camera attributes. They are the noise equivalent temperature difference (NEΔT), temperature resolution in terms of degrees/count or counts/degree, and emissivity resolution in terms of counts/%. We will use band 1 of the CTI camera for illustration. Other detection bands can be similarly obtained. Eq. (7.17) to Eq. (7.19) listed below are the different flight functions with fitting parameters we obtained in Section 7.4.1, where Eq. (7.17) is from Eq. (7.06), Eq. (7.18) is from Eq. (7.15), and Eq. (7.19) is the inverse function of Eq. (7.18).

$$y(T) = c_2 + \frac{a_2}{\exp\left(\frac{hc}{\lambda_0 k(273.15+T)}\right) - 1}, \tag{7.17}$$

$$T(N) = \frac{hc}{\lambda_0 k}\frac{1}{\ln\left(\frac{a_2}{m_r N - c_2} + 1\right)} - 273.15, \tag{7.18}$$

$$N(T) = \frac{a_2}{m_r\left[\exp\left(\frac{hc}{\lambda_0 k(T+273.15)}\right) - 1\right]} + \frac{c_2}{m_r} \tag{7.19}$$

In determining NEΔT, we note that arbitrary background subtraction and rescaling τ_{int} to achieve a larger dynamic range do not change the temporal noise of the detection. The variance of the collected electrons, σ_y^2, remains the mean value of the physically collected y for photodiodes with shot noise. NEΔT then only depends on the y value before scaling. From Eq. (7.17) and the specified readout noise $n_r = 1800\,\text{e}^-$, the temporal NEΔT limited by the detector current noise and readout noise is given by

$$\text{NE}\Delta\text{T} = \frac{\sqrt{\sigma_y^2 + n_r^2}}{\frac{dy}{dT}} = \frac{\sqrt{y + n_r^2}}{\frac{dy}{dT}}$$
$$= \frac{\sqrt{c_2 + \frac{a_2}{\exp\left(\frac{hc}{\lambda_0 k(273.15+T)}\right)-1} + n_r^2}}{a_2 \frac{hc}{\lambda_0 k} \frac{1}{(T+273.15)^2} \frac{\exp\left(\frac{hc}{\lambda_0 k(273.15+T)}\right)}{\left[\exp\left(\frac{hc}{\lambda_0 k(273.15+T)}\right)-1\right]^2}}. \tag{7.20}$$

Similarly, from Eq. (7.18),

$$\frac{dT(N)}{dN} = a_2 m_r \frac{hc}{\lambda_0 k} \frac{1}{\left[\ln\left(\frac{a_2}{m_r N - c_2} + 1\right)\right]^2} \frac{1}{\left(\frac{a_2}{m_r N - c_2} + 1\right)} \frac{1}{(m_r N - c_2)^2}, \tag{7.21}$$

and from Eq. (7.19),

$$\frac{dN(T)}{dT} = \frac{a_2 \frac{hc}{\lambda_0 k} \exp\left(\frac{hc}{\lambda_0 k(T+273.15)}\right)}{m_r \left[\exp\left(\frac{hc}{\lambda_0 k(T+273.15)}\right) - 1\right]^2 (T + 273.15)^2} \tag{7.22}$$

Finally, by noticing that $d(a_2)/d\varepsilon = a_2/\varepsilon \approx a_2$ when ε is close to 1,

$$\frac{dN(\varepsilon)}{d\varepsilon} = \frac{dN}{da_2}\frac{da_2}{d\varepsilon} \approx \frac{a_2}{m_r \left[\exp\left(\frac{hc}{\lambda_0 k(T+273.15)}\right) - 1\right]}. \tag{7.23}$$

Other quantities such as $dN/d\tau_{\text{atm}}$ can be deduced if the average τ_{atm} is known in the $T(N)$ function. Figure 7.18 is the plot of Eq. (7.20) to Eq. (7.23). Figure 7.18(a) shows that at $-10°$C, NEΔT is 160 mK. It decreases to 37 mK at 40°C since more photoelectrons are being collected at higher T. The 16-bit ADC provides sufficient T resolution

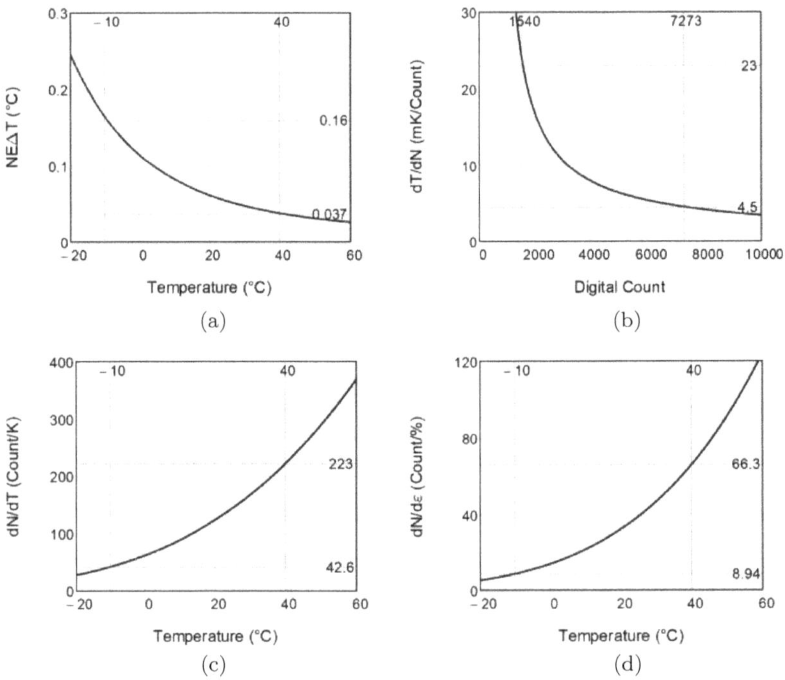

Figure 7.18. The CTI camera attributes onboard the ISS. In (b), $N = 1540$ corresponds to $T = -10°\text{C}$, and $N = 7273$ corresponds to $T = 40°\text{C}$. Reproduced with permission from [13].

of 23 mK/count at −10°C and 4.5 mK/count at 40°C in Fig. 7.18(b) because it is smaller than the corresponding NEΔT. As a result, the temporal noise rather than the ADC resolution limits the thermal sensitivity of the camera.

From 7.18(c), we expect a ΔT of one degree in the scene will result in a difference in N of 43 counts at −10°C and 223 counts at 40°C, a lower value at a lower T because of the lower radiation intensity. Figure 7.18(d) similarly shows that a percent change in the ground ε will give 9 counts at −10°C and 66 counts at 40°C. Since 66 counts is only equivalent to $66/223 = 0.30°\text{C}$ of T change at 40°C, detection in band 1 thus has a smaller ε effect by a factor of ~2 than the detection with the LWIR laboratory camera where 1% of ε change is equivalent to 0.65°C deduced from Eq. (7.13). It is because band 1 detects in MW, and MW has a larger thermal contrast $\equiv hc/(\lambda\,\mathrm{kT}^2)$ than LW by a factor of ~2. This camera information will help to understand the camera performance and to interpret the intensity image taken by the camera. It is important to note that the estimates for NEΔT consider only the temporal noise from three different noise sources, which are the dark current noise, the photocurrent noise, and the ROIC noise. The presence of the spatial noise and other temporal noises such as those arising from lens, cooler temperature fluctuations and electrical interferences are not included. The NEΔT of the overall system can be expected to be a larger value.

In summary, this section described a relatively simple and reliable calibration scheme to correlate the infrared camera's digital output N with the object temperature T being detected. By knowing the temperature analytical conversion function, one can generate a program script to provide useful temperature information in real time for any applications. Aided by the information on the detector and ROIC characteristics, very few calibration points are needed to establish the conversion functions in laboratory and in flight. These conversion functions are verified by the end-to-end experimental data within and outside the calibration range. Incorporating a physical model in the data calibration enables the effects of atmosphere transmission, scene emissivity, and different background subtractions to be readily accounted for and modified from the laboratory calibration. Hence, one can use the laboratory calibration to project reliably the temperature conversion in flight.

The temperature conversion of the present CTI camera, however, is limited by the finite dynamic range of the ROIC and the spatial resolution of the FPA. The converted temperature in the extreme conditions tends to deviate from its true value. This deficiency points to the need for a larger dynamic range and smarter sensors/processers for a wider range of temperatures. The deduced functions in the calibration can also be used to identify different camera capabilities and limitations under various scenarios such as the thermal sensitivity, temperature resolution, and emissivity resolution, which are important information for different space missions such as the study of low-temperature clouds and high-temperature fires.

7.6 Thermal Infrared Imager Onboard Landsat-8 Satellite

Rather than working with corrected images, where a single $T(N)$ function is applied to the entire array, individual conversion functions can also be applied directly to uncorrected raw images. In this approach, each detector will have its own $T(N)$ curve. An obvious disadvantage of this approach is the need for $n \times n$ distinct $T(N)$ curves for an $n \times n$ array, which makes it more appealing to linear arrays. On the other hand, one can convert any raw image in one step.

To illustrate this approach, we choose a linear array made of QWIPs, which is known as the Thermal Infrared Imager (TIRS), onboard Landsat-8 satellite. The TIRS consists of three individual QWIP arrays. A single row of 640 pixels in each array is used as a push-broom scanner to acquire images along the flight path. The ROIC is ISC 9803, which has 11 Me^- charge wells. Integrating with a 14-bit ADC, the nominal ADC $y(N) = m_r N = 11 \cdot N/2^{14} = 6.7139 \times 10^{-4} N\,\mathrm{Me}^-$. The detector array has a rigorous pre-flight calibration. The mean responsivity spectrum $R(\lambda)$ is shown in Fig. 7.19(a) while the peak responsivities of individual pixels in the center array are shown in Fig. 7.19(b).

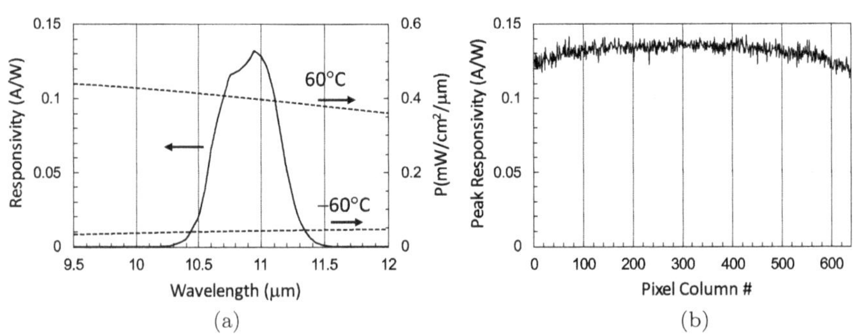

Figure 7.19. (a) The measured average responsivity of the linear QWIP array (solid curve) and the power radiance at two temperatures. (b) The peak responsivity of individual pixels. Reproduced with permission from [13].

Since $R(\lambda)$ is in A/W, the power radiance is adopted, with which the $y_i(T)$ measured in flight is

$$y_i(T) = y_{di} + A\frac{\tau_{\text{int}}\tau_{\text{atm}}\tau_{\text{opt}}\pi}{(4F^2+1)e}\int_{\lambda_1}^{\lambda_2}\frac{2hc^2}{\lambda^5}\frac{R_i(\lambda)}{\exp\left(\frac{hc}{\lambda k(273.15+T)}\right)-1}d\lambda, \tag{7.24}$$

where the index $i = 0\ldots639$ is the pixel column number, y_{di} is the dark electrons, $A = 25\,\mu\text{m} \times 25\,\mu\text{m}$, $F = 1.64$, and $\tau_{\text{int}} = 5.5\,\text{ms}$. The dark electrons are obtained in flight by pointing the camera to deep space, and their magnitudes in counts N_{di} are shown in Fig. 7.20, from which $y_{di} = y(N_{di})$ can be obtained by using the nominal ADC $y(N)$.

The parameter $\tau_{\text{atm}}\tau_{\text{opt}} = 0.52$ is obtained from one $T(N)$ calibration point in flight on the average pixel response from a conventional measurement, in which the nominal ADC $y(N)$ is assumed. With these detector characteristics, the individual $y_i(T)$ curves can be calculated from Eq. (7.24), fitted to Eq. (7.4) with parameters a_{1i}, b_{1i}, and c_{1i}, and then inverted to yield $T_i(y)$. With the single nominal ADC $y(N)$, $T_i(N)$ for each pixel i can be established in the form of Eq. (7.5), *i.e.*,

$$T_i(N) = \left(\frac{m_r N - c_{1i}}{a_{1i}}\right)^{1/b_{1i}} - 273.15. \tag{7.25}$$

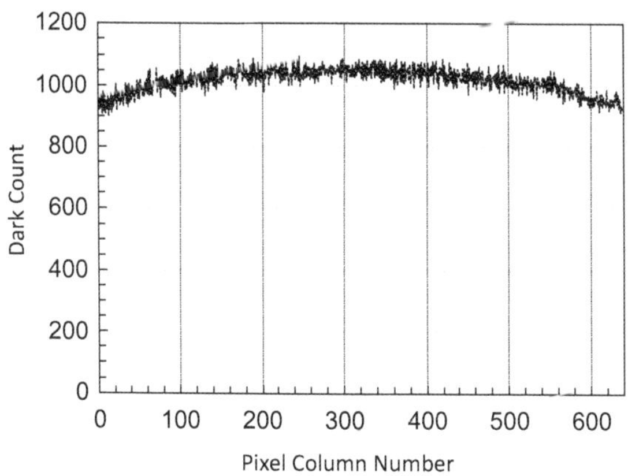

Figure 7.20. The dark counts of the array. Reproduced with permission from [13].

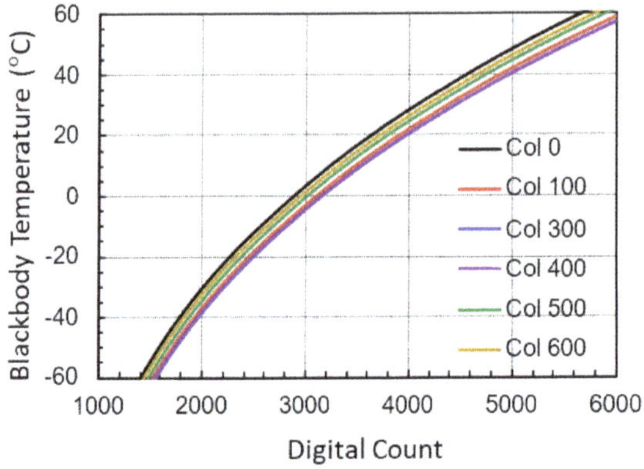

Figure 7.21. The representative $T_i(N)$ curves at different column numbers i. Reproduced with permission from [13].

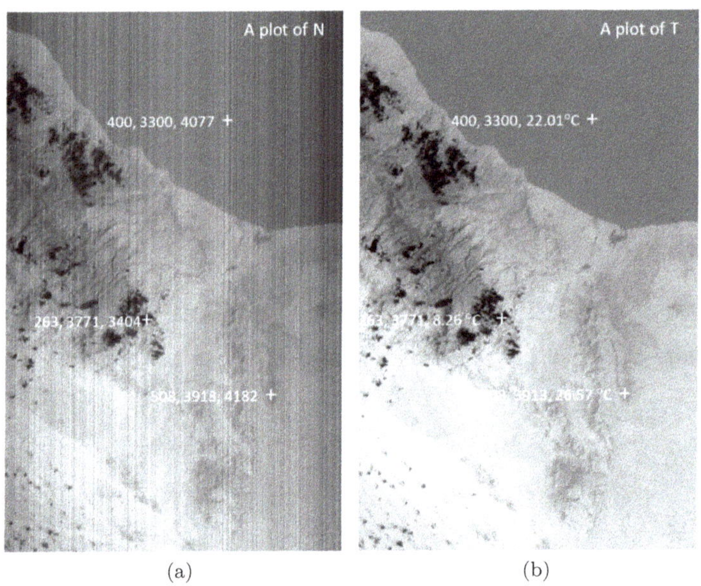

Figure 7.22. (a) The (x, y, N) in the N plot and (b) (x, y, T) in the T plot. Reproduced with permission from [13].

A few representatives are shown in Fig. 7.21. With these individual transfer functions, the raw images can be processed directly. Figure 7.22 shows a plot of raw image taken by the TIRS imager and the converted T plot. By applying the individual $T(N)$ transfer

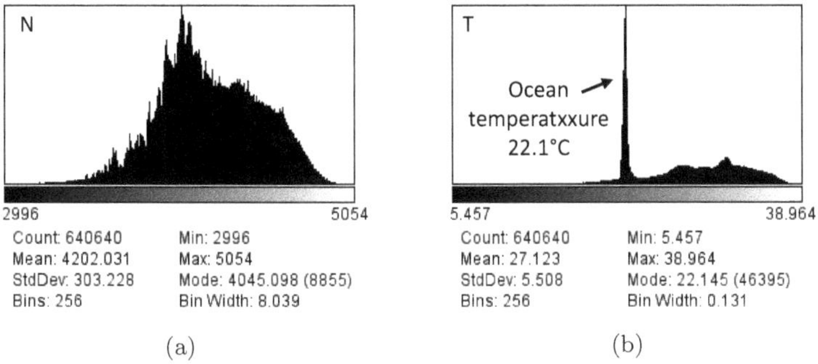

Figure 7.23. The histograms of (a) counts and (b) brightness temperature of the scene. Reproduced with permission from [13].

functions, the intensity image can be corrected and converted in a single process.

The histograms in Fig. 7.23(a) show that the N population of the scene. When it is converted to the T population, the distribution is significantly changed because the pixel nonuniformity is corrected at the same time. The resultant temperature distribution is a correct representation of the brightness temperature of the area. In particular, the ocean shows a relative constant T of 22.1°C.

In the TIRS program, the grating FPAs are supplied by two companies. Their typical pixel nonuniformity is shown in Fig. 7.24, in which a C-QWIP FPA from the same program and a R-QWIP FPA detecting at 8.4 μm from another program are also shown. The pixel nonuniformity is defined as the weighted root mean square values relative to the mean values of the rows [8]. The R-QWIP FPA shows a slightly higher pixel uniformity in this particular set of samples.

Finally, based on the same grayscale adopted in this chapter, which is more suitable for data analysis, we show again two infrared images from the LC5 C-QWIP FPA in Fig. 7.25 and from the DW1 R-QWIP FPA in Fig. 7.26.

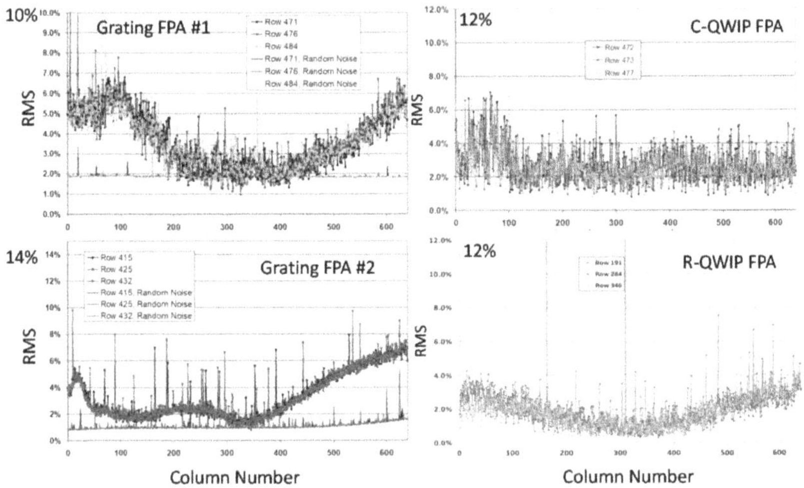

Figure 7.24. The root mean square of the pixel response, in which the mean is normalized to 1 for four different FPAs.

Figure 7.25. An infrared image taken by a 9.5 μm cutoff, 1024 $\times$ 1024 25 μm pitch LC5 C-QWIP FPA.

Figure 7.26. An infrared image taken by a 10.5 μm cutoff, 1024 $\times$ 1024 25 μm pitch DW1 R-QWIP FPA.

References

1. Chen, Y., Sun-Mack, S., Minnis, P., Young, D. F. and Smith, W. L. (2004). Seasonal surface spectral emissivity derived from TERRA MODIS data, 13th AMS Conference on Satellite Oceanography and Meteorology, Norfolk, VA.
2. Berk, A., Conforti, P. and Hawes, F. (2015). An accelerated line-by-line option for MODTRAN combining on-the-fly generation of line center absorption with 0.1 cm^{-1} bins and pre-computed line tails, *Proc. SPIE*, 9471, 947217.
3. Dash, P., Göttsche, F.-M., Olesen, F.-S. and Fischer, H. (2002). Land surface temperature and emissivity estimation from passive sensor data: Theory and practice-current trends, *Int. J. Remote Sens.*, 23, pp. 2563–2594.
4. Chen, W., Pinker, R. T., Ma, Y., Hulley, G., *et al.* (2021). Land surface temperature from GOES-East and GOES-West, *J. Atmos. Ocean. Technol.*, 38, pp. 843–858.

5. Pinker, R. T., Ma, Y., Chen, W., Hulley, G., *et al.* (2019). Towards a unified and coherent land surface temperature Earth system data record from geostationary satellites, *Remote Sens.*, 11, p. 1399.
6. Jhabvala, M., Reuter, D., Choi, K., Jhabvala, C. and Sundaram, M. (2009). QWIP-based thermal infrared sensor for the Landsat Data Continuity Mission, *Infrared Phys. Technol.*, 52, pp. 424–429.
7. Jhabvala, M., Reuter, D., Choi, K., Sundaram, M., *et al.* (2010). The QWIP focal plane assembly for NASA's Landsat Data Continuity Mission, *Proc. SPIE*, 7660, 76603J-1–13.
8. Jhabvala, M., Choi, K., Waczynski, A., La, A., *et al.* (2011). Performance of the QWIP focal plane arrays for NASA's Landsat Data Continuity Mission, *Proc. SPIE*, 8012, 80120Q-1–14.
9. Jhabvala, M. and Choi, K. (2012). Evolution of QWIP focal plane development at the NASA/Goddard space flight center, *Proc. SPIE*, 8268, 82682P-1–14.
10. Jhabvala, M., Jennings, D., Tucker, C., Ahn, A., *et al.* (2019). Strained-layer-superlattice-based compact thermal imager for the International Space Station, *Appl. Opt.*, 58, pp. 5432–5442.
11. Jhabvala, M., Choi, K. K., Gunapala, S., Razeghi, M. and Sundaram, M. (2020). QWIPs, SLS, Landsat and the international space station, *Proc. SPIE*, 11288, 1128802-1–13.
12. Wu, D. L., Jennings, D. E., Choi, K. K., Jhabvala, M. D., *et al.*, (2021). Compact thermal imager (CTI) for atmospheric remote sensing, *Remote Sens.*, 13, p. 4578.
13. Choi, K. K., Jhabvala, M., Jennings, D., Turck, K., *et al.* (2011). Remote temperature sensing by the compact thermal imager from the International Space Station, *Appl. Opt.*, 60, pp. 10390–10401.
14. Jennings, D. E., Jhabvala, M. D., Tucker, C. J., Lunsford, A. W., *et al.* (2022). Compact thermal imager: a flight demonstration of infrared technology for Earth observations, *Appl. Opt.*, 61, pp. 4215–4225.
15. Jhabvala, M., Jennings, D., Turck, L., Tucker, C., *et al.* (2023). Dual and multi-spectral band SLS infrared camera systems, *Proc. SPIE*, 12514, 1251404-1–22.
16. Sundaram, M., Bundas, J. and Reisinger, A. (2014). Dualband MW/LW strained layer superlattice focal plane arrays for satellite-based wildfire detection, SBIR Phase I final report, NASA Contract NNX14CG38P, report # QMV0002, p. 1.

Index

Milton Keynes UK
Ingram Content Group UK Ltd.
UKHW051547200624
444511UK00002B/4

9 789811 286513